A PRIMER ON
LINEAR ALGEBRA

A PRIMER ON
LINEAR ALGEBRA

I. N. HERSTEIN
University of Chicago

DAVID J. WINTER
University of Michigan, Ann Arbor

Macmillan Publishing Company
NEW YORK

Collier Macmillan Publishers
LONDON

Macmillan Publishing Company
866 Third Avenue, New York, New York 10022

Collier Macmillan Canada, Inc.

LIBRARY OF CONGRESS CATALOGING-IN-PUBLICATION DATA

Herstein, I. N.
    A primer on linear algebra.

    Includes index.
    1. Algebras, Linear.   I. Winter, David J.   III. Title.
QA184.H57   1988   512'.5   87-18489
ISBN 0-02-353953-4

Printing: 1 2 3 4 5 6 7 8   Year 8 9 0 1 2 3 4 5 6 7

ISBN 0-02-353953-4

PREFACE

When we first decided to write a book on linear algebra, we were somewhat ambivalent as to the level at which the book should be pitched. The material can be presented at a variety of different levels. What is appropriate for one group of readers may very well be inappropriate for another. Because of the importance of the techniques and results of matrix theory and linear algebra in such a diverse set of fields aside from mathematics—physics, chemistry, engineering, economics, and so forth—we wanted the book to be useful and to fill the needs of those readers who needed some, a little, or a lot of linear algebra in their work.

What are the needs of the prospective readers? This all depends on what are the intended areas of use of the subject matter. For some readers what is primarily needed is merely an introduction to the elementary operations with matrices and some of the beginning notions and theorems. For others a much deeper knowledge is required. Not only are the manipulative aspects of the subject needed, but also a fairly good acquaintance with the deeper ideas and results. To be able to adjust arguments for special use in their own disciplines, such readers would have to have a fairly good idea of how the proofs are carried out. Finally, for prospective mathematicians, a thorough control of the subject matter in all its aspects is virtually a must. This means that it is not enough to have just a compendium of results; it is necessary to know why such results are true. In other words, the nature of the proofs is essential for such a thorough understanding.

The outcome, after giving much thought to the matter, was our book *Matrix Theory and Linear Algebra*. In that book we endeavored to present the material in such a way that, by a proper selection of topics and points of emphasis, the needs of any of the three groups of students mentioned above could be accommodated.

During the writing of *Matrix Theory and Linear Algebra* we were confronted with a dilemma. On one hand, we wanted to keep our presentation as simple as possible. On the other hand, we wanted to cover the subject fully and without compromise, even when one hits the rough spots that exist. To solve this dilemma, we decided to prepare two versions of the book. One book, designed for the more experienced students, should be very thorough and should go into almost all aspects of linear algebra, even to what might be called the esoteric. That book is *Matrix Theory and Linear Algebra*. The second book—that is, this book—was designed for students who want to get a rapid introduction and good first look at linear algebra, but who do not want to go into it in its *n*th fine point, the first time around.

These two books are almost identical in many respects, but there are important differences. The fundamental difference is not so much in the topics covered (although there is quite a difference there) but, possibly more, in the philosophy and viewpoint. This aspect is characterized by the pace of the development, by the nature of the arguments used, and by the kind of problems set out for the reader.

In *Matrix Theory and Linear Algebra* we went into greater depth for some of the topics covered in both books and included material that is not included here. The decision of what material should be excluded from this version was made using a simple criteria: Exclude advanced material that would complicate a simple first development of the subject and hinder the less experienced students; exclude material that is not absolutely necessary for a good overall understanding of what the subject is about; and exclude the esoteric, especially the highly abstract.

Although large parts of this book are brought over in the exact form in which they appear in *Matrix Theory and Linear Algebra*, there are many places where the arguments used and the writing are quite different, even for topics included in both. Aside from portions of the text material, this difference is notable in the problem sets.

Toward the end of the preface we lay out some possible courses for which this book can be used. At the same time, the other material is there, for the students so inclined to be studied or looked over.

One of the intact pieces carried over from *Matrix Theory and Linear Algebra* is the rest of this preface, with the exception of the possible courses outlined.

Our approach is to start slowly, setting out at the level of 2×2 matrices. These matrices have the great advantage that everything about them is open to the eye, that the students can get their hands on the material, experiment with it, see what any specific theorem says about them. Furthermore, all this can be done by performing some simple calculations.

However, in treating the 2×2 matrices, we try to handle them as we would the general $n \times n$ case. In this microcosm of the larger matrix world, virtually every concept that will arise for $n \times n$ matrices, general vector spaces, and linear transformations make its appearance. This appearance is usually in a form ready for extension to the general situation. Probably the only exception to this is the theory of determinants, for which the 2×2 case is far too simplistic.

With the background acquired in playing with the 2×2 matrices in this general manner, the results for the most general case, as they unfold, are not as surprising, mystifying or mysterious to the students as they might otherwise be. After all, these results are almost old friends whose acquaintance we made in our earlier 2×2 incarnation. So this simplified context serves both as a laboratory and as a motivation for what is to come.

From the fairly concrete world of the $n \times n$ matrices, we pass to the more abstract realm of vector spaces and linear transformations. Here the basic strategy is to prove that an n-dimensional vector space is isomorphic to the space of n-tuples. With this isomorphism established, the whole corpus of concepts and results that we had obtained in the context of n-tuples and $n \times n$ matrices is readily transferred to the setting of arbitrary vector spaces and linear transformations. Moreover, this transfer is accomplished with little or no need of proof. Because of the nature of isomorphism, it is enough merely to cite the proof or result obtained earlier for n-tuples or $n \times n$ matrices.

The vector spaces we treat in the book are only over the fields of real or complex numbers. A little is lost in imposing this restriction, but much is gained. For instance, our vector spaces can always be endowed with an inner product. Using this inner product, we can always decompose the space as a direct sum of any subspace and its orthogonal complement. This direct sum decomposition is then exploited to the hilt to obtain very simple and illuminating proofs of many of the theorems.

There is an attempt made to give some nice, characteristic applications of the material of the book. Some of these can be integrated into a course almost from the beginning, where we discuss 2×2 matrices.

Least square methods are discussed to show how linear equations having no solutions can always be solved approximately in a very neat and efficient way. These methods are then used to show how to find functions that approximate given data.

Finally, in the last chapter, we discuss how to translate some of our methods into *linear algorithms*, that is, finite numerical step-by-step versions of methods of linear algebra. The emphasis is on linear algorithms that can be used in writing computer programs for finding exact and approximate solutions of linear equations. We then illustrate how some of these algorithms are used in such a computer program, written in the programming language PASCAL.

There are many exercises in the book. These are usually divided into categories entitled *Numerical, More Theoretical: Easier, Middle-Level,* and *Harder*. It goes without saying that the problems are an intrinsic part of any course. They are the best means for checking on one's understanding and mastery of the material. Included are exercises that are treated later in the text itself, or can be solved easily using later results. This gives you, the reader, a chance to try your own hand at developing important tools and compare your approach in an early context to our approach in the later context. An answer manual is available from the publisher.

We mentioned earlier that the book can serve as a textbook for several different levels. Of course, how this is done is up to the individual instructor. We present here some possible sample courses.

1. *One-term course emphasizing computational aspects*

Chapters 1 through 4, Chapter 5 with an emphasis on methods and a minimum of proofs. One might merely do determinants in the 3×3 case with a statement that the $n \times n$ case follows similar rules. Sections marked "optional" should be skipped. The sections in Chapter 9 entitled "Fibonacci Numbers" and "Equations of Curves" could be integrated into the course. Problems should primarily be all the numerical ones and a sampling of the easier and middle-level theoretical problems.

2. *One-term course for users of matrix theory and linear algebra in allied fields*

Chapters 1 through 5, with a possible deemphasis on the proofs of the properties of determinants and with an emphasis on computing with determinants. Some introduction to vector spaces and linear transformations would be desirable. Chapters 10 and 11, which deal with least square methods and computing, could play an important role in the course. Each of the applications in Chapter 9 could be touched on. As for problems, again all the numerical ones, most of the middle-level ones, and a few of the harder ones should be appropriate for such a course.

3. *One-term course for mathematics majors*

Most of Chapter 1 can be done very quickly, with much left for the students to read on their own. All of Chapters 2 through 6, including some of the optional topics. Definitely, some emphasis should be given to abstract vector spaces and linear transformations as in Chapters 7 and 8. The whole gamut of problems should be assignable to the students.

4. *Two-term course for users of matrix theory and linear algebra*

The entire book, but going easy on proofs for determinants, on the material on abstract vector spaces and linear transformations, plus a fairly thorough treatment of Chapters 10 and 11. The problems can be chosen from all parts of the problem sections.

5. *Two-term course for mathematics majors*

The entire book, with perhaps less emphasis on Chapters 10 and 11.

We should like to thank the many people who have looked at the manuscript, have commented on it, and have made some useful suggestions. We want to thank Bill Blair and Lynne Small for their extensive and valuable analysis of the book at its different stages, which had a very substantial effect on its final form. We should also like to thank Gary Ostedt, Bob Clark, and Elaine Wetterau of the Macmillan Publishing Company for their help in bringing this book into being. We should like to thank Lee Zukowski for the excellent typing job he did on the manuscript. And we should like to thank Pedro Sanchez for his valuable help with the computer program and the last chapter.

I. N. H.
D. J. W.

CONTENTS

SYMBOL LIST

$s \in S$	the element s is in the set S
$S \subset T$	the set S is contained in the set T
$S \cup T$	the union of the sets S and T
$S_1 \cup \cdots \cup S_n$	the union of the sets $S_1, \ldots, S_n$
$S \cap T$	the intersection of the sets S and T
$S_1 \cap \cdots \cap S_n$	the intersection of the sets $S_1, \ldots, S_n$,
$\mathbb{R}$	the set of all real numbers, 3
$M_2(\mathbb{R})$	the set of all 2×2 matrices over $\mathbb{R}$, 3
$A + B$	the sum of matrices A and B, 4, 127
$A - B$	the difference of matrices A and B, 4, 127
0	the zero matrix, 4, 127, 130
$-A$	the negative of the matrix A, 5, 127
uA	the matrix obtained by multiplying the matrix A by the scalar u, 5, 129
AB	the product of matrices A and B, 5, 129
I	the identity matrix, 7, 128
A^{-1}	the inverse of the matrix A, when it exists, 8, 132

aI	the scalar matrix corresponding to the scalar a, 9, 128
$\displaystyle\sum_{r=1}^{n} a_r$	the summation of a_r from 1 to n, 13
$\displaystyle\sum_{r=1}^{m}\sum_{a=1}^{n} a_{rs}$	the double summation of a_{rs}, 14
A^m	the product of the matrix A with itself m times, 16, 132
A^0	the identity matrix, 17, 132
A^{-m}	the product of the matrix A^{-1} with itself m times, when it exists, 17, 132
$\mathrm{tr}(A)$	the trace of the matrix A, 23, 150
A'	the transpose of the matrix A, 25, 154
$\det(A)$, $|A|$	the determinant of the matrix A, 30, 278
fg	the product or composite of functions f, g, 42
f^{-1}	the inverse of a 1-1 onto function f, 44
f^n	product of the function f with itself n times, 44
$\mathbb{R}^{(2)}$	the set of all vectors in the Cartesian plane, 49
$v + w$	the sum of vectors in $\mathbb{R}^{(2)}$, 50
dv	the vector v in $\mathbb{R}^{(2)}$ multiplied by the scalar d, 51
Av	the vector obtained by applying $A \in M_2(\mathbb{R})$ to $v \in \mathbb{R}^{(2)}$, 51
$P_A(x)$	the characteristic polynomial of the matrix A, 59, 333
$\alpha = a + bi$	the complex number, α, with real part a and pure imaginary part bi, 68
$\mathbb{C}$	the set of all complex numbers, 3, 68
$\bar{\alpha} = a - bi$	the conjugate of the complex number $\alpha = a + bi$, 72
$|\alpha|$	the absolute value of the complex number α, 72
$M_2(\mathbb{C})$	the set of all 2×2 matrices over $\mathbb{C}$, 77
(v, w)	the inner product of column vectors v, w, 80
A^*	the Hermitian adjoint of A, 82, 87, 155
$\|v\|$	the length of vector v, 85, 167
Add $(r, s; u)$	the operation of adding u times row s to row r, 113, 323
Interchange (r, s)	the operation of interchanging rows r and s, 113, 323
Multiply $(r; u)$	the operation of multiplying row r by the scalar u, 113, 323
F	the set of all scalars: $F = \mathbb{R}$ or $F = \mathbb{C}$, 126

E_{rs}	the matrix whose (r, s) entry is 1 and all of whose other entries are 0, 130
$M_n(F)$	the set of all $n \times n$ matrices over F, 126
$F^{(n)}$	the set of all column vectors with n coordinates from F, 136
$v + w$	the sum of vectors in $F^{(n)}$, 136
0	the zero vector in $F^{(n)}$, 137
$-v$	the negative of the vector v in $F^{(n)}$, 137
tv	the vector v multiplied by the scalar t, 137
Av	the vector obtained by applying the matrix or transformation $A \in M_n(F)$ to the vector $v \in F^{(n)}$, 139
$v^\perp$	the set of all vectors orthogonal to v, 165
e_s	the vector whose r entry is 0 if $r \neq s$ and 1 if $r = s$, 171
$A \sim B$	the matrices A and B are similar, 203
$cl(A)$	the set of all matrices B such that $A \sim B$, 203
$\dim (V)$	the dimension of V, 218, 383
$V + W$	the sum of subspaces V, W, 222
$V \oplus W$	the direct sum of subspaces V, W, 223, 401
$W^\perp$	the subspace of vectors orthogonal to W, 225, 401
$n(A)$	the nullity of the matrix A, 234
$< v_1, \ldots, v_r >$	the subspace spanned by the vectors $v_1, \ldots, v_r$, 176, 362, 366
$r(A)$	the rank of the matrix A, 235
$q_A(x)$	the minimum polynomial of the matrix A, 238
V_a	the set of characteristic vectors of A associated with a, 252
$x'(t)$	the derivative of the vector function $x(t)$, 270
e^A	the exponential of an $n \times n$ real or complex matrix A, 267
M_{rs}	the (r, s) minor of the matrix A, 277
A_{rs}	the (r, s) cofactor of the matrix A, 314
$A(r, s; q)$	the elementary matrix corresponding to Add $(r, s; q)$, 323
$M(r; q)$	the elementary matrix corresponding to Multiply (r, q), 323
$I(r, s)$	the elementary matrix corresponding to Interchange (r, s), 323
$v + w$	the sum of vectors v, w in a vector space V, 359
0	the zero vector in a vector space V, 359

av	the vector v multiplied by the scalar a in a vector space V, 359
Ker ϕ	the kernel of the homomorphism ϕ from V to W, 375
$V \cong W$	there exists an isomorphism from V to W, 376
(v, w)	the inner product of elements v, w in an inner product space, 388
$\|v\|$	the length of a vector in an inner product space, 393
$T_1 + T_2$	the sum of linear transformations T_1, T_2, 413
aT	the linear transformation obtained by multiplying the linear transformation T by the scalar a, 413
0	the zero linear transformation, 414
$-T$	the negative of the linear transformation T, 415
$L(V)$	the set of all linear transformations of V over F, 414
$T_1 T_2$	the product of linear transformations T_1, T_2, 418
I	the identity linear transformation, 418
T^{-1}	the inverse of the linear transformation T, when it exists, 419
$m(T)$	the matrix of a linear transformation in a given basis, 423, 424
$\text{tr}(T)$	the trace of the linear transformation T, 428
$\det(T)$	the determinant of the linear transformation T, 428
$P_T(x)$	the characteristic polynomial of the linear transformation T, 429
$[v, w]$	the inner product $(\phi(v), \phi(w))$ corresponding to a given isomorphism ϕ, 433
T^*	the Hermitian adjoint of the linear transformation T, 436
$S^{(k)}$	the kth state in a Markov process, 457
$V_a(T)$	the generalized characteristic space of T at a, 466
$\text{Proj}_W(y)$	the projection of the vector y on the subspace W, 472
A^-	the approximate inverse of the $m \times n$ matrix A, 484, 485

A PRIMER ON
LINEAR ALGEBRA

THE 2 × 2 MATRICES

1.1. INTRODUCTION

The subject whose study we are about to undertake—matrix theory and linear algebra—is one that crops up in a large variety of places. Needless to say, from its very name, it has an important role to play in algebra and, in fact, in virtually every part of mathematics. Not too surprisingly, it finds application in a much wider arena; there are few areas of science and engineering in which it does not make an appearance. What is more surprising is the extent to which results and techniques from this theory are also used increasingly in such fields as economics, psychology, business management, and sociology.

What is the subject all about? Some readers will have made a beginning acquaintance with it in courses in multivariable calculus where 2 × 2 and 3 × 3 matrices are frequently introduced. For those readers it is not so essential to explain what the subject is about. For others who have had no exposure to this sort of thing, it might be helpful if we explain, in a few words, something about the subject matter treated in this book.

We start things off with a very special situation, namely, that of 2 × 2 matrices whose entries are real numbers. Shortly after that we enlarge our universe slightly by introducing complex numbers as entries.

What will these 2 × 2 matrices be? At the outset they will be merely some formal symbols, arrays of real numbers a, b, c, d in two rows and columns such as

$$\begin{bmatrix} a & b \\ c & d \end{bmatrix}.$$

To these we assign methods of combining them, called "addition" and "multiplication," and we show that these combinations are subject to certain rules of behavior. At first, with no experience behind us, we might find these operations and rules to be arbitrary, unmotivated, and even, perhaps, contrived. Later, we shall see that they do have a natural life in geometry and solving equations.

After the initial steps of introducing the formal symbols and how to operate with these, we go about experimenting to see what is true for these formal symbols, which we call 2 × 2 matrices. In doing so, we shall prove a series of results all of which will be very special prototypes of what is to come for the far more general situation of $n \times n$ matrices for any positive integer n.

Why start with the 2 × 2 case rather than plunge immediately into the general $n \times n$ case? There are several good reasons for this. Because the 2 × 2 case is small, everything can be done very explicitly and visibly, with rather simple computations. Thus, when we come to the general case, we still have some idea of what may be true there. Hence the explicit case of the 2 × 2 matrices will be our guideline for the approach to the general case.

But even this passage to the $n \times n$ matrices is still rather formal. It would be nice to be able to see these matrices as living beings. Strangely enough, the best way to do this is to go even more abstract and to pass to the abstract notion of a vector space. Matrices will then assume a clearer role—that of objects that do something, that act by *transformations*, that is mappings, on a desirable set in a very attractive way. In making this transition we are going over from matrix theory to that of linear algebra. In this new framework we shall redo many of the things done before in a smoother, easier, and more conceptual form. Aside from redoing old things, we shall be able to press on to newer and different notions and results.

With these vague words we have given an even vaguer description of what will transpire. When we come to the more abstract approach— linear algebra—the subject will acquire a greater unity and cohesion. It will even, we hope, have a great deal of aesthetic appeal.

1.2. DEFINITIONS AND OPERATIONS

As we mentioned in the introduction, we begin everything with a simple case—perhaps even an oversimplified case—of the 2 × 2 matrices. Even though special, this case offers us the opportunity to be concrete and to do everything with our hands from the ground up.

Let $\mathbb{R}$ be the set of all real numbers. By $M_2(\mathbb{R})$, the set of 2 × 2 *matrices over* $\mathbb{R}$, we shall mean the set of all square arrays

$$\begin{bmatrix} a & b \\ c & d \end{bmatrix},$$

where a, b, c, d are real numbers. We define how to add and multiply them, and we then single out certain classes of matrices that are of special interest.

Although we shall soon enlarge our setup to matrices over the set $\mathbb{C}$ of complex numbers, we restrict our initial efforts to matrices involving real numbers.

Before doing anything with these matrices, we have to have some criterion for declaring when two of them are equal. Although this definition of equality is most natural to make, it is up to us actually to make it, so that there is no possible ambiguity as to what we mean.

Definition. The two matrices $A = \begin{bmatrix} a & b \\ c & d \end{bmatrix}$ and $B = \begin{bmatrix} e & f \\ g & h \end{bmatrix}$ in $M_2(\mathbb{R})$ are *defined* to be *equal* if and only if $a = e$, $b = f$, $c = g$, and $d = h$.

Thus, for instance, in order that the matrix $A = \begin{bmatrix} 1 & 2 \\ -6 & \pi \end{bmatrix}$ be equal to the matrix $B = \begin{bmatrix} 1 & b \\ -6 & d \end{bmatrix}$, it is necessary and sufficient that $b = 2$ and $d = \pi$.

If $A = \begin{bmatrix} a & b \\ c & d \end{bmatrix}$, then a, b, c, and d are called the *entries* of A. In these terms, we declare two matrices to be equal if their *corresponding entries* are equal.

With this definition of equality behind us, we proceed to introduce ways of combining these matrices. The first operations that we introduce in $M_2(\mathbb{R})$ are *addition* and *subtraction* of two matrices, denoted naturally enough by $+$ and $-$. Here, as in the definition of equality, the definitions strike one as reasonable and natural. We do the obvious in defining these operations as follows, to reflect the corresponding operations on their entries.

Definition. If $A = \begin{bmatrix} a & b \\ c & d \end{bmatrix}$ and $B = \begin{bmatrix} e & f \\ g & h \end{bmatrix}$ are in $M_2(\mathbb{R})$, then

$$A + B \quad \text{and} \quad A - B$$

are *defined* by

$$A + B = \begin{bmatrix} a + e & b + f \\ c + g & d + h \end{bmatrix}$$

$$A - B = \begin{bmatrix} a - e & b - f \\ c - g & d - h \end{bmatrix}.$$

Thus, to add or subtract two matrices, simply add or subtract, respectively, their corresponding entries. For instance, if A and B are the matrices

$$A = \begin{bmatrix} 1 & 7 \\ -6 & 3 \end{bmatrix}$$

$$B = \begin{bmatrix} \frac{1}{2} & \frac{2}{3} \\ 0 & 4 \end{bmatrix},$$

then $A + B$ and $A - B$ are the matrices

$$A + B = \begin{bmatrix} 1 + \frac{1}{2} & 7 + \frac{2}{3} \\ -6 + 0 & 3 + 4 \end{bmatrix} = \begin{bmatrix} \frac{3}{2} & \frac{23}{3} \\ -6 & 7 \end{bmatrix}$$

$$A - B = \begin{bmatrix} 1 - \frac{1}{2} & 7 - \frac{2}{3} \\ -6 - 0 & 3 - 4 \end{bmatrix} = \begin{bmatrix} \frac{1}{2} & \frac{19}{3} \\ -6 & -1 \end{bmatrix},$$

both of which are again in $M_2(\mathbb{R})$. In fact, from the definitions themselves, we see that for any A, B, in $M_2(\mathbb{R})$, both $A + B$ and $A - B$ are also in $M_2(\mathbb{R})$.

Note that the particular matrix

$$\begin{bmatrix} 0 & 0 \\ 0 & 0 \end{bmatrix},$$

which we shall simply denote by 0, plays a very special role with respect to the addition operation. As is clear, $A + 0 = 0 + A = A$ for every matrix A in $M_2(\mathbb{R})$. Thus this special matrix 0 plays a role in $M_2(\mathbb{R})$ very much like that played by the real number 0 in $\mathbb{R}$. It is called the *zero matrix*.

Note, too, that if $A = \begin{bmatrix} a & b \\ c & d \end{bmatrix}$ is in $M_2(\mathbb{R})$, then the matrix

$$B = \begin{bmatrix} -a & -b \\ -c & -d \end{bmatrix},$$

which is also in $M_2(\mathbb{R})$, satisfies the equations $A + B = B + A = 0$. So B acts as the "negative" of A. We denote it, simply, by $-A$.

Given $\mu \in \mathbb{R}$ and $A = \begin{bmatrix} a & b \\ c & d \end{bmatrix}$ in $M_2(\mathbb{R})$, we define the *multiplication of A by the scalar u* as follows:

$$uA = \begin{bmatrix} ua & ub \\ uc & ud \end{bmatrix}.$$

Observe that here, too, uA is again in $M_2(\mathbb{R})$.

Before considering the behavior of the addition and subtraction that we have introduced, we want still another operation for the elements of $M_2(\mathbb{R})$, namely the *multiplication* or *product* of two matrices. Unlike the earlier definitions, this multiplication will probably seem somewhat un-natural to those seeing it for the first time. As we proceed, we shall see natural patterns emerge, as well as a rationale for choosing this as our desired product.

Definition. If the matrices $A = \begin{bmatrix} a & b \\ c & d \end{bmatrix}$ and $B = \begin{bmatrix} e & f \\ g & h \end{bmatrix}$ are in $M_2(\mathbb{R})$, then their *product AB* is *defined* by the equation

$$AB = \begin{bmatrix} ae + bg & af + bh \\ ce + dg & cf + dh \end{bmatrix}.$$

Let's look at a few examples of this product starting, for example, with

$$\begin{bmatrix} 1 & 3 \\ -2 & 4 \end{bmatrix}\begin{bmatrix} 4 & 3 \\ 2 & 1 \end{bmatrix} = \begin{bmatrix} 1 \cdot 4 + 3 \cdot 2 & 1 \cdot 3 + 3 \cdot 1 \\ (-2)4 + 4 \cdot 2 & (-2) \cdot 3 + 4 \cdot 1 \end{bmatrix}$$

$$= \begin{bmatrix} 10 & 6 \\ 0 & -2 \end{bmatrix}.$$

Perhaps more interesting are the products

$$\begin{bmatrix} 0 & 1 \\ 0 & 0 \end{bmatrix}\begin{bmatrix} 1 & 0 \\ 0 & 0 \end{bmatrix} = \begin{bmatrix} 0 & 0 \\ 0 & 0 \end{bmatrix} = 0$$

and

$$\begin{bmatrix} 1 & 0 \\ 0 & 0 \end{bmatrix}\begin{bmatrix} 0 & 1 \\ 0 & 0 \end{bmatrix} = \begin{bmatrix} 0 & 1 \\ 0 & 0 \end{bmatrix} \neq 0.$$

Note several things about the product of matrices:

1. If A, B are in $M_2(\mathbb{R})$, then AB is also in $M_2(\mathbb{R})$.
2. In $M_2(\mathbb{R})$, it is possible that $AB = 0$ with $A \neq 0$ and $B \neq 0$.
3. In $M_2(\mathbb{R})$, it is possible that $AB \neq BA$.

These last two behaviors both run counter to our prior experience with number systems, where we know that

2′. In $\mathbb{R}$, $ab = 0$ if and only if $a = 0$ or $b = 0$.
3′. In $\mathbb{R}$, $ab = ba$ for all a and b.

Here (2′) is, in effect, the *cancellation law of multiplication* for real numbers:

   If $ab = 0$ and $a \neq 0$, then $b = 0$.

Thus (2) says that

   The cancellation law of multiplication does not hold in $M_2(\mathbb{R})$.

At the same time, cancellation is possible in $M_2(\mathbb{R})$ under certain circumstances, as we observe in Problem 10. Similarly, (3′) says that real numbers *commute* under multiplication. Thus (3) says that

   Matrices in $M_2(\mathbb{R})$ do not necessarily commute under multiplication.

   Matrices in $M_2(\mathbb{R})$ satisfy the *associative law* that

   $(AB)C = A(BC)$

as you can see by multiplying out the expressions on both sides of the equation. We leave this as an easy, though tedious exercise.
A matrix that plays a very special role in multiplication is the matrix

$I = \begin{bmatrix} 1 & 0 \\ 0 & 1 \end{bmatrix}$. This matrix is called the *identity matrix*, because it has the following properties:

$$AI = \begin{bmatrix} a & b \\ c & d \end{bmatrix}\begin{bmatrix} 1 & 0 \\ 0 & 1 \end{bmatrix}$$

$$= \begin{bmatrix} a \cdot 1 + b \cdot 0 & a \cdot 0 + b \cdot 1 \\ c \cdot 1 + d \cdot 0 & c \cdot 0 + d \cdot 1 \end{bmatrix} = \begin{bmatrix} a & b \\ c & d \end{bmatrix} = A.$$

Similarly,

$$IA = \begin{bmatrix} 1 & 0 \\ 0 & 1 \end{bmatrix}\begin{bmatrix} a & b \\ c & d \end{bmatrix}$$

$$= \begin{bmatrix} 1 \cdot a + 0 \cdot c & 1 \cdot b + 0 \cdot d \\ 0 \cdot a + 1 \cdot c & 0 \cdot b + 1 \cdot d \end{bmatrix} = \begin{bmatrix} a & b \\ c & d \end{bmatrix} = A.$$

Thus, multiplying any matrix A by I on either side does not change A at all. In other words, the matrix I in $M_2(\mathbb{R})$ behaves very much like the number 1 does in $\mathbb{R}$ when one multiplies.

For every nonzero real number a we can find a real number, written as $a^{-1} = 1/a$, such that $aa^{-1} = 1$. Is something similar true here in the system $M_2(\mathbb{R})$? The answer is "no." More specifically, we cannot find, for every nonzero matrix A, a matrix A^{-1} such that $AA^{-1} = I$.

Consider, for instance, the matrix $A = \begin{bmatrix} 1 & 0 \\ 0 & 0 \end{bmatrix}$. Can we find a matrix

$$B = \begin{bmatrix} e & f \\ g & h \end{bmatrix}$$

such that $AB = I$? Let's see what is needed. What we require for $AB = I$ is that

$$\begin{bmatrix} 1 & 0 \\ 0 & 1 \end{bmatrix} = AB = \begin{bmatrix} 1 & 0 \\ 0 & 0 \end{bmatrix}\begin{bmatrix} e & f \\ g & h \end{bmatrix} = \begin{bmatrix} e & f \\ 0 & 0 \end{bmatrix}.$$

This would require that $e = 1$, $f = 0$ and the absurdity that $1 = 0$. So no such B exists for this particular A. Let's try another one, the matrix

$A = \begin{bmatrix} 1 & 1 \\ 0 & 1 \end{bmatrix}$, where the result is quite different. Again we ask whether

we can find a matrix $B = \begin{bmatrix} e & f \\ g & h \end{bmatrix}$ such that $AB = I$. Again, let's see

what is needed. What is required in this case is that

$$\begin{bmatrix} 1 & 0 \\ 0 & 1 \end{bmatrix} = AB = \begin{bmatrix} 1 & 1 \\ 0 & 1 \end{bmatrix} \begin{bmatrix} e & f \\ g & h \end{bmatrix} = \begin{bmatrix} e + g & f + h \\ g & h \end{bmatrix}.$$

This requires that $g = 0$, $h = 1$, $e = 1$, $f = -1$, so that the matrix

$$B = \begin{bmatrix} 1 & -1 \\ 0 & 1 \end{bmatrix}$$

does satisfy $AB = I$. Moreover, this matrix B also satisfies the equation $BA = I$, which you can easily verify.

We have seen that for some matrices A we can find a matrix B such that $AB = BA = I$, and that for some A no such B can be found. We single out these "good" ones in our growing terminology.

Definition. A matrix A is said to be *invertible* if we can find a matrix B such that $AB = BA = I$.

A matrix A which is not invertible is called *singular*. If A is invertible, we claim that the B above is *unique*. What exactly does this mean? It means merely that if $AC = CA = I$ for a (possibly) different matrix C, then $B = C$. To see that $AB = BA = I$ and $AC = CA = I$ imply that $B = C$, just equate $AB = AC$ and cancel A by multiplying each side on the left by B. We leave the details as an exercise. You will have to use the associative law for this (see Problem 3).

If A is invertible and $AB = BA = I$ as above, we call B the *inverse* of A and in analogy to what we do in the real numbers, we write B as A^{-1}. We stress again that *not all* matrices are invertible. In a short while we shall see how the entries of A determine whether or not A is invertible.

We now come to two particular, easy-looking classes of matrices.

Definition. The matrix $A = \begin{bmatrix} a & 0 \\ 0 & d \end{bmatrix}$ is called a *diagonal matrix*.

Definition. The matrix $A = \begin{bmatrix} a & 0 \\ 0 & a \end{bmatrix}$ is called a *scalar matrix*.

So a matrix is diagonal if its off-diagonal entries are 0. And it is a scalar matrix if, in addition, the diagonal entries are equal.

Let $A = \begin{bmatrix} a & 0 \\ 0 & a \end{bmatrix}$ be a scalar matrix. Then, if we multiply any matrix B on either side by A, we get as a result a new matrix, each of whose entries are those of B merely multiplied by the number a. We *shall write* the scalar matrix $A = \begin{bmatrix} a & 0 \\ 0 & a \end{bmatrix}$ as

$A = aI.$

So, for $B = \begin{bmatrix} e & f \\ g & h \end{bmatrix}$:

$$(aI)B = \begin{bmatrix} a & 0 \\ 0 & a \end{bmatrix}\begin{bmatrix} e & f \\ g & h \end{bmatrix} = \begin{bmatrix} ae & af \\ ag & ah \end{bmatrix}$$

$$= \begin{bmatrix} e & f \\ g & h \end{bmatrix}\begin{bmatrix} a & 0 \\ 0 & a \end{bmatrix} = B(aI).$$

Looking back at the definition of multiplying a matrix by a scalar, we see that multiplying by the scalar matrix aI and multiplying by the scalar a gives us the same results. More precisely, for any matrix A,

$$(aI)A = aA,$$

for any $a \in \mathbb{R}$.

If B is also a scalar matrix bI, the multiplication becomes

$$(aI)(bI) = \begin{bmatrix} a & 0 \\ 0 & a \end{bmatrix}\begin{bmatrix} b & 0 \\ 0 & b \end{bmatrix} = \begin{bmatrix} ab & 0 \\ 0 & ab \end{bmatrix} = (ab)I.$$

Thus scalar matrices multiply like real numbers! As you can easily verify, scalar matrices also add like real numbers in the sense that

$$aI + bI = (a + b)I.$$

Thus the set $\mathbb{R}$ of real numbers a together with the operations of addition and multiplication of real numbers is duplicated, in a sense, by the set $\mathbb{R}I$ of scalar matrices aI and the operations of addition and multiplication for scalar matrices.

> *The function $f(a) = aI$ from $\mathbb{R}$ to $\mathbb{R}I$ gives a* one-to-one correspondence *between $\mathbb{R}$ and this duplicate $\mathbb{R}I$.*

The addition and multiplication of diagonal matrices is slightly more complicated than that of scalar matrices, because a scalar matrix depends on only one real number, whereas a diagonal matrix depends on two. Addition and multiplication for diagonal matrices goes as follows, as you can easily check:

$$\begin{bmatrix} a & 0 \\ 0 & b \end{bmatrix} + \begin{bmatrix} c & 0 \\ 0 & d \end{bmatrix} = \begin{bmatrix} a + c & 0 \\ 0 & b + d \end{bmatrix}$$

$$\begin{bmatrix} a & 0 \\ 0 & b \end{bmatrix} \begin{bmatrix} c & 0 \\ 0 & d \end{bmatrix} = \begin{bmatrix} ac & 0 \\ 0 & bd \end{bmatrix}.$$

We come to two other classes of special matrices.

Definition. The matrix $A = \begin{bmatrix} a & b \\ c & d \end{bmatrix}$ is called *upper triangular* if c is 0, and is called *lower triangular* if b is 0.

So, in an upper triangular matrix, the lower left entry is 0, and in a lower triangular matrix the upper right entry is 0.

PROBLEMS

In the following problems, capital Latin letters denote matrices in $M_2(\mathbb{R})$.

NUMERICAL PROBLEMS

1. Prove that $0A$ and $A0 = 0$ for every $A \in M_2(\mathbb{R})$.

2. Evaluate the following matrices.

(a) $\begin{bmatrix} 1 & 6 \\ 7 & 2 \end{bmatrix} \begin{bmatrix} -4 & 1 \\ 3 & 5 \end{bmatrix}$.

(b) $\begin{bmatrix} -1 & 4 \\ 0 & 5 \end{bmatrix} \begin{bmatrix} 6 & 2 \\ 0 & 0 \end{bmatrix} + \begin{bmatrix} -1 & 5 \\ 1 & 1 \end{bmatrix}$.

(c) $\left(\begin{bmatrix} 1 & 2 \\ 3 & 4 \end{bmatrix} \begin{bmatrix} 5 & 6 \\ 7 & 8 \end{bmatrix} \right) \begin{bmatrix} -1 & -2 \\ 3 & 4 \end{bmatrix}$

and

$\begin{bmatrix} 1 & 2 \\ 3 & 4 \end{bmatrix} \left(\begin{bmatrix} 5 & 6 \\ 7 & 8 \end{bmatrix} \begin{bmatrix} -1 & -2 \\ 3 & 4 \end{bmatrix} \right)$.

(d) How do the two resulting matrices compare?

(e) $\begin{bmatrix} \frac{1}{2} & \frac{2}{3} \\ \frac{1}{4} & -6 \end{bmatrix} \begin{bmatrix} 5 & 0 \\ 0 & -5 \end{bmatrix} - \begin{bmatrix} 5 & 0 \\ 0 & -5 \end{bmatrix} \begin{bmatrix} \frac{1}{2} & \frac{2}{3} \\ \frac{1}{4} & -6 \end{bmatrix}$.

(f) $\begin{bmatrix} \frac{1}{2} & \frac{1}{2} \\ \frac{1}{2} & \frac{1}{2} \end{bmatrix} \begin{bmatrix} \frac{1}{2} & \frac{1}{2} \\ \frac{1}{2} & \frac{1}{2} \end{bmatrix}$.

(g) $\begin{bmatrix} 1 & -1 \\ \pi & -\pi \end{bmatrix}\begin{bmatrix} 1 & -1 \\ \pi & -\pi \end{bmatrix}$.

(h) $\begin{bmatrix} 1 & 2 \\ 3 & 4 \end{bmatrix}\begin{bmatrix} -2 & 1 \\ \frac{3}{2} & -\frac{1}{2} \end{bmatrix}$.

(i) $\begin{bmatrix} -2 & 1 \\ \frac{3}{2} & -\frac{1}{2} \end{bmatrix}\begin{bmatrix} 1 & 2 \\ 3 & 4 \end{bmatrix}$.

(j) $\begin{bmatrix} a & b \\ c & d \end{bmatrix}\begin{bmatrix} d & -b \\ -c & a \end{bmatrix}$.

MORE THEORETICAL PROBLEMS

Easier Problems

3. If A, B, C are in $M_2(\mathbb{R})$, show that $(AB)C = A(BC)$. [This is known as the *associative law* of multiplication. In view of this result, we shall write $(AB)C$ without parentheses as ABC.]

4. If A, B, C are in $M_2(\mathbb{R})$, show that $A(B + C) = AB + AC$, and that $(B + C)A = BA + CA$. (These are the *distributive laws*.)

5. If A is a diagonal matrix whose diagonal entries are both nonzero, prove that A^{-1} exists and find it explicitly.

6. If A and B are upper triangular matrices, prove that AB is also upper triangular.

7. When is an upper triangular matrix invertible?

8. If A is invertible, prove that its inverse is unique.

9. If A and B are invertible, show that AB is invertible and express $(AB)^{-1}$ in terms of A^{-1} and B^{-1}.

10. If A is invertible and $AC = 0$, prove that $C = 0$; and also, show that if $DA = 0$, then $D = 0$. Then prove the more general *cancellation laws,* which state that $AB = AC$ implies $B = C$ and $BA = CA$ implies that $B = C$ if A is invertible.

Middle-Level Problems

11. If $A, B \in M_2(\mathbb{R})$, show that $(AB - BA)(AB - BA)$ is a scalar matrix.

12. If A is such that $AB = 0$ for some $B \neq 0$, show that A cannot be invertible.

13. Find two matrices A and B such that $AB = 0$ but $BA \neq 0$. What can you say about $(BA)(BA)$?

14. Prove that we cannot find two matrices A and B such that
$$AB - BA = aI$$
where $a \neq 0$ is in $\mathbb{R}$.

15. Find all matrices A that commute with *all* matrices. (That is, find all A such that $AB = BA$ for all $B \in M_2(\mathbb{R})$.)

16. What matrices commute with $A = \begin{bmatrix} a & 0 \\ 0 & b \end{bmatrix}$ if $a \neq b$?

Harder Problems

17. Find a necessary and sufficient condition on a, b, c, d in order that $A = \begin{bmatrix} a & b \\ c & d \end{bmatrix}$ be invertible.

18. If A is not invertible, show that we can find a $B \neq 0$ such that $AB = 0$.

19. If $AB = I$, prove that $BA = I$, hence A is invertible.

20. Find all $a \in \mathbb{R}$ such that $\begin{bmatrix} -1 & 2 \\ -2 & 0 \end{bmatrix} - aI$ is *not* invertible. How many such values a are there? What polynomial equation do these values a satisfy?

21. Do Problem 20 for the general $A = \begin{bmatrix} a & b \\ c & d \end{bmatrix}$, that is, find all $u \in \mathbb{R}$ such that $A - uI$ is not invertible.

22. If $A = \begin{bmatrix} a & b \\ c & d \end{bmatrix}$ satisfies $AA = 0$, show that $a + d = 0$ and $ad - bc = 0$.

23. If $A = \begin{bmatrix} a & b \\ c & d \end{bmatrix}$ satisfies $AA = A$, show that $a + d = 0$, 1, or 2.

24. For $A = \begin{bmatrix} 1 & 7 \\ 0 & -1 \end{bmatrix}$, show that we can find a matrix B such that B is invertible and $B^{-1}AB = \begin{bmatrix} u & 0 \\ 0 & v \end{bmatrix}$. What must the values of u and v be?

25. Show that we cannot find an invertible matrix B such that $B^{-1}\begin{bmatrix} 1 & 0 \\ 1 & 1 \end{bmatrix}B$ is a diagonal matrix.

26. If A is a matrix such that $B^{-1}AB$ is a scalar matrix for some invertible B in $M_2(\mathbb{R})$, prove that A must be a scalar matrix and that $AB = BA$.

27. Define the *transpose* of $A = \begin{bmatrix} a & b \\ c & d \end{bmatrix}$ as $A' = \begin{bmatrix} a & c \\ b & d \end{bmatrix}$ and show

that the transpose operation satisfies the properties $(AB)' = B'A'$, $(aA)' = aA'$, and $(A + B)' = A' + B'$

28. Find formulas for the sum and product of upper triangular matrices. Then use Problem 27 to transfer these to formulas for the sum and product of lower triangular matrices.

1.3. SOME NOTATION

This section is intended to introduce some notational devices. In reading it, the reader may very well ask: "Why go to all this trouble for the 2×2 case where everything is so explicit and can be written out in full?" Indeed, if our concerns were only the 2×2 case, it certainly would be absurd to fuss about notation. However, what we do is precisely what will be needed for the general $n \times n$ case. The 2×2 matrices afford us a good arena for acquiring some skill and familiarity with this symbolism.

Our starting point is the symbol Σ (the Greek letter for S, standing for *summation*) which some readers may have encountered before in the calculus. By

$$\sum_{r=1}^{n} a_r,$$

n being an integer greater or equal to 1, we shall mean the sum

$$\sum_{r=1}^{n} a_r = a_1 + \cdots + a_n$$

of *terms* a_r, where r varies over the set of integers from 1 to n. Note that the subscript r over which we are summing is a "dummy" index. We could easily call it k, or α, or indeed anything else. Thus

$$\sum_{r=1}^{n} a_r = \sum_{\alpha=1}^{n} a_\alpha = \sum_{\square=1}^{n} a_\square.$$

As an example of this Σ notation,

$$\sum_{r=1}^{5} r^2 = 1^2 + 2^2 + 3^2 + 4^2 + 5^2.$$

We may occasionally need a slight variant of this notation, namely the variant

$$\sum_{r=m}^{n} a_r,$$

m and n being integers with $m \leq n$, which denotes

$$a_m + a_{m+1} + \cdots + a_n.$$

So, for example,

$$\sum_{r=-2}^{3} r^2 = (-2)^2 + (-1)^2 + 0^2 + 1^2 + 2^2 + 3^2 = \sum_{\alpha=-2}^{3} \alpha^2.$$

Another, and more complicated summation symbol which will be used, although less frequently, is the *double summation*, $\Sigma\Sigma$. We express a double summation in terms of single summations:

$$\sum_{r=1}^{m} \sum_{s=1}^{n} a_{rs} = \sum_{r=1}^{m} b_r, \qquad \text{where} \quad b_r = \sum_{s=1}^{n} a_{rs}.$$

An example of this is

$$\sum_{r=1}^{4} \sum_{s=1}^{3} rs = \sum_{\alpha=1}^{4} \sum_{\beta=1}^{3} \alpha\beta = \sum_{\alpha=1}^{4} b_\alpha,$$

where

$$b_\alpha = \sum_{\beta=1}^{3} \alpha\beta = \alpha \cdot 1 + \alpha \cdot 2 + \alpha \cdot 3 = 6\alpha$$

for $\alpha = 1, 2, 3$ and 4. For instance,

$$b_3 = 3 \cdot 1 + 3 \cdot 2 + 3 \cdot 3 = 18.$$

The reader can easily verify that the values for b_1, b_2, b_3, b_4 add up to $6(1 + 2 + 3 + 4) = 60$, so that the value of the double sum $\sum_{r=1}^{4} \sum_{s=1}^{3} rs$ is 60.

We now return to the 2×2 matrices, where we want to introduce some names and a compact means of writing matrices. Consider the matrix $A = \begin{bmatrix} a & b \\ c & d \end{bmatrix}$. The *entry a* occurs in the first row and first column of A, so we call a the $(1, 1)$ *entry* of A; and b occurs in the first row and second column of A, so we call it the $(1, 2)$ *entry* of A. Similarly, c is called the $(2, 1)$ and d the $(2, 2)$ entry of A. If we denote the (r, s) entry of A by a_{rs}—which means that the entry is in the rth row and sth column of A—we shall write A as $[a_{rs}]$. To repeat, $A = [a_{rs}]$ denotes the array

$$A = \begin{bmatrix} a_{11} & a_{12} \\ a_{21} & a_{22} \end{bmatrix}.$$

If $a_{rs} = r + s$, then the matrix $A = [a_{rs}]$ is the matrix $\begin{bmatrix} 2 & 3 \\ 3 & 4 \end{bmatrix}$ since

the terms $r + s$ are arranged in the matrix A as follows:

$$A = \begin{bmatrix} 1 + 1 & 1 + 2 \\ 2 + 1 & 2 + 2 \end{bmatrix}.$$

There is a symmetry in A, in this case across the *main diagonal* of A running from its upper left corner to its lower right corner. This symmetry is due to the fact that $a_{rs} = a_{sr}$ for all values of r and s. As another example where the matrix A does not have such a condition on its entries, suppose that the entries are given by

$$a_{rs} = r + rs$$

for r and s equal to 1 and 2. Then the matrix $A = [a_{rs}]$ is the matrix $\begin{bmatrix} 2 & 3 \\ 4 & 6 \end{bmatrix}$ since the terms $r + rs$ are arranged in the matrix A as follows:

$$A = \begin{bmatrix} 1 + 1 \cdot 1 & 1 + 1 \cdot 2 \\ 2 + 2 \cdot 1 & 2 + 2 \cdot 2 \end{bmatrix}.$$

There is no such symmetry in this case, due to the fact that $a_{12} = 3$, whereas $a_{21} = 4$.

The entries of a matrix can be any real numbers. *There need not be any formula expressing a_{rs} in terms of r and s.*

What do our operations in $M_2(\mathbb{R})$ look like using these notational devices? If $A = [a_{rs}]$ and $B = [b_{rs}]$, then the sum is

$$A + B = [c_{rs}],$$

where $c_{rs} = a_{rs} + b_{rs}$ for each r and s. What about the product of the two matrices above? By our rule of multiplication, $AB = [a_{rs}][b_{rs}]$ is

$$\begin{bmatrix} a_{11} & a_{12} \\ a_{21} & a_{22} \end{bmatrix} \begin{bmatrix} b_{11} & b_{12} \\ b_{21} & b_{22} \end{bmatrix}$$

$$= \begin{bmatrix} a_{11}b_{11} + a_{12}b_{21} & a_{11}b_{12} + a_{12}b_{22} \\ a_{21}b_{11} + a_{22}b_{21} & a_{21}b_{12} + a_{22}b_{22} \end{bmatrix} = \begin{bmatrix} c_{11} & c_{12} \\ c_{21} & c_{22} \end{bmatrix},$$

where

$$c_{rs} = a_{r1}b_{1s} + a_{r2}b_{2s} = \sum_{t=1}^{2} a_{rt}b_{ts}.$$

Thus the entry in the second row and first column is $c_{21} = \sum_{t=1}^{2} a_{2t}b_{t1}$ $= a_{21}b_{11} + a_{22}b_{21}$.

In this new symbolism, the matrix $A = \begin{bmatrix} a_{11} & a_{12} \\ a_{21} & a_{22} \end{bmatrix}$ is a diagonal matrix if $a_{rs} = 0$ for $r \neq s$. And $\begin{bmatrix} a_{11} & a_{12} \\ a_{21} & a_{22} \end{bmatrix}$ is a scalar matrix if it is diagonal and $a_{rr} = a_{ss}$ for all r and s, $\begin{bmatrix} a_{11} & a_{12} \\ a_{21} & a_{22} \end{bmatrix}$ is upper triangular if $a_{rs} = 0$ for $r > s$, and $\begin{bmatrix} a_{11} & a_{12} \\ a_{21} & a_{22} \end{bmatrix}$ is lower triangular if $a_{rs} = 0$ for $r < s$. For instance, $\begin{bmatrix} 1 & 0 \\ 0 & 2 \end{bmatrix}$ is diagonal since $a_{12} = a_{21} = 0$, $\begin{bmatrix} 2 & 0 \\ 0 & 2 \end{bmatrix}$ is a scalar matrix since it is diagonal and $a_{11} = a_{22}$, $\begin{bmatrix} 1 & 3 \\ 0 & 2 \end{bmatrix}$ is upper triangular since $a_{21} = 0$ and $\begin{bmatrix} 1 & 0 \\ 2 & 2 \end{bmatrix}$ is lower triangular since $a_{12} = 0$.

The last piece of notation we want to introduce here is certainly more natural to most of us than the formalisms above. This is the notion of *exponent*. We want to define A^m for any 2 × 2 matrix A and any *positive* integer m. We do so by following the definition familiar for real numbers:

$$A^m = AAAAA \ldots A.$$
$$\text{(}m\text{ times)}$$

More formally, the two equations

$$A^1 = A, \qquad A^{m+1} = A^m A$$

define A^m *recursively* by defining it first for $m = 1$ and then for $m + 1$ after it has been defined for m, for all $m \geq 1$. We also define $A^0 = I$, since we'll need to use expressions such as $5A^0 + 6A^3 + 4A^4$.

We cannot define A^{-m} for *every* matrix A and positive integer m *unless A is invertible*. Why? Because A^{-1}, which would denote the inverse of A, makes no sense if A has no inverse, that is, if A is not invertible. However, if A is invertible, we *define A^{-m} to be* $(A^{-1})^m$ for every positive integer m.

The usual rules of exponents, namely $A^m A^n = A^{m+n}$ and $(A^m)^n = A^{mn}$, do hold for all matrices if m and n are *positive* integers, and for all invertible matrices if m and n are integers.

PROBLEMS

Make free use of the associative law $A(BC) = (AB)C$.

NUMERICAL PROBLEMS

1. Evaluate.

(a) $\displaystyle\sum_{r=1}^{6} \frac{r}{r+1}$.

(b) $\displaystyle\sum_{t=1}^{4} \frac{t+1}{t+2}$.

(c) $\displaystyle\sum_{f=-1}^{3} \frac{f}{f+5}$.

(d) $\displaystyle\sum_{g=-1}^{3} g^2$, $\displaystyle\sum_{h=-1}^{3} h^3$, and $\displaystyle\sum_{k=-2}^{3} k^3$.

(e) $\displaystyle\sum_{r=1}^{5} \sum_{s=0}^{3} \frac{r}{s+1}$ and $\displaystyle\sum_{s=0}^{3} \sum_{r=1}^{5} \frac{r}{s+1}$.

(f) $\displaystyle\sum_{b=-2}^{4} \sum_{c=-1}^{2} b(c+1)$.

2. Show that $\displaystyle\sum_{r=1}^{n} \frac{1}{r(r+1)} = 1 - \frac{1}{n+1}$.

3. Calculate the following matrices.

(a) $\begin{bmatrix} 0 & 1 \\ -4 & 0 \end{bmatrix}^2$.

(b) $\begin{bmatrix} 1 & 2 \\ 2 & 4 \end{bmatrix}^3$.

(c) $\begin{bmatrix} 1 & 1 \\ 0 & 1 \end{bmatrix}^{-1}$ (if it exists).

(d) $\begin{bmatrix} 1 & 1 \\ 1 & 1 \end{bmatrix}^{-1}$ (if it exists).

(e) $\begin{bmatrix} 1 & 0 \\ 1 & 1 \end{bmatrix}^{-3}$ (if it exists).

(f) $\begin{bmatrix} a & -a \\ b & -b \end{bmatrix}^2$.

4. If $A = \begin{bmatrix} 1 & 2 \\ 3 & 4 \end{bmatrix}$ and $B = \begin{bmatrix} 1 & -1 \\ 1 & -1 \end{bmatrix}$, calculate $(AB)^2$ and A^2B^2.

Are they equal?

5. If $A = \begin{bmatrix} 1 & 7 \\ -6 & -43 \end{bmatrix}$, show that $A^2 + 42A - I = 0$.

6. Find A^n for *all* positive integers n for the following matrices.

(a) $\begin{bmatrix} 0 & \alpha \\ \beta & 0 \end{bmatrix}$.

(b) $\begin{bmatrix} 1 & 1 \\ 0 & 1 \end{bmatrix}$.

(c) $\begin{bmatrix} 1 & 1 \\ 1 & 1 \end{bmatrix}$.

(d) $\begin{bmatrix} 1 & -1 \\ 1 & -1 \end{bmatrix}$.

MORE THEORETICAL PROBLEMS

Easier Problems

7. Let $A = \begin{bmatrix} 0 & 0 \\ 0 & 1 \end{bmatrix}$. Show that for all $B \in M_2(\mathbb{R})$,

$(AB - ABA)^2 = (BA - ABA)^2 = 0$.

8. If A is invertible and $B, C \in M_2(\mathbb{R})$, show that

$(ABA^{-1})(ACA^{-1}) = A(BC)A^{-1}$.

9. Prove the laws of exponents, namely, if A is in $M_2(\mathbb{R})$, then for all nonnegative integers m and n, $A^m A^n = A^{m+n}$ and $(A^m)^n = A^{mn}$.

Middle-Level Problems

10. Find A^n for all positive integers n for

    (a) $A = \begin{bmatrix} 1 & 2 \\ 2 & 4 \end{bmatrix}$.

    (b) $A = \begin{bmatrix} 0 & 1 \\ 1 & 1 \end{bmatrix}$.

11. If A and B are in $M_2(\mathbb{R})$:

    (a) Calculate $(AB - BA)^2$.

    (b) Find $(AB - BA)^n$ for all positive integers $n > 2$.

12. If A and B are in $M_2(\mathbb{R})$ and A is invertible, find $(ABA^{-1})^n$ for all positive integers n, expressed "nicely" in terms of A and B.

13. Find matrices B and C such that $B^2 = \begin{bmatrix} 2 & 3 \\ 0 & 2 \end{bmatrix}$ and $C^2 = \begin{bmatrix} 1 & -1 \\ 2 & 4 \end{bmatrix}$.

14. If $A \in M_2(\mathbb{R})$ satisfies $A^5 - 4A^4 + 7A^2 - 14I = 0$, show that A is invertible, and express its inverse in terms of A.

Harder Problems

15. Find a matrix $B \neq I$ such that $B^3 = I$.

16. If $A \in M_2(\mathbb{R})$ satisfies $A^2 - A + I = 0$, find A^{3n} explicitly in terms of A for every positive n.

17. If $A \in M_2(\mathbb{R})$ satisfies $A^2 + uA + vI = 0$, where $u, v \in \mathbb{R}$, find necessary and sufficient conditions on u and v that A be invertible.

18. Let A be invertible in $M_2(\mathbb{R})$ and let $B = \begin{bmatrix} u & v \\ w & x \end{bmatrix}$. Show that if $C = ABA^{-1} = \begin{bmatrix} a & b \\ c & d \end{bmatrix}$, then $a + d = u + x$.

19. Let A and B be as in Problem 18. If $C = ABA^{-1} = \begin{bmatrix} a & c \\ c & d \end{bmatrix}$, show that $ad - bc = ux - vw$.

20. If $A \in M_2(\mathbb{R})$ is invertible, show that $A^{-1} = aA + bI$ for some real numbers a and b.

21. If A is the matrix

$$
\begin{bmatrix}
\cos \dfrac{2\pi}{k} & \sin \dfrac{2\pi}{k} \\[2ex]
-\sin \dfrac{2\pi}{k} & \cos \dfrac{2\pi}{k}
\end{bmatrix},
$$

where k is a positive integer, find A^m for all m. What is the matrix A^m in the special case where $m = k$?

1.4. TRACE, TRANSPOSE, AND ODDS AND ENDS

We start off with odds and ends. Heretofore, we have not singled out any result and called it a lemma or theorem. We become more formal now and designate some results as lemmas (lemmata) or theorems. This not only provides us with a handy way of referring back to a result when we need it, but it emphasizes the result and gives us a parallel path to follow when we treat matrices of any size.

Although it may complicate what is, at present, a simple and straightforward situation, we do everything using the summation notation, Σ, throughout. In this way, we can get some practice and facility in playing around with this important symbol. The reader might find it useful to rewrite each proof in this section without using summation notation, to see how it looks written out.

Theorem 1.4.1. If A, B, and C are in $M_2(\mathbb{R})$, then:

    1. $A + B = B + A$ *Commutative Law*

    2. $(A + B) + C = A + (B + C)$ *Associative Law for Sums*

    3. $(AB)C = A(BC)$ *Associative Law for Products*

    4. $A(B + C) = AB + AC$ and

          $(B + C)A = BA + CA$ *Distributive Laws*

Proof. The proof of Part (1) amounts to letting $A = [a_{rs}]$, $B = [b_{rs}]$ and observing that the (r, s) entry $a_{rs} + b_{rs}$ of $A + B$ equals the

(r, s) entry $b_{rs} + a_{rs}$ of $B + A$ for all r, s. The proofs of Parts (2) and (4) are just as easy and are left to the reader.

We carry out the proof of Part (3) in detail. If $A = [a_{rs}]$, $B = [b_{rs}]$, and $C = [c_{rs}]$, then if $AB = [d_{rs}]$, we know that $d_{rs} = \sum_{t=1}^{2} a_{rt}b_{ts}$. Thus

$$(AB)C = [d_{rs}][c_{rs}] = [f_{rs}],$$

where

$$f_{rs} = \sum_{u=1}^{2} d_{ru}c_{us} = \sum_{u=1}^{2} \left(\sum_{t=1}^{2} a_{rt}b_{tu} \right) c_{us} = \sum_{u=1}^{2} \sum_{t=1}^{2} (a_{rt}b_{tu})c_{us}.$$

On the other hand, $BC = [g_{rs}]$, where $g_{rs} = \sum_{u=1}^{2} b_{ru}c_{us}$. Thus

$$A(BC) = [a_{rs}][g_{rs}] = [h_{rs}],$$

where

$$h_{rs} = \sum_{t=1}^{2} a_{rt}g_{ts} = \sum_{t=1}^{2} a_{rt}\left(\sum_{u=1}^{2} b_{tu}c_{us} \right) = \sum_{t=1}^{2} \sum_{u=1}^{2} a_{rt}(b_{tu}c_{us}).$$

Since $f_{rs} = h_{rs}$, we see that

$$(AB)C = [f_{rs}] = [h_{rs}] = A(BC). \qquad \blacksquare$$

In the proof above, the equations

$$d_{rs} = \sum_{t=1}^{2} a_{rt}b_{ts}$$

$$f_{rs} = \sum_{u=1}^{2} d_{ru}c_{us} = \sum_{u=1}^{2} \left(\sum_{t=1}^{2} a_{rt}b_{tu} \right) c_{us} = \sum_{u=1}^{2} \sum_{t=1}^{2} (a_{rt}b_{tu})c_{us}$$

$$g_{rs} = \sum_{u=1}^{2} b_{ru}c_{us}$$

$$h_{rs} = \sum_{t=1}^{2} a_{rt}g_{ts} = \sum_{t=1}^{2} a_{rt}\left(\sum_{u=1}^{2} b_{tu}c_{us} \right) = \sum_{t=1}^{2} \sum_{u=1}^{2} a_{rt}(b_{tu}c_{us})$$

written in summation notation become

$$d_{rs} = a_{r1}b_{1s} + a_{r2}b_{2s}$$

$$f_{rs} = d_{r1}c_{1s} + d_{r2}c_{2s}$$

$$= (a_{r1}b_{11} + a_{r2}b_{21})c_{1s} + (a_{r1}b_{12} + a_{r2}b_{22})c_{2s}$$

$$= (a_{r1}b_{11})c_{1s} + (a_{r2}b_{21})c_{1s} + (a_{r1}b_{12})c_{2s} + (a_{r2}b_{22})c_{2s}$$

$$g_{rs} = b_{r1}c_{1s} + b_{r2}c_{2s}$$

$$h_{rs} = a_{r1}g_{1s} + a_{r2}g_{2s}$$

$$= a_{r1}(b_{11}c_{1s} + b_{12}c_{2s}) + a_{r2}(b_{21}c_{1s} + b_{22}c_{2s})$$

$$= a_{r1}(b_{11}c_{1s}) + a_{r1}(b_{12}c_{2s}) + a_{r2}(b_{21}c_{1s}) + a_{r2}(b_{22}c_{2s})$$

when they are written out.

Now that you have seen in detail why the associative law

$$(AB)C = A(BC)$$

is true, you can see that it just amounts to showing that the (r, s) entries of the matrices $(AB)C$, $A(BC)$ are the double sums

$$\sum_{u=1}^{2} \sum_{t=1}^{2} (a_{rt}b_{tu})c_{us}, \qquad \sum_{t=1}^{2} \sum_{u=1}^{2} a_{rt}(b_{tu}c_{us}).$$

These double sums are equal because the order of summation does not affect the sum and *because the terms $(a_{rt}b_{tu})c_{us}$, $a_{rt}(b_{tu}c_{us})$ are equal by the associative law for numbers!*

In view of the fact that $(AB)C = A(BC)$, we can leave off the parentheses in products of three matrices—for either bracketing gives the same result—and write either product as ABC, for A, B, C in $M_2(\mathbb{R})$. In fact, we can leave off the parentheses in products of any number of matrices.

Associative operations abound in mathematics and, of course,

we can leave off the parentheses for such other associative operations as well.

We'll do this from now on without fanfare!

We know that we can add and multiply matrices. We now introduce another function on matrices—a function from $M_2(\mathbb{R})$ to $\mathbb{R}$—which plays an important role in matrix theory.

Definition. If $A = [a_{rs}]$, then the *trace* of A, denoted by tr (A), is

$$\text{tr } (A) = \sum_{r=1}^{2} a_{rr} = a_{11} + a_{22}.$$

In short, the trace of A is simply the sum of the elements on its diagonal.

What properties does this function, trace, enjoy?

Theorem 1.4.2. If $A, B \in M_2(\mathbb{R})$ and $a \in R$, then

1.   tr $(A + B) =$ tr $(A) +$ tr (B);
2.   tr $(aA) = a(\text{tr } (A))$;
3.   tr $(AB) =$ tr (BA);
4.   tr $(aI) = 2a$.

Proof.   The proofs of Parts (1), (2), and (4) are easy and direct. We leave this to you. For the sake of practice, try to show them using the summation symbol.

We prove Part (3), not because it is difficult, but to get a further acquaintance with the summation notation.

If $A = [a_{rs}]$ and $B = [b_{rs}]$, then $AB = [c_{rs}]$, where

$$c_{rs} = \sum_{t=1}^{2} a_{rt}b_{ts}.$$

Thus

$$\text{tr } (AB) = \sum_{r=1}^{2} c_{rr} = \sum_{r=1}^{2} \sum_{t=1}^{2} a_{rt}b_{tr}.$$

If $BA = [d_{rs}]$, then $d_{rs} = \sum_{u=1}^{2} b_{ru}a_{us}$, whence

$$\text{tr } (BA) = \sum_{r=1}^{2} d_{rr} = \sum_{r=1}^{2} \sum_{u=1}^{2} b_{ru}a_{ur}.$$

Since the u and r in the latter sum are dummies, we may call them what we will. Replacing u by r, and r by t, we get

$$\text{tr } (BA) = \sum_{t=1}^{2} \sum_{r=1}^{2} b_{tr}a_{rt}.$$

Since our evaluation for AB was

$$\text{tr } (AB) = \sum_{r=1}^{2} \sum_{t=1}^{2} a_{rt}b_{tr},$$

we have tr (AB) = tr (BA).   ■

In Problem 18 of Section 1.3 you were asked to show that if A is invertible and $B = \begin{bmatrix} u & v \\ w & x \end{bmatrix}$, then if $C = A^{-1}BA = \begin{bmatrix} a & b \\ c & c \end{bmatrix}$, we must have $u + x = a + d$. In terms of traces, this equality,

$$u + x = a + d,$$

becomes the equality

$$\text{tr } (B) = \text{tr } (A^{-1}BA).$$

We now formally state and prove this important result.

Corollary 1.4.3.   If A is invertible, then tr (B) = tr $(A^{-1}BA)$.

Proof.   By the theorem, tr $(A^{-1}(BA))$ = tr $((BA)A^{-1})$ = tr (B).

■

Although we know that this corollary holds in general, it is nice to see that it is true for particular matrices, If $A = \begin{bmatrix} 1 & -2 \\ 0 & 1 \end{bmatrix}$ and $B = \begin{bmatrix} -1 & 1 \\ 3 & 3 \end{bmatrix}$, then A is invertible with inverse $A^{-1} = \begin{bmatrix} 1 & 2 \\ 0 & 1 \end{bmatrix}$. Thus

$$A^{-1}BA = \begin{bmatrix} 1 & 2 \\ 0 & 1 \end{bmatrix}\begin{bmatrix} -1 & 1 \\ 3 & 3 \end{bmatrix}\begin{bmatrix} 1 & -2 \\ 0 & 1 \end{bmatrix} = \begin{bmatrix} 5 & 7 \\ 3 & 3 \end{bmatrix}\begin{bmatrix} 1 & -2 \\ 0 & 1 \end{bmatrix}$$

$$= \begin{bmatrix} 5 & -3 \\ 3 & -3 \end{bmatrix},$$

hence

$$\text{tr } (A^{-1}BA) = 5 + (-3) = 2 = (-1) + 3 = \text{tr } (B).$$

The trace is a function (or mapping) from $M_2(\mathbb{R})$ into the reals $\mathbb{R}$. We now introduce a mapping from $M_2(\mathbb{R})$ into $M_2(\mathbb{R})$ itself which exhibits a kind of internal reverse symmetry of the system $M_2(\mathbb{R})$.

Definition. If $a = [a_{rs}] \in M_2(\mathbb{R})$, then the *transpose* of A, denoted by A', is the matrix $A' = [b_{rs}]$, where $b_{rs} = a_{sr}$ for each r and s.

The transpose of a matrix A is, from the definition, that matrix which is obtained from A by interchanging the rows and columns of A. So if A is the matrix

$$\begin{bmatrix} -3 & 2 \\ 4 & 7 \end{bmatrix},$$

then its transpose A' is the matrix

$$\begin{bmatrix} -3 & 4 \\ 2 & 7 \end{bmatrix}.$$

What are the basic properties of transpose? You can easily guess some of them by looking at examples. For instance, the transpose A'' of the transpose A' of A is A itself. Because of their importance, we now list several of these properties formally.

Theorem 1.4.4. If A and B are in $M_2(\mathbb{R})$ and a and b are in $\mathbb{R}$, then

1. $(A')' = A'' = A$;
2. $(aA + bB)' = aA' + bB'$;
3. $(AB)' = B'A'$.

Proof. Both Parts (1) and (2) are very easy and are left to the reader. We do Part (3).

Let

$$A = [a_{rs}] \quad \text{and} \quad B = [b_{rs}],$$

so that $AB = [c_{rs}]$, where

$$c_{rs} = \sum_{t=1}^{2} a_{rt} b_{ts}.$$

Then $(AB)' = [c_{rs}]' = [d_{rs}]$, where $d_{rs} = c_{sr}$ for each r and s.

On the other hand, if $A' = [u_{rs}]$ and B' $[v_{rs}]$, then $u_{rs} = a_{sr}$, $v_{rs} = b_{sr}$, and $B'A' = [v_{rs}][u_{rs}] = [w_{rs}]$, where

$$w_{rs} = \sum_{t=1}^{2} v_{rt} u_{ts} = \sum_{t=1}^{2} b_{tr} a_{st} = \sum_{t=1}^{2} a_{st} b_{tr} = c_{sr} = d_{rs}.$$

Thus $(AB)' = B'A'$.   ∎

It is important to note in the formula

$$(AB)' = B'A'$$

that the order of A and B are reversed. Thus the transpose mapping from $M_2(\mathbb{R})$ to itself gives a kind of reverse symmetry of the multiplicative structure of $M_2(\mathbb{R})$. At the same time, it has the property,

$$(A + B)' = A' + B',$$

so that it is also a kind of symmetry of the additive structure of $M_2(\mathbb{R})$.

Again, let's verify our result for specific matrices A and B. If $A = \begin{bmatrix} 1 & 2 \\ -3 & 4 \end{bmatrix}$ and $B = \begin{bmatrix} 0 & 1 \\ 2 & -1 \end{bmatrix}$, then

$$AB = \begin{bmatrix} 1 & 2 \\ -3 & 4 \end{bmatrix}\begin{bmatrix} 0 & 1 \\ 2 & -1 \end{bmatrix} = \begin{bmatrix} 4 & -1 \\ 8 & -7 \end{bmatrix},$$

so that $(AB)' = \begin{bmatrix} 4 & 8 \\ -1 & -7 \end{bmatrix}$. On the other hand, $A' = \begin{bmatrix} 1 & -3 \\ 2 & 4 \end{bmatrix}$ and $B' = \begin{bmatrix} 0 & 2 \\ 1 & -1 \end{bmatrix}$, whence

$$B'A' = \begin{bmatrix} 0 & 2 \\ 1 & -1 \end{bmatrix}\begin{bmatrix} 1 & -3 \\ 2 & 4 \end{bmatrix} = \begin{bmatrix} 4 & 8 \\ -1 & -7 \end{bmatrix} = (AB)'.$$

We call a matrix A *symmetric* if $A' = A$, and a matrix B *skew-symmetric* or *skew* if $B' = -B$. Thus $\begin{bmatrix} 1 & \pi \\ \pi & 0 \end{bmatrix}$ is symmetric, while $\begin{bmatrix} 0 & \pi \\ -\pi & 0 \end{bmatrix}$ is skew.

PROBLEMS

Use the summation notation in your solutions for these problems.

NUMERICAL PROBLEMS

1. If $A = \begin{bmatrix} 1 & 2 \\ 3 & 4 \end{bmatrix}$ and $B = \begin{bmatrix} 0 & 1 \\ 1 & 2 \end{bmatrix}$, show by a direct calculation that tr $(AB) =$ tr (BA).

2. For the matrices A and B in Problem 1, calculate $(AB)'$ and $B'A'$ and show that they are equal.

3. Let $A = \begin{bmatrix} 1 & 0 \\ 0 & 3 \end{bmatrix}$, $B = \begin{bmatrix} 0 & 1 \\ 2 & 3 \end{bmatrix}$, and $C = \begin{bmatrix} 1 & 1 \\ 0 & 2 \end{bmatrix}$. Calculate both tr (ABC) and tr (CBA). Are they equal?

4. Calculate tr (AB') if $A = \begin{bmatrix} 0 & \frac{1}{2} \\ -\frac{1}{3} & 1 \end{bmatrix}$ and $B = \begin{bmatrix} 1 & 2 \\ 3 & 0 \end{bmatrix}$.

5. Show that ABA' is symmetric if $A = \begin{bmatrix} 1 & 1 \\ 4 & 2 \end{bmatrix}$ and $B = \begin{bmatrix} 1 & -1 \\ -1 & 7 \end{bmatrix}$.

6. Show that $AB - B'A'$ is skew-symmetric where $A = \begin{bmatrix} 2 & -3 \\ 4 & 5 \end{bmatrix}$ and $B = \begin{bmatrix} 5 & \frac{1}{2} \\ 6 & 0 \end{bmatrix}$.

MORE THEORETICAL PROBLEMS

Easier Problems

7. Prove Parts (1), (2), and (4) of Theorem 1.4.1.
8. Prove Parts (1), (2), and (4) of Theorem 1.4.2.
9. Prove Parts (1) and (2) of Theorem 1.4.4.
10. If B is any matrix in $M_2(\mathbb{R})$, show that $B = C + aI$, where tr $(C) = 0$ and $a \in \mathbb{R}$.
11. If A and B are symmetric, prove that
    - (a) A^n is symmetric for all positive integers n.
    - (b) $AB + BA$ is symmetric.
    - (c) $AB - BA$ is skew-symmetric.
    - (d) ABA is symmetric.
    - (e) If A is invertible, then A^{-1} is symmetric.
12. If A is skew-symmetric, show that A^n is skew-symmetric if n is a positive odd integer and A^n is symmetric if n is a positive even integer.
13. If A and B are skew-symmetric, show that
    - (a) $AB + BA$ is symmetric.
    - (b) $AB - BA$ is skew-symmetric.
    - (c) $A(AB + BA) - (AB + BA)A$ is symmetric.

(d) *ABA* is skew-symmetric.

(e) If A is invertible, then A^{-1} is skew-symmetric.

14. If A and B *commute* (i.e., $AB = BA$) show that

(a) If A is symmetric and B skew-symmetric, then AB is skew-symmetric.

(b) If A and B are skew-symmetric, then AB is symmetric.

(c) If A and B are symmetric, then AB is symmetric.

15. Produce specific matrices A and B which do not commute for which Parts (a) and (b) of Problem 14 are false.

16. If $A = A'$, show that BAB' is symmetric for all matrices B.

17. If $A' = -A$, show that BAB' is skew-symmetric for all matrices B.

18. Show that tr $(A) = $ tr (A').

Middle-Level Problems

19. If $A \neq 0$, prove that tr $(AA') > 0$.

20. Using Problem 19, prove that if A is symmetric and $A^n = 0$ for some positive integer n, then $A = 0$.

21. If $A_1, A_2, \ldots, A_n$ are symmetric and

$$\text{tr } (A_1^2 + A_2^2 + \cdots + A_n^2) = 0,$$

show that $A_1 = A_2 = \cdots = A_n = 0$.

22. Given any matrix A, show that we can write A as $A = B + C$, where B is symmetric and C is skew-symmetric.

23. Show that the B and C in Problem 22 are uniquely determined by A.

24. If $AA' = I = BB' = I$, shows that $(AB)(AB)' = I$.

25. If $AA' = I$, prove that

(a) ABA^{-1} is symmetric whenever B is symmetric.

(b) ABA^{-1} is skew-symmetric whenever B is skew-symmetric.

Harder Problems

26. If $A' = -A$, show that $A + aI$ is invertible for all $a \neq 0$ in $\mathbb{R}$.

27. If tr $(AB) = 0$ for *all* matrices B, prove that $A = 0$.

28. If A is any matrix, prove that AA' and $A'A$ are both symmetric.

29. If both tr $(A) = 0$ and tr $(A^2) = 0$, show that $A^2 = 0$.

30. If $A^2 = 0$, prove that tr $(A) = 0$.

31. If $A^2 = A$, prove that tr (A) is an integer. What integers can arise this way?

32. Show that if A is any matrix, then, for $a < 0$ in $\mathbb{R}$, $AA' - aI$ is invertible.

Very Hard Problems

33. If tr $(A) = 0$, prove that there exists an invertible matrix B such that $B^{-1}AB = \begin{bmatrix} 0 & u \\ v & 0 \end{bmatrix}$ for some u, v in $\mathbb{R}$.

34. If tr $(A) = 0$, show that we can find a B and C such that

$$A = BC - CB.$$

35. If tr $(ABC) = $ tr (CBA) for all matrices C, prove that $AB = BA$.

36. Let A be an invertible matrix. Define the mapping * from $M_2(\mathbb{R})$ to $M_2(\mathbb{R})$ by $B* = AB'A^{-1}$ for every B in $M_2(\mathbb{R})$. What are the necessary and sufficient conditions on A in order that * satisfy the three rules:

(1) $B** = B$ [where $B**$ denotes $(B*)*$];

(2) $(B + C)* = B* + C*$;

(3) $(BC)* = C*B*$;

for all B, C in $M_2(\mathbb{R})$?

1.5. DETERMINANTS

So far, while the 2×2 matrices have served as a very simple model of the various definitions, concepts, and results that will be seen to hold in the more general context, one could honestly say that these special matrices do not present an oversimplified case. The pattern outlined for them is the same pattern that we shall employ, later, in general. Even the method of proof, with only a slight variation, will go over.

Now, for the first time, we come to a situation where the 2×2 case is vastly oversimplified—in the notion of the *determinant*. Even the very definition of the determinant in the $n \times n$ matrices is of a fairly complicated nature. Moreover, the proofs of many, if not most of the results will be of a different flavor, and of a considerably different degree of difficulty. While the general case will be a rather sticky affair, for the 2×2 matrices there are no particularly nasty points.

Keeping this in mind, we proceed to define the determinant for elements of $M_2(\mathbb{R})$.

Definition. If $A = \begin{bmatrix} a & b \\ c & d \end{bmatrix}$, then the *determinant* of A, written det (A), is defined by det $(A) = ad - bc$.

We sometimes denote det $\begin{bmatrix} a & b \\ c & d \end{bmatrix}$ by $\begin{vmatrix} a & b \\ c & d \end{vmatrix}$.

For instance, det $(0) = 0$, det $I = 1$, det $\begin{bmatrix} a & b \\ 0 & d \end{bmatrix} = ad$,

det $\begin{bmatrix} 5 & 6 \\ -1 & 2 \end{bmatrix} = 5 \cdot 2 - (-1)6 = 16$, and det $\begin{bmatrix} 1 & 3 \\ 5 & 15 \end{bmatrix}$

$= \begin{vmatrix} 1 & 3 \\ 5 & 15 \end{vmatrix} = 1 \cdot 15 - 3 \cdot 5 = 0$.

The first result that we prove for the determinant is a truly trivial one.

Lemma 1.5.1. det $(A) = $ det (A').

Proof. If $A = \begin{bmatrix} a & b \\ c & d \end{bmatrix}$, then $A' = \begin{bmatrix} a & c \\ b & d \end{bmatrix}$ and det $(A) = $ $ad - bc = ad - cb = $ det (A').   ■

Trivial though it be, Lemma 1.5.1 assures us that whatever result holds for the rows of a matrix holds equally for the columns.

Now for some of the basic properties of the determinant.

Theorem 1.5.2. In changing from one matrix to another, the following properties describe how the determinant changes:

1. If we interchange two rows of a matrix A, the determinant of the resulting matrix is $-$det (A) (only the sign of the determinant is changed).

2. If we add any multiple of any row of a matrix A to another, the determinant of the resulting matrix is just det (A) (the determinant does not change).

3. If we multiply a given row of a matrix A by a real number u, then the determinant of the resulting matrix is u(det (A)) (the determinant is multiplied by u).

4. If two rows of A are equal, det $(A) = 0$.

Proof. Let A be the matrix

$$A = \begin{bmatrix} a & b \\ c & d \end{bmatrix}.$$

If we interchange the rows of A, we get the matrix

$$\begin{bmatrix} c & d \\ a & b \end{bmatrix},$$

so that the first property simply asserts that

$$\det \begin{bmatrix} c & d \\ a & b \end{bmatrix} = -\det \begin{bmatrix} a & b \\ c & d \end{bmatrix},$$

which is true because $cb - ad = -(ad - bc)$. For the second assertion, suppose that the multiple (ua, ub) of the first row is added to the second row, changing A from the matrix $\begin{bmatrix} a & b \\ c & d \end{bmatrix}$ to the matrix

$$\begin{bmatrix} a & b \\ ua + c & ub + d \end{bmatrix}.$$

Then the determinant of the new matrix is $aub + ad - (bua + bc) = ad - bc$, which is the same as the determinant of the original matrix. A similar argument works if, instead, a multiple (uc, ud) of the second row is added to the first row.

The third assertion states that

$$\det \begin{bmatrix} ua & ub \\ c & d \end{bmatrix} = \det \begin{bmatrix} a & b \\ uc & ud \end{bmatrix} = u \begin{vmatrix} a & b \\ c & d \end{vmatrix},$$

which is true because $uad - ubc = aud - buc = u(ad - bc)$.

The remaining assertion simply states the obvious, namely that the $\begin{vmatrix} a & b \\ a & b \end{vmatrix} = ab - ba$ is 0. Nevertheless, we also prove this as a conse-

quence of Part (1) of the theorem, because this is how it is proved in the $n \times n$ case. Namely, observe that since two different rows of A are equal, when these two rows are interchanged, the following two things happen:

1. The matrix A does not change, so neither does its determinant.
2. By Part (1), its determinant is multiplied by a factor of -1.

So

$$\det (A) = -\det (A),$$

whence

$$\det (A) = 0. \quad \blacksquare$$

Since the determinant does not change when we replace a matrix A by its transpose, that is, we replace its rows by its columns, we have

Corollary 1.5.3.   If we replace the word "row" by the word "column" in Theorem 1.5.2, all the results remain valid.

Proof.   The determinant of a matrix and its transpose are the same, so that row properties of determinants imply corresponding column properties and conversely. $\blacksquare$

We interrelate the algebraic property of the invertibility of a matrix with properties of the determinant.

Theorem 1.5.4. A is invertible if and only if $\det (A) \neq 0$. If $A = \begin{bmatrix} a & b \\ c & d \end{bmatrix}$ is invertible, its inverse is $\dfrac{1}{\det (A)} \begin{bmatrix} d & -b \\ -c & a \end{bmatrix}$.

Proof.   If $A = \begin{bmatrix} a & b \\ c & d \end{bmatrix}$, then for $B = \begin{bmatrix} d & -b \\ -c & a \end{bmatrix}$, we have

$$AB = \begin{bmatrix} ad - bc & 0 \\ 0 & ad - bc \end{bmatrix} = (\det (A))I.$$

So if $\alpha = \det (A) \neq 0$, then

$$A(\alpha^{-1}B) = \alpha^{-1}AB = (\alpha^{-1} \det (A))I = \alpha^{-1}\alpha I = I.$$

Similarly,

$$(\alpha^{-1}B)A = I.$$

Hence $\alpha^{-1}B$ is the inverse of A.

If $\alpha = \det (A) = 0$, then, for the B above, $AB = 0$. Thus if A were invertible, we would have

$$A^{-1}(AB) = A^{-1}0 = 0.$$

But

$$A^{-1}(AB) = (A^{-1}A)B = IB = B.$$

So $B = 0$. Thus $d = b = c = a = 0$. But this implies that $A = 0$, and 0 is certainly not invertible. Thus, if det $(A) = 0$, then A is not invertible. ■

The key result—and for the general case it will be the most difficult one—intertwines the determinant of the product of matrices with that of their determinants.

Theorem 1.5.5. If A and B are matrices, then

  det $(AB) = $ det (A) det (B).

  Proof. Suppose that $A = \begin{bmatrix} a & b \\ c & d \end{bmatrix}$ and $B = \begin{bmatrix} u & v \\ w & x \end{bmatrix}$. Therefore,

det $(A) = ad - bc$ and det $(B) = ux - wv$. Since

$$AB = \begin{bmatrix} a & b \\ c & d \end{bmatrix} \begin{bmatrix} u & v \\ w & x \end{bmatrix} = \begin{bmatrix} au + bw & av + bx \\ cu + dw & cv + dx \end{bmatrix},$$

we can compute as follows:

$$\begin{aligned} \det (AB) &= (au + bw)(cv + dx) - (av + bx)(cu + dw) \\ &= aucv + audx + bwcv + bwdx - avcu - avdw \\ &\quad - bxcu - bxdw \\ &= adux + bcwv - adwv - bcux \\ &= (ad - bc)(ux - wv) = \det (A) \det (B). \end{aligned}$$

This proves the assertion. ■

An immediate consequence of the theorem is

Corollary 1.5.6. If A is invertible, then det (A) is nonzero and

$$\det (A^{-1}) = \frac{1}{\det (A)}.$$

  Proof. Since $AA^{-1} = I$, we know that det $(AA^{-1}) = $ det $(I) = 1$. But by the theorem, det $(AA^{-1}) = $ det (A) det (A^{-1}), in consequence of which, det (A) det $(A^{-1}) = 1$. Therefore, det (A) is nonzero and

$$\det (A^{-1}) = \frac{1}{\det (A)}. \quad ■$$

There is yet another important consequence of Theorem 1.5.5, namely

Corollary 1.5.7. If A is invertible and B is any matrix, then

det $(A^{-1}BA)$ = det (B).

Proof. Since $A^{-1}BA = (A^{-1}B)A$, by the theorem we have

$$\text{det } (A^{-1}BA) = \text{det } ((A^{-1}B)A) = \text{det } (A^{-1}B) \text{ det } (A)$$
$$= \text{det } (A^{-1}) \text{ det } (B) \text{ det } (A)$$
$$= \text{det } (A^{-1}) \text{ det } (A) \text{ det } (B)$$
$$= \text{det } (B).$$

(We have made another use of the theorem, and also of Corollary 1.5.6, in this chain of equalities.)   ■

Note the similarity of Corollary 1.5.7 to that of Corollary 1.4.3.

It is helpful, even when we know a general result, to see it in a specific instance. We do this for Corollary 1.5.7. Let A be the matrix $\begin{bmatrix} 1 & -2 \\ 0 & 1 \end{bmatrix}$ and B be the matrix $\begin{bmatrix} 3 & 4 \\ 5 & 6 \end{bmatrix}$. Then A is invertible, with $A^{-1} = \begin{bmatrix} 1 & 2 \\ 0 & 1 \end{bmatrix}$. Therefore, we have

$$A^{-1}BA = \begin{bmatrix} 1 & 2 \\ 0 & 1 \end{bmatrix} \begin{bmatrix} 3 & 4 \\ 5 & 6 \end{bmatrix} \begin{bmatrix} 1 & -2 \\ 0 & 1 \end{bmatrix}$$
$$= \begin{bmatrix} 13 & 16 \\ 5 & 6 \end{bmatrix} \begin{bmatrix} 1 & -2 \\ 0 & 1 \end{bmatrix} = \begin{bmatrix} 13 & -10 \\ 5 & -4 \end{bmatrix}.$$

Thus det $(A^{-1}BA) = 13 \cdot (-4) - (5)(-10) = -52 + 50 = -2$ and

$$\text{det } (B) = \text{det } \begin{bmatrix} 3 & 4 \\ 5 & 6 \end{bmatrix} = 3 \cdot 6 - 4 \cdot 5 = -2.$$

We remind you for the nth time—probably to the point of utter boredom—that although the results for the determinant of the 2 × 2 matrices are established easily and smoothly, this will not be the case for the analogous results in the case of $n \times n$ matrices.

PROBLEMS

NUMERICAL PROBLEMS

1. Compute the following determinants.

   (a) $\det \begin{bmatrix} 1 & 6 \\ -6 & 1 \end{bmatrix}$.

   (b) $\det \left(\begin{bmatrix} 5 & 6 \\ 7 & 8 \end{bmatrix} \begin{bmatrix} 4 & 3 \\ -3 & 4 \end{bmatrix} \right)$.

   (c) $\det \left(\begin{bmatrix} 5 & 6 \\ 7 & 8 \end{bmatrix} + \begin{bmatrix} 4 & 3 \\ -3 & 4 \end{bmatrix} \right)$.

   (d) $\det \left(\begin{bmatrix} 1 & 2 \\ 2 & 4 \end{bmatrix} + \begin{bmatrix} 5 & -6 \\ -7 & 8 \end{bmatrix} \right)$.

2. In Part (c) of Problem 1, is it true that

   $$\det \left(\begin{bmatrix} 5 & 6 \\ 7 & 8 \end{bmatrix} + \begin{bmatrix} 4 & 3 \\ -3 & 4 \end{bmatrix} \right) = \det \begin{bmatrix} 5 & 6 \\ 7 & 8 \end{bmatrix} + \det \begin{bmatrix} 4 & 3 \\ -3 & 4 \end{bmatrix}?$$

3. For the matrix $A = \begin{bmatrix} 1 & 2 \\ 3 & 4 \end{bmatrix}$, calculate tr $(A^2) - (\text{tr } (A))^2$.

4. Find all real numbers a such that $\det \left(aI - \begin{bmatrix} 4 & 5 \\ 0 & -\pi \end{bmatrix} \right) = 0$.

5. Find all real numbers a such that $\det \left(aI - \begin{bmatrix} 3 & 4 \\ 1 & 3 \end{bmatrix} \right) = 0$.

6. Find all real numbers a such that $\det \left(aI - \begin{bmatrix} 5 & 6 \\ 6 & 7 \end{bmatrix} \right) = 0$.

7. Is there a real number a such that $\det \left(aI - \begin{bmatrix} 0 & 6 \\ -6 & 0 \end{bmatrix} \right) = 0?$

MORE THEORETICAL PROBLEMS

Easier Problems

8. Complete the proof of Theorem 1.5.2.

9. If det (A) and det (B) are rational numbers and C is invertible, determine whether det (AB) and det (CAC^{-1}) are rational numbers.

10. If A and B are matrices such that $AB = I$, then prove—using the results of this section on determinants—that $BA = I$.

11. If A and B are invertible, show that det $(ABA^{-1}B^{-1}) = 1$.

12. Prove for any matrix in $M_2(\mathbb{R})$ that 2 det $(A) = $ tr $(A^2) - ($tr $(A))^2$.

13. Show that every matrix A in $M_2(\mathbb{R})$ satisfies the relation

$$A^2 - (\text{tr }(A))A + (\det (A))I = 0.$$

Middle-Level Problems

14. From Problem 13, find the inverse of A, if A is invertible, in terms of A.

15. Using Problem 13, show that if $C = AB - BA$, then $C^2 = aI$ for some real number a.

16. Verify the results of Problem 15 for the matrix $A = \begin{bmatrix} 1 & -6 \\ \frac{2}{3} & -5 \end{bmatrix}$ and all B.

17. Calculate det $(xI - A)$ for any matrix A.

18. If det $(aI - A) = 0$, show that det $(a^nI - A^n) = 0$ for all positive integers n.

19. Show that if a is any real number, then

$$\det (aI - A) = a^2 - (\text{tr } A)a + \det (A).$$

20. Using determinants, show for any real numbers a and b that if $A = \begin{bmatrix} a & b \\ -b & a \end{bmatrix} \neq 0$, then A is invertible.

Harder Problems

21. If det $(A) < 0$, show that there are exactly two real numbers a such that $aI - A$ is not invertible.

22. If $A = \begin{bmatrix} a & b \\ b & a \end{bmatrix}$ with $b \neq 0$, prove that there exists exactly two real numbers u such that $uI - A$ is not invertible.

23. If $A = \begin{bmatrix} a & b \\ -b & a \end{bmatrix}$ with $b \neq 0$, show that for all real numbers $u \neq 0$,

$uI - A$ is invertible.

1.6. CRAMER'S RULE

The determinant was introduced in a rather ad hoc manner as a number associated with a given matrix. The meaning or relationship of the determinant of a matrix to that matrix was not clear or motivated. Aside from providing a criterion for the invertibility of a matrix, the determinant seemed to have nothing much to do with anything.

It should be somewhat illuminating if we could see the determinant arise naturally, and with some bite, in a concrete context. This context will be in the solution of simultaneous linear equations.

Suppose that we want to solve for x and y in the equations

$$ax + by = g \qquad\qquad (1)$$
$$cx + dy = h.$$

The method of solution is to eliminate one of x or y between these two equations. To eliminate y, we multiply the first equation by d, and the second one by b, and subtract. The outcome of all this is that $(ad - bc)x = dg - bh$, so, provided that $ad - bc \neq 0$, we obtain the solution

$$x = \frac{dg - bh}{ad - bc}.$$

Similarly, we get

$$y = \frac{ah - cg}{ad - bc}.$$

We can recognize the expressions

$$ad - bc, \ dg - bh, \ ah - cg$$

as the determinants of the matrices

$$\begin{bmatrix} a & b \\ c & d \end{bmatrix}, \begin{bmatrix} g & b \\ h & d \end{bmatrix}, \begin{bmatrix} a & g \\ c & h \end{bmatrix}$$

respectively. What are these matrices? The matrix

$$\begin{bmatrix} a & b \\ c & d \end{bmatrix}$$

is simply the matrix of the coefficients of x and y in the system of equations (1). Furthermore, the matrix

$$\begin{bmatrix} g & b \\ h & d \end{bmatrix}$$

is the matrix obtained from $\begin{bmatrix} a & b \\ c & d \end{bmatrix}$ by replacing the first column—the

one associated with x—by the right-hand sides of (1). Similarly, $\begin{bmatrix} a & g \\ c & h \end{bmatrix}$ is obtained from $\begin{bmatrix} a & b \\ c & d \end{bmatrix}$ by replacing the second column—the

one associated with y—by the right-hand sides of (1). *This is no fluke*, as we shall see in a moment. More important, when we have the determinant of an $n \times n$ matrix defined and under control, the exact analog of this procedure for finding the solution for systems of linear equations will carry over. This method of solution is called *Cramer's Rule*.

Consider the matrix $\begin{bmatrix} a & b \\ c & d \end{bmatrix}$ and its determinant. If x is any real number, then

$$x \left(\det \begin{bmatrix} a & b \\ c & d \end{bmatrix} \right) = \det \begin{bmatrix} ax & b \\ cx & d \end{bmatrix}$$

by one of the properties we showed for determinants. Furthermore, by another of the properties shown for determinants, we have

$$\det \begin{bmatrix} ax & b \\ cx & d \end{bmatrix} = \det \begin{bmatrix} ax + by & b \\ cx + dy & d \end{bmatrix}$$

for any real number y. Therefore, if x and y are such that $ax + by = g$ and $cx + dy = h$, the discussion above leads us to the equations

$$x \left(\det \begin{bmatrix} a & b \\ c & d \end{bmatrix} \right) = \det \begin{bmatrix} ax + by & b \\ cx + cy & d \end{bmatrix} = \det \begin{bmatrix} g & b \\ h & d \end{bmatrix},$$

hence to the equation

$$x = \frac{\det \begin{bmatrix} g & b \\ h & d \end{bmatrix}}{\det \begin{bmatrix} a & b \\ c & d \end{bmatrix}}.$$

A similar argument works for y, showing that

$$y = \frac{\det \begin{bmatrix} a & g \\ c & h \end{bmatrix}}{\det \begin{bmatrix} a & b \\ c & d \end{bmatrix}}.$$

From what we just did, two things are clear:

1. Matrices and their determinants have something to do with solving linear equations; and

2. The vanishing or nonvanishing of the determinant of the coefficients of a system of linear equations is a deciding criterion for carrying out the solution.

Let's try out Cramer's rule for a system of two specific linear equations. According to the recipe provided us by Cramer's rule, the solutions x and y to the system of two simultaneous equations

$$\begin{aligned} x - y &= 7 \\ 3x + 4y &= 5 \end{aligned}$$

are

$$x = \frac{\begin{vmatrix} 7 & -1 \\ 5 & 4 \end{vmatrix}}{\begin{vmatrix} 1 & -1 \\ 3 & 4 \end{vmatrix}} = \frac{33}{7} \quad \text{and} \quad y = \frac{\begin{vmatrix} 1 & 7 \\ 3 & 5 \end{vmatrix}}{\begin{vmatrix} 1 & -1 \\ 3 & 4 \end{vmatrix}} = -\frac{16}{7}.$$

You can verify that these values of x and y satisfy both of the given equations.

Cramer's rule has a nice geometric interpretation. The condition that the determinant

$$\begin{vmatrix} a & b \\ c & d \end{vmatrix}$$

be nonzero is equivalent to the condition that the lines

$$\begin{aligned} ax + by &= g \\ cx + dy &= h \end{aligned}$$

not be parallel, that is, are two different intersecting lines. Their point of intersection, by Cramer's rule, is then

$$(x, y) = \left(\frac{\begin{vmatrix} g & b \\ h & d \end{vmatrix}}{\begin{vmatrix} a & b \\ c & d \end{vmatrix}}, \frac{\begin{vmatrix} a & g \\ c & h \end{vmatrix}}{\begin{vmatrix} a & b \\ c & d \end{vmatrix}} \right).$$

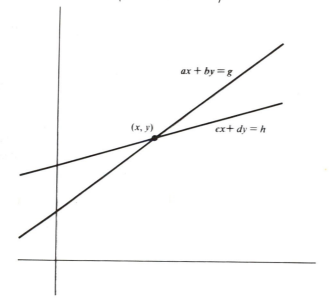

So, in the above example, the point of intersection of the lines

$$x - y = 7$$
$$3x + 4y = 5$$

is

$$\left(\frac{33}{7}, -\frac{16}{7} \right).$$

PROBLEMS

NUMERICAL PROBLEM

1. Using determinants, solve the following equations for x and y.

(a) $x + 7y = 11$
       $\frac{3}{2}x - 4y = 6.$

(b) $\pi x - 3y = \pi^2$
        $x + 5y = \pi.$

MORE THEORETICAL PROBLEMS

Easier Problems

2. Show that the solution for x and y in

$$ax + by = g$$
$$cx + dy = h$$

is the same as the values x and y for which

$$\begin{bmatrix} a & b \\ c & d \end{bmatrix} \begin{bmatrix} x & 0 \\ y & 0 \end{bmatrix} = \begin{bmatrix} g & 0 \\ h & 0 \end{bmatrix}.$$

3. For the situation in Problem 2, show that if det $\begin{bmatrix} a & b \\ c & d \end{bmatrix} \neq 0$, then

$\begin{bmatrix} x & 0 \\ y & 0 \end{bmatrix} = \begin{bmatrix} a & b \\ c & d \end{bmatrix}^{-1} \begin{bmatrix} g & 0 \\ h & 0 \end{bmatrix}$. From this, computing $\begin{bmatrix} a & b \\ c & d \end{bmatrix}^{-1}$

find x and y.

1.7.   MAPPINGS

If there is one notion that is central in every part of mathematics, it that of a *mapping* or *function* from one set S to another set T. Let's rec a few things about mappings. Our main concerns will be with mappin of a given set S to itself (so where $S = T$).

By a *mapping* f from S to T—which we denote by $f: S \rightarrow T$—v mean a rule or mechanism that assigns to every element of S anothe unique element of T. If $s \in S$ and $f: S \rightarrow T$, we write $t = f(s)$ as th *element of T into which s is carried by f*. We call $t = f(s)$ the *ima* of s under f. Thus, if $S = T = \mathbb{R}$ and $f: S \rightarrow T$ is defined by $f(s)$ s^2, then $f(4) = 4^2 = 16$, $f(-\pi) = (-\pi)^2$, and so on.

When do we declare two mappings to be equal? A natural way to define the two mappings f and g of S to T to be equal if they do t same thing to the same objects of S. More formally,

     $f = g$ *if and only if* $f(s) = g(s)$ *for every* $s \in S$.

Among the possible hordes of mappings from a set S to T, we sing out some particularly decent types of mappings.

Definition.   The mapping $f: S \rightarrow T$ is said to be *one-to-one*, written as 1-1, if $f(s_1) = f(s_2)$ implies that $s_1 = s_2$ for all s_1, s_2 in S.

In other words,

a mapping f is 1-1 if it takes distinct elements of S into distinct images.

If we look at $S = \mathbb{R}$ and the mapping $f: S \rightarrow S$ defined by

$$f(s) = s + 1,$$

we readily see that f is 1-1. For if $f(s) = f(t)$, we have $s + 1 = f(s) = f(t) = t + 1$, which implies that $s = t$. On the other hand, the mapping $g: \mathbb{R} \rightarrow \mathbb{R}$ defined by $g(s) = s^2$ is *not* 1-1 since $f(-2) = (-2)^2 = 2^2 = f(2)$, yet $-2 \neq 2$.

Another type of good mapping is defined in the

Definition.   A mapping $f: S \rightarrow T$ is said to be *onto* if given any $t \in T$, there is an $s \in S$ such that $t = f(s)$.

In other words, the mapping f is onto if the images of S under f *fill out all of T*. Again, returning to the two examples of f and g above, where $S = T = \mathbb{R}$, since $f(s) = s + 1$, given the real number t, then $t = f(t - 1) = (t - 1) + 1$, so that f is onto, while $g(s) = s^2$ is *not* onto. [If g were onto, then the negative number -1 could be expressed as a square $g(s) = s^2$ for some s, which is clearly impossible.] Note, however, that if S is the set of all *positive* real numbers, then the mapping $g(s) = s^2$ maps S onto S, since every positive real number has a positive (and also, negative) square root.

The advantage of considering mappings of a set S into itself, rather than a mapping of one set S to a different set T, lies in the fact that

in the setup where S = T, we can define a product for two mappings of S into itself.

How? If f, g are mappings of S into S and $s \in S$, then $g(s)$ is again in S. As such, $g(s)$ is a candidate for action by f, that is, $f(g(s))$ is again an element of S. This prompts the

Definition.   If f and g are mappings of S to itself, then the *product* or *composition* of f and g, written as fg, is the mapping defined by $(fg)(s) = f(g(s))$ for every $s \in S$.

So fg is that mapping obtained by first applying g and then, to the result of this, applying f. Let's look at a couple of examples. Suppose that $S = \mathbb{R}$ and $f: S \to S$ and $g: S \to S$ are defined by

$$f(s) = -4s + 3 \text{ and } g(s) = s^2.$$

What is fg? Computing,

$$(fg)(s) = f(g(s)) = f(s^2) = -4s^2 + 3;$$

so, for instance, $(fg)(1) = -4(1)^2 + 3 = -1$, $f(\pi) = -4\pi^2 + 3$, and $f(0) = -4(0)^2 + 3 = 3$. While we are at it, what is gf? Computing,

$$\begin{aligned}(gf)(s) &= g(f(s)) = g(-4s + 3) = (-4s + 3)^2 \\ &= 16s^2 - 24s + 9;\end{aligned}$$

hence, for instance, $(gf)(1) = 16(1)^2 - 24(1) + g = 1$. *Notice that*

$$(gf)(1) = 1 \neq -1 = (fg)(1).$$

Since fg and gf do not agree on $s = 1$, they are not equal as mappings, so that $fg \neq gf$. In other words,

the commutative law does not necessarily hold for two mappings of S to itself.

However, another basic law, which we should like to hold for the products of mappings, does indeed hold true. This is the *associative law*.

Lemma 1.7.1. (*Associative Law*). If f, g, and h are mappings of S to itself, then $f(gh) = (fg)h$.

Proof. Note first that since g and h are mappings of S to itself, then so is gh a mapping of S to itself. Therefore, $f(gh)$ is a mapping from S to itself. Similarly, $(fg)h$ is a mapping from S to itself. Hence it at least makes sense to ask if $f(gh) = (fg)h$.

To prove the result, we must merely show that $f(gh)$ and $(fg)h$ agree on every element of S. So if $s \in S$, then

$$(f(gh))(s) = f((gh)(s)) = f(g(h(s))),$$

while, on the other hand,

$$((fg)h)(s) = (fg)(h(s)) = f(g(h(s))),$$

by successive uses of the definition of the product of mappings. Since we thus see that $(f(gh))(s) = ((fg)h)(s)$ for every $s \in S$, by the definition of the equality of mappings, we have that $f(gh) = (fg)h$, the desired result. ∎

By virtue of the lemma, we can dispense with a particular bracketing for the product of three or more mappings, and we write such products simply as fgh, $fgghfr$ (product of mappings f, g, g, h, f, r), and so forth.

A very nice mapping is always present, that is, the *identity mapping* $e: S \rightarrow S$, that is, the mapping e which disturbs no element of S:

$e(s) = s$ for all $s \in S$.

We leave as an exercise the proof of

Lemma 1.7.2.   If e is the identity mapping on S, then $ef = fe = f$ for every mapping $f: S \rightarrow S$.

Before leaving this short discussion of mappings, there is one more fact about certain mappings that should be highlighted. Let $f: S \rightarrow S$ be *both* 1-1 and onto. Thus, given $t \in S$, then $t = f(s)$ for some $s \in S$, since f is onto. Furthermore, because f is 1-1, there is *one and only one s* that does the trick. So if we define $g: S \rightarrow S$ by the rule $g(t) = s$ if and only if $t = f(s)$, we do indeed get a mapping of S to itself. What properties does this g enjoy? If we compute $(gf)(s)$ for any $s \in S$, what do we get? Suppose that $t = f(s)$. Then, by the very definition of g, $s = g(t) = g(f(s)) = (gf)(s)$. So $(gf)(s) = s = e(s)$ for every $s \in S$, so that $gf = e$. We call g the *inverse* of f and denote it by f^{-1}. So,

$t = f(s)$ if and only if $s = f^{-1}(t)$.

We summarize what we have just done in

Theorem 1.7.3.   If f is 1-1 and onto from S to S, then there exists a mapping $f^{-1}: S \rightarrow S$ such that $ff^{-1} = f^{-1}f = e$.

We compute f^{-1} for some sample f's. Let $S = \mathbb{R}$ and let f be the function from S to S defined by $f(s) = 6s + 5$. Then f is a 1-1 mapping of S *onto* itself, as is readily verified. What, then, is f^{-1}? We claim that $f^{-1}(s) = \dfrac{1}{6} s - \dfrac{5}{6} = \dfrac{s - 5}{6}$ for every s in $\mathbb{R}$. (Verify!). *Do not confuse* $f^{-1}(s)$ with $(f(s))^{-1} = \dfrac{1}{f(s)}$. In our example, $f^{-1}(s) = \dfrac{s - 5}{6}$, whereas $(f(s))^{-1} = \dfrac{1}{f(s)} = \dfrac{1}{6s + 5}$ and these are not equal as functions on S.

For instance, $f^{-1}(1) = \dfrac{1-5}{6} = -\dfrac{4}{6}$ while $(f(1))^{-1} = \dfrac{1}{6+5} = \dfrac{1}{11}$.

Since $t = f(s) = 6s + 5$ if and only if $s = f^{-1}(t) = \dfrac{t-5}{6}$, we can illustrate the mappings f and f^{-1} as follows:

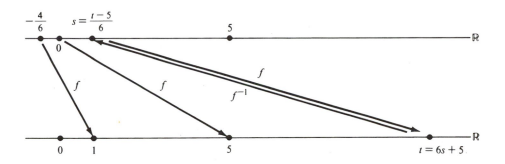

We do one more example. Let S be the set of all *positive* real numbers and suppose that $f: S \to S$ is defined by $f(s) = 1/s$. Clearly, f is a 1-1 mapping of S onto itself, hence f^{-1} exists. What is f^{-1}? We leave it to the reader to prove that $f^{-1}(s) = 1/s = f(s)$ for all s, so that $f^{-1} = f$.

One last bit. To have a handy notation, we use exponents for the product of a function (several times) with itself. Thus

$$f^2 = ff, \ldots , f^n = f^{n-1}f$$

for every positive integer n. So, by our definition,

$$f^n = \underset{(n \text{ times})}{ff f \ldots f}.$$

As we might expect, the usual rules of exponents hold, namely

$$f^n f^m = f^{n+m}$$

and

$$(f^m)^n = f^{mn}$$

for m and n positive integers. By f^0 we shall mean the identity mapping of S onto itself.

PROBLEMS

NUMERICAL PROBLEMS

1. Determine whether the following mappings are 1-1, onto, or both.

   (a) $f: \mathbb{R} \to \mathbb{R}$, where f is defined by $f(s) = \dfrac{s}{s^2 + 1}$.

   (b) $f: \mathbb{R} \to \mathbb{R}$, where f is defined by $f(s) = s^2 + s + 1$.

   (c) $f: \mathbb{R} \to \mathbb{R}$, where f is defined by $f(s) = s^3$.

2. Let S be the set of integers and let f be a mapping from S to S. Then determine, for the following f's, whether they are 1-1, onto, or both.

   (a) $f(s) = -s + 1$.

   (b) $f(s) = 6s$.

   (c) $f(s) = s$ if s is even and $f(s) = s + 2$ if s is odd.

3. If $f: \mathbb{R} \to \mathbb{R}$ is defined by $f(s) = 3s + 2$, find the formula for $f^2(s)$.

4. Is the "mapping" $f: \mathbb{R} \to \mathbb{R}$ defined by $f(s) = \dfrac{s}{1 + s}$ really a mapping?

5. If S is the set of *positive* real numbers, verify that $f: S \to S$ defined by $f(s) = \dfrac{s}{1 + s}$ really is a mapping. (Compare with Problem 4.)

6. In Problem 5, is f 1-1? Is f onto?

7. If S is the set of all real numbers s such that $0 < s < 1$, verify that f defined by $f(s) = \dfrac{s}{1 + s}$ is a mapping of S into itself. (Compare with Problem 4.)

8. In Problem 7, is f 1-1? Is f onto?

9. In Problem 7, compute f^2 and f^3.

10. In Problem 7, find f^{-1}.

MORE THEORETICAL PROBLEMS

Easier Problems

11. If S is the set of positive integers, give an example of a mapping $f: S \to S$ such that

(a) f is 1-1 but not onto.

(b) f is onto but not 1-1.

12. If f maps S onto itself and is 1-1, show that if $fg = fh$, then $g = h$.

13. If f and g are 1-1 mappings of S onto itself, show that

(a) fg is also 1-1 and onto.

(b) f^{-1} and g^{-1} exist.

(c) $(fg)^{-1} = g^{-1}f^{-1}$.

14. Let $S = \mathbb{R}$ and consider the mapping $t_{a,b}: S \to S$, where $a \neq 0$ and b are real numbers, defined by $t_{a,b}(s) = as + b$. Then

(a) Show that $t_{a,b}$ is 1-1 and onto.

(b) Find $t_{a,b}^{-1}$ in the form $t_{u,v}$.

(c) If $t_{c,d}: S \to S$ where $t_{c,d}(s) = cs + d$, and $c \neq 0$, d are real numbers, find the formula for $t_{a,b}t_{c,d}$.

15. For what values of a and b does $t_{a,b}$ in Problem 14 satisfy $t_{a,b}^2 = e$?

16. If $t_{1,2}$ and $t_{a,b}$ are as in Problem 14, then find all $t_{a,b}$ such that $t_{1,2}t_{a,b} = t_{a,b}t_{1,2}$.

17. Let S be the x-y plane and let f be defined on S by the rule that f of any point (x, y) is the corresponding point $(0, y)$ on the y-axis. This mapping f is called *projection onto the y-axis*. Show that $f^2 = f$. Letting g be the analogous *projection onto the x-axis*, show that fg and gf both are the zero function that maps s to $(0, 0)$ for all $s \in S$.

Middle-Level Problems

18. Let S be a set having at least three elements. Show that we can find 1-1 mappings of S into itself, f and g, such that $fg \neq gf$.

19. If f is a mapping of the set of positive real numbers into $\mathbb{R}$ such that $f(ab) = f(a) + f(b)$, and f is not identically 0, find

(a) $f(1)$.

(b) $f(a^3)$.

(c) $f(a^n)$ for every positive integer n.

20. Show that $f(a^r) = rf(a)$ for any *positive rational* number r in Problem 19.

21. Let S be the x-y plane $\mathbb{R} \times \mathbb{R} = \{(a, b) | a, b \in \mathbb{R}\}$ and let f be the reflection of the plane about the x-axis and g the rotation of the plane about the origin through $120°$ in the counterclockwise direction. Prove:

    (a) f and g are 1-1 and onto.

    (b) $f^2 = g^3 = e$, the identity mapping on S.

    (c) $fg \neq gf$.

    (d) $fg = g^{-1}f$.

22. Let S be the set of all integers and define, for a, b integers and $a \neq 0$, the functions $u_{a,b}: S \to S$ by $u_{a,b}(s) = as + b$. Then

    (a)   When is the function $u_{a,b}$ 1-1?

    (b)   When is the function $u_{a,b}$ onto?

    (c)   When is $u_{a,b}$ both 1-1 and onto? And when it is both 1-1 and onto, describe its inverse.

23. Define $f(s)$ as $\dfrac{s^2 + 1}{s^2 + 2}$ for $s \in \mathbb{R}$. Describe the set of all images $f(s)$ as s ranges throughout the set $\mathbb{R}$.

24. Define $f: \mathbb{R} \to \mathbb{R}$ by $f(s) = s^3 + 6$. Then

    (a)   Show that f is 1-1 and onto.

    (b)   Find f^{-1}.

25. Define $f: \mathbb{R} \to \mathbb{R}$ by $f(s) = s^2$. Then determine whether f is 1-1 and onto.

26. "Prove" the law of exponents: $f^m f^n = f^{m+n}$ and $(f^m)^n = f^{mn}$, where m, n are nonnegative integers and f maps some set S to itself.

Harder Problems

27. (a)   Is $(fg)^2$ necessarily equal to f^2g^2 for f, g mappings of S into S? Prove or give a counterexample.

    (b)   If f, g are 1-1 mappings of S onto S, is $(fg)^2$ always equal to f^2g^2? Prove or give a counterexample.

28. If f, g are 1-1 mappings of S onto itself, show that a necessary and sufficient condition that $(fg)^n = f^n g^n$ for all $n > 1$ is that $fg = gf$.

29. If S is a *finite* set and $f: S \to S$ is 1-1, show that f is onto.

30. If S is a *finite* set and $f: S \to S$ is onto, show that f is 1-1.

Generalization of terminology to two sets S and T: If S and T are two sets, then $f: S \rightarrow T$ is said to be 1-1 if $f(s) = f(s')$ only if $s = s'$ and f is said to be *onto* if, given $t \in T$, there is an $s \in S$ such that $t = f(s)$.

31. If $f: S \rightarrow T$ is 1-1 and $fg = fh$ where $g, h: T \rightarrow S$ are two functions, show that $g = h$.

32. If $g: S \rightarrow T$ is *onto* and p and q are mappings of T to S such that $pg = qg$, prove that $p = q$.

33. If $f: S \rightarrow T$ is 1-1 and onto, show that we can find $g: T \rightarrow S$ such that $fg = e_S$, the identity mapping on S, and $gf = e_T$, the identity mapping on T.

34. Let $S = \mathbb{R}$ and let T be the set of all *positive* real numbers. Define $f: S \rightarrow T$ by $f(s) = 10^s$. Then

  (a) Prove that f is 1-1 and onto.

  (b) Find the function g of Problem 33 for our f.

35. Let S be the set of all positive real numbers, and let $T = \mathbb{R}$. Define $f: S \rightarrow T$ by $f(s) = \log_2 s$. Then

  (a) Show that f is 1-1 and onto.

  (b) Find the g of Problem 33 for f.

1.8. MATRICES AS MAPPINGS

It would be nice if we could strip matrices of their purely formal definitions and, instead, could find some way in which they would *act as transformations*, that is, mappings, on a reasonable set. A reasonable sort of *action* would be in terms of a nice kind of mapping of a set of familiar objects into itself.

For our set of familiar objects, we use the usual Cartesian plane $\mathbb{R}^{(2)}$. Traditionally, we have written the points of the plane $\mathbb{R}^{(2)}$ as ordered pairs (x, y) with $x, y \in \mathbb{R}$. For a purely technical reason, we now begin to write the points as $\begin{bmatrix} x \\ y \end{bmatrix}$. The entries x and y are referred to as the *coordinates* of $\begin{bmatrix} x \\ y \end{bmatrix}$.

We now let V denote $\mathbb{R}^{(2)} = \left\{ \begin{bmatrix} x \\ y \end{bmatrix} \middle| x, y \in \mathbb{R} \right\}$. We refer to the

elements (points) of V as *vectors* and to V as the *space of vectors* $\begin{bmatrix} x \\ y \end{bmatrix}$

with $x,\ y \in \mathbb{R}$ or, simply, as the *vector space of column vectors over* $\mathbb{R}$.
In the subject of *vector analysis*, two operations are introduced in V,
namely, the *addition* of vectors and the *multiplication* of a vector by a

scalar (element of $\mathbb{R}$). We explain. If $\begin{bmatrix} a \\ b \end{bmatrix}$ and $\begin{bmatrix} c \\ d \end{bmatrix}$ are in V, we *define*

$\begin{bmatrix} a \\ b \end{bmatrix} + \begin{bmatrix} c \\ d \end{bmatrix} = \begin{bmatrix} a + c \\ b + d \end{bmatrix}$ and $s \begin{bmatrix} a \\ b \end{bmatrix} = \begin{bmatrix} sa \\ sb \end{bmatrix}$ for any s in $\mathbb{R}$. Pictori-

ally, these two operations look like this where $P = \begin{bmatrix} a \\ b \end{bmatrix}$, $Q =$

$\begin{bmatrix} c \\ d \end{bmatrix}$, and $R = \begin{bmatrix} a + c \\ b + d \end{bmatrix}$:

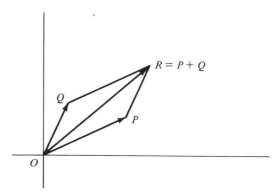

Addition: R is the sum of P and Q.

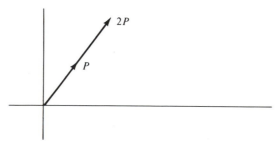

Multiplication: $2P$ is the scalar multiple $2\begin{bmatrix} a \\ b \end{bmatrix}$ of P.

So addition is performed *component-wise*, that is, by adding the coordinates (components) individually. Multiplication by a scalar a is a stretching (or shrinking) of the line by a factor of a. The stretching represented in the picture as the effect of stretching the line segment OP into the segment OQ.

With these operations, V satisfies many of the readily verified rules such as: if $u, v, w \in V$, $a, b \in \mathbb{R}$, then

$$(u + v) + w = u + (v + w),$$
$$u + v = v + u,$$
$$a(u + v) = au + av,$$
$$a(bv) = (ab)v,$$

and so on.

The structure V (the set V together with the operations of addition and multiplication by scalars $a \in \mathbb{R}$) is a particular case of the abstract notion of a *vector space*, a topic treated later in the book. The 2×2 matrices which we view in this section as mappings from this vector space V to itself are the *linear transformations* from V to V. Since the subject *linear algebra* is, more or less, simply the study of linear transformations of vector spaces, what we do in this section sets the stage for what is to come later in the book.

We shall let the matrices, that is, the elements of $M_2(\mathbb{R})$, *act* as mappings from V to itself. We do this by *defining*

$$\begin{bmatrix} a & b \\ c & d \end{bmatrix} \begin{bmatrix} x \\ y \end{bmatrix} = \begin{bmatrix} ax + by \\ cx + dy \end{bmatrix}$$

and saying that

the action *of* $\begin{bmatrix} a & b \\ c & d \end{bmatrix}$ *on V is the mapping that sends* $\begin{bmatrix} x \\ y \end{bmatrix}$ *to* $\begin{bmatrix} ax + by \\ cx + dy \end{bmatrix}$.

Geometrically,

the mapping sending $\begin{bmatrix} x \\ y \end{bmatrix}$ *to* $\begin{bmatrix} ax + by \\ cx + dy \end{bmatrix}$ *can be given the*

interpretation of a change of coordinates *or* coordinate transformation *on V*.

If $v \in V$ and $A \in M_2(\mathbb{R})$, we shall write Av as a shorthand for the specific action of A on v as described above.

Let's see two particular actions of this sort, which we can interpret geometrically. Let

$$A = \begin{bmatrix} 1 & 0 \\ 0 & -1 \end{bmatrix}.$$

Then

$$A \begin{bmatrix} x \\ y \end{bmatrix} = \begin{bmatrix} 1 & 0 \\ 0 & -1 \end{bmatrix} \begin{bmatrix} x \\ y \end{bmatrix} = \begin{bmatrix} x \\ -y \end{bmatrix}.$$

So, A carries the point $\begin{bmatrix} x \\ y \end{bmatrix}$ into its mirror image $\begin{bmatrix} x \\ -y \end{bmatrix}$ under reflection about the x-axis. So A merely is this reflection.

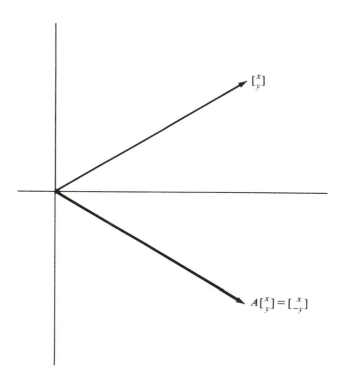

If $B = \begin{bmatrix} 0 & -1 \\ 1 & 0 \end{bmatrix}$, then $B \begin{bmatrix} x \\ y \end{bmatrix} = \begin{bmatrix} 0 & -1 \\ 1 & 0 \end{bmatrix} \begin{bmatrix} x \\ y \end{bmatrix} = \begin{bmatrix} -y \\ x \end{bmatrix}$. So B can

be interpreted as the rotation of the plane, through 90° in the counter-clockwise direction, about the origin.

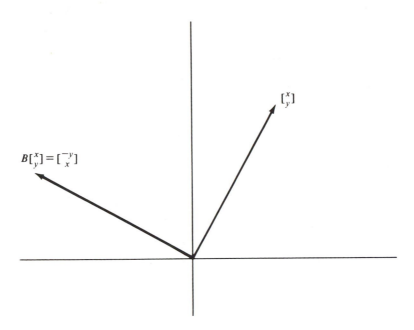

From the definition of the action of a matrix on a vector, we easily can show the following basic properties of this action.

Lemma 1.8.1. If $A, B \in M_2(\mathbb{R})$, $v, w \in V$ and $a, b \in \mathbb{R}$, then:

1. $A(v + w) = Av + Aw$;
2. $A(av) = a(Av)$;
3. $(A + B)v = Av + Bv$;
4. $(aA + bB)v = aAv + bBv$.

We leave the verification of these properties as an exercise for the reader.

Mappings on an abstract vector space (whatever that is) that satisfy (a) and (b) are called *linear transformations* on V, and A is said to act *linearly* on V. So, in these terms, $M_2(\mathbb{R})$ consists of linear transformations on V. Pictorially, the *linearity conditions*

$$A(v + w) = Av + Aw$$

and

$$A(av) = a(Av)$$

for the linear transformation $B\begin{bmatrix} x \\ y \end{bmatrix} = \begin{bmatrix} 0 & -1 \\ 1 & 0 \end{bmatrix}\begin{bmatrix} x \\ y \end{bmatrix} = \begin{bmatrix} -y \\ x \end{bmatrix}$

described above look like this:

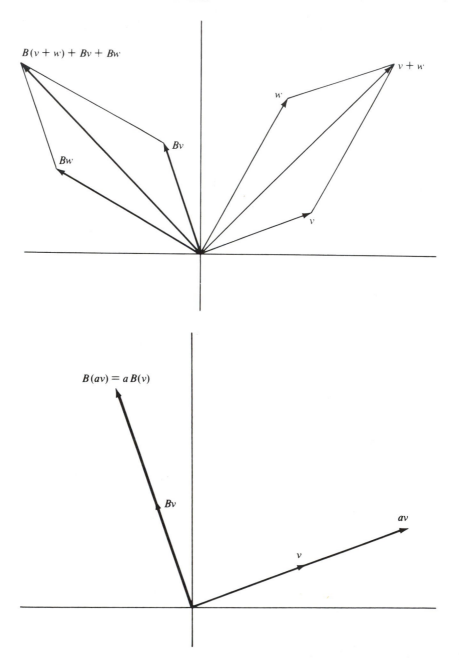

How does the composition of two matrices A and B as mappings jibe with the formal matrix product we defined for matrices? Letting A

and B be the matrices $A = \begin{bmatrix} a & b \\ c & d \end{bmatrix}$ and $B = \begin{bmatrix} r & s \\ t & u \end{bmatrix}$, then by the composition of mappings we have

$$(A \cdot B)\begin{bmatrix} x \\ y \end{bmatrix} = A\left(B\begin{bmatrix} x \\ y \end{bmatrix}\right) = A\left(\begin{bmatrix} r & s \\ t & u \end{bmatrix}\begin{bmatrix} x \\ y \end{bmatrix}\right) = A\begin{bmatrix} rx + sy \\ tx + uy \end{bmatrix}$$

$$= \begin{bmatrix} a(rx + sy) + b(tx + uy) \\ c(rx + sy) + d(tx + uy) \end{bmatrix}$$

$$= \begin{bmatrix} (ar + bt)x + (as + bu)y \\ (cr + dt)x + (cs + du)y \end{bmatrix}$$

$$= \begin{bmatrix} ar + bt & as + bu \\ cr + dt & cs + du \end{bmatrix}\begin{bmatrix} x \\ y \end{bmatrix} = (AB)\begin{bmatrix} x \\ y \end{bmatrix},$$

where $A \cdot B$ is the product of A and B as mappings, and AB is the matrix product of A and B. So $A \cdot B = AB$. Thus *we see that treating matrices as mappings, or treating them as formal objects and multiplying by some weird multiplication leads to the same result for what we want the product to be.* In fact, the reason that matrix multiplication was defined as it was it precisely this fact, namely, that this formal matrix multiplication corresponds to the more natural notion of the product of mappings. Note one little by-product of all this. Since the product of mappings is associative (Lemma 1.7.1), we get, free of charge (without any messy computation), that the product of matrices satisfies the associative law.

For emphasis, we state what we just discovered as

Theorem 1.8.2. If A, B are matrices in $M_2(\mathbb{R})$ then their product as mappings, $A \cdot B$, coincides with their product as matrices in $M_2(\mathbb{R})$.

If A in $M_2(\mathbb{R})$ is an invertible matrix with inverse B, then we know from Theorem 1.8.2 that $A \cdot B$ and $B \cdot A$ are both the identity mapping on $\mathbb{R}^{(2)}$. This means that

as a mapping from $\mathbb{R}^{(2)}$ to itself, the effect of the inverse B of A is just to reverse the effect of A.

For example, if A is the matrix $\begin{bmatrix} 0 & 1 \\ -1 & 0 \end{bmatrix}$, then A can be interpreted as the rotation of the plane, through 90° in the clockwise direction, about the origin. Its inverse B is the matrix $B = \begin{bmatrix} 0 & -1 \\ 1 & 0 \end{bmatrix}$, which acts as the

rotation of the plane, through $90°$ in the counterclockwise direction.

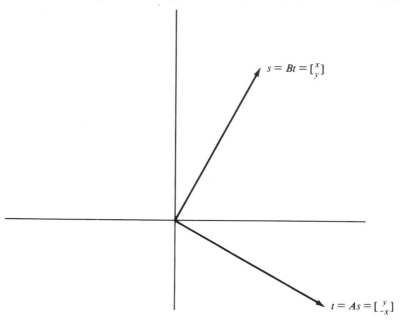

PROBLEMS

NUMERICAL PROBLEMS

1. Show that every element in V can be written in a unique way as
$ae_1 + be_2$, where $a, b \in \mathbb{R}$ and $e_1 = \begin{bmatrix} 1 \\ 0 \end{bmatrix}$ and $e_2 = \begin{bmatrix} 0 \\ 1 \end{bmatrix}$
are in V.

2. Let the matrices E_{ij} be defined as follows: E_{ij} is a matrix such that the (i, j) entry is 1 and all other entries are 0.

(a) Find a formula for $E_{ij}E_{k1}$.

(b) Interpret the formula you get geometrically.

3. Find all matrices A such that $AE_{11} = E_{11}A$.

4. Find all matrices A such that $AE_{12} = E_{21}A$.

5. If A is any matrix, compute $E_{11}AE_{11}$.

6. If A is any matrix, compute $E_{12}AE_{21}$.

MORE THEORETICAL PROBLEMS

Easier Problems

7. Prove Lemma 1.8.1 in detail.

8. Give a geometric interpretation of the action of the matrix
$\begin{bmatrix} -1 & 0 \\ 0 & 1 \end{bmatrix}$ on V.

9. Give a geometric interpretation of the action of the matrix
$\begin{bmatrix} 1 & c \\ 0 & 1 \end{bmatrix}$ on V.

10. Give a geometric interpretation of the action of the matrix
$\begin{bmatrix} \cos \theta & -\sin \theta \\ \sin \theta & \cos \theta \end{bmatrix}$ on V.

11. Under what condition on the matrix $A = \begin{bmatrix} a & b \\ c & d \end{bmatrix}$ will the new axes

$x_1 = ax + by$, $y_1 = cx + dy$ intersect at $90°$?

Middle-Level Problems

12. Show that every matrix A in $M_2(\mathbb{R})$ can be written uniquely as $A = \sum_{r=1}^{2} \sum_{s=1}^{2} a_{rs} E_{rs}$.

13. Let T be a linear transformation on V [i.e., T satisfies $T(v + w) = Tv + Tw$ and $T(av) = aT(v)$ for $v, w \in V, a \in \mathbb{R}$]. Show that we
can realize T as some matrix $T = \begin{bmatrix} t_{11} & t_{12} \\ t_{21} & t_{22} \end{bmatrix}$. Can you see a way
of determining the entries of this matrix $[t_{rs}]$?

14. Let $\varphi: V \to M_2(\mathbb{R})$ be defined by $\varphi \begin{bmatrix} x \\ y \end{bmatrix} = \begin{bmatrix} x & 0 \\ y & 0 \end{bmatrix}$. Then prove:

   (a) $\varphi(v + w) = \varphi(v) + \varphi(w)$ for $v, w \in V$.

   (b) $\varphi(\alpha v) = \alpha\varphi(v)$ for $\alpha \in \mathbb{R}, v \in V$.

   (c) If $A \in M_2(\mathbb{R})$, then $\varphi(Av) = A(\varphi(v))$, where Av is the result of the action of A on v and $A\varphi(v)$ is the usual matrix product of the matrices A and $\varphi(v)$.

   (d) If $\varphi(v) = 0$, then $v = 0$.

15. If $\det(A) = 0$, show that there is a vector $v \neq 0$ in V such that $Av = 0$.

1.9. THE CAYLEY–HAMILTON THEOREM

Given a matrix A in $M_2(\mathbb{R})$, we can attach to A a polynomial; called the *characteristic polynomial* of A, the roots of which determine, to a large extent, the behavior of A as a linear transformation on V. This result will be shown later to hold quite generally for square matrices of all sizes. Interestingly enough, although this result is attributed to Cayley, he really showed it only for the 2×2 and 3×3 matrices, where it is quite easy and straightforward. It took later mathematicians to prove it for the general case.

Consider a polynomial

$$p(x) = a_0 x^n + a_1 x_1^{n-1} + \cdots + a_{n-1} x^1 + a_n x^0$$

with real coefficients $a_0, a_1, \ldots, a_n$. We usually will write 1 in place of x^0, but it is convenient for the sake of the next definition to write it as x^0.

Definition.   A matrix A is said to *satisfy* the polynomial $p(x)$ if $p(A) = 0$, where $p(A)$ denotes the matrix obtained by replacing each occurrence of x in $p(x)$ by the matrix A:

$$p(A) = a_0 A^n + a A_1^{n-1} + \cdots + a_{n-1} A^1 + a_n A^0.$$

(Here, it is convenient to adopt the convention that A^0 is the identity matrix I.)

For instance, the matrices $I = \begin{bmatrix} 1 & 0 \\ 0 & 1 \end{bmatrix}$, $A = \begin{bmatrix} 1 & 0 \\ 0 & -1 \end{bmatrix}$, and $B = \begin{bmatrix} 0 & 1 \\ -1 & 0 \end{bmatrix}$ satisfy the polynomials $p(x) = x - 1$, $p(x) = x^2 - 1$, and $p(x) = x^2 + 1$, respectively. To see the last one, we must show that

$$B^2 + I = 0.$$

But

$$B^2 = \begin{bmatrix} 0 & 1 \\ -1 & 0 \end{bmatrix} \begin{bmatrix} 0 & 1 \\ -1 & 0 \end{bmatrix} = \begin{bmatrix} -1 & 0 \\ 0 & -1 \end{bmatrix} = -I,$$

hence $B^2 + I = 0$.

We now introduce the very important polynomial associated with a given matrix.

Definition. If $A \in M_2(\mathbb{R})$, then the *characteristic polynomial* of A, denoted $P_A(x)$, is defined by

$$P_A(x) = \det (xI - A).$$

The characteristic polynomial comes up often in various sciences, such as physics and chemistry. There it is often referred to as the *secular determinant*.

For instance, going back to the matrix $B = \begin{bmatrix} 0 & 1 \\ -1 & 0 \end{bmatrix}$, we have

$$P_B(x) = \det [xI - B] = \det \begin{bmatrix} x - 0 & -1 \\ -(-1) & x - 0 \end{bmatrix}$$

$$= \det \begin{bmatrix} x & -1 \\ 1 & x \end{bmatrix} = x^2 + 1.$$

Note that we showed above that B satisfies $x^2 + 1 = P_B(x)$, that is, *B satisfies its characteristic polynomial.* This is no accident, as we proceed to show in the

Theorem 1.9.1 (Cayley–Hamilton Theorem). A matrix satisfies its characteristic polynomial.

Proof. Since we are working with 2×2 matrices, the proof will be quite easy, for we can write down $P_A(x)$ explicitly. Let $A = \begin{bmatrix} a & b \\ c & d \end{bmatrix}$. Then we have

$$xI - A = \begin{bmatrix} x & 0 \\ 0 & x \end{bmatrix} - \begin{bmatrix} a & b \\ c & d \end{bmatrix} = \begin{bmatrix} x - a & -b \\ -c & x - d \end{bmatrix}.$$

Hence we have

$$P_A(x) = \det [xI - A] = \det \begin{bmatrix} x - a & -b \\ -c & x - d \end{bmatrix}$$

$$= (x - a)(x - d) - bc = x^2 - (a + d)x + ad - bc$$

$$= x^2 - (\text{tr } (A))x + (\det (A))1.$$

What we must verify is, therefore, that A satisfies the polynomial

$$x^2 - (\text{tr } (A))x + (\det (A))1,$$

that is, that

$$A^2 - (\text{tr } (A))A + (\det (A))I = 0.$$

This was actually given as Problem 13 of Section 1.5. However, we do it in detail here. Because

$$A^2 = \begin{bmatrix} a & b \\ c & d \end{bmatrix}^2 = \begin{bmatrix} aa + bc & ab + bd \\ ca + cd & cb + dd \end{bmatrix},$$

we have

$$A^2 - (\text{tr } (A))A + (\det (A))I$$

$$= \begin{bmatrix} aa + bc & ab + bd \\ cd + ca & dd + bc \end{bmatrix} - (a + d)\begin{bmatrix} a & b \\ c & d \end{bmatrix} + (ad - bc)\begin{bmatrix} 1 & 0 \\ 0 & 1 \end{bmatrix}$$

$$= \begin{bmatrix} aa + bc - (a + d)a + ad - bc & ab + bd - (a + d)b \\ cd + ca - (a + d)c & dd + bc - (a + d)d + ad - bc \end{bmatrix}$$

$$= 0.$$

Thus A satisfies the polynomial $P_A(x) = x^2 - (\text{tr } (A))x + (\det (A))1$. ■

Given a polynomial $f(x)$ with real coefficients, associated with $f(x)$ are certain "numbers," called the *roots* of $f(x)$. The number a is a root of f if $f(a) = 0$. For instance, 3 is a root of $f(x) = x^2 - 9x + 18$, since $f(3) = 3^2 - 9 \cdot 3 + 18 = 0$.

We used the vague word "numbers" above in defining a root of a polynomial. There is a very good reason for this vagueness. If we insist that the roots be real numbers, we get into a bit of a bind, for the very nice polynomial $f(x) = x^2 + 1$ has no real root, since $a^2 + 1 \neq 0$ for every real number a. On the other hand, we want our polynomials to have roots. This necessitates the enlargement of the concept of real number to some wider arena: that of *complex numbers*. We do this in the next section. When this has been accomplished, we shall not have to worry about the presence or absence of roots for the polynomials that will arise. To be precise, the powerful so-called *Fundamental Theorem of Algebra* ensures that

> *every polynomial f(x) with complex coefficients has enough complex roots $a_1, \ldots, a_n$ that it factors as a product $f(x) = (x - a_1) \cdots (x - a_n)$.*

For the moment, let's act as if the polynomials we are dealing with had real roots—something that is often false. At any rate, let's see where this will lead us.

Definition. A *characteristic root* of the matrix A is a number a such that $P_A(a) = 0$.

Here, too, we're still being vague as to the meaning of the word "number."

If $A = \begin{bmatrix} 1 & 6 \\ 0 & -3 \end{bmatrix}$, we readily find that $P_A(x) = \det(xI - A) = (x - 1)(x + 3) = x^2 + 2x - 3$, so 1 and -3 are characteristic roots of A. On the other hand, if $B = \begin{bmatrix} 0 & 1 \\ -1 & 0 \end{bmatrix}$ then, as we saw, $P_B(x) = x^2 + 1$, which has no real roots, so *for the moment* has no characteristic root.

Incidentally, what we call a characteristic root of A is often called an *eigenvalue* of A. In fact, this hybrid English–German word is the one most often used by physicists.

Suppose that a is a characteristic root of A; thus $\det(aI - A) = P_A(a) = 0$. By Theorem 1.5.4, since $\det(aI - A) = 0$, $aI - A$ is *not invertible*. On the other hand, if b is a number such that $bI - A$ is not invertible, then, again by Theorem 1.5.4, $\det(bI - A) = 0$. Therefore,

> *the characteristic roots of A are precisely those numbers a for which $aI - A$ fails to have an inverse.*

Since $P_A(x) = \det(xI - A)$ is a polynomial of degree 2, it can have at most two real roots (and, in fact, will have two real or complex roots). So $aI - A$ can fail to be invertible for *at most* two real numbers. Thus $aI - A$ is *almost always* invertible.

We summarize all of this in

Theorem 1.9.2. The characteristic roots of A are precisely those numbers a for which the matrix $aI - A$ is not invertible. If $A \in M_2(\mathbb{R})$, then A has at most two real characteristic roots.

In Problem 15 of Section 1.8, we asked to show that if $\det B = 0$, then there is some $v \neq 0$ in V such that $Bv = 0$. Since we need this result now, we prove it. If $B = 0$, there is nothing to prove, for *every* vector v satisfies $0v = 0$. So, suppose that $B = \begin{bmatrix} r & s \\ t & u \end{bmatrix} \neq 0$. Then some row of B does not consist entirely of zeros, say the first row. Then, if $v = \begin{bmatrix} s \\ -r \end{bmatrix}$, we have

$$Bv = \begin{bmatrix} r & s \\ t & u \end{bmatrix} \begin{bmatrix} s \\ -r \end{bmatrix} = \begin{bmatrix} rs - sr \\ ts - ur \end{bmatrix} = \begin{bmatrix} 0 \\ -\det B \end{bmatrix} = \begin{bmatrix} 0 \\ 0 \end{bmatrix}.$$

Hence $Bv = 0$ and $v \neq 0$. The same argument works if it is the other row that does not consist entirely of zeros.

Suppose, now, that the real number a is a characteristic root of A. Then $aI - A$ is not invertible and $\det (aI - A) = 0$. Thus, by the above, there is a nonzero $v \in V$ such that $(aI - A)v = 0$, which is to say, $Av = (aI)v = av$. The converse is also true, so we have

Theorem 1.9.3.   The real number a is a characteristic root of A if and only if for some $v \neq 0$ in V, $Av = av$.

Definition.   An element $v \neq 0$ in V such that $Av = av$ is called a *characteristic vector associated with a*.

Physicists often use a hybrid English–German word *eigenvector* for a characteristic vector.

Let $A = \begin{bmatrix} 5 & 12 \\ 12 & -5 \end{bmatrix}$. Then $P_A(x) = \det \begin{bmatrix} x - 5 & -12 \\ -12 & x + 5 \end{bmatrix} = $ $(x - 5)(x + 5) - 12^2 = x^2 - 169$, whose roots are ± 13. What v are characteristic vectors associated with 13, and which with -13? If $\begin{bmatrix} 5 & 12 \\ 12 & -5 \end{bmatrix} \begin{bmatrix} x \\ y \end{bmatrix} = 13 \begin{bmatrix} x \\ y \end{bmatrix}$, then $5x + 12y = 13x$, hence $8x = 12y$, so $2x = 3y$. If we let $x = 3$, then $y = 2$. So the vector $\begin{bmatrix} 3 \\ 2 \end{bmatrix}$ is a characteristic vector of $\begin{bmatrix} 5 & 12 \\ 12 & -5 \end{bmatrix}$ associated with 13. In fact, all such vectors are of the form $\begin{bmatrix} 3s \\ 2s \end{bmatrix}$ for any $s \neq 0$. For -13, we similarly can see that $\begin{bmatrix} 2 \\ -3 \end{bmatrix} \left(\text{and, in fact, all of its nonzero multiples } s \begin{bmatrix} 2 \\ -3 \end{bmatrix} \right)$ is a characteristic vector associated with -13.

(Don't expect the characteristic roots always to be "nice" numbers. For instance, the characteristic roots of $\begin{bmatrix} 1 & 2 \\ 3 & 4 \end{bmatrix}$ are the roots of the polynomial $x^2 - 5x - 2$, which are $\dfrac{5 \pm \sqrt{33}}{2}$. Can you find the respective characteristic vectors associated with these?)

Return to $A = \begin{bmatrix} 5 & 12 \\ 12 & -5 \end{bmatrix}$, whose characteristic roots were 13 and

-13, with corresponding characteristic vectors $\begin{bmatrix} 3 \\ 2 \end{bmatrix}$ and $\begin{bmatrix} 2 \\ -3 \end{bmatrix}$. Note

the following about these characteristic vectors: If $a \begin{bmatrix} 2 \\ 2 \end{bmatrix} +$

$b \begin{bmatrix} 2 \\ -3 \end{bmatrix} = \begin{bmatrix} 0 \\ 0 \end{bmatrix}$, then $a = b = 0$. (Prove this!) This is a rather im-

portant attribute of $\begin{bmatrix} 3 \\ 2 \end{bmatrix}$ and $\begin{bmatrix} 2 \\ -3 \end{bmatrix}$ which we treat in great detail later.

Note that if the characteristic roots of a matrix are real, we need not have two distinct ones. The easiest example is the matrix I, which has only the number 1 as a characteristic root. This illustrates the following:

> *The characteristic polynomial $P_A(x)$ need not be a polynomial of lowest degree satisfied by A.*

For instance, the characteristic polynomial of I is $P_I(x) = (x - 1)^2$, yet I satisfies the polynomial $q(x) = x - 1$. Is there only one *monic* (leading coefficient is 1) polynomial of lowest degree satisfied by A? Yes: The difference of any two such has lower degree and is still satisfied by A, so the difference is 0.

This prompts the

Definition. The *minimal polynomial* for the matrix A is the nonzero polynomial of lowest degree satisfied by A having 1 as the coefficient of its highest term.

For 2×2 matrices, this distinction between the minimal polynomial and the characteristic polynomial of A is not very important, for if these are different, then A must be a scalar. For the general case, this is important. This is also true because of

Theorem 1.9.4. If A satisfies a polynomial $f(x)$, then every characteristic root of A is a root of $f(x)$. So every characteristic root of A is a root of the minimal polynomial of A.

Proof. Suppose that A satisfies

$$f(x) = a_0 x^n + a_1 x^{n-1} + \cdots + a_n x^0.$$

If λ is a characteristic root of A, then we can find a vector $v \neq 0$ such that $Av = \lambda v$. Thus we have

$$A^2 v = A(Av) = A(\lambda v) = \lambda(Av) = \lambda(\lambda v) = \lambda^2 v.$$

Continuing in this way, we obtain

$$A^k v = \lambda^k v \qquad \text{for all } k > 0.$$

So, since $f(A) = 0$, we have

$$\begin{aligned}
0 = f(A)v &= (a_0 A^n + a_1 A^{n-1} + \cdots + a_n I)v \\
&= (a_0 A^n)v + (a_1 A^{n-1})v + \cdots + (a_n I)v \\
&= a_0 \lambda^n v + a_1 \lambda^{n-1} v + \cdots + a_n v \\
&= (a_0 \lambda^n + a_1 \lambda^{n-1} + \cdots + a_n)v.
\end{aligned}$$

Because

$$\mu = a_0 \lambda^n + a_1 \lambda^{n-1} + \cdots + a_{n-1}\lambda + a_n = f(\lambda)$$

is a scalar and $\mu v = 0$ with $v \neq 0$, we end up with $0 = \mu = f(\lambda)$. So λ is a root of f.

Since A satisfies its minimal polynomial, every characteristic root is a root of this minimal polynomial. ■

PROBLEMS

NUMERICAL PROBLEMS

1. Calculate the characteristic polynomial for the given matrix.

   (a) $\begin{bmatrix} 1 & -5 \\ -6 & 2 \end{bmatrix}$.

   (b) $\begin{bmatrix} a & 0 \\ b & c \end{bmatrix}$.

   (c) $\begin{bmatrix} \frac{1}{2} & \frac{1}{2} \\ \frac{1}{2} & \frac{1}{2} \end{bmatrix}$.

   (d) $\begin{bmatrix} \frac{1}{2} & -\frac{1}{2} \\ \frac{1}{2} & -\frac{1}{2} \end{bmatrix}$.

2. In Problem 1, find the characteristic roots of the matrices given.

3. In Problem 2, find the characteristic vector for each characteristic root of each matrix given.

4. If a is *not* a characteristic root of $A = \begin{bmatrix} 1 & -5 \\ -6 & 2 \end{bmatrix}$, find the explicit form of $(aI - A)^{-1}$. What goes wrong with your calculation if a is a characteristic root?

5. In Part (c) of Problem 1 find an invertible matrix A such that
$$A \begin{bmatrix} \frac{1}{2} & \frac{1}{2} \\ \frac{1}{2} & \frac{1}{2} \end{bmatrix} A^{-1} = \begin{bmatrix} 1 & 0 \\ 0 & 0 \end{bmatrix}.$$

MORE THEORETICAL PROBLEMS

Easier Problems

6. If B is invertible and $C = BAB^{-1}$, show that $P_A(x) = P_C(x)$.

7. If A and C are as in Problem 6, and if a is a characteristic root of A with associated characteristic vector v, show that a is a characteristic root of C and exhibit its associated characteristic vector.

8. If $A^2 = 0$, find the possible forms for $P_A(x)$.

9. If $A^2 = A$, find the possible forms for $P_A(x)$.

10. Calculate $P_A(x)$ if $A = \begin{bmatrix} a & b \\ -b & a \end{bmatrix}$.

11. If A is a matrix and A' its transpose, show that $P_A(x) = P_{A'}(x)$.

12. Using Problem 11, show that A and A' have the same characteristic roots.

13. If A is an invertible matrix, show that the characteristic roots of A^{-1} are the inverses of the characteristic roots of A, and vice versa.

Middle-Level Problems

14. If a is a characteristic root of A and $f(x)$ is a polynomial with coefficients in F, show that $f(a)$ is a characteristic root of $f(A)$.

15. If A is invertible, express $P_{A^{-1}}(x)$ in terms of $P_A(x)$.

16. Define for A, B in $M_2(F)$ that $A \sim B$ if there exists an invertible U in $M_2(F)$ such that $B = UAU^{-1}$. (Matrices A, B such that $A \sim B$ are called *similar*.) Prove for all A, B, $C \in M_2(F)$:

    (a) $A \sim A$.

    (b) If $A \sim B$ then $B \sim A$.

    (c) If $A \sim B$ and $B \sim C$, then $A \sim C$.

17. Show that $\begin{bmatrix} a & b \\ b & c \end{bmatrix}$ has real characteristic roots.

18. If det $(A) < 0$, show that A has distinct, real characteristic roots.

19. If $A^2 = A$, show that $(I - A)^2 = I - A$.

Harder Problems

20. If $A^2 = A$ is a matrix, show that every vector $v \in V$ can be written in a unique way as $v = v_1 + v_2$, where $Av_1 = v_1$ and $Av_2 = 0$.

21. Show that a necessary and sufficient condition that $A = \begin{bmatrix} a & b \\ c & d \end{bmatrix}$ have two distinct, real characteristic roots u and v is that for some invertible matrix C, $CAC^{-1} = \begin{bmatrix} u & 0 \\ 0 & v \end{bmatrix}$.

22. Prove that for any two matrices A and B, the characteristic polynomials for the products AB and BA are equal; that is, $P_{AB}(x) = P_{BA}(x)$.

23. Are the minimal polynomials of AB and BA always equal? If yes, prove; and if no, give a *counterexample*.

Very Hard Problems

24. Find all matrices A in $M_2(\mathbb{R})$ whose characteristic roots are real and which satisfy the equation $AA' = I$.

1.10. COMPLEX NUMBERS

As we saw in looking at the matrix

$$B = \begin{bmatrix} 0 & 1 \\ -1 & 0 \end{bmatrix},$$

whose characteristic polynomial

$$P_B = x^2 + 1$$

has no real roots, if we restrict ourselves to working only with real numbers, we easily run into matrices that have no real characteristic roots. Yet it would be desirable if all matrices had characteristic roots— even a full complement of them. In order to reach this state of nirvana we must widen the concept of number. In doing so the set of complex numbers crops up naturally. This enlarged set of numbers play a prominent part not only in every phase of mathematics but in all sort of other fields—chemistry, physics, engineering, and so on. The analysis of electrical circuits becomes easy once one introduces the nature of complex impedance. So, aside from whatever role they will play in matrix theory, it would be highly useful to know someting about them.

The right way of defining complex numbers is by formally introducing a set of symbols—constructed over the reals—and to show that with the proper definition of addition and multiplication this set of new objects behaves very well. This we shall do very soon. But to motivate the definitions we adopt a backdoor kind of approach and introduce the complex numbers via a certain set of matrices in $M_2(\mathbb{R})$.

Let $\mathscr{C}$ be the set of all matrices of the form $\begin{bmatrix} a & b \\ -b & a \end{bmatrix}$, where a and b are real numbers. If we denote the matrix $\begin{bmatrix} 0 & 1 \\ -1 & 0 \end{bmatrix}$ by J, then all the matrices in $\mathscr{C}$ look like $aI + bJ$, where a and b are any real numbers:

$$\begin{bmatrix} a & b \\ -b & a \end{bmatrix} = a\begin{bmatrix} 1 & 0 \\ 0 & 1 \end{bmatrix} + b\begin{bmatrix} 0 & 1 \\ -1 & 0 \end{bmatrix} = aI + bJ$$

How do elements of $\mathscr{C}$ behave under the matrix operations? Since

$$J^2 = \begin{bmatrix} 0 & 1 \\ -1 & 0 \end{bmatrix}\begin{bmatrix} 0 & 1 \\ -1 & 0 \end{bmatrix} = \begin{bmatrix} -1 & 0 \\ 0 & -1 \end{bmatrix} = -I$$

we see that:

1. $(aI + bJ) + (cI + dJ) = (a + c)I + (b + d)J$, so is again in $\mathscr{C}$.

2. $(aI + bJ)(cI + dJ) = (ac - bd)I + (ad + bc)J$, so it too is again in $\mathscr{C}$. Here we have used the fact that $J^2 = -I$.

3. $(aI + bJ)(cI + dJ) = (ac - bd)I + (ad + bc)J = (ca - db)I + (da + cb)J = (cI + dJ)(aI + bJ)$. Thus the multiplication of elements of $\mathscr{C}$ is *commutative*.

But possibly the most important property of $\mathscr{C}$ is yet to come. If $aI + bJ \neq 0$, then one of $a \neq 0$, $b \neq 0$ holds. Hence $c = a^2 + b^2 \neq 0$. Now $(aI + bJ)(aI - bJ) = (a^2 + b^2)I$; therefore,

$$(aI + bJ)\left(\frac{a}{c}I - \frac{b}{c}J\right) = \frac{a^2 + b^2}{c}I = I.$$

So, the *inverse* of $aI + bJ$ is

$$\frac{a}{c}I - \frac{b}{c}J$$

where $c = a^2 + b^2$, which is also in $\mathscr{C}$. Thus we have:

4.   If $aI + bJ \neq 0$ is in $\mathscr{C}$, then $(aI + bJ)^{-1}$ is also in $\mathscr{C}$.

Finally, if we *identify* the matrix aI with the real number a, we get that $\mathbb{R}$ (or something quite like it) is in $\mathscr{C}$. Thus

5. $\mathbb{R}$ is contained in $\mathscr{C}$.

So in many ways $\mathscr{C}$, which is larger than $\mathbb{R}$, behaves very much like $\mathbb{R}$ with respect to addition, multiplication, and division. But $\mathscr{C}$ has the advantage that in it $J^2 = -I$, so we have something like $\sqrt{-1}$ in $\mathscr{C}$.

With this little discussion of $\mathscr{C}$ and the behavior of its elements relative to addition and multiplication as our guide, we are ready formally to define the set of complex numbers $\mathbb{C}$. Of course, the complex numbers did not arise historically via matrices—matrices arose much later in the game than the complex numbers. But since we do have the 2 × 2 matrices over $\mathbb{R}$ in hand it does not hurt to motivate the complex numbers in the fashion we did above.

Definition.   The set of complex numbers $\mathbb{C}$ is the set of all *formal symbols* $a + bi$, where a and b are any real numbers, where we define:

1. $a + bi = c + di$ if and only if $a = c$ and $b = d$.
2. $(a + bi) + (c + di) = (a + c) + (b + d)i$.
3. $(a + bi)(c + di) = (ac - bd) + (ad + bc)i$.

This last property comes from the fact that we want i^2 to be equal to -1.

If we merely write $a + 0i$ as a and $0 + bi$ as bi, then the rule of multiplication in (3) gives us

$$i^2 = (0 + i)(0 + i) = (0 \cdot 0 - 1 \cdot 1) + (0 \cdot 1 + 1 \cdot 0)i$$
$$= -1 + 0i = -1.$$

If $\alpha = a + bi$, then a is called the *real part* of α and bi the *imaginary part* of α. If $\alpha = bi$, we call *a pure imaginary*.

We assume that $i^2 = -1$ and multiply out formally to get the multiplication rule of (3). This is the best way to remember how complex numbers multiply.

As we pointed out, we *identify* $a + 0i$ with a. This implies that $\mathbb{R}$ is contained in $\mathbb{C}$.

Also, as we did for matrices, rule (3) implies that we have $(a + bi)(a - bi) = a^2 + b^2$, so

if $a + bi \neq 0$, then

$$(a + bi)^{-1} = \frac{a}{a^2 + b^2} - \frac{b}{a^2 + b^2} i$$

and is again in $\mathbb{C}$.

If $\alpha = 2 + 3i$, we see that

$$\alpha^{-1} = \frac{2}{2^2 + 3^2} - \frac{3}{2^2 + 3^2} i = \frac{2}{13} - \frac{3}{13} i.$$

If $\beta = \dfrac{1}{\sqrt{2}} - \dfrac{1}{\sqrt{2}} i$, we leave it to the reader to show that $\beta^{-1} = \dfrac{1}{\sqrt{2}} + \dfrac{1}{\sqrt{2}} i.$

Before going any further, we document other basic rules of behavior in $\mathbb{C}$. We shall use lowercase Greek letters to denote elements of $\mathbb{C}$.

Theorem 1.10.1. $\mathbb{C}$ satisfies the following rules:

1. If $\alpha, \beta \in \mathbb{C}$, then $\alpha + \beta \in \mathbb{C}$.
2. If $\alpha, \beta \in \mathbb{C}$, then $\alpha + \beta = \beta + \alpha$.
3. For all $\alpha \in \mathbb{C}$, $\alpha + 0 = \alpha$.
4. Given $\alpha = a + bi \in \mathbb{C}$, then $-\alpha = -a - bi \in \mathbb{C}$ and $\alpha + (-\alpha) = 0$.
5. If $\alpha, \beta, \gamma \in \mathbb{C}$, then $\alpha + (\beta + \gamma) = (\alpha + \beta) + \gamma$.

These specify how $\mathbb{C}$ behaves relative to $+$. We now specify how it behaves relative to multiplication.

6. If $\alpha, \beta \in \mathbb{C}$, then $\alpha\beta \in \mathbb{C}$.
7. If $\alpha, \beta \in \mathbb{C}$, then $\alpha\beta = \beta\alpha$.
8. If $\alpha, \beta, \gamma \in \mathbb{C}$, then $\alpha(\beta\gamma) = (\alpha\beta)\gamma$.
9. The complex number $1 = 1 + 0i$ satisfies $\alpha 1 = \alpha$ for all $\alpha \in \mathbb{C}$.

10.   Given $\alpha \neq 0 \in \mathbb{C}$, then $\alpha^{-1} \in \mathbb{C}$, where $\alpha^{-1}\alpha = 1$.

The final rule weaves together addition and multiplication.

11.   If $\alpha, \beta, \gamma \in \mathbb{C}$, then $\alpha(\beta + \gamma) = \alpha\beta + \alpha\gamma$.

Proof.   We leave all parts of Theorem 1.10.1, except for (8), as an exercise for the reader. How do we prove (8)? Let

$$\alpha = a + bi, \beta = c + di, \gamma = g + hi.$$

Then, by our rule for the product,

$$\alpha\beta = (a + bi)(c + di) = (ac - bd) + (ad + bc)i;$$

hence

$$
\begin{aligned}
(\alpha\beta)\gamma &= ((ac - bd) + (ad + bc)i)(g + hi) \\
&= (ac - bd)g - (ad + bc)h \\
&\quad + ((ac - bd)h + (ad + bc)g)i \\
&= (a(cg - dh) - b(dg + ch)) \\
&\quad + (a(ch + dg) + b(-dh + cg)i) \\
&= (a + bi)((cg - dh) + (ch + dg)i) \\
&= (a + bi)((c + di)(g + hi)) = \alpha(\beta\gamma).
\end{aligned}
$$

This long chain is easy but looks a little frightening.   ■

The properties specified for $\mathbb{C}$ by (1)–(11) define what is called a *field* in mathematics. So $\mathbb{C}$ is an example of a field. Other examples of fields are $\mathbb{R}$, $\mathbb{Q}$, the rational numbers, and $T = \{a + b\sqrt{2} \mid a, b \text{ rational}\}$. Notice that rules (1)–(11), while forming a long list, are really a reminder of the usual rules of arithmetic to which we are so accustomed.

$\mathbb{C}$ has the property that $\mathbb{R}$ fails to have, namely, we can find roots in $\mathbb{C}$ of any quadratic polynomials having their coefficients in $\mathbb{C}$ (something which is all too false in $\mathbb{R}$). To head toward this goal we first establish that every element in $\mathbb{C}$ has a *square root* in $\mathbb{C}$.

Theorem 1.10.2.   If α is in $\mathbb{C}$, then there exists a β in $\mathbb{C}$ such that $\alpha = \beta^2$.

Proof.   Let $\alpha = a + bi$. If $b = 0$, we leave to the reader to show that there is a complex number β such that $\beta^2 = a = \alpha$. So we may assume that $b \neq 0$. We want to find *real* numbers x and y such that

$$\alpha = a + bi = (x + yi)^2.$$

Now, by our rule of multiplication, $(x + yi)^2 = (x^2 - y^2) + 2xyi$. So we want

$$a + bi = (x^2 - y^2) + 2xyi,$$

that is, $a = x^2 - y^2$ and $b = 2xy$. Substituting $y = b/2x$ in $a = x^2 - y^2$, we come up with $a = x^2 - b^2/4x^2$, so $4x^4 - 4ax^2 - b^2 = 0$. This is a quadratic equation in x^2, so we can solve for x^2 using the *quadratic formula*. This yields that

$$x^2 = \frac{4a \pm \sqrt{16a^2 + 16b^2}}{8} = \frac{a \pm \sqrt{a^2 + b^2}}{2}.$$

Since $b \neq 0$, $a^2 + b^2 > a^2$ and $\sqrt{a^2 + b^2}$ is therefore larger in size than a. So x^2 *cannot* be $\dfrac{a - \sqrt{a^2 + b^2}}{2}$ for a real value of x, since $\dfrac{a - \sqrt{a^2 + b^2}}{2} < 0$. So we are forced to conclude that

$$x = \pm \sqrt{\frac{a + \sqrt{a^2 + b^2}}{2}}$$

are the *only real solutions* for x. The formula for x makes sense since $a + \sqrt{a^2 + b^2}$ is *positive*. (Prove!) So we have our required x. What is y? Since $y = b/2x$ we get the value for y from this using the value of x obtained above. The x and y, which are real numbers, then satisfy $(x + yi)^2 = a + bi = \alpha$. Thus Theorem 1.10.2 is proved. Note that if $\alpha \neq 0$, then we can find exactly two square roots of α in $\mathbb{C}$. ∎

An immediate consequence of Theorem 1.10.2 is that any quadratic equation with coefficients in $\mathbb{C}$ has two solutions in $\mathbb{C}$. This is

Theorem 1.10.3. Given α, β, γ in $\mathbb{C}$, with $\alpha \neq 0$, we can always find a solution of $\alpha x^2 + \beta x + \gamma = 0$ for some x in $\mathbb{C}$.

Proof. The usual quadratic formula, namely

$$x = \frac{-\beta \pm \sqrt{\beta^2 - 4\alpha\gamma}}{2\alpha},$$

holds as well here as it does for real coefficients. (Prove!) Now, by Theorem 1.10.2, since $\beta^2 - 4\alpha\gamma$ is in $\mathbb{C}$, its square root, $\sqrt{\beta^2 - 4\alpha\gamma}$, is also in $\mathbb{C}$. Thus x is in $\mathbb{C}$. Since x is a solution to $\alpha x^2 + \beta x + \gamma = 0$, the theorem is proved. ∎

In fact, Theorem 1.10.3 actually shows that $\alpha x^2 + \beta x + \gamma = 0$ has two *distinct* solutions in $\mathbb{C}$ provided that $\beta^2 - 4\alpha\gamma \neq 0$.

We have another nice operation, that of taking the *comlex conjugate* of a complex number.

Definition. If $\alpha = a + bi$ is in $\mathbb{C}$, then the *complex conjugate* of α, denoted by $\bar{\alpha}$, is $\bar{\alpha} = a - bi$.

The properties governing the behavior of "conjugacy" are contained in

Lemma 1.10.4. For any complex numbers α and β we have:

1. $\overline{(\bar{\alpha})} = \bar{\bar{\alpha}} = \alpha$;
2. $\overline{(\alpha + \beta)} = \bar{\alpha} + \bar{\beta}$;
3. $\overline{(\alpha\beta)} = \bar{\alpha}\bar{\beta}$;
4. $\alpha\bar{\alpha}$ is real and nonnegative. Moreover, $\alpha\bar{\alpha} > 0$ if $\alpha \neq 0$.

Proof. The proofs of all the parts are immediate from the definitions of the sum, product, and complex conjugate. We leave Parts (1), (2), and (4) as exercises, but do prove Part (3).

Let $\alpha = a + bi$, $\beta = c + di$. Thus

$$\alpha\beta = (a + bi)(c + di) = (ac - bd) + (ad + bc)i,$$

so

$$\overline{(\alpha\beta)} = \overline{(ac - bd) + (ad + bc)i} = (ac - bd) - (ad + bc)i.$$

On the other hand, $\bar{\alpha} = a - bi$ and $\bar{\beta} = c - di$, whence

$$\begin{aligned}
\bar{\alpha}\bar{\beta} &= (a - bi)(c - di) \\
&= (ac - (-b)(-d)) + (a(-d) + (-b)c)i \\
&= (ac - bd) - (ad + bc)i = \overline{(\alpha\beta)}. \quad \blacksquare
\end{aligned}$$

Yet another notion plays an important role in dealing with complex numbers. This is the notion of *absolute value*, which, in a sense, measures the size of a complex number.

Definition. If $\alpha = a + bi$, then the *absolute value* of α, $|\alpha|$, is defined by $|\alpha| = \sqrt{a^2 + b^2}$.

Note that $|\alpha| = \sqrt{a^2 + b^2} = \sqrt{\alpha\bar{\alpha}}$.

The properties of $|\cdot|$ are outlined in

Lemma 1.10.5.   If α and β are complex numbers, then

1. $\alpha + \bar{\alpha}$ is real and $\alpha + \bar{\alpha} \le 2|\alpha|$;
2. $|\alpha\beta| = |\alpha| \, |\beta|$;
3. $|\alpha + \beta| \le |\alpha| + |\beta|$ (triangle inequality).

Proof.   (a) Let $\alpha = a + bi$. Then

$$\alpha + \bar{\alpha} = 2a \le 2\sqrt{a^2 + b^2} = 2|\alpha|.$$

2. $|\alpha\beta| = \sqrt{\alpha\beta\overline{\alpha\beta}} = \sqrt{\alpha\bar{\alpha}\beta\bar{\beta}} = \sqrt{\alpha\bar{\alpha}} \sqrt{\beta\bar{\beta}} = |\alpha| \, |\beta|$ since $\alpha\bar{\alpha}$ is real and nonnegative.

3. $\begin{aligned}|\alpha + \beta|^2 &= (\alpha + \beta)(\overline{\alpha + \beta}) \\ &= \alpha\bar{\alpha} + \alpha\bar{\beta} + \bar{\alpha}\beta + \beta\bar{\beta} \\ &= |\alpha|^2 + \alpha\bar{\beta} + \overline{(\alpha\bar{\beta})} + |\beta|^2 \\ &\le |\alpha|^2 + 2|\alpha\bar{\beta}| + |\beta|^2 \\ &= |\alpha|^2 + 2|\alpha| \, |\beta| + |\beta|^2 = (|\alpha| + |\beta|)^2.\end{aligned}$

since $(\alpha\bar{\beta}) + \overline{(\alpha\bar{\beta})} \le 2|a\bar{\beta}| = 2|\alpha| \, |\beta|$. Taking square roots, we obtain $|\alpha + \beta| \le |\alpha| + |\beta|$.   ∎

Why the name "triangle inequality" for the inequality in Part (3)? We can identify the complex number $\alpha = a + bi$ with the point (a, b) in the Euclidean plane:

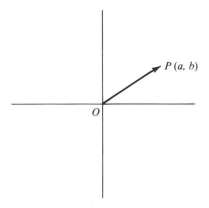

Then $|\alpha| = \sqrt{a^2 + b^2}$ is the length of the line segment OP. If we recall the diagram for the addition of complex numbers as points of the plane,

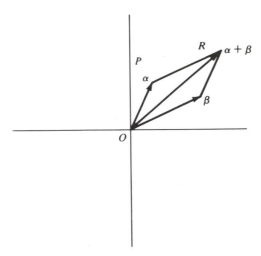

then the point R, $(a + c, b + d)$, is that point associated with $\alpha + \beta$, where $\alpha = a + bi$, $\beta = c + di$.

The triangle inequality then merely asserts that the length of OR is at most that of OP plus that of PR, in other words, in a triangle the length of any side is at most the sum of the lengths of the other two sides.

Again turning to the geometric interpretation of complex numbers, if $\alpha = a + bi$ it corresponds to the point P in

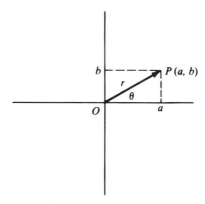

Thus

$$r = \sqrt{a^2 + b^2}$$

is the *length* of OP, and the angle θ—called the *argument* of α—is determined by

$$\sin (\theta) = b/r$$

if we restrict θ to be in the range $-\pi$ to π. Note that $a = r \cos (\theta)$ and

$b = r \sin (\theta)$; thus $\alpha = a + bi = r \cos (\theta) + r \sin (\theta) i$ and

$$\alpha = r (\cos (\theta) + i \sin (\theta)).$$

This is called the *polar form* of the complex number α.

Before leaving the subject we make a remark related to the result of Theorem 1.10.3. There it was shown that any quadratic polynomial having complex coefficients has a complex root. This is but a special case of a beautiful and very important result due to Gauss which is known as the *Fundamental Theorem of Algebra*. This theorem asserts that

> *any polynomial, of whatever positive degree, with complex coefficients always has a complex root.*

We shall need this result when we treat the $n \times n$ matrices.

PROBLEMS

NUMERICAL PROBLEMS

1. Carry out the following.
   - **(a)** $(6 - 5i)(7 + \frac{3}{2}i)$.
   - **(b)** $(16 + 4i)(\frac{1}{2} - \frac{2}{3}i)^{-1}$.
   - **(c)** $(4 - 3i)(16 + 12i)$.
   - **(d)** $(4 + 7i)\overline{(8 - 6i)}$.

2. Find $\sqrt{i}$ in the form $a + bi$.

3. Find $\sqrt{6 + 7i}$.

4. Solve the equation $x^2 - ix + 5 = 0$.

5. Compute $|\cos (\theta) + i \sin (\theta)|$.

6. Show that $\sqrt{\cos (\theta) + i \sin (\theta)} = \cos \left(\dfrac{\theta}{2}\right) + i \sin \left(\dfrac{\theta}{2}\right)$.

7. Use the result of Problem 6 to find $\sqrt{\dfrac{1}{2} + \dfrac{\sqrt{3}}{2} i}$.

MORE THEORETICAL PROBLEMS

Easier Problems

8. Show that $T = \{a + b\sqrt{2} \mid a, b$ rational numbers$\}$ is a field.

9. Prove all the various parts of Theorem 1.10.1.

10. Verify that the quadratic formula gives you the roots of the polynomial $\alpha x^2 + \beta x + \gamma$ even when α, β, and γ are complex numbers.

11. Prove Parts (a), (b), and (d) of Lemma 1.10.4. Show that $\alpha + \bar{\alpha}$ is twice the real part of α, and $\alpha - \bar{\alpha}$ is twice the imaginary part of α.

12. Let $\alpha \neq 0$ be a complex number. Suppose that β is a complex number such that $\alpha\beta$ is real. Show that $\beta = u\bar{\alpha}$ for some real number u.

13. If $\alpha = r(\cos(\theta) + i\sin(\theta))$ and $\beta = s(\cos(\phi) + i\sin(\phi))$ are two complex numbers in polar form, show that

$$\alpha\beta = rs(\cos(\theta + \phi) + i\sin(\theta + \phi)).$$

(Thus the argument of the product of two complex numbers is the sum of their arguments.)

14. Show that $T = \{a + bi \mid a, b \text{ rational numbers}\}$ is a field.

Middle-Level Problems

15. If α is a root of the polynomial $p(x) = a_0 x^n + a_1 x^{n-1} + \cdots + a_n$, where the a_j are all *real*, show that $\bar{a}$ is also a root of $p(x)$.

16. From the result of Problem 13 prove De Moivre's theorem, namely that $(\cos(\theta) + i\sin(\theta))^n = \cos(n\theta) + i\sin(n\theta)$ for every integer $n \geq 0$.

17. Prove De Moivre's theorem even if $n < 0$.

18. Express α^{-1}, where $\alpha = r[\cos(\theta) + i\sin(\theta)] \neq 0$, in polar form.

19. Show that

$$\left(\cos\left(\frac{2\pi}{k}\right) + \sin\left(\frac{2\pi}{k}\right)\right)^k = 1$$

and

$$\left(\cos\left(\frac{2\pi}{k}\right) + \sin\left(\frac{2\pi}{k}\right)\right)^m \neq 1$$

for $0 < m < k$.

Harder Problems

20. Given a complex number $a \neq 0$ show that for every integer $n > 0$ we can find n distinct complex numbers $b_1, \ldots, b_n$ such that $a = b_r^n$ for $1 \leq r \leq n$. (So complex numbers have nth roots.)

21. Find necessary and sufficient conditions on α, $\beta \in \mathbb{C}$ so that $|\alpha + \beta| = |\alpha| + |\beta|$.

1.11. $M_2(\mathbb{C})$

We have discussed $M_2(\mathbb{R})$, the 2×2 matrices over the real numbers, $\mathbb{R}$. Now that we have the complex numbers, $\mathbb{C}$, at our disposal we can enlarge our domain of discourse from $M_2(\mathbb{R})$ to $M_2(\mathbb{C})$. What is the advantage of passing form $\mathbb{R}$ to $\mathbb{C}$? For one thing, since the characteristic polynomial of a matrix in $M_2(\mathbb{R})$ (and in $M_2(\mathbb{C})$) is quadratic, by Theorem 1.10.3, we know that this quadratic polynomial has roots in $\mathbb{C}$. Therefore, *every matrix in $M_2(\mathbb{C})$ will have characteristic roots, and these will be in $\mathbb{C}$.*

How do we carry out this passage from $M_2(\mathbb{R})$ to $M_2(\mathbb{C})$? In the most obvious way! *Define equality, addition, and multiplication of matrices* in $M_2(\mathbb{C})$, that is, matrices having complex numbers as their entries, *exactly as we did for matrices having real entries.* Everything we have done, so far, carries over in its entirety. We leave this verification to the reader. We will have some of these verifications as problems.

We shall let $M_2(\mathbb{C})$ act on something—namely the set of 2-tuples having complex coordinates. As before, let

$$W = \left\{ \begin{bmatrix} \xi \\ \vartheta \end{bmatrix} \,\middle|\, \xi, \vartheta \in \mathbb{C} \right\}.$$

We let $M_2(\mathbb{C})$ act on W as follows:

If $A = \begin{bmatrix} \alpha & \beta \\ \gamma & \delta \end{bmatrix} \in M_2(\mathbb{C})$

and if $\begin{bmatrix} \xi \\ \vartheta \end{bmatrix} \in W$; then

$$Aw = \begin{bmatrix} \alpha & \beta \\ \gamma & \delta \end{bmatrix} \begin{bmatrix} \xi \\ \vartheta \end{bmatrix} = \begin{bmatrix} \alpha\xi + \beta\vartheta \\ \gamma\xi + \delta\vartheta \end{bmatrix}.$$

Because of the nice properties of $\mathbb{C}$ relative to addition, multiplication, and division, what we did for $M_2(\mathbb{R})$ carries over *verbatim* to $M_2(\mathbb{C})$.

PROBLEMS

NUMERICAL PROBLEMS

1. Multiply.

(a) $\begin{bmatrix} 1 & i \\ -i & i \end{bmatrix} \begin{bmatrix} 1 & -i \\ i & 1 \end{bmatrix}.$

(b) $\begin{bmatrix} 3 + i & 0 \\ \frac{1}{2} - \frac{1}{3}i & 1 \end{bmatrix} \begin{bmatrix} 3 - i & 0 \\ 6 + 7i & 1 \end{bmatrix}.$

(c) $\begin{bmatrix} i & 1 \\ 1 & i \end{bmatrix} \begin{bmatrix} 4 & 5 \\ 6 & 3 - 2i \end{bmatrix} \begin{bmatrix} i & 1 \\ 1 & i \end{bmatrix}^{-1}.$

2. Calculate the determinant of the matrices in Problem 1 directly and also by using the fact that $\det (AB) = \det (A) \det (B)$.

3. Find the characteristic roots of

(a) $\begin{bmatrix} 1 & i \\ -i & 1 \end{bmatrix}.$

(b) $\begin{bmatrix} 5 & 6 \\ -6 & 5 \end{bmatrix}.$

(c) $\begin{bmatrix} (4 - i)/2 & 4 \\ (i + 3)/3 & 0 \end{bmatrix}.$

4. In Problem 3, find the characteristic vectors for the matrices for each characteristic root.

5. Find all matrices A in $M_2(\mathbb{C})$ such that

$$A \begin{bmatrix} 1 & 2 \\ -2 & 1 \end{bmatrix} = \begin{bmatrix} 1 & 2 \\ -2 & 1 \end{bmatrix} A.$$

6. Find all matrices A in $M_2(\mathbb{C})$ such that $AB = BA$ for all $B \in M_2(\mathbb{C})$.

7. Find all α such that for every invertible A in $M_2(\mathbb{C})$, $\det$

$$\left(A \begin{bmatrix} 2 + i & \alpha \\ 1 - i & 1 + i \end{bmatrix} A^{-1} \right) = 0.$$

MORE THEORETICAL PROBLEMS

Easier Problems

8. For what values of $\beta \in \mathbb{C}$ is $\begin{bmatrix} u & \beta \\ -\beta & u \end{bmatrix}$ invertible for all *real* $u \ne 0$?

9. Prove that in $M_2(\mathbb{C})$:

(a) $\det (AB) = \det (A) \det (B)$

(b) $\operatorname{tr} (AB) = \operatorname{tr} (BA)$

(c) $\operatorname{tr} (\alpha A + B) = \alpha (\operatorname{tr} (A)) + \operatorname{tr} (B)$

(d) $(AB)' = B'A'$

for all $A, B \in M_2(\mathbb{C})$ and all $\alpha \in \mathbb{C}$.

10. Define the mapping $^-$ in $M_2(\mathbb{C})$ by $\overline{\begin{bmatrix} \alpha & \beta \\ \gamma & \delta \end{bmatrix}} = \begin{bmatrix} \bar{\alpha} & \bar{\beta} \\ \bar{\gamma} & \bar{\delta} \end{bmatrix}$. Prove:

   (a) $\overline{\overline{A}} = \overline{(\overline{A})} = A$

   (b) $\overline{(A + B)} = \overline{A} + \overline{B}$

   (c) $\overline{(\alpha A)} = \bar{\alpha}\overline{A}$

   (d) $\overline{(AB)} = \overline{A}\,\overline{B}$

   for all A, B, $\in M_2(\mathbb{C})$ and all $\alpha \in \mathbb{C}$.

11. Show that $\overline{(AB)'} = \overline{B'}\overline{A'}$.

12. If $A \in M_2(\mathbb{C})$ is invertible, show that $\overline{A^{-1}} = \overline{A}^{-1}$.

13. If $A \in M_2(\mathbb{C})$, define A^* by $A^* = \overline{A}'$. Prove:

   (a) $A^{**} = (A^*)^* = A$

   (b) $(\alpha A + B)^* = \bar{\alpha}A^* + B^*$

   (c) $(AB)^* = B^*A^*$

   for all A, $B \in M_2(\mathbb{C})$ and $\alpha \in \mathbb{C}$.

Middle-Level Problems

14. Show that $AA^* = 0$ if and only if $A = 0$.

15. Show that $(A + A^*)^* = A + A^*$, $(A - A^*)^* = -(A - A^*)$, and $(AA^*)^* = AA^*$ for $A \in M_2(\mathbb{C})$.

16. If $B \in M_2(\mathbb{C})$, show that $B = A + C$, where $A^* = A$, $C^* = -C$. Moreover, show that A and C are uniquely determined by B.

17. If $A^* = \pm A$ and $A^2 = 0$, show that $A = 0$.

Harder Problems

18. If $A^* = A$, prove that the characteristic roots of A are real.

19. If $A^* = -A$, prove that the characteristic roots of A are pure imaginaries.

20. If $A^*A = I$ and α is a characteristic root of A, what can you say about $|\alpha|$?

21. If $A^*A = I$ and $B^*B = I$, what is a simple form for $(AB)^{-1}$?

22. If $A^* = -A$, show that:

   (a) $I - A$ and $I + A$ are invertible.

   (b) If $B = (I - A)(I + A)^{-1}$, then $B^*B = I$.

23. If $AB = 0$, is $A^*B^* = 0$? Either prove or give a counterexample.

24. If $A \in M_2(\mathbb{C})$, show that $\text{tr}\,(AA^*) > 0$ for $A \neq 0$.

Very Hard Problems

25. Prove that for $A \in M_2(\mathbb{C})$ the characteristic roots of A^*A are *real* and *nonnegative*.

26. If $A(AA^* - A^*A) = (AA^* - A^*A)A$, show that $AA^* = A^*A$.

27. If $A^* = A$, show that we can find a B such that $B^*B = I$ and $BAB^{-1} = \begin{bmatrix} \alpha & 0 \\ 0 & \beta \end{bmatrix}$. What are α and β in relation to A?

1.12. INNER PRODUCTS

For those readers who have been exposed to vector analysis, the concept of *dot product* is not a new one. Given

$$W = \left\{ \begin{bmatrix} \xi \\ \vartheta \end{bmatrix} \,\middle|\, \xi, \vartheta \in \mathbb{C} \right\},$$

if $v, w \in W$ and $v = \begin{bmatrix} \alpha \\ \beta \end{bmatrix}$ and $w = \begin{bmatrix} \gamma \\ \delta \end{bmatrix}$, we define the *inner product* of v and w.

Definition. If $v = \begin{bmatrix} \alpha \\ \beta \end{bmatrix}$ and $w = \begin{bmatrix} \gamma \\ \delta \end{bmatrix}$ are in W, then their *inner product*, denoted by (v, w), is defined by $(v, w) = \alpha\bar{\gamma} + \beta\bar{\delta}$.

The reason we use the complex conjugate, rather than defining (v, w) as $\alpha\gamma + \beta\delta$, is because we want (v, v) to be nonzero if $v \neq 0$. For instance, if $v = \begin{bmatrix} 1 \\ i \end{bmatrix}$, then, if we defined the inner product without complex conjugates, (v, v) *would* be $(1)(1) + (i)(i) = 1 - 1 = 0$, yet $v \neq 0$. You might very well ask why we want to avoid such a possibility. The answer is that we want (v, v) to determine the length (actually, it is the square of the length) of v and we want a nonzero element to have positive length.

EXAMPLE

Let $v = w = \begin{bmatrix} 1 + i \\ 2i \end{bmatrix}$. Then $(v, w) = (1 + i)(1 - i) +$

$(2i)(-2i) = 2 + 4 = 6$.

We summarize some of the basic behavior of $(\cdot, \cdot)$ in

Lemma 1.12.1. If $u, v, w \in W$ and $\sigma \in \mathbb{C}$, then

1. $(v + w, u) = (v, u) + (w, u)$.
2. $(u, v + w) = (u, v) + (u, w)$.
3. $(v, w) = \overline{(w, v)}$.
4. $(\sigma v, w) = \sigma(v, w) = (v, \overline{\sigma} w)$.
5. $(v, v) \geq 0$ is real and $(v, v) = 0$ if and only if $v = 0$.

Proof. The proof is straightforward, following directly from the definition of the inner product. We do verify Parts (3), (4), and (5).

To see Part (3), let $v = \begin{bmatrix} \alpha \\ \beta \end{bmatrix}$ and let $w = \begin{bmatrix} \gamma \\ \delta \end{bmatrix}$. Then, by definition,

$(v, w) = \alpha\overline{\gamma} + \beta\overline{\delta}$ while $(w, v) = \gamma\overline{\alpha} + \delta\overline{\beta}$. Noticing that

$$\overline{\gamma\overline{\alpha} + \delta\overline{\beta}} = \overline{\gamma}\overline{\overline{\alpha}} + \overline{\delta}\overline{\overline{\beta}} = \alpha\overline{\gamma} + \beta\overline{\delta},$$

we get that $(v, w) = \overline{(w, v)}$.

To see Part (4), note that

$$(\sigma v, w) = (\sigma\alpha)\overline{\gamma} + (\sigma\beta)\overline{\delta} = \sigma(\alpha\overline{\gamma} + \beta\overline{\delta}) = \sigma(v, w)$$

because $\sigma v = \sigma \begin{bmatrix} \alpha \\ \beta \end{bmatrix} = \begin{bmatrix} \sigma\alpha \\ \sigma\beta \end{bmatrix}$.

Finally, if $v = \begin{bmatrix} \alpha \\ \beta \end{bmatrix} \neq 0$, then at least one of α or β is not 0. Thus

$(v, v) = \alpha\overline{\alpha} + \beta\overline{\beta} = |\alpha|^2 + |\beta|^2 \neq 0$, and is, in fact, positive. This proves Part (5). ∎

We call v *orthogonal* to w if $(v, w) = 0$. For example, the vector

$v = \begin{bmatrix} 2 \\ 3 \end{bmatrix}$ is orthogonal to $w = \begin{bmatrix} -3 \\ 2 \end{bmatrix}$:

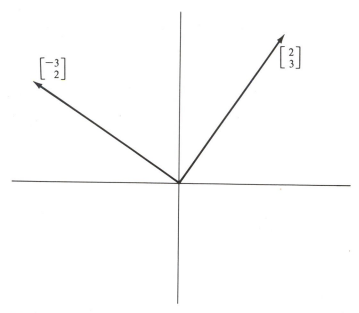

Note that if v is orthogonal to w, then w is orthogonal to v because v is orthogonal to w if and only if $(v, w) = 0$. But then, by Part (3) of Lemma 1.12.1, $(w, v) = \overline{(v, w)} = \overline{0} = 0$. So w is orthogonal to v.

Clearly, $v = 0$ is orthogonal to every element of W. Is it the only vector (element of W) with this property? Suppose that $(v, w) = 0$ for all w in W; then certainly $(v, v) = 0$. By Part (5) of Lemma 1.12.1 we obtain that $v = 0$. So, we've proved parts (1) and (2) of

Lemma 1.12.2.
1. If v is orthogonal to w, then w is orthogonal to v.
2. If v is orthogonal to all of W, then $v = 0$.
3. If $v^{\perp} = \{w \in W \mid (v, w) = 0\}$, then:
    (a) $w_1, w_2 \in v^{\perp}$ implies that $w_1 + w_2 \in v^{\perp}$.
    (b) $\alpha w \in v^{\perp}$ for all $\alpha \in \mathbb{C}$ and $w \in v^{\perp}$.

We leave the proof of Part (3) to the reader.

A natural question is to investigate the relationship between acting on a vector by matrix and the inner product. We do this now.

Definition. If $A \in M_2(\mathbb{C})$, we let $A^* = (\overline{A})'$.

Theorem 1.12.3. If $A \in M_2(\mathbb{C})$ and $v, w \in W$, then $(Av, w) = (v, A^*w)$.

Proof. We verify this result by computing (Av, w) and (v, A^*w) and comparing the results.

Let $A = \begin{bmatrix} \alpha & \beta \\ \gamma & \delta \end{bmatrix}$ and $v = \begin{bmatrix} \varepsilon \\ \phi \end{bmatrix}$, $w = \begin{bmatrix} \zeta \\ \eta \end{bmatrix}$. Then

$$Av = \begin{bmatrix} \alpha & \beta \\ \gamma & \delta \end{bmatrix}\begin{bmatrix} \varepsilon \\ \phi \end{bmatrix} = \begin{bmatrix} \alpha\varepsilon + \beta\phi \\ \gamma\varepsilon + \delta\phi \end{bmatrix},$$

hence

$$(Av, w) = (\alpha\varepsilon + \beta\phi)\bar{\zeta} + (\gamma\varepsilon + \delta\phi)\bar{\eta}$$
$$= \alpha\varepsilon\bar{\zeta} + \beta\phi\bar{\zeta} + \gamma\varepsilon\bar{\eta} + \delta\phi\bar{\eta}.$$

On the other hand, $A^* = \begin{bmatrix} \bar{\alpha} & \bar{\gamma} \\ \bar{\beta} & \bar{\delta} \end{bmatrix}$, so

$$A^*w = \begin{bmatrix} \bar{\alpha} & \bar{\gamma} \\ \bar{\beta} & \bar{\delta} \end{bmatrix}\begin{bmatrix} \zeta \\ \eta \end{bmatrix} = \begin{bmatrix} \bar{\alpha}\zeta + \bar{\gamma}\eta \\ \bar{\beta}\zeta + \bar{\delta}\eta \end{bmatrix},$$

where

$$(v, A^*w) = \varepsilon\overline{(\bar{\alpha}\zeta + \bar{\gamma}\eta)} + \phi\overline{(\bar{\beta}\zeta + \bar{\delta}\eta)}$$
$$= \varepsilon(\bar{\bar{\alpha}}\bar{\zeta} + \bar{\bar{\gamma}}\bar{\eta}) + \phi(\bar{\bar{\beta}}\bar{\zeta} + \phi\bar{\bar{\delta}}\bar{\eta})$$
$$= \varepsilon(\alpha\bar{\zeta} + \gamma\bar{\eta}) + \phi(\beta\bar{\zeta} + \delta\bar{\eta})$$
$$= \alpha\varepsilon\bar{\zeta} + \gamma\varepsilon\bar{\eta} + \beta\phi\bar{\zeta} + \delta\phi\bar{\eta},$$

which we see is exactly what we obtained for (Av, w). Thus Theorem 1.12.3 is proved. ∎

Note a few consequences of the theorem. These are things that were done directly—in the problems in the preceding section—by a direct computation. For the tactical reason that the proof we are about to give works in general, we derive these consequences more abstractly from Theorem 1.12.3.

Theorem 1.12.4. If $A, B \in M_2(\mathbb{C})$ and $\alpha \in \mathbb{C}$, then

1. $A^{**} = (A^*)^* = A$;
2. $(A + B)^* = (A^* + B^*)$;
3. $(\alpha A)^* = \bar{\alpha}A^*$;
4. $(AB)^* = B^*A^*$.

Proof. The proofs of all the parts will depend on something we worked out earlier, namely, if $(v, w) = 0$ for all $w \in W$, then $v = 0$.

We start with Part (1). If $v, w \in W$ then

$$(Av, w) = (v, A^*w) = \overline{(A^*w, v)} = \overline{(w, A^{**}v)} = (A^{**}v, w),$$

using Lemma 1.12.1 and the definition of A^*. Thus

$$(Av - A^{**}v, w) = (Av, w) - (A^{**}v, w) = 0$$

hence $((A - A^{**})v, w) = 0$ for all $w \in W$. Therefore, $(A - A^{**})v = 0$ for all $v \in W$. This imples that

$$A^{**} = A.$$

For Part (2) notice that

$$(Av, w) + (Bv, w) = ((A + B)v, w) = (v, (A + B)^*w).$$

However,

$$(Av, w) + (Bv, w) = (v, A^*w) + (v, B^*w) = (v, (A^* + B^*)w).$$

Comparing these evaluations leads us to

$$(A + B)^* = A^* + B^*.$$

Part (3) we leave to the reader.

Finally, for Part (4), we have

$$\begin{aligned}(ABv, w) &= (A(Bv), w) = (Bv, A^*w) \\ &= (v, B^*(A^*w)) = (v, B^*A^*w);\end{aligned}$$

however, $(ABv, w) = (v, (AB)^*w)$. Comparing the results yields that

$$(AB)^* = B^*A^*. \quad \blacksquare$$

Another interesting question about the interaction of multiplication by a matrix and the inner product is: For what A is (v, w) *preserved*, that is, for what A does

$$(Av, Aw) = (v, w)$$

hold for all v, w? Clearly, if $(Av, Aw) = (v, w)$, then $(v, A^* Aw) = (v, w)$, hence $A^*Aw = w$ for all $w \in W$. This forces

$$A^*A = I.$$

On the other hand, if $A^*A = I$, then for all $v, w \in W$,

$$(A^*Av, w) = (Iv, w) = (v, w).$$

But $(A^*Av, w) = (Av, A^{**}w) = (Av, Aw)$, by Theorem 1.12.4. Thus

$$(Av, Aw) = (v, w).$$

We have proved

Theorem 1.12.5. For $A \in M_2(\mathbb{C})$, $(Av, Aw) = (v, w)$ for all $v, w \in W$ if and only if $A*A = I$.

Definition. A matrix $A \in M_2(\mathbb{C})$ is called *unitary* if $A*A = I$.

Unitary matrices are very nice in that they preserve orthogonality and also (v, w).

Definition. If v is in W, then the *length* of v, denoted by $\|v\|$, is defined by $\|v\| = \sqrt{(v, v)}$.

So if $v = \begin{bmatrix} \xi \\ \vartheta \end{bmatrix} \in W$, then $\|v\| = \sqrt{\xi\bar{\xi} + \vartheta\bar{\vartheta}}$. Note that if ξ and ϑ are *real*, then $\|v\| = \sqrt{\xi^2 + \vartheta^2}$, the usual length of the line segment join the origin to the point (ξ, ϑ) in the plane.

We know that any unitary matrix preserves length, for if A is unitary then $\|Av\|$ equals $\sqrt{(Av, Av)} = \sqrt{(v, v)} = \|v\|$. What about the other way around? Suppose that $(Av, Av) = (v, v)$ for all $v \in W$; is A unitary? The answer, as we see in the next theorem, is "yes."

Theorem 1.12.6. If $A \in M_2(\mathbb{C})$ is such that $(Aw, Aw) = (w, w)$ for all $w \in W$, then A is unitary [hence $(Av, Aw) = (v, w)$ for all $v, w \in W$].

Proof. Let $v, w \in W$. Then

$(A(v + w), A(v + w))$
$= (Av + Aw, Av + Aw)$
$= (Av, Av) + (Aw, Aw) + (Av, Aw) + (Aw + Av).$

Since $(Av, Av) = (v, v)$ and $(Aw, Aw) = (w, w)$ we get

$$(Av, Aw) + (Aw, Av) = (v, w) + (w, v). \qquad (1)$$

Since (1) is true for all $v, w \in W$ and since $iw \in W$, if we replace w by iw in (1) we get

$$(Av, iAw) + (iAw, Av) = (v, iw) + (iw, v), \qquad (2)$$

and so

$$\bar{i}(Av, Aw) + i(Aw, Av) = \bar{i}(v, w) + i(w, v). \qquad (3)$$

However, $\bar{i} = -i$; thus (3) becomes

$$i\{-(Av, Aw) + (Aw, Av)\} = i\{-(v, w) + (w, v)\}, \qquad (4)$$

which gives us

$$(Aw, Av) - (Av, Aw) = (w, v) - (v, w). \tag{5}$$

If we add the result in (1) to that in (5), we end up with $2(Aw, Av)$ $= 2(w, v)$, hence $(Aw, Av) = (w, v)$. Therefore, A is unitary. ∎

Thus to express that a matrix in $M_2(\mathbb{C})$ is unitary, we can say it succinctly as: A *preserves length.*

For example, the matrix $A = \begin{bmatrix} \cos(\theta) & -\sin(\theta) \\ \sin(\theta) & \cos(\theta) \end{bmatrix}$ is unitary, and

it preserves length because it is the rotation through the angle θ in the counterclockwise direction.

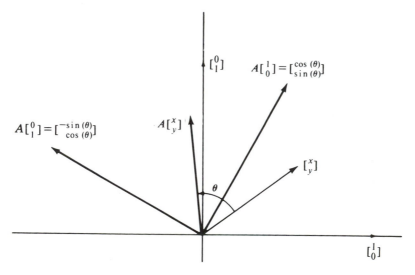

Let's see how easy and noncomputational the proof of the following fact becomes:

If $A^* = A$, *then the characteristic roots of A are all real.*

To prove it by computing it directly can be very messy. Let's see how it goes using inner products.

Let $\alpha \in \mathbb{C}$ be a characteristic root of $A \in M_2(\mathbb{C})$, where $A^* = A$. Thus $Av = \alpha v$ for some $v \neq 0 \in V$. Therefore,

$$\alpha(v, v) = (\alpha v, v) = (Av, v) = (v, A^*v)$$
$$= (v, Av) = (v, \alpha v) = \bar{\alpha}(v, v)$$

Since $(v, v) > 0$ we get $\alpha = \bar{\alpha}$, hence α is real. We have proved

Theorem 1.12.7. If $A^* = A$, then the characteristic roots of A are real. Using a similar technique, we prove

Theorem 1.12.8. If $A \in M_2(\mathbb{C})$ is unitary and if α is a characteristic root of A, then $|\alpha| = 1$.

Proof. Let α be a characteristic root of A and $v \neq 0$ a corresponding characteristic vector. Thus $Av = \alpha v$. Hence

$$\alpha(v, v) = (Av, v) = (v, A^*v) = (v, A^{-1}v).$$

But if $Av = \alpha v$, then $A^{-1}(Av) = \alpha A^{-1}v$, that is, $A^{-1}v = \dfrac{1}{\alpha} v$. There-fore, returning to our calculation above, $\alpha(v, v) = (v, A^{-1}v) = \left(v, \dfrac{1}{\alpha} v\right) = \dfrac{1}{\bar{\alpha}}(v, v)$. Thus $\alpha\bar{\alpha}(v, v) = (v, v)$ and since $(v, v) \neq 0$, we get $\alpha\bar{\alpha} = 1$. Thus $|\alpha| = \sqrt{\alpha\bar{\alpha}} = \sqrt{1} = 1$. ∎

Before closing this section, we should introduce the names for some of these things.

1. A^* is called the *Hermitian adjoint*.
2.   If $A^* = A$, then A is called *Hermitian*.
3.   If $A^* = -A$, then A is called *skew-Hermitian*.
4.   If $A^*A = AA^*$, then A is called *normal*.

The word "Hermitian" comes from the name of the French mathematician Hermite. We use these terms freely in the problems.

PROBLEMS

NUMERICAL PROBLEMS

1. Determine the values for a (if any) such that the vectors $\begin{bmatrix} 1 + \alpha \\ 1 - \alpha \end{bmatrix}$ and $\begin{bmatrix} 1 - \alpha \\ 3 \end{bmatrix}$ are orthogonal.

2. Describe the set of vectors $\begin{bmatrix} a \\ b \end{bmatrix}$ orthogonal to $\begin{bmatrix} 3 \\ 4 \end{bmatrix}$.

3. Describe the set of vectors $\begin{bmatrix} a \\ b \end{bmatrix}$ orthogonal to $\begin{bmatrix} 1 \\ 4 \end{bmatrix}$.

4. Describe the set of vectors $\begin{bmatrix} a \\ b \end{bmatrix}$ which are orthogonal to $\begin{bmatrix} 3 \\ 4 \end{bmatrix}$ as well as to $\begin{bmatrix} 1 \\ 4 \end{bmatrix}$.

MORE THEORETICAL PROBLEMS

Easier Problems

5. Given the matrix $A = \begin{bmatrix} \alpha & \beta \\ \gamma & \delta \end{bmatrix} \in M_2(\mathbb{C})$, if $v = \begin{bmatrix} \xi \\ \vartheta \end{bmatrix} \in W$, find the formula for (Av, v) in terms of α, β, γ, δ, ξ, and ϑ.

6. If A is unitary and has *real* entries, show that $\det(A) = \pm 1$.

7. If A and B are unitary, show that AB is unitary.

Middle-Level Problems

8. If α is a characteristic root of A, show that α^n is a characteristic root of A^n for all $n \geq 1$.

Harder Problems

9. If A is unitary and has real entries, show that $\det(A) = \pm 1$ and if $\det(A) = 1$, then $A = \begin{bmatrix} \cos(\theta) & -\sin(\theta) \\ \sin(\theta) & \cos(\theta) \end{bmatrix}$ for some θ.

10. If $(Av, v) = 0$ for all v with real coordinates, where $A \in M_2(\mathbb{C})$, determine the form of A if $A \neq 0$.

11. Show that if A is as in Problem 10, then $A \neq BB^*$ for all $B \in M_2(\mathbb{C})$.

12. If $A \in M_2(\mathbb{R})$ is such that for all $v = \begin{bmatrix} \xi \\ \vartheta \end{bmatrix}$, with ξ and ϑ real, we have $(Av, Av) = (v, v)$, is A^*A necessarily equal to I? Either prove or give a counterexample.

13. Prove Parts (1) and (2) of Lemma 1.12.1.

14. Prove Part (3) of Lemma 1.12.2.

15. If $B \in M_2(\mathbb{C})$, show that $B = C + D$, where C is Hermitian and D is skew-Hermitian.

16. If A is both Hermitian and unitary, what possible values can the characteristic roots of A have?

17. If A is Hermitian and $A^n = I$ for some $n \geq 1$, show that $A^2 = I$.

18. If A is normal and if for some $v \neq 0$ in W, $Av = \alpha v$, prove that $A^*v = \bar{\alpha}v$.

19. If tr $(AA^*) = 0$, show that $A = 0$.

20. Show that det $(A^*) = \overline{\det (A)}$.

21. If $A^* = -A$ and $A^n = I$ for some $n > 0$, show that $A^4 = I$.

22. If A is skew-Hermitian, show that its characteristic roots are pure imaginaries, making use of the inner product on W.

23. If A is normal and invertible, show that $B = A^*A^{-1}$ is unitary.

24. Is the result of Problem 23 valid if A is not normal? Either prove or give a counterexample.

25. Define, for $v, w \in W$, $<v, w> = (Av, w)$, where $A \in M_2(\mathbb{C})$. What conditions must A satisfy in order that $<\cdot, \cdot>$ satisfies the five properties in Lemma 1.12.1?

26. Prove that if $A \in M_2(\mathbb{C})$ and if $p(x) = \det (xI - A)$, then $p(A) = 0$. (This is the Cayley–Hamilton Theorem for $M_2(\mathbb{C})$.)

27. If U is unitary, show that UBU^{-1} is Hermitian if B is Hermitian and skew-Hermitian if B is skew-Hermitian.

28. If A and B are in $M_2(\mathbb{R})$ and $B = CAC^{-1}$, where $C \in M_2(\mathbb{C})$, show that we can find a $D \in M_2(\mathbb{R})$ such that $B = DAD^{-1}$.

29. If α is a characteristic root of A and if $p(x)$ is a polynomial with complex coefficients, show that $p(\alpha)$ is a characteristic root of $p(A)$.

30. If $p(A) = 0$ for some polynomial $p(x)$ with complex coefficients, show that if α is a characteristic root of A, then $p(\alpha) = 0$.

2

SYSTEMS OF
LINEAR EQUATIONS

2.1. INTRODUCTION

If you should happen to glance at a variety of books on matrix theory or linear algebra, you would find that many of them, if not most, start things rolling by considering systems of linear equations and their solutions. A great number of these books go even further and state that this subject—systems of linear equations—is the most important part and central focus of linear algebra. We certainly do not subscribe to this point of view. We can think of many topics in linear algebra which play an even more important role in mathematics and allied fields. To name a few such (and these will come up in the course of the book): the theory of characteristic roots, the Cayley–Hamilton Theorem, the diagonalization of Hermitian matrices, vector spaces and linear transformations on them, inner product spaces, and so on. These are areas of linear algebra which find everyday use in a host of theoretical and applied problems.

Be that as it may, we do not want to minimize the importance of the study of systems of linear equations. Not only does this topic stand firmly on its own feet, but the results and techniques we develop in this chapter will crop up often as tools in the subsequent material of this book.

Many of the things that we do with linear equations lend themselves readily to an algorithmic description. In fact, it seems that a recipe for attack is available for almost all the results we shall obtain here. Wherever possible, we shall stress this algorithmic approach and summarize what was done into a "Method for . . . ."

In starting with 2×2 matrices and their properties, one is able to get to the heart of linear algebra easily and quickly, and to get an idea of how, and in what direction, things will run in general. The results on systems of linear equations will provide us with a rich set of techniques to study a large cross section of ideas and results in matrix theory and linear algebra.

We start things off with some examples.

1. The system of equations

$$2x + 3y = 5$$
$$4x + 5y = 9$$

can be solved by using the first equation to express y in terms of x as

$$y = (\tfrac{1}{3})(5 - 2x),$$

then substituting this expression for y into the second equation and solving for x, then solving for y:

$$4x + 5(\tfrac{1}{3})(5 - 2x) = 9$$
$$12x + 25 - 10x = 27$$
$$x = 1$$
$$y = (\tfrac{1}{3})(5 - 2) = 1.$$

For this system of equations, the solution $x = 1$, $y = 1$ is the only solution. Geometrically, these equations are represented by lines in the plane which are not parallel and so intersect in a single point. The point of intersection is the solution $\begin{bmatrix} x \\ y \end{bmatrix} = \begin{bmatrix} 1 \\ 1 \end{bmatrix}$ to the system of equations.

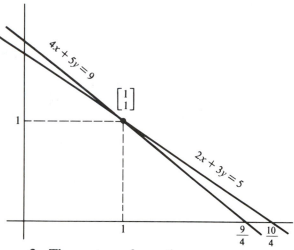

2. The system of equations

$$2x + 3y = 5$$
$$4x + 6y = 10$$

is easy to solve because any solution to the first equation is also a solution to the second. Geometrically, these equations are represented by lines, and it turns out that these lines are really the same line. The points on this line are the solutions $\begin{bmatrix} x \\ y \end{bmatrix}$ to the system of equations.

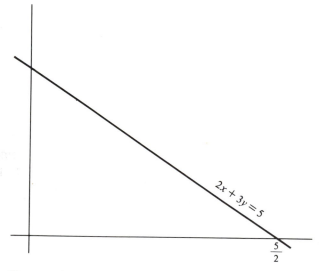

If we take any value for x, we can solve for y and we get the corresponding solution

x (any value)

$y = (\frac{1}{3})(5 - 2x)$.

Because we give the variable x any value we want and then give y the corresponding value $(\frac{1}{3})(5 - 2x)$, which depends on the value we give to x.

we call x an independent variable and y a dependent variable.

However, if instead we had chosen any value for y and set $x = (\frac{1}{2})(5 - 3y)$, we would have called y the independent variable any x the dependent variable.

3. The system of equations

$2x + 3y = 5$
$4x + 6y = 20$

has no solution, since if x and y satisfy the first equation, they cannot satisy the second. Geometrically, these equations represent parallel lines that do not intersect.

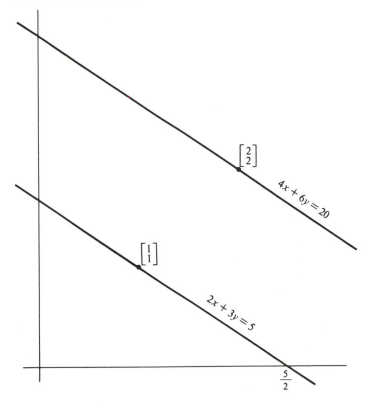

The examples above illustrate the principle that

the set of solutions $\begin{bmatrix} x \\ y \end{bmatrix} \in \mathbb{R}^{(2)}$ *to a system of linear equations in two variables x and y is either a point, a line, the empty set or all of $\mathbb{R}^{(2)}$.*

This principle even takes care of cases where all coefficients are 0. For example, the set of solutions of the system consisting of the one equation $0x + 0y = 2$ is the empty set, and the set of solutions of the system consisting of the one equation $0x + 0y = 0$ is all of $\mathbb{R}^{(2)}$. A similar principle holds for any number of variables. If the number of variables is three, for instance, this principle is that

the set of solutions $\begin{bmatrix} x \\ y \\ z \end{bmatrix}$ *to a system of linear equations in three variables x, y, z is one of the following: a point, a line, a plane, the empty set or all of $\mathbb{R}^{(3)}$.*

The reason for this is quite simple. If there is only one equation $ax + by + cz = d$, then if a, b, and c are 0, the set of solutions $\begin{bmatrix} x \\ y \\ z \end{bmatrix}$ is $\mathbb{R}^{(3)}$ if $d = 0$ and the empty set if $d \neq 0$. If, on the other hand, the equation is *nonzero* (meaning that at least one coefficient on the left hand side is nonzero), the set of solutions is a plane. Suppose that we now add another nonzero equation $fx + gy + hz = e$, ending up with the system of equations

$$ax + by + cz = d$$
$$fx + gy + hz = e.$$

The set of solutions to the second equation is also a plane, so *the set of solutions to the system of two equations is the intersection of two planes.* This will be a line or a plane or the empty set. Each time we add another nonzero equation, the new solution set is the intersection of the old solution set with the plane of solutions to the new equation. If the old solution set was a line, then the intersection of that line with this plane is either a point, a line, or the empty set. In the case of a system

$$ax + by + cz = d$$
$$fx + gy + hz = k$$
$$rx + sy + tz = u$$

of three nonzero equations, each equation is represented by a plane. We leave it as an exercise for the reader to show that if two of these planes are parallel, then the set of solutions is a line, a plane, or the empty set, and each of these possibilities really can happen. Assuming, on the other hand, that no two of these planes are parallel, their intersection is a point, a line, or the empty set:

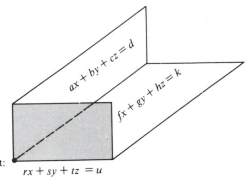

Intersection is a point:

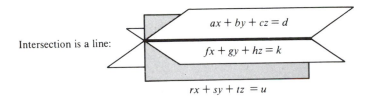

Intersection is a line:

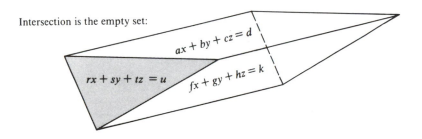

Intersection is the empty set:

EXAMPLE

Suppose that we have a supply of 6000 units of S and 8000 units of T, materials used in manufacturing products P, Q, R. If each unit of P uses 2 units of S and 0 units of T, each unit of Q uses 3 units of S and 4 units of T and each unit of R uses 1 unit of S and 4 units of T, how many units p, q, and r of P, Q, and R should we make if we want to use up the entire supply?

To answer this, we write down the sysem of linear equations

$$2p + 3q + 1r = 6000$$
$$0p + 4q + 4r = 8000,$$

which represents the given information. Why do we call these equations *linear*? Because the variables p, q, r occur in these equations only to the *first* power.

The rectangular 2×3 array of numbers $\begin{bmatrix} 2 & 3 & 1 \\ 0 & 4 & 4 \end{bmatrix}$ is called

the *matrix of coefficients* (or coefficient matrix) of this system of equations.

Since there are as many variables as equations (in fact, there is an extra *independent* variable r), and since the *coefficient matrix* $\begin{bmatrix} 2 & 3 & 1 \\ 0 & 4 & 4 \end{bmatrix}$ has nonzero entries 2 and 4 on the diagonal going down

from the top left hand corner of the matrix with only 0 below it, we can solve this system of equations. How do we do this? We use a process of *back substitution*.

What do we mean by this? We work backward. From the second equation

$$4q + 4r = 8000$$

we get that

$$q = 2000 - r;$$

this expresses q in terms of r. We now substitute this expression for q in the first equation to obtain

$$2p + 3(2000 - r) + 1r = 6000 \text{ or } 2p = 2r,$$

hence $p = r$. Thus we have also expressed p in terms of r. Assigning any value to r gives us values for p and q which are solutions to our system of equations. For instance, if we use $r = 500$, we get the solution

$$p = 500$$
$$q = 2000 - 500 = 1500.$$

Since p and q are determined by r we call r an *independent variable* of this system. The set of solutions

$$\begin{bmatrix} r \\ 2000 - r \\ r \end{bmatrix}$$

is a line, which is just the intersection of the two planes defined by the two equations in the system.

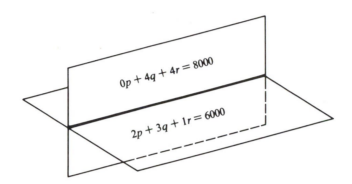

We let a matrix A—even if only rectangular instead of square— operate on column vectors v using a rule similar to what we used for 2×2 matrices. If A is an $m \times n$ matrix, that is, has m row and n columns, it operates on column vectors v with n entries. Namely, we "multiply" the first row of A by v to get the first entry of Av, the second row of A by v to get the second entry of Av, and so on. What does it mean to multiply a row of A times the column vector v? Just multiply the first entries together, then the second entries, and so on. When this has been done, add them up. Of course, the number of columns of A must equal the number of entries of v for it to make sense when we "multiply" a row of A times v in this fashion. And, of course, the vector Av we end up with has m (the number of rows of A) entries, instead of the number n of entries of v. So when we let the 2×3 matrix

$\begin{bmatrix} 2 & 3 & 1 \\ 0 & 4 & 4 \end{bmatrix}$ operate on the vector $\begin{bmatrix} p \\ q \\ r \end{bmatrix}$ of three entries, we perform the product

$$2p + 3q + 1r = 6000$$

to get the first entry and the product

$$0p + 4q + 4r = 8000$$

to get the second. So we get a vector $\begin{bmatrix} 6000 \\ 8000 \end{bmatrix}$ with two entries. In this way we can represent the system of equations

$$2p + 3q + 1r = 6000$$
$$0p + 4q + 4r = 8000$$

by a corresponding *matrix equation* $\begin{bmatrix} 2 & 3 & 1 \\ 0 & 4 & 4 \end{bmatrix} \begin{bmatrix} p \\ q \\ r \end{bmatrix} = \begin{bmatrix} 6000 \\ 8000 \end{bmatrix}.$

Now we can represent the information given in the system of equations in the *matrix* $\begin{bmatrix} 2 & 3 & 1 \\ 0 & 4 & 4 \end{bmatrix}$ and *column vector* $\begin{bmatrix} 6000 \\ 8000 \end{bmatrix}$. We then can represent the information found after solving the system as the column vector $\begin{bmatrix} p \\ q \\ r \end{bmatrix} = \begin{bmatrix} r \\ 2000 - r \\ r \end{bmatrix}$. *We equate solving the system of equations*

$$2p + 3q + 1r = 6000$$
$$0p + 4q + 4r = 8000$$

for p, q, r with solving the matrix equation

$$\begin{bmatrix} 2 & 3 & 1 \\ 0 & 4 & 4 \end{bmatrix} \begin{bmatrix} p \\ q \\ r \end{bmatrix} = \begin{bmatrix} 6000 \\ 8000 \end{bmatrix}$$

for $\begin{bmatrix} p \\ q \\ r \end{bmatrix}.$

We can use systems of equations and their corresponding matrix equations interchangeably, since each contains all the entries needed to write down the other. Let's use subscripts to keep track of equations and variables in the most general *system*

$$a_{11}x_1 + \cdots + a_{1n}x_n = y_1$$
$$\vdots \qquad\qquad \vdots \qquad \vdots$$
$$a_{m1}x_1 + \cdots + a_{mn}x_n = y_m$$

of m *linear equations in n variables.* The $m \times n$ array or *matrix*

$$\begin{bmatrix} a_{11} & \cdots & a_{1n} \\ \cdot & & \cdot \\ \cdot & & \cdot \\ \cdot & & \cdot \\ a_{m1} & \cdots & a_{mn} \end{bmatrix}$$ is called the *coefficient matrix* of the system of equa-

tions. As in our example, we equate the system of equations with the corresponding *matrix equation*

$$\begin{bmatrix} a_{11} & \cdots & a_{1n} \\ \vdots & & \vdots \\ a_{m1} & \cdots & a_{mn} \end{bmatrix} \begin{bmatrix} x_1 \\ \vdots \\ x_n \end{bmatrix} = \begin{bmatrix} y_1 \\ \vdots \\ y_m \end{bmatrix}.$$

We can abbreviate this equation as $Ax = y$ where

$$A = \begin{bmatrix} a_{11} & \cdots & a_{1n} \\ \vdots & & \vdots \\ a_{m1} & \cdots & a_{mn} \end{bmatrix}$$

and x and y are the column vectors

$$x = \begin{bmatrix} x_1 \\ \vdots \\ x_n \end{bmatrix}, y = \begin{bmatrix} y_1 \\ \vdots \\ y_m \end{bmatrix}.$$

The *main diagonal* of A holds the *diagonal entries* $a_{11}, \ldots, a_{dd}$, where $d = m$ if $m < n$ and $d = n$ otherwise. For example, the diagonal entries

of $\begin{bmatrix} 1 & 2 & 3 \\ 5 & 6 & 7 \\ 9 & 8 & 7 \\ 5 & 4 & 3 \end{bmatrix}$ and $\begin{bmatrix} 1 & 2 & 3 \\ 5 & 6 & 7 \\ 9 & 8 & 7 \end{bmatrix}$ are 1, 6, and 7, whereas those of

$\begin{bmatrix} 1 & 2 & 3 \\ 5 & 6 & 7 \end{bmatrix}$ are 1 and 6.

We say that the matrix A is *upper triangular* if the entries a_{rs} with

$r > s$ (those below the main diagonal) are all 0. So $\begin{bmatrix} 2 & 3 & 1 \\ 0 & 4 & 4 \\ 0 & 0 & 7 \\ 0 & 0 & 0 \end{bmatrix}$ and

$\begin{bmatrix} 2 & 3 & 1 \\ 0 & 4 & 4 \\ 0 & 0 & 7 \end{bmatrix}$ are upper triangular and $\begin{bmatrix} 2 & 3 & 1 \\ 0 & 4 & 4 \\ 0 & 0 & 7 \\ 0 & 0 & 3 \end{bmatrix}$ and $\begin{bmatrix} 2 & 3 & 1 \\ 5 & 4 & 4 \\ 0 & 0 & 7 \end{bmatrix}$ are

not.

Just as for the system of two equations in three variables represented

by the matrix equation $\begin{bmatrix} 2 & 3 & 1 \\ 0 & 4 & 4 \end{bmatrix} \begin{bmatrix} p \\ q \\ r \end{bmatrix} = \begin{bmatrix} 6000 \\ 8000 \end{bmatrix}$, we can solve the

matrix equation $\begin{bmatrix} 2 & 3 & 1 \\ 0 & 4 & 4 \\ 0 & 0 & 7 \end{bmatrix} \begin{bmatrix} p \\ q \\ r \end{bmatrix} = \begin{bmatrix} 6000 \\ 8000 \\ 7000 \end{bmatrix}$ by back substitution, get-

ting $r = 1000$ from the last equation, then $q = 1000$ from the next-to-

last equation, and $p = 1000$ from the first equation, that is, $\begin{bmatrix} p \\ q \\ r \end{bmatrix} =$

$\begin{bmatrix} 1000 \\ 1000 \\ 1000 \end{bmatrix}$. In this case, however, the number of equations equals the

number of variables, and $\begin{bmatrix} p \\ q \\ r \end{bmatrix} = \begin{bmatrix} 1000 \\ 1000 \\ 1000 \end{bmatrix}$ is the *only* solution.

When the coefficient matrix of a matrix equation is an upper triangular matrix having as many columns as rows and the entries on the
main diagonal are all nonzero, it is very easy to solve the equation.

EXAMPLE

To solve $\begin{bmatrix} 1 & 0 & 2 & 4 & 5 \\ 0 & 2 & 2 & 4 & 4 \\ 0 & 0 & 1 & 3 & 4 \end{bmatrix} \begin{bmatrix} a \\ b \\ c \\ d \\ e \end{bmatrix} = \begin{bmatrix} 8 \\ 8 \\ 2 \end{bmatrix}$, let's first write down the corresponding system of linear equations

$$1a + 0b + 2c + 4d + 5e = 8$$
$$2b + 2c + 4d + 4e = 8$$
$$1c + 3d + 4e = 2.$$

To solve this system, let's rewrite it with the last $5 - 3 = 2$ variables on the right-hand side:

$$1a + 0b + 2c = 8 - 4d - 5e$$
$$2b + 2c = 8 - 4d - 4e$$
$$1c = 2 - 3d - 4e.$$

We solve these equations starting with the last equation and then working backward, getting

$$c = 2 - 3d - 4e$$
$$2b + 2c = 8 - 4d - 4e$$
$$b = 4 - 2d - 2e - c$$
$$= 4 - 2d - 2e - 2 + 3d + 4e$$
$$= 2 + d + 2e$$
$$1a + 0b + 2c = 8 - 4d - 5e$$
$$a = 8 - 4d - 5e - 2c$$
$$= 8 - 4 + 6d + 8e - 4d - 5e,$$
$$= 4 + 2d + 3e,$$

that is, getting
$$\begin{bmatrix} a \\ b \\ c \\ d \\ e \end{bmatrix} = \begin{bmatrix} 4 + 2d + 3e \\ 2 + d + 2e \\ 2 - 3d - 4e \\ d \\ e \end{bmatrix}.$$

We now give the general result.

Theorem 2.1.1. Suppose that $m \le n$ and the coefficient matrix of a system of m linear equations in n variables is upper triangular with only nonzero entries on the main diagonal. Then we can always solve the system by back substitution. If $m < n$, there are $n - m$ independent variables in the solution. On the other hand, if $m = n$, there is exactly one solution.

Proof. Instead of using the matrix equation, we shall play with the problem as a problem on linear equations. Since we have assumed that the coefficient matrix is upper triangular with nonzero entries on the main diagonal, the system of linear equations that we are dealing with looks like

$$
\begin{aligned}
a_{11}x_1 + a_{12}x_2 + a_{13}x_3 + \cdots + \quad a_{1,m}x_m + \cdots + \quad a_{1,n}x_n &= y_1 \\
a_{22}x_2 + a_{23}x_3 + \cdots + \quad a_{2,m}x_m + \cdots + \quad a_{2,n}x_n &= y_2 \\
a_{33}x_3 + \cdots + \quad a_{3,m}x_m + \cdots + \quad a_{3,n}x_n &= y_3 \\
&\vdots \\
a_{m-1,m-1}x_{m-1} + a_{m-1,m}x_m + \cdots + a_{m-1,n}x_n &= y_{m-1} \\
a_{m,m}x_m + \cdots + \quad a_{m,n}x_n &= y_m,
\end{aligned}
$$

(1)

where $a_{11} \neq 0$, $a_{22} \neq 0$, ..., $a_{m,m} \neq 0$, and $y_1, \ldots, y_m$ are given numbers.

In all the equations above, move the terms involving

$$x_{m+1}, \ldots, x_n$$

to the right-hand side. The result is a system of m linear equations

$$
\begin{aligned}
a_{11}x_1 + a_{12}x_2 + a_{13}x_3 + \cdots + a_{1,m}x_m &= y_1 - (a_{1,m+1}x_{m+1} + \cdots + a_{1,n}x_n) \\
a_{22}x_2 + a_{23}x_3 + \cdots + a_{2,m}x_m &= y_2 - (a_{2,m+1}x_{m+1} + \cdots + a_{2,n}x_n) \\
a_{33}x_3 + \cdots + a_{3,m}x_m &= y_3 - (a_{3,m+1}x_{m+1} + \cdots + a_{3,n}x_n) \\
&\vdots \\
a_{m-1,m-1}x_{m-1} + a_{m-1,m}x_m &= y_{m-1} - (a_{m-1,m+1}x_{m+1} + \cdots + a_{m-1,n}x_n) \\
a_{mm}x_m &= y_m - (a_{m,m+1}x_{m+1} + \cdots + a_{m,n}x_n).
\end{aligned}
$$

(2)

The solutions of the systems (1) and (2) are the same. To solve system (2) is not hard. Start from the last equation,

$$a_{mm}x_m = y_m - (a_{m,m+1}x_{m+1} + \cdots + a_{m,n}x_n).$$

Since $a_{mm} \neq 0$, we can solve for x_m merely by dividing by a_{mm}. *Notice that x_m is expressed in terms of $x_{m+1}, \ldots, x_n$. Now feed this value of x_m into the second-to-last equation,*

$$a_{m-1,m-1}x_{m-1} + a_{m-1,m}x_m = y_{m-1} - (a_{m-1,m+1}x_{m+1} + \cdots + a_{m-1,n}x_n).$$

Since $a_{m-1,m-1} \neq 0$ we can solve for x_{m-1} in terms of $x_m, x_{m+1}, \ldots,$ x_n. *Since we now have x_m expressed in terms of $x_{m+1}, \ldots, x_n$, we have x_{m-1} also expressed in such terms.*

Continue the process. Feed the values of x_{m-1} and x_m just found into the third-to-last equation. Because $a_{m-2,m-2} \neq 0$ we can solve for x_{m-2} in terms of $x_{m-1}, x_m, x_{m+1}, \ldots, x_n$. But since x_m, x_{m-1} are expressible in terms of $x_{m+1}, \ldots, x_n$, we get that x_{m-2} is also so expressible.

Repeat the procedure, climbing up through the equations to reach x_1. In this way we get that each of $x_1, \ldots, x_m$ is expressible in terms of $x_{m+1}, \ldots, x_n$, and we are free to assign values to $x_{m+1}, \ldots, x_n$

at will. Each such assignment of values for $x_{m+1}, \ldots, x_n$ leads to a solution of the systems (1) and (2).

Because $x_{m+1}, \ldots, x_n$ can take on any values independently we call them the *independent variables*. So if $m < n$, we get what we claimed, namely that the solutions to system (1) are given by $n - m$ independent variables whose values we can assign arbitrarily.

We leave the verification to the reader that if $m = n$, there is one and only one solution to the system (1).   ■

Because of the importance of the method outlined in the proof of this theorem, we summarize it here.

Method to solve a system of linear equations by back substitution if the coefficient matrix is upper triangular with nonzero diagonal entries

1. If $m < n$, let the $n - m$ extra variables $x_{m+1}, \ldots, x_n$ have any values, and move them to the right-hand side of the equations.
2. Find the value for the last dependent variable x_m from the last equation.
3. Eliminate that variable by substituting its value back in place of it in all of the preceding equations.
4. Repeat (2), (3), and (4) for the new last remaining dependent variable using the new last remaining equation to get its value and eliminate it and the equation. Continue until the values for all dependent variables have been found.

PROBLEMS

NUMERICAL PROBLEMS

1. Represent each of the following systems of equations by a matrix equation.

   (a) $1u + 4k + 6h = 6$
            $0u + 3k + 5h = 5.$

   (b) $3s + 5t + 8h = \pi$
            $10t + 3r + 5w = \frac{1}{5}.$

  (c) $2a + 3b + 6c = 6$
            $0a + 4b + 5c = 5$
            $1a + 4b + 6c = 6$
                 $3b + 5c = 5.$

  (d) $2a + 3b + 6c = 3$
                 $4b + 5c = 5$
            $1d + 4a + 6e = 2$
                 $6b + 5e = 5.$

  (e) $1u + 4k + 6h = 3$
            $2u + 3h + 4x = 8.$

2. Find all solutions for each of the following triangular systems of equations.

  (a) $2x + 3w = 6$
               $4w = 5.$

  (b) $2a + 3b + 6c = 6$
                 $4b + 5c = 5$
                     $6c = 6.$

  (c) $4a + 3b + 6c = 6$
                 $4b + 5c = 5$
                     $6c = 6.$

  (d) $2a + 3b + 6c = 6$
                 $4b + 5c = 5$
                     $6c = 6$
            $4d + 6e = 2.$

3. Describe the system of equations corresponding to the following matrix equations, and for each say how many independent variables must appear in its general solution.

  (a) $\begin{bmatrix} 1 & 2 & 1 & 1 & 5 \\ 0 & 2 & 1 & 2 & 1 \\ 0 & 0 & 1 & 3 & 4 \end{bmatrix} \begin{bmatrix} a \\ b \\ c \\ d \\ e \end{bmatrix} = \begin{bmatrix} 4 \\ 4 \\ 4 \end{bmatrix}.$

  (b) $\begin{bmatrix} 1 & 1 & 5 \\ 0 & 1 & 1 \\ 0 & 1 & 4 \end{bmatrix} \begin{bmatrix} w \\ q \\ d \end{bmatrix} = \begin{bmatrix} 554 \\ 1084 \\ 54 \end{bmatrix}.$

  (c) $\begin{bmatrix} 2 & 1 & 1 & 5 \\ 2 & 1 & 2 & 1 \\ 0 & 1 & 3 & 4 \end{bmatrix} \begin{bmatrix} a \\ s \\ f \\ g \end{bmatrix} = \begin{bmatrix} 44 \\ 41 \\ 14 \end{bmatrix}.$

4. Solve the system of equations in Part (a) of Problem 3 by back substitution.

5. Find a nontrivial solution to the system

$$x - y - x - w = 0$$
$$- 2y - 4z - 5w = 0.$$

MORE THEORETICAL PROBLEMS

Easier Problems

6. Show that the system of equations

$$ax + by + cz = u$$
$$0x + fy + gz = v$$
$$0x + hy + kz = w$$

has a unique solution for every u, v, w if and only if $a(fk - gh)$ is nonzero.

7. Give a formula for the solution x, y, z to the system of equations in Problem 6 which works as long as $a(fk - gh)$ is nonzero.

8. Prove that if $m = n$ in Theorem 2.1.1, then there is one and only one solution to the system (1).

2.2. EQUIVALENT SYSTEMS

Suppose that $Ax = y$ represents a system of m linear equations in n variables. We ask:

1. Are there solutions x to the equation $Ax = y$?

2. If so, how do we find x using the entries of A and y?

If $m \le n$ and A is upper triangular with only nonzero entries on the main diagonal, we know by Theorem 2.1.1 that the system can be solved by back substitution. Otherwise, we find the answers to these questions by *eliminating variables* and replacing the given system by an *equivalent* system which we can use to get the answers. Let's look at an example.

EXAMPLE

To solve the system

$$2p + 3q + r = 6000$$
$$- 4p - 2q + 2r = - 4000$$

represented by the matrix equation $\begin{bmatrix} 2 & 3 & 1 \\ -4 & -2 & 2 \end{bmatrix} \begin{bmatrix} p \\ q \\ r \end{bmatrix} =$

$\begin{bmatrix} 6000 \\ -4000 \end{bmatrix}$, we use the first equation to modify the second equation

by adding two times $(2p + 3q + r)$ to its left-hand side and two times 6000 to its right-hand side, getting

$$0p + 4q + 4r = 8000.$$

So, we can replace our original system by the system

$$2p + 3q + \ \ r = 6000$$
$$0p + 4q + 4r = 8000$$

obtained from it by adding 2 times the first equation to the second equation. Since the second system is obtained from the first by adding 2 times the first equation to the second, and since the first system could by obtained from the second by subtracting 2 times the first equation from the second, *the two systems have the same solutions*. The second system, represented by the matrix equation

$$\begin{bmatrix} 2 & 3 & 1 \\ 0 & 4 & 4 \end{bmatrix} \begin{bmatrix} p \\ q \\ r \end{bmatrix} = \begin{bmatrix} 6000 \\ 8000 \end{bmatrix},$$

has an upper triangular coefficient matrix, so it can be solved by back substitution, by Theorem 2.1.1. In fact, we already did this in our first example, getting the solutions $\begin{bmatrix} r \\ 2000 - r \\ r \end{bmatrix}$. Checking,

we see that

$$\begin{bmatrix} 2 & 3 & 1 \\ -4 & -2 & 2 \end{bmatrix} \begin{bmatrix} r \\ 2000 - r \\ r \end{bmatrix} = \begin{bmatrix} 6000 \\ -4000 \end{bmatrix},$$

as expected.

Of course, if we modify a system of equations by multiplying all terms of one of its equations by a nonzero value or by interchanging two of the equations, the solutions again will be the same for the original system and the modified system.

EXAMPLE

The system

$$0p + 4q + 4r = 8000$$
$$2p + 3q + r = 6000$$

has the same solutions as the system

$$2p + 3q + r = 6000$$
$$0p + 4q + 4r = 8000$$

obtained from it by interchanging rows 1 and 2. This system, in turn has the same solutions as the system

$$2p + 3q + r = 6000$$
$$0p + q + r = 2000$$

obtained from it by multiplying row 2 by $\frac{1}{4}$. So, since the solutions

to the last system are $\begin{bmatrix} r \\ 2000 - r \\ r \end{bmatrix}$, where $r \in F$, these are the

solutions to the first system as well.

These operations come up often enough that we give them a name in the

Definition. The following three kinds of operations are called *elementary operations*:

1. The operation of adding u times equation s to equation r (where $r \neq s$).
2. The operation of interchanging equations r and s (where $r \neq s$);
3. The operation of multiplying equation r by u (where $u \neq 0$).

Each elementary operation can be reversed by a corresponding *inverse* operation. The inverse operations are

1' The operation of adding $-u$ times equation s to equation r (where $r \neq s$);

2' The operation of interchanging equations s and r (where $r \neq s$);

3' The operation of by multiplying equation r by $1/u$ (where $u \neq 0$).

EXAMPLE

In order to solve the system

$$2p + 3q + r = 6000$$
$$-4p - 2q + 2r = -4000,$$

we first got the system

$$2p + 3q + r = 6000$$
$$0p + 4q + 4r = 8000$$

by adding 2 times the first equation to the second equation. If we now were to apply the inverse operation of adding -2 times the first equation to the second equation, we would get back the system,

$$2p + 3q + r = 6000$$
$$-4p - 2q + 2r = -4000,$$

with which we started.

Since each elementary operation can be reversed in this fashion,

the solutions to a system obtained from a given system by any elementary operation are the same as the solutions to the original system.

To release the full power of this observation, we make the

Definition.   Two systems of equations are *equivalent* if the second can be obtained from the first by applying elementary operations successively a finite number of times.

The relation of equivalence satisfies the following properties:

1.   Any system of equations is equivalent to itself.
2.   If a system of equations represented by $Ax = y$ is equivalent to a system of equations represented by $Bx = z$, then the system of equations represented by $Bx = z$ is equivalent to the system of equations represented by $Ax = y$.
3.   If the system of equations represented by $Ax = y$ is equivalent to the system of equations represented by $Bx = z$ and if the system of equations represented by $Bx = z$ is equivalent to the system of equations represented by $Cx = w$, then the system of equations represented by $Ax = y$ is equivalent to $Cx = w$.

Why? Multiplying by 1 each equation of the system of equations represented by the equation $Ax = y$, we find that there is no change and we end up with the same system of equations that we started with. This proves (1). For (2), suppose that the system of equations represented by the equation $Ax = y$ is equivalent to the system of equations represented by the equation $Bx = z$. Then the system of equations represented by the equation $Bx = z$ is obtained from the system of equations represented by the equation $Ax = y$ by performing, successively, a finite number of elementary operations. Let's now *reverse* each of these elementary operations and apply the reverse operations successively in *reverse order* starting with the system of equations represented by the equation $Bx = z$. Then we get back the system of equations that we started with. Since the reverse operations are also elementary operations, it follows that the system of equations represented by the equation $Bx = z$ is equivalent to the system of equations represented by the equation $Ax = y$, which proves (2). For (3), suppose that starting with the system of equations represented by the equation $Ax = y$, we apply elementary operations successively a finite number of times, ending up with the system of equations represented by the equation $Bx = z$. And then, starting with the equation $Bx = z$, we apply elementary operations successively a finite number of times, ending up with the system of equations represented by the equation $Cx = w$. Then, starting with the system of equations represented by the equation $Ax = y$, we can apply all of these elementary operations in succession, first applying those used to get to the system represented by $Bx = z$ and then those used to get to $Cx = w$. So if the system of equations represented by $Ax = y$ is equivalent to the system of equations represented by $Bx = z$ and if the system of equations represented by $Bx = z$ is equivalent to the system of equations represented by $Cx = w$, then the system of equations represented by $Ax = y$ is equivalent to $Cx = w$, which proves (3).

     These properties of equivalence become very important when we take into account

Theorem 2.2.1.   If two matrix equations $Ax = y$ and $Bx = z$ are equivalent, then they have the same solutions x.

     Proof.   We already know that if a matrix equation is obtained from another matrix equation by a single elementary operation, the new equation has the same solutions as the old. So, using this time and time again, the new equation still has the same solutions as the old.   ■

PROBLEMS

NUMERICAL PROBLEMS

1. For each matrix A listed below, given the system of equations represented by $Ax = 0$, find an equivalent system of equations represented by $Bx = 0$, where B is upper triangular, and solve (find all solutions of) the latter.

(a) $A = \begin{bmatrix} 1 & 4 & 6 \\ 1 & 7 & 11 \end{bmatrix}$.

(b) $A = \begin{bmatrix} 2 & 4 & 6 \\ 5 & 5 & 61 \end{bmatrix}$.

(c) $A = \begin{bmatrix} 1 & 2 & 3 \\ 4 & 5 & 6 \\ 7 & 8 & 9 \end{bmatrix}$.

2. For each matrix A and vector y listed below, given the system of equations represented by $Ax = y$, find an equivalent system of equations represented by $Bx = z$, where B is upper triangular, and solve the latter.

(a) $A = \begin{bmatrix} 1 & 6 \\ 1 & 11 \end{bmatrix}$ and $y = \begin{bmatrix} 6 \\ 11 \end{bmatrix}$.

(b) $A = \begin{bmatrix} 1 & 4 & 6 \\ 1 & 7 & 11 \end{bmatrix}$ and $y = \begin{bmatrix} 6 \\ 11 \end{bmatrix}$.

(c) $A = \begin{bmatrix} 1 & 2 & 3 \\ 4 & 5 & 6 \\ 7 & 9 & 9 \end{bmatrix}$ and $y = \begin{bmatrix} 16 \\ 1 \\ 0 \end{bmatrix}$.

3. Show that the system of equations

$$\begin{aligned} x - y + z - w &= 0 \\ 2x + 3y + 4z + 5w &= 0 \\ x - y + 2z - 2w &= 0 \\ 5x + 5y + 9z + 9w &= 0 \end{aligned}$$

has a nontrivial solution, by finding a system of three equations having the same solutions.

MORE THEORETICAL PROBLEMS

Easier Problems

4. Show by three examples that the set of solutions x to an equation $Ax = y$ with upper triangular coefficient matrix can be any one of the following: empty (i.e., no solutions), infinite, or consisting of exactly one solution.

Middle-Level Problems

5. Let A be an upper triangular $m \times n$ matrix with real entries and let

$$y = \begin{bmatrix} y_1 \\ \cdot \\ \cdot \\ \cdot \\ y_m \end{bmatrix}.$$ Show that the set of solutions x to the equation $Ax =$

y must be one of the following: empty, infinite, or consisting of exactly one solution.

2.3. ELEMENTARY ROW OPERATIONS. ECHELON MATRICES

Elementary operations on systems of equations can be represented very efficiently by operations on matrices.

EXAMPLE

To solve the system

$$\begin{aligned} 2p + 3q + \;\; r &= \;\;\;\; 6000 \\ -4p - 2q + 2r &= -4000, \end{aligned}$$

we first got the system

$$\begin{aligned} 2p + 3q + \;\; r &= 6000 \\ 0p + 4q + 4r &= 8000 \end{aligned}$$

by adding 2 times the first equation to the second equation. Instead, we can represent the system

$$\begin{aligned} 2p + 3q + \;\; r &= \;\;\;\; 6000 \\ -4p + 2q + 2r &= -4000 \end{aligned}$$

by the matrix $\begin{bmatrix} 2 & 3 & 1 \\ -4 & -2 & 2 \end{bmatrix}$ and vector $\begin{bmatrix} 6000 \\ -4000 \end{bmatrix}$, which we

combine into the *augmented matrix* $\begin{bmatrix} 2 & 3 & 1 & 6000 \\ -4 & -2 & 2 & -4000 \end{bmatrix}$.

We then add 2 times the first row to the second, getting
$\begin{bmatrix} 2 & 3 & 1 & 6000 \\ 0 & 4 & 4 & 8000 \end{bmatrix}$, which represents the system

$$2p + 3q + r = 6000$$
$$0p + 4q + 4r = 8000.$$

Since the coefficient matrix is now upper triangular, we can solve the system by back substitution.

Taking this point of view, we can concentrate on how the operations affect the matrix representing the system. When the system of equations is *homogeneous*, that is, when the corresponding matrix equation is of the form $Ax = 0$, we do not even bother to augment the coefficient matrix A, so we then just conventrate on how the operations affect the coefficient matrix.

EXAMPLE

The homogeneous equations $\begin{bmatrix} 2 & 3 & 1 \\ -4 & -2 & 2 \end{bmatrix} \begin{bmatrix} p \\ q \\ r \end{bmatrix} = \begin{bmatrix} 0 \\ 0 \end{bmatrix}$

and $\begin{bmatrix} 2 & 3 & 1 \\ 0 & 4 & 4 \end{bmatrix} \begin{bmatrix} p \\ q \\ r \end{bmatrix} = \begin{bmatrix} 0 \\ 0 \end{bmatrix}$ have the same solutions since adding

2 times row 1 of $\begin{bmatrix} 4 & 3 & 1 \\ -4 & -2 & 2 \end{bmatrix}$ to row 2 gives us the new

coefficient matrix $\begin{bmatrix} 2 & 3 & 1 \\ 0 & 4 & 4 \end{bmatrix}$. Solving the latter, we get the solu-

tion $\begin{bmatrix} p \\ q \\ r \end{bmatrix} = \begin{bmatrix} r \\ -r \\ r \end{bmatrix}$, which has $3 - 2 = 1$ independent variables.

These operations on matrices are just as important as are the corresponding operations on systems of equations. We now give them names.

Definition. The following three kinds of operations on $m \times n$ matrices are called *elementary row operations*:

1. The operation of adding u times row s to row r (where $r \neq s$), which we denote Add $(r, s; u)$.

2. The operation of interchanging rows r and s (where $r \neq s$), which we denote Interchange (r, s).

3. The operation of multiplying row r by u (where $u \neq 0$), which we denote Multiply $(r; u)$.

In our notation for the elementary row operations, we have arranged the subscripts first, followed by a semicolon, followed by a scalar in (1) and (3). Let's try out this notation by looking at what we get when we apply the operations Add $(r, s; u)$, Interchange (r, s), Multiply $(r; u)$ to the $n \times n$ identity matrix:

1. If we apply the operation Add $(r, s; u)$ to the $n \times n$ identity matrix I, we get the matrix which has 1's on the diagonal, (r, s) entry u and all other entries 0. For example, if $n = 3$, the operation Add $(2, 3; u)$ changes the identity matrix to the matrix $\begin{bmatrix} 1 & 0 & 0 \\ 0 & 1 & u \\ 0 & 0 & 1 \end{bmatrix}$ whose $(2, 3)$ entry is u.

2. If we apply the operation Interchange (r, s) to the $n \times n$ identity matrix I, it interchanges rows r and s. So we get the matrix which has 0 as (r, r) entry, 0 as (s, s) entry, 1 as (r, s) entry, 1 as (s, r) entry; all other diagonal entries 1; and all other entries 0. For example, Interchange $(2, 3)$ changes the 3×3 identity matrix to the matrix $\begin{bmatrix} 1 & 0 & 0 \\ 0 & 0 & 1 \\ 0 & 1 & 0 \end{bmatrix}$.

3. If we apply the operation Multiply $(r; u)$ to the $n \times n$ identity matrix I, we get the matrix which has (r, r) entry u, all other diagonal entries 1 and all other entries 0. For example, the operation Multiply $(2; u)$ changes the 3×3 identity matrix to the matrix $\begin{bmatrix} 1 & 0 & 0 \\ 0 & u & 0 \\ 0 & 0 & 1 \end{bmatrix}$, whose $(2, 2)$ entry is u.

Matrices such as $\begin{bmatrix} 1 & 0 & 0 \\ 0 & 1 & u \\ 0 & 0 & 1 \end{bmatrix}$, $\begin{bmatrix} 1 & 0 & 0 \\ 0 & 0 & 1 \\ 0 & 1 & 0 \end{bmatrix}$, $\begin{bmatrix} 1 & 0 & 0 \\ 0 & u & 0 \\ 0 & 0 & 1 \end{bmatrix}$ obtained by applying elementary row operations to the identity matrix are called *elementary matrices*. These matrices turn out to be very useful.

Each of the three elementary row operations on matrices can be reversed by a corresponding *inverse operation*, in the same way that we reversed the elementary operations on equations. These inverse operations are:

1′. The operation of adding $-u$ times row s to row r (where $r \neq s$), denoted Add $(r, s; -u)$;

2′. The operation of interchanging rows s and r (where $r \neq s$), denoted Interchange (s, r);

3′. The operation of multiplying row r by $1/u$ (where $u \neq 0$), denoted Multiply $(r; 1/u)$.

Of course, Interchange (r, s) and Interchange (s, r) are equal and the net effect of carrying out Interchange (r, s) twice is not to change the matrix at all. So the inverse of the operation Interchange (r, s) is just Interchange (r, s) itself.

EXAMPLE

When we apply the elementary row operations Add $(2, 3; 5)$, Interchange $(2, 3)$, Multiply $(3; 7)$ to the matrix $\begin{bmatrix} 1 & 3 & 4 \\ 3 & 2 & 8 \\ 2 & 3 & 1 \end{bmatrix}$,

we get the matrices $\begin{bmatrix} 1 & 3 & 4 \\ 3 + 5 \cdot 2 & 2 + 5 \cdot 3 & 8 + 5 \cdot 1 \\ 2 & 3 & 1 \end{bmatrix}$,

$\begin{bmatrix} 1 & 3 & 4 \\ 2 & 3 & 1 \\ 3 & 2 & 8 \end{bmatrix}$, and $\begin{bmatrix} 1 & 3 & 4 \\ 3 & 2 & 8 \\ 7 \cdot 2 & 7 \cdot 3 & 7 \cdot 1 \end{bmatrix}$, respectively. Then when

we apply the inverse operation Add $(2, 3; -5)$ to $\begin{bmatrix} 1 & 3 & 4 \\ 3 + 5 \cdot 2 & 2 + 5 \cdot 3 & 8 + 5 \cdot 1 \\ 2 & 3 & 1 \end{bmatrix}$, we get back the matrix

$$\begin{bmatrix} 1 & 3 & 4 \\ 3 + 5 \cdot 2 + (-5) \cdot 2 & 2 + 5 \cdot 3 + (-5) \cdot 3 & 8 + 5 \cdot 1 + (-5) \cdot 1 \\ 2 & 3 & 1 \end{bmatrix} = \begin{bmatrix} 1 & 3 & 4 \\ 3 & 2 & 8 \\ 2 & 3 & 1 \end{bmatrix}.$$

Similarly, we get back the matrix $\begin{bmatrix} 1 & 3 & 4 \\ 3 & 2 & 8 \\ 2 & 3 & 1 \end{bmatrix}$ when we apply the

inverse operations Interchange (3, 2), Multiply $(3; \frac{1}{7})$ to

$$\begin{bmatrix} 1 & 3 & 4 \\ 2 & 3 & 1 \\ 3 & 2 & 8 \end{bmatrix}, \quad \begin{bmatrix} 1 & 3 & 4 \\ 3 & 2 & 8 \\ 7 \cdot 2 & 7 \cdot 3 & 7 \cdot 1 \end{bmatrix} :$$

$$\text{Interchange (3, 2)} \begin{bmatrix} 1 & 3 & 4 \\ 2 & 3 & 1 \\ 3 & 2 & 8 \end{bmatrix} = \begin{bmatrix} 1 & 3 & 4 \\ 3 & 2 & 8 \\ 2 & 3 & 1 \end{bmatrix}$$

$$\text{Multiply } (3; \tfrac{1}{7}) \begin{bmatrix} 1 & 3 & 4 \\ 3 & 2 & 8 \\ 7 \cdot 2 & 7 \cdot 3 & 7 \cdot 1 \end{bmatrix}$$

$$= \begin{bmatrix} 1 & 3 & 4 \\ 3 & 2 & 8 \\ (\tfrac{1}{7})7 \cdot 2 & (\tfrac{1}{7})7 \cdot 3 & (\tfrac{1}{7})7 \cdot 1 \end{bmatrix} = \begin{bmatrix} 1 & 3 & 4 \\ 3 & 2 & 8 \\ 2 & 3 & 1 \end{bmatrix}.$$

The counterpart for matrices of our definition of equivalence for systems of equations is

Definition. Two matrices are *row equivalent* if the second can be obtained from the first by successively applying elementary row operations a finite number of times.

The relation of row equivalence of matrices satisfies the following properties, which you can verify by following in the footsteps that we made in verifying the properties equivalence of systems of equations:

1. Any matrix A is row equivalent to itself.

2. If A is row equivalent to B, then B is row equivalent to A.

3. If A is row equivalent to B and if B is row equivalent to C, then A is row equivalent to C.

We now get the following counterpart of Theorem 2.2.1 for matrices. We leave it as an exercise for the reader to prove it, along the same lines as we proved Theorem 2.2.1.

Theorem 2.3.1. If two matrices A and B are row equivalent, then the homogeneous equations $Ax = 0$ and $Bx = 0$ have the same solutions x.

This theorem provides us with an efficient way of solving the homogeneous equation $Ax = 0$. How? The coefficient matrix A is row equivalent to an echelon matrix B, defined below, and the equation $Bx = 0$ can be solved by a variation of the method of back substitution described in Section 2.1.

Definition. An $m \times n$ matrix B is an *echelon matrix* if

1.   The *leading entry* (first nonzero entry) of each nonzero row is 1; and

2.   For any two consecutive rows, either the second of them is 0 or both rows are nonzero and the leading entry of the second of them is to the right of the leading entry of the first of them.

EXAMPLE

The matrices

$$\begin{bmatrix} 1 & 2 & 3 & 4 \\ 0 & 0 & 1 & 3 \\ 0 & 0 & 0 & 1 \\ 0 & 0 & 0 & 0 \end{bmatrix}, \begin{bmatrix} 1 & 4 & 0 & 0 & 0 \\ 0 & 0 & 0 & 0 & 1 \\ 0 & 0 & 0 & 0 & 0 \\ 0 & 0 & 0 & 0 & 0 \end{bmatrix}$$

are echelon matrices. On the other hand, the matrices

$$\begin{bmatrix} 1 & 2 & 3 & 4 \\ 0 & 0 & 1 & 3 \\ 0 & 0 & 1 & 1 \\ 0 & 0 & 0 & 1 \end{bmatrix}, \begin{bmatrix} 1 & 4 & 0 & 0 & 0 \\ 0 & 0 & 0 & 1 & 0 \\ 0 & 0 & 0 & 0 & 0 \\ 0 & 0 & 0 & 0 & 1 \end{bmatrix}$$

are not. Although the leading entries of their nonzero rows are 1, each has two consecutive rows, where the leading entry 1 of the second does not occur to the right of a leading entry 1 of the first.

For $\begin{bmatrix} 1 & 2 & 3 & 4 \\ 0 & 0 & 1 & 3 \\ 0 & 0 & 1 & 1 \\ 0 & 0 & 0 & 1 \end{bmatrix}$, these are the second and third rows; and for

$\begin{bmatrix} 1 & 4 & 0 & 0 & 0 \\ 0 & 0 & 0 & 1 & 0 \\ 0 & 0 & 0 & 0 & 0 \\ 0 & 0 & 0 & 0 & 1 \end{bmatrix}$, the third and fourth.

An $m \times n$ matrix A can be row reduced to an echelon matrix as follows.

Method to row reduce an $m \times n$ matrix A to an echelon matrix

1. Reduce A to a matrix B with 0's in the $(r, 1)$ position for $r = 2, \ldots, n$ as follows:
   (a) If the first column of A is zero, let $B = A$.
   (b) If the first column of A is not zero, then
       (i) If the a_{11} entry is 0, look for the first nonzero entry a_{r1} and interchange rows 1 and r.
       (ii) Multiply row 1 by a_{11}^{-1} and add $-a_{r1}$ times row 1 to row r for all $r > 1$.
2. If done, then stop. Otherwise, pass to the matrix C obtained from B by ignoring the first row and column of B.
3. Replace n by $n - 1$ and A by the $(n - 1) \times (n - 1)$ matrix C. Then repeat this entire process for the new n and A.

We state this result for future reference.

Theorem 2.3.2. Any $m \times n$ matrix is row equivalent to an echelon matrix.

EXAMPLE

To solve

$$\begin{bmatrix} 2 & 3 & 1 \\ -2 & -3 & 3 \end{bmatrix} \begin{bmatrix} a \\ b \\ c \end{bmatrix} = 0,$$

we reduce $\begin{bmatrix} 2 & 3 & 1 \\ -2 & -3 & 3 \end{bmatrix}$ to an echelon matrix by a variation of the algorithm above. Starting with $\begin{bmatrix} 2 & 3 & 1 \\ -2 & -3 & 3 \end{bmatrix}$, we add the first row to the second to get $\begin{bmatrix} 2 & 3 & 1 \\ 0 & 0 & 4 \end{bmatrix}$. We continue on to multiply the first row by $\frac{1}{2}$, then the second row by $\frac{1}{4}$, getting the echelon matrix

$$\begin{bmatrix} 1 & \frac{3}{2} & \frac{1}{2} \\ 0 & 0 & 1 \end{bmatrix}.$$

So the solutions to $\begin{bmatrix} 2 & 3 & 1 \\ -2 & 1 & 3 \end{bmatrix}\begin{bmatrix} a \\ b \\ c \end{bmatrix} = 0$ are the solutions to the

homogeneous equation

$$\begin{bmatrix} 1 & \frac{3}{2} & \frac{1}{2} \\ 0 & 0 & 1 \end{bmatrix}\begin{bmatrix} a \\ b \\ c \end{bmatrix} = 0,$$

which can be found by back substitution to be $\begin{bmatrix} a \\ b \\ c \end{bmatrix} = \begin{bmatrix} -(\frac{3}{2})b \\ b \\ 0 \end{bmatrix}$.

PROBLEMS

NUMERICAL PROBLEMS

1. Reduce the following matrices to echelon matrices using elementary row operations.

(a) $\begin{bmatrix} 1 & 2 & 3 & 4 \\ 1 & 2 & 1 & 3 \\ 1 & 3 & 1 & 1 \end{bmatrix}$.

(b) $\begin{bmatrix} 1 & 2 & 3 & 4 \\ 2 & 2 & 2 & 2 \\ 4 & 3 & 2 & 1 \end{bmatrix}$.

(c) $\begin{bmatrix} 1 & 2 & 3 & 1 & 2 & 3 \\ 2 & 3 & 2 & 3 & 2 & 3 \\ 3 & 5 & 5 & 4 & 4 & 6 \end{bmatrix}$.

(d) $\begin{bmatrix} 3 & 4 \\ 2 & 2 \\ 2 & 1 \end{bmatrix}$.

2. Find some nonzero solution to $Ax = 0$ (or show that none exists) for each matrix A listed in Problem 1.

MORE THEORETICAL PROBLEM

Easier Problem

3. Prove the following properties of row equivalence.

(1) Any matrix A is row equivalent to itself.

(2) If A is row equivalent to B, then B is row equivalent to A.

(3) If A is row equivalent to B and if B is row equivalent to C, then A is row equivalent to C.

2.4. SOLVING SYSTEMS OF LINEAR EQUATIONS

Let A be an $m \times n$ matrix and suppose that we have computed an echelon matrix B row equivalent to A, by the reduction algorithm given above. Then the solutions to the homogeneous equation $Ax = 0$ are the same as the solutions to the homogeneous equation $Bx = 0$, and the latter can be gotten by back substitution. Since B is an echelon matrix, the number of independent variables in the general solution to $Bx = 0$ by back substitution is $n - m'$, where m' is the number of nonzero rows of B. Why? Let's look at an example before we give the answer in general. The solution to

$$
\begin{bmatrix} 1 & 2 & 3 & 4 \\ 0 & 0 & 1 & 3 \\ 0 & 0 & 0 & 0 \end{bmatrix}
\begin{bmatrix} a \\ b \\ c \\ d \end{bmatrix}
=
\begin{bmatrix} 0 \\ 0 \\ 0 \end{bmatrix}
$$

is

$$
\begin{bmatrix} a \\ b \\ c \\ d \end{bmatrix}
=
\begin{bmatrix} -2b + 9d - 4d \\ b \\ -3d \\ d \end{bmatrix},
$$

which has $2 = 4 - 2$ independent variables b and d. Note that b and d correspond to columns 2 and 4, which are the columns that do not contain leading entries 1 of nonzero rows. This gives the answer for our example. In the general case, the answer is the same. The independent variables correspond to those columns that do not contain a leading entry of 1. So there are $n - m'$ of them, where m' is the number of nonzero rows of B. This proves

Theorem 2.4.1. Let A be an $m \times n$ matrix and suppose that A is row equivalent to an echelon matrix B. Then the general solution x to the homogeneous equation $Ax = 0$ obtained by solving the homogeneous equation $Bx = 0$ by back substitution contains $n - m'$ independent variables, where m' is the number of nonzero rows of B.

A useful special case of Theorem 2.4.1 is

Corollary 2.4.2.　Let A be an $m \times n$ matrix and suppose that A is row equivalent to an echelon matrix B. Then 0 is the only solution to the homogeneous equation $Ax = 0$ if and only if B has n nonzero rows.

This corollary, in turn, leads to the

Corollary 2.4.3.　Let A be an $m \times n$ matrix with $m < n$. Then the homogeneous equation $Ax = 0$ has a nonzero solution.

Proof.　Take B to be an echelon matrix which is row equivalent to A. Since $m < n$ and m is the number of rows, B cannot have n nonzero rows. So $Ax = 0$ has a nonzero solution by Corollary 2.4.2.　　■

Equivalently, if the number of variables exceeds the number of equations in a system of homogeneous equations, then this system has a nontrivial solution.

From the discussion above we see that the method of back substitution to solve the matrix equation $Bx = z$ when the coefficient matrix B is an $m \times n$ echelon matrix is just the following slight refinement of the method of back substitution that we gave in a special case in Section 2.1. Of course, *this method works only when $z_r = 0$ whenever row r of B is 0, since otherwise $Bx = z$ has no solution x.*

Method to solve a system of linear equations by back substitution if the coefficient matrix is an echelon matrix

1. Let the $n - m$ extra variables $x_{s_1}, \ldots, x_{s_{n-m}}$ corresponding to the columns $s_1, \ldots, s_{n-m}$ which do not contain leading entries 1 of nonzero rows have any values, and move them to the right-hand side of the equations.
2. Find the value for the last remaining dependent variable from the last remaining equation and discard that equation.
3. Eliminate that variable by substituting its value back in place of it in all of the preceding equations.
4. Repeat (2), (3), and (4) for the next last remaining dependent variable using the next last remaining equation to get its value

and eliminate it and the equation. Continue until the values for all dependent variables have been found.

EXAMPLE

To solve the equation

$$\begin{bmatrix} 1 & 2 & 3 & 4 \\ 0 & 0 & 1 & 3 \\ 0 & 0 & 0 & 0 \end{bmatrix} \begin{bmatrix} a \\ b \\ c \\ d \end{bmatrix} = \begin{bmatrix} 8 \\ 4 \\ 0 \end{bmatrix}$$

by back substitution, we look for the columns that do not contain leading entries 1 of nonzero rows, namely columns 2 and 4. The corresponding *free* variables are b and d, so we move them to the right-hand side, getting the system of equations

$$1a + 3c = 8 - 2b - 4d$$
$$0a + 1c = 4 - \qquad 3d$$
$$0a + 0a = 0.$$

We now solve this system by back substitution, first getting

$$c = 4 - 3d$$

and then

$$a = 8 - 2b - 4d - 3(4 - 3d) = -4 - 2b + 5d.$$

Our solution is

$$\begin{bmatrix} a \\ b \\ c \\ d \end{bmatrix} = \begin{bmatrix} -4 - 2b + 5d \\ b \\ 4 - 3d \\ d \end{bmatrix}.$$

How can we use this method to solve a *nonhomogeneous* matrix equation $Ax = y$ for x? We find an equivalent matrix equation $Bx = z$, where B is an echelon matrix. This is done by using the method of the Section 2.3 to row reduce A to an echelon matrix B *with the variation that each row operation during the reduction process is applied both to the coefficient matrix and to the column vector, ultimately changing A to B and y to z.*

In practice, this can be done by forming the *augmented matrix*

$$[A, y]$$

and reducing it to an echelon matrix

$$[B, z].$$

Then the equations $Ax = y$ and $Bx = z$ are equivalent and the solutions to $Ax = y$ are found as the solutions to $Bx = z$ using the method of back substitution. So, we have

Method to solve $Ax = y$ for x

1. Form the augmented matrix $[A, y]$ and reduce it to an echelon matrix $[B, z]$.
2. Find x as the general solution to $Bx = z$ by back substitution.

EXAMPLE

To find the solutions to $\begin{bmatrix} 1 & 2 & 3 & 4 \\ 1 & 2 & 4 & 7 \\ 2 & 4 & 6 & 8 \end{bmatrix} \begin{bmatrix} a \\ b \\ c \\ d \end{bmatrix} = \begin{bmatrix} 8 \\ 12 \\ 16 \end{bmatrix}$, reduce the augmented matrix

$$\begin{bmatrix} 1 & 2 & 3 & 4 & 8 \\ 1 & 2 & 4 & 7 & 12 \\ 2 & 4 & 6 & 8 & 16 \end{bmatrix}$$

to the echelon matrix

$$\begin{bmatrix} 1 & 2 & 3 & 4 & 8 \\ 0 & 0 & 1 & 3 & 4 \\ 0 & 0 & 0 & 0 & 0 \end{bmatrix}.$$

Since the third entry of the vector $\begin{bmatrix} 8 \\ 4 \\ 0 \end{bmatrix}$ on the right-hand side is 0,

we can solve the equation $\begin{bmatrix} 1 & 2 & 3 & 4 \\ 0 & 0 & 1 & 3 \\ 0 & 0 & 0 & 0 \end{bmatrix} \begin{bmatrix} a \\ b \\ c \\ d \end{bmatrix} = \begin{bmatrix} 8 \\ 4 \\ 0 \end{bmatrix}$ by back sub-

stitution. We did this in the example above, getting

$$\begin{bmatrix} a \\ b \\ c \\ d \end{bmatrix} = \begin{bmatrix} -4 - 2b + 5d \\ b \\ 4 - 3d \\ d \end{bmatrix}.$$

How do we know when $Ax = y$ has a solution? Because $Bx = z$ has a solution if and only if $z_r = 0$ whenever row r of B consists of 0's, we have

Method to determine whether $Ax = y$ has a solution

1. Form the augmented matrix $[A, y]$ and reduce it to an echelon marix $[B, z]$.
2. Then $Ax = y$ has a solution if and only if $z_r = 0$ whenever row r of B consists of 0's.

When $Ax = y$ does have a solution, the discussion at the beginning of this section shows that Theroem 2.4.1 has the following counterpart for nonhomogeneous equations.

Theorem 2.4.4. Let A be an $m \times n$ matrix and suppose that the equation $Ax = y$ is equivalent to the equation $Bx = z$, where B is an echelon matrix. If $Ax = y$ has a solution, then the general solution x to the equation $Ax = y$ obtained by solving the equation $Bx = z$ by back substitution contains $n - m'$ independent variables, where m' is the number of nonzero rows of B.

The number of nonzero rows of an $m \times n$ echelon matrix B is called its *rank*, denoted rank (B). Our theorem gives the formula $n = $ rank (B) + number of independent variables in the solution of $Bx = 0$. We discuss this formula in more depth in Section 4.4.

PROBLEMS

NUMERICAL PROBLEMS

1. Show that each of the following equations has a nonzero solution.

(a) $\begin{bmatrix} 1 & 2 & 3 & 4 \\ 1 & 2 & 1 & 3 \\ 1 & 3 & 1 & 1 \end{bmatrix} x = 0.$

(b) $\begin{bmatrix} 5 & 3 & 4 \\ 1 & 1 & 2 \\ 6 & 4 & 6 \end{bmatrix} x = 0.$

(c) $\begin{bmatrix} 2 & 5 & 5 & 4 & 2 & 2 & 2 & 2 & 3 & 4 \\ 1 & 2 & 4 & 5 & 8 & 7 & 7 & 7 & 6 & 3 \\ 1 & 1 & 1 & 1 & 3 & 2 & 2 & 3 & 4 & 4 \end{bmatrix} x = 0.$

2. Solve the equation $\begin{bmatrix} 1 & 4 & 3 & 2 \\ 0 & 0 & 1 & 3 \\ 0 & 0 & 0 & 0 \end{bmatrix} x = \begin{bmatrix} 3 \\ 3 \\ 0 \end{bmatrix}$ using the method of

back substitution.

3. Solve the equation $\begin{bmatrix} 1 & 2 & 3 & 2 \\ 2 & 4 & 1 & 3 \\ 3 & 6 & 4 & 5 \end{bmatrix} x = \begin{bmatrix} 4 \\ 7 \\ 11 \end{bmatrix}$ by forming the aug-

mented matrix to find an equivalent equation whose coefficient matrix is an echelon matrix and solving the latter by the method of
back substitution.

4. Determine whether the equation $\begin{bmatrix} 1 & 2 & 3 & 2 \\ 2 & 4 & 6 & 2 \\ 3 & 6 & 9 & 3 \end{bmatrix} x = \begin{bmatrix} 4 \\ 7 \\ 1 \end{bmatrix}$ has a

solution.

MORE THEORETICAL PROBLEMS

Easier Problems

5. Represent the system of equations

$$1u + 4k + 6h = 6$$
$$0u + 3k + 5h = 5$$

by a matrix equation and show that the set of all solutions is the set

of vectors $v + w$ where v is one solution and $w = \begin{bmatrix} u \\ k \\ h \end{bmatrix}$ is any

solution to the homogeneous system of equations

$$1u + 4k + 6h = 0$$
$$0u + 3k + 5h = 0.$$

C H A P T E R

3

THE $n \times n$ MATRICES

With the experience of the 2×2 matrices behind us the transition to the $n \times n$ case will be rather straightforward and should not present us with many large difficulties. In fact, for a large part, the development of the $n \times n$ matrices will closely follow the line used in the 2×2 case. So we have some inkling of what results to expect and how to go about establishing these results. All this will stand us in good stead now.

3.1. THE OPENING

What is a better place to start than the beginning? So we start things off with our cast of players. If $\mathbb{R}$ is the set of real numbers and $\mathbb{C}$ that of complex numbers, by $M_n(\mathbb{R})$ and $M_n(\mathbb{C})$ we shall mean the set of all square arrays, with n rows and n columns, of real or complex numbers, respectively.

We could speak about $m \times n$ matrices, even when $m \neq n$, but we shall restrict our attention for now to $n \times n$ matrices. We call them *square matrices*.

125

What are they? An $n \times n$ matrix is a square array

$$\begin{bmatrix} a_{11} & a_{12} & \cdots & a_{1n} \\ a_{21} & a_{22} & \cdots & a_{2n} \\ \vdots & \vdots & & \vdots \\ a_{n1} & a_{n2} & \cdots & a_{nn} \end{bmatrix}$$

where all the a_{rs} are in $\mathbb{C}$. We use $M_n(\mathbb{C})$ to denote the set of all $n \times n$ matrices over $\mathbb{C}$, that is, the set of all such square arrays. Similarly $M_n(\mathbb{R})$ is the set of such arrays where all the a_{rs} are restricted to being in $\mathbb{R}$. Since $\mathbb{R} \subset \mathbb{C}$ we have $M_n(\mathbb{R}) \subset M_n(\mathbb{C})$.

We don't want to keep specifying whether we are working over $\mathbb{R}$ or $\mathbb{C}$ at any given time. So we shall merely speak about matrices. For most things it will not be important if these matrices are in $M_n(\mathbb{R})$ or $M_n(\mathbb{C})$. If we simply say that we are working with $M_n(F)$, where F is understood to be either $\mathbb{C}$ or $\mathbb{R}$, then what we say applies in both cases. When the results do depend on the number system we are using, it will be so pointed out.

If A is the matrix above we call the a_{rs} the *entries* of A, and a_{rs} is called the (r, s) *entry*, meaning that it is that entry in row r and column s.

In analogy with what we did for the 2×2 matrices we want to be able to express when two matrices are equal, how to add them, and how to multiply them. We shall use the shorthand $A = (a_{rs})$ to denote the matrix A above. [Mathematicians usually—out of habit—write (a_{ij}), but to avoid confusion with the complex number i, where $i^2 = -1$, we prefer to use other subscripts.] So (a_{rs}) will be that matrix whose (r, s) entry is a_{rs}. In all that follows n is a fixed integer with $n \geq 2$ and all matrices are in $M_n(F)$.

Definition. The two matrices $A = (a_{rs})$ and $B = (b_{rs})$ are defined to be *equal*, $A = B$, if and only if $a_{uv} = b_{uv}$ for all u and v.

So, for instance, $\begin{bmatrix} x & y & 1 \\ 0 & 5 & z \\ 0 & 1 & 1 \end{bmatrix} = \begin{bmatrix} 1 & 3 & 1 \\ 0 & 5 & \pi \\ 0 & 1 & 1 \end{bmatrix}$ if and only if

$x = 1$, $y = 3$, and $z = \pi$. Note that for *no* choice of x, y, z can $\begin{bmatrix} x & y & 1 \\ 0 & 5 & 3 \\ 0 & 1 & 1 \end{bmatrix}$ equal $\begin{bmatrix} 1 & 3 & 1 \\ 0 & 5 & \pi \\ 0 & 0 & 1 \end{bmatrix}$, for these differ in the $(3, 2)$ entry.

Now that we have equality of two matrices defined we pass on to the next notion, that of the *addition* of matrices.

Definition. Given the matrices $A = (a_{rs})$ and $B = (b_{rs})$, then $A + B = (c_{rs})$, where $c_{uv} = a_{uv} + b_{uv}$ for all u and v.

By this definition it is clear that the sum of matrices is again a matrix, that is, if A, $B \in M_n(\mathbb{C})$, then $A + B \in M_n(\mathbb{C})$, and if A, $B \in M_n(\mathbb{R})$, then $A + B \in M_n(\mathbb{R})$. We add matrices by adding their corresponding entries.

As before, the matrix all of whose entries are 0—which we shall write merely as 0—has the property that $A + 0 = A$ for all matrices A. Similarly, if $A = (a_{rs})$, then the matrix (b_{rs}), where $b_{rs} = -a_{rs}$, is written as $-A$ and has the property that $A + (-A) = 0$. Note that the addition of matrices satisfies the *commutative law*, namely, that $A + B = B + A$. For example,

$$
\begin{bmatrix} 2 & 0 & 0 \\ 1 & 3 & 1 \\ 0 & 0 & 9 \end{bmatrix} + \begin{bmatrix} 2 & 0 & 0 \\ 2 & 3 & 4 \\ 0 & 0 & 8 \end{bmatrix} = \begin{bmatrix} 4 & 0 & 0 \\ 3 & 6 & 5 \\ 0 & 0 & 17 \end{bmatrix}
$$

$$
= \begin{bmatrix} 2 & 0 & 0 \\ 2 & 3 & 4 \\ 0 & 0 & 8 \end{bmatrix} + \begin{bmatrix} 2 & 0 & 0 \\ 1 & 3 & 1 \\ 0 & 0 & 9 \end{bmatrix}.
$$

If $A = \begin{bmatrix} 1 & i & 0 & 2 \\ 3+i & 0 & \frac{1}{2} & -1 \\ i & 0 & 0 & -3i \\ 1 & 1 & 1 & 0 \end{bmatrix}$ and $B = \begin{bmatrix} 0 & -i & 1 & 3 \\ i & i & i & -i \\ 1 & -i & 1 & -1 \\ 0 & i & 0 & -i \end{bmatrix}$ are

two 4×4 matrices, then $A + B$ is the matrix

$$
\begin{bmatrix} 1 & 0 & 1 & 5 \\ 3+2i & i & \frac{1}{2}+i & -1-i \\ 1+i & -1 & 1 & -1-3i \\ 1 & 1+i & 1 & -i \end{bmatrix}.
$$

We single out some nice-looking matrices, as we did in the 2×2 case.

Definition. The matrix $A = (a_{rs})$ is a *diagonal matrix* if $a_{uv} = 0$ for $u \neq v$.

So a diagonal matrix is one in which the entries off the main diagonal are all 0. For example, the matrices $\begin{bmatrix} 2 & 0 & 0 \\ 0 & 3 & 0 \\ 0 & 0 & 9 \end{bmatrix}$ and $\begin{bmatrix} 99 & 0 \\ 0 & 99 \end{bmatrix}$

are diagonal matrices, whereas the matrix $A = \begin{bmatrix} 0 & 0 & 1 \\ 0 & 1 & 0 \\ 1 & 0 & 0 \end{bmatrix}$ is not, since

a_{13} and a_{31} are not 0.

Among the diagonal matrices there are some that are nicer than others. They are the so-called scalar matrices.

Definition.   A diagonal matrix is called a *scalar matrix* if all the entries on the main diagonal are equal.

Thus a scalar matrix looks like

$$\begin{bmatrix} a & 0 & 0 & \cdots & 0 \\ 0 & a & 0 & \cdots & 0 \\ 0 & 0 & a & \cdots & 0 \\ \vdots & & & \cdots & \vdots \\ 0 & 0 & 0 & \cdots & a \end{bmatrix},$$

where a is a number. We shall write this matrix as aI, where I is the matrix

$$I = \begin{bmatrix} 1 & 0 & 0 & \cdots & 0 \\ 0 & 1 & 0 & \cdots & 0 \\ 0 & 0 & 1 & \cdots & 0 \\ \vdots & & & \cdots & \vdots \\ 0 & 0 & 0 & \cdots & 1 \end{bmatrix}.$$

Here, in writing I, wherever we wrote no entry that entry is understood to be 0. So the matrix aI is

$$aI = \begin{bmatrix} a & 0 & 0 & \cdots & 0 \\ 0 & a & 0 & \cdots & 0 \\ 0 & 0 & a & \cdots & 0 \\ \vdots & & & \cdots & \vdots \\ 0 & 0 & 0 & \cdots & a \end{bmatrix}.$$

Notice that $aI + bI = (a + b)I$; for instance,

$$1I + \pi I = \begin{bmatrix} 1 & 0 \\ 0 & 1 \end{bmatrix} + \begin{bmatrix} \pi & 0 \\ 0 & \pi \end{bmatrix} = \begin{bmatrix} 1 + \pi & 0 \\ 0 & 1 + \pi \end{bmatrix}$$

$$= (1 + \pi)I.$$

So the scalar matrices aI behave like the scalar (numbers) a themselves—real or complex as the case may be—under addition.

Definition. We define *multiplication by a scalar* (i.e., an element of F) as follows. If $A = (a_{rs})$ and b is a number, then $bA = (c_{rs})$, where for each u and v, $c_{uv} = ba_{uv}$.

So, for example,

$$(-5)\begin{bmatrix} 1 & 2 & 3 \\ 0 & 1 & 6 \\ 1 & -1 & 4 \end{bmatrix} = \begin{bmatrix} -5 & -10 & -15 \\ 0 & -5 & -30 \\ -5 & 5 & -20 \end{bmatrix}.$$

The result of multiplying A by the scalar -1 is the matrix $(-1)A = -A$ which we introduced earlier.

In a moment we shall see that this type of multiplication is merely a special case of the product defined for any matrices.

We now come to the trickiest of the operations, that of the *product* of two matrices. With what we saw for the 2×2 matrices, the definition we are about to make should not be as mysterious or confusing as it might otherwise be.

Definition. Given the matrices $A = (a_{rs})$ and $B = (b_{rs})$, then $AB = (c_{rs})$, where $c_{uw} = \sum_{\varphi=1}^{n} a_{u\varphi}b_{\varphi w}$.

Note that the subscript φ over which the sum above is taken is a dummy index. We could equally well call it anything. So

$$c_{uw} = \sum_{\varphi=1}^{n} a_{u\varphi}b_{\varphi w} = \sum_{\square=1}^{n} a_{u\square}b_{\square w} = \sum_{v=1}^{n} a_{uv}b_{vw}.$$

To highlight the dummy variables while you get used to them, we'll use Greek letters for them in this section and part of the next. After that, we'll just use Latin letters, to keep the notation simple.

Note also that if $A, B \in M_n(\mathbb{C})$, then $AB \in M_n(\mathbb{C})$, and also if $A, B \in M_n(\mathbb{R})$, then $AB \in M_n(\mathbb{R})$. We express this by saying that $M_n(\mathbb{C})$ (and $M_n(\mathbb{R})$) is *closed* with respect to products.

Let's compute the product of some matrices.

If $A = \begin{bmatrix} 1 & 2 & 3 \\ 3 & 2 & 1 \\ 0 & -1 & 0 \end{bmatrix}$ and $B = \begin{bmatrix} i & 0 & 0 \\ 0 & -i & 0 \\ 0 & 0 & 1 \end{bmatrix}$, then AB is

$$\begin{bmatrix} 1(i) + 2(0) + 3(0) & 1(0) + 2(-i) + 3(0) & 1(0) + 2(0) + 3(1) \\ 3(i) + 2(0) + 1(0) & 3(0) + 2(-i) + 1(0) & 3(0) + 2(0) + 1(1) \\ 0(i) + (-1)0 + 0(0) & 0(0) + (-1)(-i) + 0(0) & 0(0) + (-1)(0) + 0(1) \end{bmatrix}$$

which equals $\begin{bmatrix} i & -2i & 3 \\ 3i & -2i & 1 \\ 0 & i & 0 \end{bmatrix}$.

As a way of remembering how to form the product AB:

The (s, t) entry is the "dot product" of the row s of A by column t of B.

In the case above, for example, the $(2, 3)$ entry is the dot product $3(0) + 2(0) + 1(1)$ of row $2 = (3, 2, 1)$ of A and $(0, 0, 1)$, which is column 3 written horizontally.

Note that the matrix I introduced above satisfies $AI = IA = A$ for every matrix A. So I acts in matrices as the number 1 does in numbers. Note also that $(aI)A = aA$, so multiplication by a scalar is merely a special case of matrix multiplication. (Prove!) For this reason, we call I the *identity matrix*.

Similarly, the matrix $0I$, which we denote simply by 0, satisfies $A0 = 0A = 0$ for all A. We call $0 = 0I$ the *zero matrix*.

As we already saw in the 2×2 matrices, matrices do not, in general, satisfy the *commutative law* of multiplication; that is, AB *need not equal* BA. Remember also that in matrices it is possible that $AB = 0$ but $A \neq 0$ and $B \neq 0$.

In the $n \times n$ matrices there are n^2 particular matrices that play a key role. These are called the *matrix units*, E_{rs}, which are defined as follows: E_{rs} is the matrix whose (r, s) entry is 1 and all of whose other entries are 0. So, for instance, in $M_3(\mathbb{C})$,

$$E_{32} = \begin{bmatrix} 0 & 0 & 0 \\ 0 & 0 & 0 \\ 0 & 1 & 0 \end{bmatrix} \quad \text{and} \quad E_{22} = \begin{bmatrix} 0 & 0 & 0 \\ 0 & 1 & 0 \\ 0 & 0 & 0 \end{bmatrix}.$$

One of the important properties of the E_{rs} is that

every matrix can be expressed as a combination of them with scalar coefficients, that is, $A = (a_{rs}) = \Sigma_\rho \Sigma_\sigma a_{\rho\sigma} E_{\rho\sigma}$.

For example,

$$\begin{bmatrix} 2 & 0 & 0 \\ 0 & 5 & 0 \\ 0 & 8 & 0 \end{bmatrix} = 2\begin{bmatrix} 1 & 0 & 0 \\ 0 & 0 & 0 \\ 0 & 0 & 0 \end{bmatrix} + 5\begin{bmatrix} 0 & 0 & 0 \\ 0 & 1 & 0 \\ 0 & 0 & 0 \end{bmatrix} + 8\begin{bmatrix} 0 & 0 & 0 \\ 0 & 0 & 0 \\ 0 & 1 & 0 \end{bmatrix}$$

$$= 2E_{11} + 5E_{22} + 8E_{32}.$$

Another very important thing about the E_{rs}'s is how they multiply. The equations

$$E_{rs}E_{uv} = 0 \qquad \text{if } s \neq u$$
$$E_{rs}E_{sv} = E_{rv}$$

hold for all r, s, u, v. Instances of this are

$$E_{31}E_{22} = \begin{bmatrix} 0 & 0 & 0 \\ 0 & 0 & 0 \\ 1 & 0 & 0 \end{bmatrix}\begin{bmatrix} 0 & 0 & 0 \\ 0 & 1 & 0 \\ 0 & 0 & 0 \end{bmatrix} = 0$$

and

$$E_{32}E_{22} = \begin{bmatrix} 0 & 0 & 0 \\ 0 & 0 & 0 \\ 0 & 1 & 0 \end{bmatrix}\begin{bmatrix} 0 & 0 & 0 \\ 0 & 1 & 0 \\ 0 & 0 & 0 \end{bmatrix} = \begin{bmatrix} 0 & 0 & 0 \\ 0 & 0 & 0 \\ 0 & 1 & 0 \end{bmatrix} = E_{32}.$$

We leave these important things for the reader to verify (they occur in the problem set to follow).

We compute—anticipating the arithmetic properties that hold for matrices—the product of some conbination of the E_{rs}'s. Let $A = 2E_{11} + 3E_{12} + E_{33}$ and $B = E_{12} - E_{31}$. Then AB is

$$(2E_{11} + 3E_{12} + E_{33})(E_{12} - E_{31})$$
$$= 2E_{11}E_{12} + 3E_{12}E_{12} + E_{33}E_{12} - 2E_{11}E_{31} - 3E_{12}E_{31} - E_{33}E_{31}$$
$$= 2E_{12} - E_{31}$$

by the rules of multiplying the matrix units. As matrices, we have

$$A = \begin{bmatrix} 2 & 3 & 0 \\ 0 & 0 & 0 \\ 0 & 0 & 1 \end{bmatrix}, \qquad B = \begin{bmatrix} 0 & 1 & 0 \\ 0 & 0 & 0 \\ -1 & 0 & 0 \end{bmatrix}$$

and

$$AB = \begin{bmatrix} 0 & 2 & 0 \\ 0 & 0 & 0 \\ -1 & 0 & 0 \end{bmatrix} = 2E_{12} - E_{31}.$$

Not surprisingly, the answer we get this way for AB is exactly the same as if we computed AB by our definition of the product of the matrices A and B.

In the 2×2 matrices we found that the introduction of exponents was a handy and useful device for representing the powers of matrix. Here, too, it will be equally useful. As before, we define for an $n \times n$ matrix A, the *power* A^m for positive integer values m to be

$$A^m = AAA \cdots A.$$
$$\quad\quad (m \text{ times})$$

We define A^0 to be

$$A^0 = I.$$

For negative integers m there is also a problem, for it cannot be defined in a sensible way for all matrices. We would want A^{-1} to have the meaning: that matrix such that $AA^{-1} = I$. However, for some matrices A, no such A^{-1} exists!

Definition.   The matrix A is said to be *invertible* if there exists a matrix B such that $AB = BA = I$.

If A is invertible, there is only one matrix B such that $AB = BA = I$ (Prove!), and we denote this matrix B as A^{-1}.

Even for 2×2 matrices we saw that A^{-1} need not exist for every matrix A. As we proceed we shall find the necessary and sufficient criterion that a given matrix have an *inverse* A^{-1}, that is, that it be invertible.

Finally, for nonnegative integers r and s the usual rules of exponents hold, namely

$$A^r A^s = A^{r+s}$$

and

$$(A^r)^s = A^{rs}.$$

(Verify!) If A is invertible, we define A^{-m}, for m a positive integer, by

$$A^{-m} = (A^{-1})^m.$$

If A is invertible then $A^r A^s = A^{r+s}$ and $(A^r)^s = A^{rs}$ hold for *all* integers r and s (if we understand A^0 to be I). (Verify!)

We close this section with an omnibus theorem about addition, multiplication, and how they interact. Its proof we leave as a massive, although generally easy, exercise for the reader, since our proof of the 2×2 version of it is a road map for the general proof.

Theorem 3.1.1. For matrices the following hold:

1. $\qquad A + B = B + A$ commutativity;
2. $(A + B) + C = A + (B + C)$ associativity of sums;
3. $\qquad\qquad A + 0 = A;$
4. $\qquad A + (-A) = 0;$
5. $\qquad\qquad\qquad A0 = 0A = 0;$
6. $\qquad\qquad\qquad AI = IA = A;$
7. $\qquad\qquad (AB)C = A(BC)$ associativity of products;
8. $\qquad A(B + C) = AB + AC;$ and
   $\qquad (B + C)A = BA + CA$ distributivity.

PROBLEMS

NUMERICAL PROBLEMS

1. Calculate.

(a) $\begin{bmatrix} 1 & -6 & 0 & \pi \\ \pi & i & 0 & i \\ 0 & 0 & 1 & 0 \\ 1 & 0 & i & i \end{bmatrix} \begin{bmatrix} 1 & 2 & 3 & 4 \\ 0 & 1 & 2 & 3 \\ 0 & 0 & 1 & 2 \\ 0 & 0 & 0 & 1 \end{bmatrix}.$

(b) $\begin{bmatrix} \frac{1}{2} & \frac{1}{3} & \frac{1}{4} \\ -\frac{1}{2} & -\frac{1}{3} & -\frac{1}{4} \\ 1 & 1 & 1 \end{bmatrix}^2.$

(c) $\begin{bmatrix} 0 & 1 & 0 & 0 \\ 0 & 0 & 1 & 0 \\ 0 & 0 & 0 & 1 \\ 1 & 0 & 0 & 0 \end{bmatrix}^4.$

(d) $\begin{bmatrix} 1 & -1 & 2 \\ 3 & 1 & 4 \\ 5 & 6 & 0 \end{bmatrix} \left(\begin{bmatrix} 1 & 2 & 0 \\ -10 & -9 & 8 \\ 7 & 6 & 4 \end{bmatrix} + \begin{bmatrix} 3 & 0 & 2 \\ 10 & 8 & -9 \\ -7 & -6 & 4 \end{bmatrix} \right).$

(e) $\begin{bmatrix} \frac{1}{3} & \frac{1}{3} & \frac{1}{3} \\ \frac{1}{3} & \frac{1}{3} & \frac{1}{3} \\ \frac{1}{3} & \frac{1}{3} & \frac{1}{3} \end{bmatrix} \begin{bmatrix} \frac{2}{3} & -\frac{1}{3} & -\frac{1}{3} \\ -\frac{1}{3} & \frac{2}{3} & -\frac{1}{3} \\ -\frac{1}{3} & -\frac{1}{3} & \frac{2}{3} \end{bmatrix}.$

(f) $\begin{bmatrix} 0 & 0 & i \\ 0 & i & 0 \\ i & 0 & 0 \end{bmatrix} \begin{bmatrix} 0 & 0 & 3 \\ 0 & 3 & 0 \\ 3 & 0 & 0 \end{bmatrix} - \begin{bmatrix} 0 & 0 & 3 \\ 0 & 3 & 0 \\ 3 & 0 & 0 \end{bmatrix} \begin{bmatrix} 0 & 0 & i \\ 0 & i & 0 \\ i & 0 & 0 \end{bmatrix}.$

2. Find all matrices M such that $M \begin{bmatrix} 1 & 2 & 3 \\ 0 & 2 & 3 \\ 0 & 0 & 3 \end{bmatrix} = \begin{bmatrix} 1 & 2 & 3 \\ 0 & 2 & 3 \\ 0 & 0 & 3 \end{bmatrix} M.$

3. (a) Find a matrix $A \neq 0$ such that $A \begin{bmatrix} 1 & 1 & 1 \\ 3 & 3 & 3 \\ 0 & 5 & 6 \end{bmatrix} = 0.$

   (b) Find the form of all matrices A such that $A \begin{bmatrix} 1 & 1 & 1 \\ 3 & 3 & 3 \\ 0 & 5 & 6 \end{bmatrix} = 0.$

   (c) Find the form of all matrices A such that $\begin{bmatrix} 1 & 1 & 1 \\ 3 & 3 & 3 \\ 0 & 5 & 6 \end{bmatrix} A = 0.$

4. Show that $A = \begin{bmatrix} 0 & 1 & 0 \\ 0 & 0 & 1 \\ -a_3 & -a_2 & -a_1 \end{bmatrix}$ satisfies

$$A^3 + a_1 A^2 + a_2 A + a_3 I = 0.$$

MORE THEORETICAL PROBLEMS

Easier Problems

5. Show that $(aI)A = aA$.
6. Prove that
   (a) $(0I)(bI) = 0I$ and $(1I)(bI) = bI$.
   (b) $(aI) + (bI) = (a + b)I$.
   (c) $(aI)(bI) = (ab)I$.
7. Show that $(aI)A = A(aI)$ for all matrices A.
8. If D is a diagonal 3×3 matrix all of whose diagonal elements are distinct, show that if $AD = DA$, then A must be a diagonal matrix.

Middle-Level Problems

9. Prove Theorem 3.1.1 in its entirety.
10. Prove the product rule for the matrix units, namely, $E_{rs}E_{uv} = 0$ if $s \neq u$ and $E_{rs}E_{sv} = E_{rv}$.

11. Show that $\sum_{\sigma=1}^{n} E_{\sigma\sigma} = I$.

12. Using the fact that $A = (a_{rs}) = \sum_{\rho=1}^{n} \sum_{\sigma=1}^{n} a_{\rho\sigma} E_{\rho\sigma}$ and $B = (b_{rs}) = \sum_{\rho=}^{n} \sum_{\sigma=1}^{n} b_{\rho\sigma} E_{\rho\sigma}$, reconstruct what the product AB must be, by exploiting the multiplication rules for the E_{rs}'s.

13. Prove the associative law $A(BC) = (AB)C$ using the representation of A, B, C as combinations of the matrix units. (This should give you a lot of practice with playing with sums as Σ's.)

14. If $AB = 0$, is $BA = 0$? Either prove or produce a counterexample.

15. If M is a matrix such that $MA = AM$ for all matrices A, show that M must be a scalar matrix.

16. If A is invertible and $AB = 0$, show that $B = 0$.

17. If A is invertible, show that its inverse is unique.

18. If A and B are invertible, show that AB is also invertible and express $(AB)^{-1}$ in terms of A^{-1} and B^{-1}.

19. If $A^m = 0$ and $AB = B$, prove that $B = 0$.

20. If $A^m = 0$, prove that $aI - A$ is invertible for all scalars $a \neq 0$.

Harder Problems

21. If $A^2 = A$, prove that $(AB - ABA)^2 = 0$ for all matrices B.

22. If $AB = I$, prove that $BA = I$.

23. If $C = BAB^{-1}$, where A is as in Problem 4 and B is invertible, prove that $C^3 + a_1 C^2 + a_2 C + a_3 I = 0$.

24. Prove that $(BAB^{-1})(BVB^{-1}) = B(AV)B^{-1}$ for all matrices A, V, and invertible B.

25. What does Problem 24 tell you about $(BAB^{-1})^k$ for $k \geq 1$ an integer?

26. Let B, C be invertible matrices. Define $\phi_B: M_n(\mathbb{C}) \to M_n(\mathbb{C})$ by $\phi_B(A) = BAB^{-1}$ for all $A \in M_n(\mathbb{C})$. Similarly, define $\phi_C(A) = CAC^{-1}$ for all $A \in M_n(\mathbb{C})$. Prove that $\phi_B \phi_C = \phi_{BC}$.

27. Given a fixed matrix A define $d: M_n(\mathbb{C}) \to M_n(\mathbb{C})$ by $d(X) = AX - XA$ for all $X \in M_n(\mathbb{C})$. Then:

    (a) Prove that $d(XY) = d(X)Y + Xd(Y)$ and $d(X + Y) = d(X) + d(Y)$ for all X, $Y \in M_n(\mathbb{C})$. (What does this remind you of?)

    (b) Determine $d(I)$, where I is the identity.

    (c) Determine $d(aI)$.

    (d) Determine $d(A^3)$.

3.2. MATRICES AS OPERATORS

For the 2×2 matrices we found that we could define an action on the set of all 2-tuples, which was rather nice and which gave some sense to the product of matrices as we had defined them. In Chapter 2 we did a similar thing in expressing a system of linear equations as $Av = w$, where A is an $m \times n$ coefficient matrix and v an element of $F^{(n)}$ (see below). We now do the same for the $n \times n$ matrices.

Let F be either the set of real or complex numbers. By $F^{(n)}$ we shall

mean the set of n-tuples $\begin{bmatrix} a_1 \\ a_2 \\ \vdots \\ a_n \end{bmatrix}$, where $a_1, a_2, \ldots, a_n$ are all in F.

In $F^{(n)}$ we introduce an algebraic structure by defining two operations—addition and multiplication by a scalar—and it is on this structure that we shall allow the $n \times n$ matrices over F to act.

But first we need a way of recognizing when two *vectors*—that is, elements of $F^{(n)}$—are equal.

Definition. If $v = \begin{bmatrix} a_1 \\ \vdots \\ a_n \end{bmatrix}$ and $w = \begin{bmatrix} b_1 \\ \vdots \\ b_n \end{bmatrix}$ are in $F^{(n)}$, then we *define*

$v = w$ if and only if $a_j = b_j$ for every $j = 1, 2, \ldots, n$.

We call a_r the rth *coordinate* or *component of v*. So two vectors are equal only when their corresponding coordinates are equal.

Now that we are able to tell when two vectors are equal, we are ready to introduce the two operations, mentioned above, in $F^{(n)}$.

Definition. If $v = \begin{bmatrix} a_1 \\ \vdots \\ a_n \end{bmatrix}$ and $w = \begin{bmatrix} b_1 \\ \vdots \\ b_n \end{bmatrix}$ are in $F^{(n)}$, we *define*

$$v + w = \begin{bmatrix} a_1 + b_1 \\ \vdots \quad \vdots \\ a_n + b_n \end{bmatrix}.$$

For example, if $v = \begin{bmatrix} 1 \\ 2 \\ 3 \end{bmatrix}$ and $w = \begin{bmatrix} \pi \\ 2.3 \\ 3 \end{bmatrix}$, then

$$v + w = \begin{bmatrix} 1 + \pi \\ 2 + 2.3 \\ 3 + 3 \end{bmatrix} = \begin{bmatrix} 1 + \pi \\ 4.3 \\ 6 \end{bmatrix}.$$

So to add two vectors, just add their corresponding coordinates. Note that the vector $\begin{bmatrix} 0 \\ \vdots \\ 0 \end{bmatrix}$, all of whose coordinates are 0, has the key

property that $v + \begin{bmatrix} 0 \\ \vdots \\ 0 \end{bmatrix} = v$ for every $v \in F^{(n)}$. We denote $\begin{bmatrix} 0 \\ \vdots \\ 0 \end{bmatrix}$ simply

as 0. Note also that the vector $u = \begin{bmatrix} -a_1 \\ \vdots \\ -a_n \end{bmatrix}$ has the property that $u +$

$v = 0$, where $v = \begin{bmatrix} a_1 \\ \vdots \\ a_n \end{bmatrix}$. We write u as $-v$.

We now come to the second operation, that of multiplying a vector in $F^{(n)}$ by an element of F. We call this *multiplication by a scalar*.

Definition.   If $v = \begin{bmatrix} a_1 \\ \vdots \\ a_n \end{bmatrix} \in F^{(n)}$ and $t \in F$, then we define

$tv = \begin{bmatrix} ta_1 \\ \vdots \\ ta_n \end{bmatrix}.$

For instance, if $w = \begin{bmatrix} \pi \\ 2.3 \\ 3 \end{bmatrix}$, then $4w = \begin{bmatrix} 4\pi \\ 9.2 \\ 12 \end{bmatrix}$.

So, to multiply a vector v by a scalar t, just multiply each component of v by t. Note that $(-1)v$ is the same as $-v$ defined above.

The basic rules governing the behavior of addition and multiplication by a scalar are contained in

Theorem 3.2.1 If u, v, w are in $F^{(n)}$ and a, b are in F then av and $v + w$ are in $F^{(n)}$ and

1. $\qquad\qquad v + w = w + v;$
2. $\quad u + (v + w) = (u + v) + w;$
3. $\qquad\quad v + 0 = v$ and $v + (-v) = 0;$
4. $\qquad a(v + w) = av + aw;$
5. $\qquad (a + b)v = av + bv;$
6. $\qquad\qquad 1v = v;$
7. $\qquad\qquad 0v = 0,\ a0 = 0;$
8. $\qquad\quad a(bv) = (ab)v.$

Proof. The proofs of all these parts of the theorem are easy and direct. We pick two sample ones, (4) and (8), to show how such a proof should run.

Suppose that $v = \begin{bmatrix} s_1 \\ \vdots \\ s_n \end{bmatrix}$ and $w = \begin{bmatrix} t_1 \\ \vdots \\ t_n \end{bmatrix}$ are in $F^{(n)}$ and $a \in F$.

Then

$$v + w = \begin{bmatrix} s_1 + t_1 \\ \vdots \\ s_n + t_m \end{bmatrix},$$

hence

$$a(v + w) = \begin{bmatrix} a(s_1 + t_1) \\ \vdots \\ a(s_n + t_n) \end{bmatrix} = \begin{bmatrix} as_1 + at_1 \\ \vdots \\ as_n + at_n \end{bmatrix} = \begin{bmatrix} as_1 \\ \vdots \\ as_n \end{bmatrix} + \begin{bmatrix} at_1 \\ \vdots \\ at_n \end{bmatrix}$$

$$= a\begin{bmatrix} s_1 \\ \vdots \\ s_n \end{bmatrix} + a\begin{bmatrix} t_1 \\ \vdots \\ t_n \end{bmatrix} = av + aw.$$

Here we have used the basic definition of addition and multiplication by a scalar several times. This proves (4).

Now to demonstrate (8). If $v = \begin{bmatrix} s_1 \\ \vdots \\ s_n \end{bmatrix}$, then

$$bv = \begin{bmatrix} bs_1 \\ \vdots \\ bs_n \end{bmatrix},$$

so

$$a(bv) = \begin{bmatrix} a(bs_1) \\ \vdots \\ a(bs_n) \end{bmatrix} = \begin{bmatrix} (ab)s_1 \\ \vdots \\ (ab)s_n \end{bmatrix} = (ab)\begin{bmatrix} s_1 \\ \vdots \\ s_n \end{bmatrix} = (ab)v.$$

This shows (8) to be proved.

The proofs of all the other parts are along the lines used for (4) and (8). ∎

Since we have two zeros running around—that of F and that of $F^{(n)}$—we might get some confusion by using the same symbol for them. But there really is no danger of this because $0v = 0$ for any $v \in F^{(n)}$, where the zero on the left is that of F and the one on the right that of $F^{(n)}$, and $a0 = 0$ for any $a \in F$, where the zero on the left is that of $F^{(n)}$ and the one on the right that of $F^{(n)}$ also. To avoid notational extravaganzas, we'll use the symbol 0 everywhere for both.

The properties described in Theorem 3.2.1 make of $F^{(n)}$ what is called in mathematics a *vector space over* F. Abstract vector spaces will play a prominent role in the last half of the book.

If $n = 3$ and $F = \mathbb{R}$ then $F^{(3)}$ is the usual 3-dimensional Euclidean space that we run into in geometry. So it is natural to call $F^{(n)}$ the *n-dimensional space* over F.

The space $F^{(n)}$ gives us a natural domain on which to define an action by matrices in $M_n(F)$, which we do now in the

Definition.   Suppose that $A = (a_{rs})$ is in $M_n(F)$ and $v = (v_s) = \begin{bmatrix} v_1 \\ \vdots \\ v_n \end{bmatrix}$

is in $F^{(n)}$. Then we define $Av = (w_r) = \begin{bmatrix} w_1 \\ \vdots \\ w_n \end{bmatrix}$, where for each $r = 1, 2, \ldots, n$, $w_r = \Sigma_{\sigma=1}^{n} a_{r\sigma}v_\sigma = a_{r1}v_1 + \cdots + a_{rn}v_n$.

Let's try this definition out for a few matrices and vectors. If $A =$

$$\begin{bmatrix} 1 & 2 & -1 \\ \pi & 1 & 1 \\ 1 & \pi & 3 \end{bmatrix} \text{ and } v = \begin{bmatrix} 0 \\ 1 \\ 0 \end{bmatrix}, \text{ then}$$

$$Av = \begin{bmatrix} 1 & 2 & -1 \\ \pi & 1 & 1 \\ 1 & \pi & 3 \end{bmatrix}\begin{bmatrix} 0 \\ 1 \\ 0 \end{bmatrix} = \begin{bmatrix} 2 \\ 1 \\ \pi \end{bmatrix}.$$

For example, the first entry of Av is $1 \cdot 0 + 2 \cdot 1 + (-1) \cdot 0 = 2$. Perhaps a little more interesting is the example $A =$

$$\begin{bmatrix} 1 & -1 & -1 & 1 \\ 1 & 0 & 0 & 0 \\ 0 & 1 & 0 & 0 \\ -5 & -5 & -5 & 5 \end{bmatrix} \text{ and } v = \begin{bmatrix} 0 \\ 0 \\ \pi+1 \\ \pi+1 \end{bmatrix}, \text{ for we notice that}$$

$Av = 0$. So it is possible for a nonzero matrix to knock-out a nonzero vector.

How do matrices in $M_n(F)$ behave as mappings on $F^{(n)}$? Given $A = (a_{rs})$ and $B = (b_{rs})$, then $A + B = (a_{rs} + b_{rs})$, so for any vector

$$v = \begin{bmatrix} v_1 \\ \vdots \\ v_n \end{bmatrix}, \quad (A + B)\begin{bmatrix} v_1 \\ \vdots \\ v_n \end{bmatrix} = \begin{bmatrix} t_1 \\ \vdots \\ t_n \end{bmatrix}, \quad \text{where}$$

$$t_r = \sum_{\sigma=1}^{n}(a_{r\sigma} + b_{r\sigma})v_\sigma = \sum_{\sigma=1}^{n} a_{r\sigma}v_\sigma + \sum_{\sigma=1}^{n} b_{r\sigma}v_\sigma$$

$$= A\begin{bmatrix} v_1 \\ \vdots \\ v_n \end{bmatrix} + B\begin{bmatrix} v_1 \\ \vdots \\ v_n \end{bmatrix}.$$

In short,

$$(A + B)v = Av + Bv.$$

A similar argument shows that

$$(aA)v = a(Av) = A(av).$$

Also,

$$A(v + w) = Av + Aw.$$

These are easy to verify. (Do so!) These last two behaviors of A on vectors (on scalar products and on sums) are very important. We use them to make the

Definition. A mapping $T: F^{(n)} \to F^{(n)}$ is a *lineaar transformation on* $F^{(n)}$ if:

     (a) $T(v + w) = T(v) + T(w)$
     (b) $T(av) \quad = \quad\quad aT(v)$

for every $a \in F$ and for any $v, w \in F^{(n)}$.

EXAMPLE

For $n = 3$, the mapping $T: F^{(n)} \to F^{(n)}$ defined by

$$T\begin{bmatrix} a \\ b \\ c \end{bmatrix} = \begin{bmatrix} 3c - b \\ 3b - a \\ a \end{bmatrix}$$

is a *linear transformation on* $F^{(n)}$ since

$$T\left(\begin{bmatrix} a \\ b \\ c \end{bmatrix} + \begin{bmatrix} g \\ h \\ k \end{bmatrix}\right) = T\begin{bmatrix} a + g \\ b + h \\ c + k \end{bmatrix}$$

$$= \begin{bmatrix} 3(c + k) - (b + h) \\ 3(b + h) - (a + g) \\ a + g \end{bmatrix}$$

$$= \begin{bmatrix} 3c - b \\ 3b - a \\ a \end{bmatrix} + \begin{bmatrix} 3k - h \\ 3h - g \\ g \end{bmatrix}$$

$$= T\begin{bmatrix} a \\ b \\ c \end{bmatrix} + T\begin{bmatrix} g \\ h \\ k \end{bmatrix}$$

and

$$T\left(d\begin{bmatrix} a \\ b \\ c \end{bmatrix}\right) = T\left(\begin{bmatrix} da \\ db \\ dc \end{bmatrix}\right) = \begin{bmatrix} 3dc - db \\ 3db - da \\ da \end{bmatrix}$$

$$= d\begin{bmatrix} 3c - b \\ 3b - a \\ a \end{bmatrix} = dT\begin{bmatrix} a \\ b \\ c \end{bmatrix}$$

for all $\begin{bmatrix} a \\ b \\ c \end{bmatrix}$, $\begin{bmatrix} g \\ h \\ k \end{bmatrix}$ in $F^{(n)}$ and all d in F.

By this definition, every $n \times n$ matrix A defines by its action $Av(v \in F^{(n)})$ a linear transformation on $F^{(n)}$. The opposite is also true, namely, a linear transformation on $F^{(n)}$ is given by a matrix, as in the following example.

EXAMPLE

Let $n = 3$ and define $T: F^{(n)} \to F^{(n)}$ by

$$T\begin{bmatrix} a \\ b \\ c \end{bmatrix} = \begin{bmatrix} 3c - b \\ 3b - a \\ a \end{bmatrix}$$

as in the example above. Then

$$T\begin{bmatrix} 1 \\ 0 \\ 0 \end{bmatrix} = \begin{bmatrix} 0 \\ -1 \\ 1 \end{bmatrix}, \quad T\begin{bmatrix} 0 \\ 1 \\ 0 \end{bmatrix} = \begin{bmatrix} -1 \\ 3 \\ 0 \end{bmatrix}, \quad T\begin{bmatrix} 0 \\ 0 \\ 1 \end{bmatrix} = \begin{bmatrix} 3 \\ 0 \\ 0 \end{bmatrix}.$$

But then it is clear that the matrix

$$\begin{bmatrix} 0 & -1 & 3 \\ -1 & 3 & 0 \\ 1 & 0 & 0 \end{bmatrix}$$

has the same effect:

$$\begin{bmatrix} 0 & -1 & 3 \\ -1 & 3 & 0 \\ 1 & 0 & 0 \end{bmatrix}\begin{bmatrix} 1 \\ 0 \\ 0 \end{bmatrix} = \begin{bmatrix} 0 \\ -1 \\ 1 \end{bmatrix},$$

$$\begin{bmatrix} 0 & -1 & 3 \\ -1 & 3 & 0 \\ 1 & 0 & 0 \end{bmatrix}\begin{bmatrix} 0 \\ 1 \\ 0 \end{bmatrix} = \begin{bmatrix} -1 \\ 3 \\ 0 \end{bmatrix},$$

$$\begin{bmatrix} 0 & -1 & 3 \\ -1 & 3 & 0 \\ 1 & 0 & 0 \end{bmatrix}\begin{bmatrix} 0 \\ 0 \\ 1 \end{bmatrix} = \begin{bmatrix} 3 \\ 0 \\ 0 \end{bmatrix}.$$

Trying it with the general vector, we get

$$\begin{bmatrix} 0 & -1 & 3 \\ -1 & 3 & 0 \\ 1 & 0 & 0 \end{bmatrix}\begin{bmatrix} a \\ b \\ c \end{bmatrix} = \begin{bmatrix} 3c - b \\ 3b - a \\ a \end{bmatrix}.$$

You may be convinced from this example that every linear transformation is given by a matrix. This is not surprising, since we can model the general proof after the example in

Theorem 3.2.2. Let T be a linear transformation on $F^{(n)}$. Then $T = (t_{rs})$ as mappings for some matrix (t_{rs}) in $M_n(F)$.

Proof. Consider the vectors

$$e_1 = \begin{bmatrix} 1 \\ 0 \\ \vdots \\ 0 \end{bmatrix}, \; e_2 = \begin{bmatrix} 0 \\ 1 \\ \vdots \\ 0 \end{bmatrix}, \ldots, e_n = \begin{bmatrix} 0 \\ 0 \\ \vdots \\ 1 \end{bmatrix},$$

that is, those vectors e_r whose rth component is 1 and all of whose other components are 0. Given any vector $v \in F^{(n)}$, if $v = \begin{bmatrix} a_1 \\ a_2 \\ \vdots \\ a_n \end{bmatrix}$, we see that

$$v = a_1 e_1 + a_2 e_2 + \cdots a_n e_n.$$

Thus

$$\begin{aligned} T(v) &= T(a_1 e_1 + a_2 e_2 + \cdots + a_n e_n) \\ &= a_1 T(e_1) + a_2 T(e_2) + \cdots + a_n T(e_n). \end{aligned}$$

Therefore, if we knew what T did to each of $e_1, \ldots, e_n$ we would know what T did to any vector. Now Te_s as an element of $F^{(n)}$ is representable as

$$Te_s = t_{1s} e_1 + t_{2s} e_2 + \cdots + t_{ns} e_n$$

for each $s = 1, 2, \ldots, n$. Consider the action of the matrix (t_{rs}) on $F^{(n)}$. As noted above, to know how (t_{rs}) behaves on $F^{(n)}$, it is enough to know what it does to $e_1, \ldots, e_n$. Computing yields

$$(t_{rs})e_1 = \begin{bmatrix} t_{11} & t_{12} & \cdots & t_{1n} \\ t_{21} & t_{22} & \cdots & t_{2n} \\ \vdots & \vdots & & \vdots \\ t_{n1} & t_{n2} & \cdots & t_{nn} \end{bmatrix} \begin{bmatrix} 1 \\ 0 \\ \vdots \\ 0 \end{bmatrix} = \begin{bmatrix} t_{11} \\ t_{21} \\ \vdots \\ t_{n1} \end{bmatrix}$$

$$= t_{11} e_1 + t_{21} e_2 + \cdots + t_{n1} e_n = Te_1.$$

Similarly, we get that $(t_{rs})e_u = Te_u$ for every $u = 1, 2, \ldots, n$. Thus T and (t_{rs}) agree on each of $e_1, \ldots, e_n$, from which it follows that T and (t_{rs}) agree on all elements of $F^{(n)}$. Hence $T = (t_{rs})$. ■

Definition. Let T be a linear transformation on $F^{(n)}$. Then the matrix (t_{rs}) in $M_n(F)$ such that $T = (t_{rs})$ as mappings of $F^{(n)}$, that is the matrix (t_{rs}) such that

$$Te_s = t_{1s}e_1 + t_{2s}e_2 + \cdots + t_{ns}e_n \text{ for } 1 \leq s \leq n$$

is called the *matrix* of T.

This is a constructive definition in the sense that to find the matrix of T it tells us to use as column s of this matrix the vector Te_s.

EXAMPLE

For instance, if T is the mapping on $F^{(4)}$ defined by

$$T\begin{bmatrix} a_1 \\ a_2 \\ a_3 \\ a_4 \end{bmatrix} = \begin{bmatrix} a_2 \\ 2a_3 \\ 3a_4 \\ 0 \end{bmatrix}, \text{ then}$$

$$Te_1 = T\begin{bmatrix} 1 \\ 0 \\ 0 \\ 0 \end{bmatrix} = \begin{bmatrix} 0 \\ 0 \\ 0 \\ 0 \end{bmatrix}, Te_2 = T\begin{bmatrix} 0 \\ 1 \\ 0 \\ 0 \end{bmatrix} = \begin{bmatrix} 1 \\ 0 \\ 0 \\ 0 \end{bmatrix},$$

$$Te_3 = T\begin{bmatrix} 0 \\ 0 \\ 1 \\ 0 \end{bmatrix} = \begin{bmatrix} 0 \\ 2 \\ 0 \\ 0 \end{bmatrix}, Te_4 = T\begin{bmatrix} 0 \\ 0 \\ 0 \\ 1 \end{bmatrix} = \begin{bmatrix} 0 \\ 0 \\ 3 \\ 0 \end{bmatrix}.$$

Therefore, the matrix of T is $\begin{bmatrix} 0 & 1 & 0 & 0 \\ 0 & 0 & 2 & 0 \\ 0 & 0 & 0 & 3 \\ 0 & 0 & 0 & 0 \end{bmatrix}$. Checking out

$$\begin{bmatrix} 0 & 1 & 0 & 0 \\ 0 & 0 & 2 & 0 \\ 0 & 0 & 0 & 3 \\ 0 & 0 & 0 & 0 \end{bmatrix}\begin{bmatrix} a_1 \\ a_2 \\ a_3 \\ a_4 \end{bmatrix} = \begin{bmatrix} a_2 \\ 2a_3 \\ 3a_4 \\ 0 \end{bmatrix},$$

we see that, indeed, the matrix of the linear transformation T is

$$\text{given by } \begin{bmatrix} 0 & 1 & 0 & 0 \\ 0 & 0 & 2 & 0 \\ 0 & 0 & 0 & 3 \\ 0 & 0 & 0 & 0 \end{bmatrix}.$$

We should like to see how the product of two matrices behaves in terms of mappings. Since matrices A and B in $M_n(F)$ define mappings of $F^{(n)}$ into itself, it is fairly natural to ask how $A \circ B$—the product (i.e., composite) of A and B as mappings—relates to AB, the product of A and B as matrices. Let $A = (a_{rs})$ and $B = (b_{rs})$; so if

$$x = \begin{bmatrix} x_1 \\ \vdots \\ x_n \end{bmatrix} \in F^{(n)},$$

then $Bx = \begin{bmatrix} y_1 \\ \vdots \\ y_n \end{bmatrix}$, where $y_s = \sum_{t=1}^{n} b_{st}x_t$ for $s = 1, 2, \ldots, n$. Thus

$$(A \circ B)(x) = A(Bx) = A\begin{bmatrix} y_1 \\ \vdots \\ y_n \end{bmatrix} = \begin{bmatrix} z_1 \\ \vdots \\ z_n \end{bmatrix},$$

where

$$z_r = \sum_{s=1}^{n} a_{rs}y_s = \sum_{s=1}^{n} a_{rs}\left(\sum_{t=1}^{n} b_{st}x_t \right) = \sum_{s=1}^{n} \sum_{t=1}^{n} a_{rs}b_{st}x_t$$

$$= \sum_{t=1}^{n} \sum_{s=1}^{n} a_{rs}b_{st}x_t = \sum_{t=1}^{n} \left(\sum_{s=1}^{n} a_{rs}b_{st} \right)x_t.$$

This implies that z_r, which was defined as the rth entry of $(A \circ B)(x)$, is also the rth entry of $(AB)x$, since AB is the matrix $(\sum_{s=1}^{n} a_{rs}b_{st})$. Thus $(A \circ B)(x)$ and $(AB)x$ have the same rth entry for all r, which implies that $(A \circ B)(x) = (AB)x$. Since this is true for all x, we can conclude that $A \circ B = AB$. In other words,

the composition of A and B as mappings agrees with their product as matrices.

In fact, this is the basic reason for which the product of matrices is defined in the way it is—which at first glance seems contrived.

The result above merits being singled out, and we do so in

Theorem 3.2.3. The mapping defined by AB on $F^{(n)}$ coincides with the mapping $A \circ B$—the composition of the mapping A and B—defined by $(A \circ B)v = A(Bv)$ for every $v \in F^{(n)}$.

Although we know it is true in general, let's check out Theorem 3.2.3 in a specific instance. Let $A = \begin{bmatrix} 1 & 2 & 4 \\ -2 & 0 & 1 \\ 3 & -1 & 0 \end{bmatrix}$ and $B = \begin{bmatrix} 0 & 1 & 0 \\ 1 & 1 & 1 \\ 0 & 0 & 1 \end{bmatrix}$. Then the *matrix product* of A and B is $AB = \begin{bmatrix} 2 & 3 & 6 \\ 0 & -2 & 1 \\ -1 & 2 & -1 \end{bmatrix}$. So if $v = \begin{bmatrix} x_1 \\ x_2 \\ x_3 \end{bmatrix} \in F^{(3)}$ then

$$(AB)v = \begin{bmatrix} 2 & 3 & 6 \\ 0 & -2 & 1 \\ -1 & 2 & -1 \end{bmatrix}\begin{bmatrix} x_1 \\ x_2 \\ x_3 \end{bmatrix} = \begin{bmatrix} 2x_1 + 3x_2 + 6x_3 \\ -2x_2 + x_3 \\ -x_1 + 2x_2 - x_3 \end{bmatrix}.$$

On the other hand, $Bv = \begin{bmatrix} 0 & 1 & 0 \\ 1 & 1 & 1 \\ 0 & 0 & 1 \end{bmatrix}\begin{bmatrix} x_2 \\ x_1 + x_2 + x_3 \\ x_3 \end{bmatrix}$; thus

$$(A \circ B)v = A(Bv)$$

$$= \begin{bmatrix} 1 & 2 & 4 \\ -2 & 0 & 1 \\ 3 & -1 & 0 \end{bmatrix}\begin{bmatrix} x_2 \\ x_1 + x_2 + x_3 \\ x_3 \end{bmatrix}$$

$$= \begin{bmatrix} x_2 + 2(x_1 + x_2 + x_3) + 4x_3 \\ -2x_2 + x_3 \\ 3x_2 - (x_1 + x_2 + x_3) \end{bmatrix}$$

$$= \begin{bmatrix} 2x_1 + 3x_2 + 6x_3 \\ -2x_2 + x_3 \\ -x_1 + 2x_2 - x_3 \end{bmatrix} = (AB)v.$$

So, in this specific example, we see that $A \circ B = AB$, as the theorem tells us that it should be.

PROBLEMS

In the following problems, $A \circ B$ will denote the product of A and B as mappings. Also, *if T is a linear transformation, we shall mean by m(T) the matrix of T.*

NUMERICAL PROBLEMS

1. Compute the vector $-\begin{bmatrix} 1 \\ 2 \\ -3 \end{bmatrix}$.

2. Compute the vector $\begin{bmatrix} 4 \\ -2 \\ -3 \end{bmatrix} - \begin{bmatrix} 6 \\ -4 \\ 5 \end{bmatrix}$.

3. Compute the vector $3 \begin{bmatrix} 4 \\ -2 \\ -3 \end{bmatrix}$.

4. Compute the vector $8 \begin{bmatrix} 4 \\ -2 \\ -3 \end{bmatrix} + (-4)\left(\begin{bmatrix} 4 \\ -2 \\ -3 \end{bmatrix} - \begin{bmatrix} 6 \\ -4 \\ 5 \end{bmatrix} \right)$.

5. Compute the vector $\begin{bmatrix} 3 & 2 & 1 \\ -2 & 3 & 1 \\ -4 & -3 & 1 \end{bmatrix} \begin{bmatrix} 6 \\ -4 \\ 5 \end{bmatrix}$.

6. Compute the vectors

$$\begin{bmatrix} 3 & 2 & 5 \\ -2 & 2 & 3 \\ -1 & -3 & 1 \end{bmatrix} \begin{bmatrix} 3 \\ -2 \\ -1 \end{bmatrix}, \quad \begin{bmatrix} 3 & 2 & 5 \\ -2 & 2 & 3 \\ -1 & -3 & 1 \end{bmatrix} \begin{bmatrix} 3 \\ 2 \\ 1 \end{bmatrix}, \quad \begin{bmatrix} 3 & 2 & 5 \\ -2 & 2 & 3 \\ -1 & -3 & 1 \end{bmatrix} \begin{bmatrix} 3 \\ 1 \\ 1 \end{bmatrix}.$$

7. Compute the matrix $\begin{bmatrix} 3 & 2 & 5 \\ -2 & 2 & 3 \\ -1 & -3 & 1 \end{bmatrix} \begin{bmatrix} 3 & 3 & 3 \\ -2 & 2 & 1 \\ -1 & -1 & 1 \end{bmatrix}$. How

does the result relate to the three vectors found in Problem 6?

8. What is the matrix of the linear transformation that maps the vector $\begin{bmatrix} a \\ b \\ c \end{bmatrix}$ to the vector $\begin{bmatrix} u \\ v \\ w \end{bmatrix}$ such that $\begin{bmatrix} 0 & 0 & 0 \\ a & b & c \\ 0 & 0 & 0 \end{bmatrix} \begin{bmatrix} 3 & 3 & 3 \\ -2 & 2 & 1 \\ -1 & -1 & 1 \end{bmatrix} =$

$\begin{bmatrix} 0 & 0 & 0 \\ u & v & w \\ 0 & 0 & 0 \end{bmatrix}$?

9. What is the matrix of the linear transformation that maps the vector
$$\begin{bmatrix} a \\ b \\ c \end{bmatrix} \text{ to the vector } \begin{bmatrix} u \\ v \\ w \end{bmatrix} \text{ such that } \begin{bmatrix} a & b & c \\ 0 & 0 & 0 \\ 0 & 0 & 0 \end{bmatrix} \begin{bmatrix} 3 & 3 & 3 \\ -2 & 2 & 1 \\ -1 & -1 & 1 \end{bmatrix} =$$
$$\begin{bmatrix} u & v & w \\ 0 & 0 & 0 \\ 0 & 0 & 0 \end{bmatrix}?$$

10. Let S be the linear transformation whose matrix is $\begin{bmatrix} 1 & 2 & 2 \\ 3 & 3 & 3 \\ 4 & 4 & 4 \end{bmatrix}$ and

    let T be the linear transformation whose matrix is $\begin{bmatrix} 1 & 3 & 4 \\ 2 & 3 & 4 \\ 2 & 3 & 4 \end{bmatrix}$. Com-

    pute the matrix of the linear transformation $S \circ T$ and verify directly
    that the vectors $(S \circ T)(v)$ and $(ST)(v)$ are the same for all v in $F^{(3)}$.

MORE THEORETICAL PROBLEMS

Easier Problems

11. Show that the identity mapping T on $F^{(n)}$ is a linear transformation
    of $F^{(n)}$ whose matrix is the unit matrix I.

12. If A is an invertible matrix and A^{-1} is its inverse, show that
    $A \circ A^{-1}$ is the identity mapping.

13. Verify that the T's given are linear transformations on the appro-
    priate $F^{(n)}$ and for each find $m(T)$, the matrix of T.

    (a) $T\begin{bmatrix} a_1 \\ a_2 \\ a_3 \\ a_4 \end{bmatrix} = \begin{bmatrix} a_2 \\ a_1 \\ a_4 \\ a_3 \end{bmatrix}$ for all $\begin{bmatrix} a_1 \\ a_2 \\ a_3 \\ a_4 \end{bmatrix}$ in $F^{(4)}$.

    (b) $T\begin{bmatrix} a_1 \\ a_2 \\ a_3 \\ a_4 \\ a_5 \end{bmatrix} = \begin{bmatrix} 0 \\ a_1 \\ \frac{1}{2}a_2 \\ a_3 \\ \frac{1}{4}a_4 \end{bmatrix}$ on $F^{(5)}$.

    (c) $T\begin{bmatrix} a_1 \\ a_2 \\ a_3 \\ a_4 \end{bmatrix} = \begin{bmatrix} ia_4 \\ -a_1 \\ 3a_2 \\ -ia_3 \end{bmatrix}$ on $F^{(4)}$ $(F = \mathbb{C})$.

(d) $T\begin{bmatrix} a_1 \\ a_2 \\ a_3 \end{bmatrix} = \begin{bmatrix} -a_3 + a_2 - a_1 \\ a_1 \\ a_2 \end{bmatrix}$ on $F^{(3)}$.

14. Let A be the T in Part (a) and B the T in Part (c) of Problem 13. Find the matrix $m(A \circ B)$ and verify directly the assertion of Theorem 3.2.3 that $m(A \circ B) = m(A)m(B)$.

15. If T on $F^{(4)}$ is defined by $T\begin{bmatrix} a_1 \\ a_2 \\ a_3 \\ a_4 \end{bmatrix} = \begin{bmatrix} a_4 \\ a_1 \\ a_2 \\ a_3 \end{bmatrix}$ for all $\begin{bmatrix} a_1 \\ a_2 \\ a_3 \\ a_4 \end{bmatrix}$ in $F^{(4)}$ show

that $T^4 = T \circ T \circ T \circ T$ is the identity mapping on $F^{(4)}$, find $m(T)$, and show by a direct calculation that $m(T)^4 = I$.

16. If $v \in F^{(n)}$ and $A \in M_n(F)$ is invertible, show that $A^{-1}(Av) = v$.

17. If A in $M_n(F)$ is invertible and $Av = 0$ for $v \in F^{(n)}$, show that $v = 0$.

18. If two columns of the matrix A are identical, show that $Av = 0$ for some $v \neq 0$, hence A cannot be invertible.

19. If A in $M_3(F)$ is not invertible, show that you can find a $v \neq 0$ in $F^{(3)}$ such that $Av = 0$.

20. If $A \in M_n(F)$ is such that the vectors $w_1, \ldots, w_n$, where w_s is column s of A, satisfy a relation of the form

$$a_1 w_1 + a_2 w_2 + \cdots + a_n w_n = 0,$$

where the $a_r \in F$ and not all of them are 0, then show that $Av = 0$ for some $v \neq 0$ in $F^{(n)}$ by describing the coordinates of such a v.

21. Let T be defined by $T\begin{bmatrix} x_1 \\ x_2 \\ x_3 \end{bmatrix} = \begin{bmatrix} x_2 \\ x_3 \\ -cx_1 - bx_2 + ax_3 \end{bmatrix}$, where

$a, b, c \in F$, for all $\begin{bmatrix} x_1 \\ x_2 \\ x_3 \end{bmatrix}$ in $F^{(3)}$. Show that T is a linear

transformation
on $F^{(3)}$ and find $A = m(T)$. Then show that

$$A^3 + aA^2 + bA + cI = 0.$$

22. Let $v_1 = \begin{bmatrix} 1 \\ 0 \\ \vdots \\ 0 \end{bmatrix}$, $v_2 = \begin{bmatrix} 0 \\ 1 \\ \vdots \\ 0 \end{bmatrix}$, . . . , $v_n = \begin{bmatrix} 0 \\ 0 \\ \vdots \\ 1 \end{bmatrix}$ be in $F^{(n)}$. Suppose

that T is defined on $F^{(n)}$ by $T(a_1 v_1 + a_2 v_2 + \cdots + a_n v_n) = a_1 v_2 + a_2 v_3 + \cdots + a_{n-1} v_n$ for all $a_1, \ldots, a_n$ in F. Show that T is a linear transformation on $F^{(n)}$, find $A = m(T)$, and show that $A^n = 0$.

3.3. TRACE

In discussing the 2×2 matrices, one of the concepts introduced there was the trace of a matrix. For the matrix $\begin{bmatrix} a & b \\ c & d \end{bmatrix}$ we defined its trace as $a + d$, the sum of its diagonal elements. This notion was a useful one for the 2×2 matrices. One would hope its extension to the $n \times n$ case would be equally useful.

Given the matrix $A = (a_{rs})$ in $M_n(F)$ we do the obvious and make the

Definition. If $A = (a_{rs}) \in M_n(F)$, then the *trace* of A, written tr (A), is defined as tr $(a) = \Sigma_{t=1}^{n} a_{tt}$.

In words, tr (A) *is the sum of the diagonal entries of A.*

Fortunately, all the theorems we proved for the trace in $M_2(F)$ go over smoothly and exactly to $M_n(F)$. Note that tr $(I) = n$.

Theorem 3.3.1. If $A, B \in M_n(F)$ and $a \in F$, then

1. tr $(A + B) =$ tr $(A) +$ tr (B);
2. tr $(aA) = a$ tr (A);
3. tr $(AB) =$ tr (BA).

Proof. Suppose that $A = (a_{rs})$ and $B = (b_{rs})$. Then tr $(A) = \Sigma_{t=1}^{n} a_{tt}$ and tr $(B) = \Sigma_{t=1}^{n} b_{tt}$. By the way we add matrices $A + B = (c_{rs})$, where $c_{rs} = a_{rs} + b_{rs}$; thus

$$\text{tr } (A + B) = \Sigma^n_{t=1} c_{tt} = \Sigma^n_{t=1} (a_{tt} + b_{tt})$$
$$= \Sigma^n_{t=1} a_{tt} + \Sigma^n_{t=1} b_{tt} = \text{tr } (A) + \text{tr } (B).$$

Similarly, tr $(aA) = a$ tr (A), whose proof we leave to the reader.

If $AB = (d_{rs})$ and $BA = (g_{rs})$, we know that $d_{rt} = \Sigma^n_{s=1} a_{rs}b_{st}$ and $g_{rt} = \Sigma^n_{s=1} b_{rs}a_{st}$ for every r and t. (Remember, the index of summation, s for d_{rt} and s for g_{rt}, is a dummy index. We can designate it however we like.) Thus

$$\text{tr } (AB) = \sum_{r=1}^{n} d_{rr} = \sum_{r=1}^{n} \sum_{s=1}^{n} a_{rs}b_{sr}.$$

Similarly,

$$\text{tr } (BA) = \sum_{r=1}^{n} g_{rr} = \sum_{r=1}^{n} \sum_{s=1}^{n} b_{rs}a_{sr}.$$

Since the r and s are dummy indices, if we interchange the letters r by s in the first expression, we get

$$\text{tr } (BA) = \sum_{r=1}^{n} \sum_{s=1}^{n} b_{rs}a_{sr} = \sum_{s=1}^{n} \sum_{r=1}^{n} a_{sr}b_{rs} = \text{tr } (AB). \qquad \blacksquare$$

If you have trouble playing around with these summations Σ, try the argument given for 3×3 matrices. It should clarify what is going on.

As always, it is nice to see that the general theorem holds for specific cases. So let us illustrate that tr $(AB) = $ tr (BA) with the example

$$A = \begin{bmatrix} 1 & 2 & 0 & -1 \\ 0 & i & 0 & i \\ i & 1 & 1 & i \\ 1+i & 0 & 0 & 1 \end{bmatrix}, \quad B = \begin{bmatrix} 0 & 1 & 0 & 0 \\ 1 & 1+i & 1 & i \\ 1 & 0 & 2 & 3 \\ 0 & 1 & 4 & -5 \end{bmatrix}.$$

Then, to know tr (AB) we don't need to know all the entries of AB, just its diagonal entries. So

$$AB = \begin{bmatrix} 2 & & & * \\ & i(1 + i) + i & & \\ & & 1 + 2 + 4i & \\ * & & & -5 \end{bmatrix},$$

where we are not interested in the $*$. So

$$\text{tr } (AB) = 2 + i(i + 1) + 1 + 1 + 2 + 4i - 5 = -1 + 6i.$$

We also have

$$
BA = \begin{bmatrix} 0 & & & * \\ & 2 + (1 + i)i + 1 & & \\ & & 2 & \\ * & & & i + 4i - 5 \end{bmatrix},
$$

whence

$$
\begin{aligned}
\mathrm{tr}\,(BA) &= 0 + 2 + (1 + i)i + 1 + 2 + i + 4i - 5 \\
&= 2 + i(1 + i) + i + 1 + 2 + 4i - 5 = -1 + 6i \\
&= \mathrm{tr}\,(AB).
\end{aligned}
$$

The third part of Theorem 3.3.1 has an equivalent statement, namely

Corollary 3.3.2. For $A, B \in M_n(F)$, tr $(AB - BA) = 0$.

Proof. By Parts (1) and (2) of the theorem,

$$
\begin{aligned}
\mathrm{tr}\,(AB - BA) \\
&= \mathrm{tr}\,(AB) + \mathrm{tr}\,(-BA) \\
&= \mathrm{tr}\,(AB) - \mathrm{tr}\,(BA),
\end{aligned}
$$

which is 0 by Part (3) of the theorem. ∎

This corollary itself has a corollary, namely

Corollary 3.3.3. If $A, B \in M_n(F)$ and A is invertible, then

$$
\mathrm{tr}\,(B) = \mathrm{tr}\,(ABA^{-1}).
$$

Proof. Let $C = BA^{-1}$. Then tr $(ABA^{-1}) = \mathrm{tr}\,(AC) = \mathrm{tr}\,(CA) = \mathrm{tr}\,(BA^{-1}A) = \mathrm{tr}\,(BI) = \mathrm{tr}\,(B)$. ∎

PROBLEMS

NUMERICAL PROBLEMS

1. Find tr (A) for

(a) $A = \begin{bmatrix} 1 & 2 & 3 \\ 4 & 5 & 6 \\ 7 & 8 & 9 \end{bmatrix} \begin{bmatrix} 7 & 8 & 9 \\ 4 & 5 & 6 \\ 1 & 2 & 3 \end{bmatrix}$.

(b) $A = \begin{bmatrix} 1 & -1 & 0 \\ 0 & 0 & 1 \\ 2 & -3 & 0 \end{bmatrix} \left(\begin{bmatrix} 1 & .5 & .5 \\ 1 & 1 & 1 \\ 3 & 0 & -2 \end{bmatrix} + \begin{bmatrix} 0 & \pi & 0 \\ \pi & 0 & 0 \\ 0 & 0 & \pi \end{bmatrix} \right).$

(c) $A = \begin{bmatrix} 1 & 0 & 3 \\ 0 & 1 & 0 \\ 0 & 0 & 1 \end{bmatrix} \begin{bmatrix} -2 & .3 & 0 \\ 0 & 5 & 1 \\ 0 & 0 & -3 \end{bmatrix} \begin{bmatrix} 1 & 0 & -3 \\ 0 & 1 & 0 \\ 0 & 0 & 1 \end{bmatrix}.$

(d) $A = \begin{bmatrix} 1 & 0 & 0 \\ 0 & 2 & 0 \\ 0 & 0 & 3 \end{bmatrix} \begin{bmatrix} -1 & 0 & -2 \\ -3 & 0 & -4 \\ 5 & 0 & 6 \end{bmatrix} \begin{bmatrix} 1 & 0 & 0 \\ 0 & .5 & 0 \\ 0 & 0 & .3 \end{bmatrix}.$

(e) $A = \begin{bmatrix} .25 & .25 & .25 & .25 \\ .25 & .25 & .25 & .25 \\ .25 & .25 & .25 & .25 \\ .25 & .25 & .25 & .25 \end{bmatrix}^2.$

2. By a direct computation find all 3×3 matrices A such that

$$\text{tr }(AB) = 0, \text{ where } B = \begin{bmatrix} 1 & 0 & 0 \\ 0 & 2 & 0 \\ 0 & 0 & 3 \end{bmatrix}.$$

3. Verify by a direct calculation of the products that tr $(AB) = $ tr (BA),

where $A = \begin{bmatrix} 0 & 1 & 0 \\ .5 & 0 & 1 \\ 3 & 0 & -1 \end{bmatrix}$ and $B = \begin{bmatrix} .2 & .3 & .4 \\ 0 & 1 & 0 \\ 0 & 6 & 5 \end{bmatrix}.$

4. For what values of a is tr $(A) = 0$, where $A = \begin{bmatrix} 0 & 1 & 0 \\ a & 0 & 1 \\ 0 & 0 & -a \end{bmatrix}$?

5. For what values of a is tr $(A) = 0$, where

$$A = \begin{bmatrix} 0 & 1 & 0 \\ a & 0 & 1 \\ 0 & 0 & -a \end{bmatrix} \begin{bmatrix} a & 1 & 0 \\ 1 & -a & 0 \\ 0 & 0 & -a \end{bmatrix}?$$

MORE THEORETICAL PROBLEMS

Easier Problems

6. Show that tr $(ABC) = $ tr (CAB) for all matrices C.

Middle-Level Problems

7. Show that tr $((AB)^m) = $ tr $((BA)^m)$ for all positive integers m.

8. If $A \in M_2(F)$ and $A^2 = 0$, show that tr $(A) = 0$.

9. If $A \in M_2(F)$ and tr $(A) = 0 = $ tr (A^2), show that $A^2 = 0$.

10. If $A = \begin{bmatrix} a_1 & b_1 & c_1 \\ 0 & a_2 & b_2 \\ 0 & 0 & a_3 \end{bmatrix}$ is such that tr $(A) = $ tr $(A^2) = $ tr $(A^3) = 0$,

prove that $a_1 = a_2 = a_3 = 0$ and that $A^3 = 0$.

11. If $A = \begin{bmatrix} a_1 & & & * \\ & a_2 & & \\ & & \ddots & \\ 0 & & & a_n \end{bmatrix}$, where every entry below the main

diagonal is 0 and where the entries above the main diagonal are arbitrary elements of F, show that if $A^k = 0$ for some k, then tr $(A) = 0$.

Harder Problems

12. If $A \in M_n(F)$, show that $A = aI + B$ for some $a \in F$, $B \in M_n(F)$, and tr $(B) = 0$.

3.4. TRANSPOSE AND HERMITIAN ADJOINT

In working with the 2 × 2 matrices with real entries, we introduced the concept of the transpose of a matrix. If you recall, the transpose of a matrix was the matrix obtained by interchanging the rows with the columns of the given matrix. Later, after we had discussed the complex numbers and were working with matrices whose entries were complex numbers, we talked about the Hermitian adjoint of a matrix, A. It was like the transpose, but with a twist. The Hermitian adjoint $A*$ of a matrix A was obtained by first taking the transpose of A and then applying the operation of complex conjugation to all the entries of this transposed matrix. Of course, for a matrix with real entries its transpose and its Hermitian adjoint are the same.

For these operations on the matrices we found that certain rules pertained. We shall find the exact analog of these rules, here, for the $n \times n$ case. But first we make the formal definitions that we need.

Definition.   Given $A = (a_{rs}) \in M_n(F)$, then the *transpose* of A denoted by A', is the matrix $A' = (b_{rs})$, where for each r and s, $b_{rs} = a_{sr}$.

So, for example, if $A = \begin{bmatrix} 1 & i & 2 \\ 3 & 7 & 0 \\ 1+i & 0 & 0 \end{bmatrix}$, then

$$A' = \begin{bmatrix} 1 & 3 & 1+i \\ i & 7 & 0 \\ 2 & 0 & 0 \end{bmatrix}.$$

We immediately pass on to the Hermitian adjoint.

Definition. If $A = (a_{rs}) \in M_n(\mathbb{C})$, then the *Hermitian adjoint* A^* of A is defined by $A^* = (c_{rs})$ where $c_{rs} = \bar{a}_{sr}$, the $^-$ indicating the complex conjugate of a_{sr}.

Thus, for the example we used above, $A = \begin{bmatrix} 1 & i & 2 \\ 3 & 7 & 0 \\ 1+i & 0 & 0 \end{bmatrix}$, we

have

$$A^* = \begin{bmatrix} \bar{1} & \bar{3} & \overline{1+i} \\ \bar{i} & \bar{7} & \bar{0} \\ \bar{2} & \bar{0} & \bar{0} \end{bmatrix} = \begin{bmatrix} 1 & 3 & 1-i \\ -i & 7 & 0 \\ 2 & 0 & 0 \end{bmatrix}.$$

When we discuss matrices with complex entries, we shall seldom—possibly never—use the transpose; instead, we shall always use the Hermitian adjoint. Accordingly, we shall prove the basic properties only for the Hermitian adjoint, leaving the corresponding results for transpose as exercises for the reader.

Theorem 3.4.1. If $A, B \in M_n(\mathbb{C})$ and $a \in \mathbb{C}$, then

1. $(A^*)^* = A^{**} = A$;
2. $(A + B)^* = A^* + B^*$;
3. $(aA)^* = \bar{a}A^*$;
4. $(AB)^* = B^*A^*$.

Proof. Before getting down to the details of the proof, note that the rules for $*$ and $'$ are the same, with the exception of (3), where for transpose $(aA') = aA'$. Note that important property of $*$ in (4), which says that on taking the Hermitian adjoint we reverse the order of the matrices involved.

Now to the proof itself. If $A = (a_{rs})$, then $A^* = (b_{rs})$, where $b_{rs} = \bar{a}_{sr}$. Thus $(A^*)^* = (c_{rs})$, where $c_{rs} = \bar{b}_{sr} = \bar{\bar{a}}_{rs} = a_{rs}$. Thus

$$(A^*)^* = A.$$

This proves (1). To get (2), if $A = (a_{rs})$ and $B = (b_{rs})$, then $A^* = (c_{rs})$, where $c_{rs} = \bar{a}_{sr}$, and $B^* = (d_{rs})$, where $d_{rs} = \bar{b}_{sr}$. Since $A + B = (u_{rs})$, where $u_{rs} = a_{rs} + b_{rs}$, we have that

$$(A + B)^* = (v_{rs}),$$

where $v_{rs} = \bar{u}_{sr} = \overline{a_{sr} + b_{sr}} = \bar{a}_{sr} + \bar{b}_{sr}$. On the other hand,

$$A^* + B^* = (c_{rs}) + (d_{rs})$$
$$= (c_{rs} + d_{rs}),$$

and $c_{rs} + d_{rs} = \bar{a}_{sr} + \bar{b}_{sr}$. Therefore, we see, on comparing these evaluations, that indeed

$$(A + B)^* = A^* + B^*.$$

To prove (3) is even easier. With the notation used above, $aA = (aa_{rs})$ and $(aA)^* = (w_{rs})$, where $w_{rs} = \overline{(aa_{sr})} = \bar{a}\bar{a}_{sr}$, whence

$$(aA)^* = (\overline{aa}_{sr}) = \bar{a}A^*.$$

Finally, we come to the hardest of the four. Again with the notation above, $AB = (a_{rs})(b_{rs}) = (t_{rs})$, where $t_{rs} = \Sigma_{k=1}^{n} a_{rk}b_{ks}$. Thus

$$(AB)^* = (f_{rs}),$$

where $f_{rs} = \bar{t}_{rs} = \sum_{k=1}^{n} \bar{a}_{sk}\bar{b}_{kr}$. On the other hand,

$$B^*A^* = (d_{rs})(c_{rs}) = (f_{rs}),$$

where $d_{rs} = \bar{b}_{sr}$, $c_{rs} = \bar{a}_{sr}$, since

$$\sum_{k=1}^{n} d_{rk}c_{ks} = \sum_{k=1}^{n} \bar{b}_{kr}\bar{a}_{sk} = \sum_{k=1}^{n} \bar{a}_{sk}\bar{b}_{kr} = a_{rs}.$$

Thus

$$B^*A^* = (AB)^*,$$

as claimed.   ∎

Let's see (4) in a particular example. If

$$A = \begin{bmatrix} i & 1+i & 3 \\ 1-i & 4 & i \\ 3 & -i & 6+3i \end{bmatrix}, \quad B = \begin{bmatrix} 1 & -i & .5i \\ 0 & 0 & 1+i \\ 5+6i & 0 & 1 \end{bmatrix},$$

then AB is

$$\begin{bmatrix} i + 3(5+6i) & -i^2 & .5i^2 + (1+i)^2 + 3 \\ 1 - i + i(5+6i) & (1-i)(-i) & .5i(1-i) + 4(1+i) + i \\ 3 + (6+3i)(5+6i) - 3i & -3i & 1.5i - i(1+i) + (6+3i) \end{bmatrix}$$

which reduces to

$$\begin{bmatrix} 15+19i & 1 & 2.5+2i \\ -5+4i & -1-i & 4.5+5.5i \\ 15+51i & -3i & 7+3.5i \end{bmatrix}.$$

(Check!). So

$$(AB)^* = \begin{bmatrix} 15-19i & -5-4i & 15-51i \\ 1 & -1+i & 3i \\ 2.5-2i & 4.5-5.5i & 7-3.5i \end{bmatrix}.$$

We now compute B^*A^*. Since

$$A^* = \begin{bmatrix} -i & 1+i & 3 \\ 1-i & 4 & +i \\ 3 & -i & 6-3i \end{bmatrix},$$

$$B^* = \begin{bmatrix} 1 & 0 & 5-6i \\ i & 0 & 0 \\ -.5i & i-i & 1 \end{bmatrix}$$

we get B^*A^* as

$$B^*A^* = \begin{bmatrix} -i + 3(5-6i) & 1(1+i) - i(5-6i) & 3 + (5-6i)(6-3i) \\ (-i)i & i(1+i) & 3i \\ (-.5i)(-i) + (1-i)^2 + 3 & -.5i(1+i) + 4(1-i) - i & -1.5i + (1-i) + (6-3i) \end{bmatrix}$$

which reduces to

$$\begin{bmatrix} 15-19i & -5-4i & 15-51i \\ 1 & -1+i & 3i \\ 2.5-2i & 4.5-5.5i & 7-3.5i \end{bmatrix} = (AB)^*.$$

Thus we see that $(AB)^* = B^*A^*$, as it should be.

As we mentioned earlier, for the transpose we have

$(A')' = A$, $(A + B)' = A' + B'$,
$(aA)' = aA'$,
$(AB)' = B'A'$.

We should like to interrelate the trace and Hermitian adjoint. Let $A = (a_{rs}) \in M_n(\mathbb{C})$. Then $A^* = (b_{rs})$, where $b_{rs} = \bar{a}_{sr}$. Thus $AA^* = (a_{rs})(b_{rs}) = (c_{rs})$, where $c_{rs} = \sum_{k=1}^{n} a_{rk}b_{ks} = \sum_{k=1}^{n} a_{rk}\bar{a}_{sk}$. In particular, the diagonal entries, c_{rr}, are given by $c_{rr} = \sum_{k=1}^{n} a_{rk}\bar{a}_{rk} = \sum_{k=1}^{n} |a_{rk}|^2$. Thus

$$\text{tr } (AA^*) = \sum_{r=1}^{n}\left(\sum_{k=1}^{n} |a_{rk}|^2\right).$$

Suppose that tr $(AA^*) = 0$; then $\sum_{r=1}^{n} (\sum_{k=1}^{n} |a_{rk}|^2) = 0$. Since $|a_{rk}|^2 \geq 0$, the only way this sum can be 0 is if each $a_{rk} = 0$ for every r and k. But this merely tells us that $A = 0$. We have proved

Theorem 3.4.2.   If A is in $M_n(\mathbb{C})$, then tr $(AA^*) = 0$ if and only if $A = 0$. In fact, tr $(AA^*) > 0$ if $A \neq 0$.

This result is false if we use transpose instead of Hermitian adjoint. For example, if $A = \begin{bmatrix} 1 & i \\ -2i & 2 \end{bmatrix}$, then

$$AA' = \begin{bmatrix} 1 & i \\ -2i & 2 \end{bmatrix}\begin{bmatrix} 1 & -2i \\ i & 2 \end{bmatrix}$$

$$= \begin{bmatrix} 1 + i^2 & -2i + 2i \\ -2i + 2i & 4i^2 + 4 \end{bmatrix} = \begin{bmatrix} 0 & 0 \\ 0 & 0 \end{bmatrix},$$

yet $A \neq 0$. Of course, if all the entries of A *are real*, then $A^* = A'$, so in this instance we do have that tr $(AA') = 0$ forces $A = 0$.

To illustrate what is going on in the proof of Theorem 3.4.2, let us look at it for a matrix of small size, namely a 3×3 matrix. Letting

$$A = \begin{bmatrix} a_{11} & a_{12} & a_{13} \\ a_{21} & a_{22} & a_{23} \\ a_{31} & a_{32} & a_{33} \end{bmatrix}$$

we have

$$A^* = \begin{bmatrix} \bar{a}_{11} & \bar{a}_{21} & \bar{a}_{31} \\ \bar{a}_{12} & \bar{a}_{22} & \bar{a}_{32} \\ \bar{a}_{13} & \bar{a}_{23} & \bar{a}_{33} \end{bmatrix}.$$

Since we want to look at tr (AA^*), our only concern will be with the diagonal entries of AA^* These are

$$a_{11}\bar{a}_{11} + a_{12}\bar{a}_{12} + a_{13}\bar{a}_{13} = |a_{11}|^2 + |a_{12}|^2 + |a_{13}|^2,$$

$$a_{21}\bar{a}_{21} + a_{22}\bar{a}_{22} + a_{23}\bar{a}_{23} = |a_{21}|^2 + |a_{22}|^2 + |a_{23}|^2,$$

and finally,

$$a_{31}\bar{a}_{31} + a_{32}\bar{a}_{32} + a_{33}\bar{a}_{33} = |a_{31}|^2 + |a_{32}|^2 + |a_{33}|^2.$$

Therefore,

$$\begin{aligned} \text{tr } (AA^*) = {} & |a_{11}|^2 + |a_{12}|^2 + |a_{13}|^2 \\ & + |a_{21}|^2 + |a_{22}|^2 + |a_{23}|^2 \\ & + |a_{31}|^2 + |a_{32}|^2 + |a_{33}|^2 \end{aligned}$$

and we see that if $A \neq 0$, then some entry is not 0; hence tr (AA^*) is positive. Also, tr $(AA^*) = 0$ forces each a_{jk} to be 0; hence $A = 0$.

Associated with the Hermitian adjoint are two particular classes of matrices.

Definition. $A \in M_n(\mathbb{C})$ is a *Hermitian* matrix if $A = A^*$, and A is a *skew-Hermitian* matrix if $A^* = -A$.

EXAMPLES

1. The matrix $\begin{bmatrix} -1 & 1 + i & 3 \\ 1 - i & 4 & +i \\ 3 & -i & 2 \end{bmatrix}$ is Hermitian.

2. The matrix $\begin{bmatrix} -i & 1 + i & 3 \\ -1 + i & 4i & +i \\ -3 & +i & 0 \end{bmatrix}$ is skew-Hermitian.

3. The matrix $\begin{bmatrix} -1 & 1 & 3 \\ 1 & 4 & 3 \\ 3 & 3 & 2 \end{bmatrix}$ is Hermitian with real entries.

In case $A \in M_n(\mathbb{R})$, A is Hermitian, as in (3) above, we just call it a (real) *symmetric matrix*. Thus a real matrix A is a symmetric matrix if and only if it equals its transpose A'.

Given $A \in M_n(\mathbb{C})$, then $A = \dfrac{A + A^*}{2} + \dfrac{A - A^*}{2}$; however,

$$(A + A^*)^* = A^* + A^{**} = A^* + A$$

and

$$(A - A^*)^* = A^* - A^{**} = A^* - A = -(A - A^*).$$

This says that $A + A^*$ is Hermitian and $A - A^*$ is skew-Hermitian. So

A is the sum of the Hermitian and skew-Hermitian matrices
$\dfrac{A + A^*}{2}$ *and* $\dfrac{A - A^*}{2}$.

Furthermore, if $A = B + C$, where B is Hermitian and C is a skew-Hermitian, then $A^* = (B + C)^* = B^* + C^* = B - C$. Thus $B = \dfrac{A + A^*}{2}$ and $C = \dfrac{A - A^*}{2}$. (Prove!) So

there is only one way in which we can decompose A as a sum of a Hermitian and a skew-Hermitian matrix.

We summarize this in

Lemma 3.4.3.   Given $A \in M_n(\mathbb{C})$, then $A = \dfrac{A + A^*}{2} + \dfrac{A - A^*}{2}$ is a decomposition of A as a sum of a Hermitian and a skew-Hermitian matrix. Moreover, this decomposition is unique in that if $A = B + C$, where B is Hermitian and C is skew-Hermitian, then $B = \dfrac{A + A^*}{2}$ and $C = \dfrac{A - A^*}{2}$.

It is easy to write down a great number of Hermitian matrices. As we saw above, $A + A^*$ is Hermitian for any A. Similarly, since $(AA^*)^* = (A^*)^*A^* = AA^*$, AA^* is Hermitian. A similar computation shows that XAX^* is Hermitian, for any matrix X, if A is Hermitian. Finally, note that if A is skew-Hermitian, then A^2 is Hermitian for $(A^2)^* = (A^*)^2 = (-A)^2 = A^2$.

For skew-Hermitian matrices we also have easy means of producing them. For instance, for any matrix A the matrix $A - A^*$ is skew-Hermitian, and for any matrix X the matrix XAX^* is skew-Hermitian if A is skew-Hermitian.

In general, the product of two Hermitian matrices is *not* Hermitian.

If $A = A^*$ and $B = B^*$ and $(AB)^* = AB$, we get, since $(AB)^* = B^*A^* = BA$, that AB is Hermitian only if A and B commute.

There are many combinations of Hermitian and skew-Hermitian matrices which end up as Hermitian or skew-Hermitian. Some of these will appear in the problems.

PROBLEMS

NUMERICAL PROBLEMS

1. For the given matrices A write down A', A^*, and $A' - A^*$.

(a) $A = \begin{bmatrix} i & 2 & \frac{1+i}{1-i} \\ (1+i)^2 & 3 & 5 \\ \frac{1}{2}i & \frac{1}{3}i & \frac{1}{4}i \end{bmatrix}$.

(b) $A = \begin{bmatrix} 1 & i & -i & 0 \\ -i & \pi & 0 & 1+i \\ i & 0 & \pi^2 & 6 \\ 0 & 1-i & 0 & \pi^3 \end{bmatrix}$.

(c) $A = \begin{bmatrix} 1 & i & -i & 0 \\ i & \pi & 0 & 1+i \\ -i & 0 & \pi^2 & 6 \\ 0 & 1+i & 6 & \pi^3 \end{bmatrix}$.

(d) $A = \begin{bmatrix} 1 & i & 2i \\ 1+i & 1-i & 0 \\ 0 & 1 & 1 \end{bmatrix}^2$.

2. For the given matrices A and B write down A^*, B^* and verify that $(AB)^* = B^*A^*$.

(a) $A = \begin{bmatrix} 1 & 2 & 1 \\ -1 & 6 & i \\ i & i & -1 \end{bmatrix}$, $B = \begin{bmatrix} 1 & 0 & 1+i \\ 0 & 2-i & 0 \\ 3i & 1 & 1 \end{bmatrix}$.

(b) $A = \begin{bmatrix} 1 & i & i^2 & i^3 \\ -i & 0 & 1 & 2 \\ i^2 & 1 & i & 0 \\ -i^3 & 2 & 0 & i \end{bmatrix}$, $B = \begin{bmatrix} 1 & 1 & 1 & 1 \\ 1 & 1 & 1 & 1 \\ 1 & 1 & 1 & 1 \\ 1 & 1 & 1 & 1 \end{bmatrix}$.

(c) $A = \begin{bmatrix} 1 & i & -1 \\ 0 & 2 & i \\ 3-2i & 0 & 1 \end{bmatrix}$, $B = \begin{bmatrix} 1 & 0 & 3-2i \\ i & 2 & 0 \\ -1 & i & 1 \end{bmatrix}$.

(d) $A = \begin{bmatrix} 1 & i & i^2 \\ -i & 2 & 1+i \\ i^2 & 1-i & 3 \end{bmatrix}$, $B = \begin{bmatrix} 1 & 0 & 0 \\ 0 & 1 & 0 \\ 0 & 0 & 1 \end{bmatrix}$.

3. For what values of a and b are the following matrices Hermitian?

(a) $A = \begin{bmatrix} 1 & \frac{1+i}{i} & \frac{1-i}{1+i} \\ a & a & 0 \\ b & 0 & b \end{bmatrix}$.

(b) $A = \begin{bmatrix} 1+i & 0 \\ b & a \end{bmatrix} \begin{bmatrix} 1-i & 0 \\ a & b \end{bmatrix}$.

(c) $A = \begin{bmatrix} a & 0 & b \\ 0 & a & a \\ i & 1 & a \end{bmatrix}$.

(d) $A = \begin{bmatrix} a & 0 & b \\ 0 & a & a \\ i & 1 & a \end{bmatrix}^2$.

4. By a direct calculation show that if $A = \begin{bmatrix} 1 & i & 0 \\ -i & 0 & i \\ 0 & -i & 0 \end{bmatrix}$ and B is Hermitian, then AB is Hermitian only if $AB = BA$.

5. Compute tr (AA^*) for

(a) $A = \begin{bmatrix} 1 & \frac{1+i}{1-i} & 2 \\ -i & 0 & 0 \\ 0 & 1 & 2 \end{bmatrix}$.

(b) $A = \begin{bmatrix} i & 0 & 1 & 0 \\ 0 & i & 2+i & 0 \\ 1 & 2-i & i & \pi \\ 0 & 0 & \pi & i \end{bmatrix}$.

(c) $A = \begin{bmatrix} 1 & 2 & 3 & 4 \\ 2i & 0 & 0 & 0 \\ 3i & 0 & 0 & 0 \\ 4i & 0 & 0 & 0 \end{bmatrix}$.

MORE THEORETICAL PROBLEMS

Easier Problems

6. Show that if $'$ is the transpose, then

(a) $(A')' = A$

(b) $(A + B)' = A' + B'$

(c) $(aA)' = aA'$

(d) $(AB)' = B'A'$

for all $A, B \in M_n(F)$ and all $a \in F$.

7. Show that the following matrices are Hermitian.

(a) $AB + BA$, where $A^* = A$, $B^* = B$.

(b) $AB - BA$, where $A^* = A$, $B^* = -B$.

(c) $AB + BA$, where $A^* = -A$, $B^* = -B$.

(d) A^{2n}, where $A^* = -A$.

8. Show that the following matrices are skew-Hermitian.

(a) $AB - BA$, where $A^* = A$, $B^* = B$.

(b) $AB + BA$, where $A^* = A$, $B^* = -B$.

(c) $AB - BA$, where $A^* = -A$, $B^* = -B$.

(d) $A^2B - BA^2$, where $A^* = -A$, $B^* = B$.

9. If X is any matrix in $M_n(\mathbb{C})$, show that XAX^* is Hermitian if A is Hermitian, and XAX^* is skew-Hermitian if A is skew-Hermitian.

Middle-Level Problems

10. If $A_1, \ldots, A_n \in M_n(\mathbb{C})$, show that
$$(A_1A_2 \cdots A_n)^* = A_n^*A_{n-1}^* \cdots A_1^*.$$

11. If $A_1, \ldots, A_n \in M_n(\mathbb{C})$ and
$$\mathrm{tr}\,(A_1A_1^* + A_2A_2^* + \cdots + A_nA_n^*) = 0,$$
prove that $A_1 = A_2 = \cdots = A_n = 0$.

12. If $A^* = A$ show that if $A^m = 0$, then $A = 0$.   (**Hint:** Use Theorem 3.4.2.)

13. If A is invertible, then $(A^*)^{-1} = (A^{-1})^*$.

3.5. INNER PRODUCT SPACES

In Section 1.12 the topic under discussion was the inner product on the space W of 2-tuples whose components were in $\mathbb{C}$. The inner product of

$$v = \begin{bmatrix} x_1 \\ x_2 \end{bmatrix} \text{ and } w = \begin{bmatrix} y_1 \\ y_2 \end{bmatrix}, \text{ where } v \text{ and } w \text{ are in } W, \text{ was defined as}$$

$(v, w) = x_1 \bar{y}_1 + x_2 \bar{y}_2$. We then saw some properties obtained by W relative to its inner product.

What we shall do for $\mathbb{C}^{(n)}$, the set of n-tuples over $\mathbb{C}$, will be the exact analogue of what we said above.

Definition. If $v = \begin{bmatrix} x_1 \\ \vdots \\ x_n \end{bmatrix}$ and $w = \begin{bmatrix} y_1 \\ \vdots \\ y_n \end{bmatrix}$ are in $\mathbb{C}^{(n)}$, then their *inner*

product, denoted by (v, w), is defined by $(v, w) = \sum_{j=1}^{n} x_j \bar{y}_j$.

For instance, the inner product of $\begin{bmatrix} 41 \\ e \\ 12 \end{bmatrix}$ and $\begin{bmatrix} 1 \\ f \\ 2 \end{bmatrix}$ is

$41 + e\bar{f} + 24$.

Everything we did before, in Chapter 1, carries over immediately to our present context. So we can be a little briefer in the discussion and the proofs. If you are perplexed by anything that comes up, go back to Section 1.12 to see how it was done there. We state, without proof, the first of these results, leaving the proof to the reader.

Lemma 3.5.1. If u, v, w are in $\mathbb{C}^{(n)}$ and $a \in \mathbb{C}$, then

1. $(v + w, u) = (v, u) + (w, u)$;
2. $(u, v + w) = (u, v) + (u, w)$;
3. $(v, w) = \overline{(w, v)}$;
4. $(av, w) = a(v, w) = (v, \bar{a}w)$;
5. $(v, v) \geq 0$ is real and $(v, v) = 0$
          if and only if $v = 0$.

Note that property 4 says we can pull a scalar a out of the symbol $(\cdot, \cdot)$ if it occurs as a multiplier of the first entry of $(\cdot, \cdot)$, but we can only pull it out as $\bar{a}$ if it occurs as a multiplier of the second one. That is, if $a \in \mathbb{C}$, then $(v, aw) = \bar{a}(v, w)$.

Definition. If $v, w \in \mathbb{C}^{(n)}$, then v is said to be *orthogonal* to w if $(v, w) = 0$.

Notice that if v is orthogonal to w, then w is orthogonal to v, since $(w, v) = \overline{(v, w)} = \bar{0} = 0$. Also, if $(v, w) = 0$ for *all* $w \in \mathbb{C}^{(n)}$, then, in particular, $(v, v) = 0$, and so, by Part (e) of Lemma 3.5.1, $v = 0$.

Definition. $v^{\perp} = \{w \in \mathbb{C}^{(n)} \mid (v, w) = 0\}$.

    EXAMPLE

$$\text{If } v = \begin{bmatrix} 2 \\ i \\ 3 \end{bmatrix}, \text{ then } v^{\perp} = \left\{ \begin{bmatrix} x \\ y \\ z \end{bmatrix} \middle| 2\bar{x} + i\bar{y} + 3\bar{z} = 0 \right\}.$$

If w_1, w_2 are in $v^{\perp}$, then $(v, w_1 + w_2) = (v, w_1) + (v, w_2) = 0 + 0 = 0$, hence $w_1 + w_2$ is again in $v^{\perp}$. Also if $a \in \mathbb{C}$ and w is in $v^{\perp}$, then $(v, aw) = \bar{a}(v, w) = \bar{a}0 = 0$, so aw is in $v^{\perp}$.

We summarize what was just done in

Lemma 3.5.2

1.   If v is orthogonal to w then w is orthogonal to v.
2.   If v is orthogonal to all elements of $\mathbb{C}^{(n)}$ then $v = 0$.
3.   If $v^{\perp} = \{w \in \mathbb{C}^{(n)} \mid (v, w) = 0\}$, then:
   (a) w_1, w_2 in $v^{\perp}$ implies that $w_1 + w_2$ is in $v^{\perp}$.
   (b) a in $\mathbb{C}$, w in $v^{\perp}$ implies that aw is in $v^{\perp}$.

In Chapter 1 we interrelated the inner product and the Hermitian adjoint. We should like to do so here as well. All the proofs given will be the exact duplicates of those in Section 1.12. So we shall give relatively few proofs and just wave our hands with a "look back in Chapter 1 for the method of proof." However, since the next result involves the summation symbol and many readers do not feel comfortable with it, we do the proof in detail.

Theorem 3.5.3. If $A \in M_n(\mathbb{C})$ and $v, w \in \mathbb{C}^{(n)}$, then $(Av, w) = (v, A^*w)$, where A^* is the Hermitian adjoint of A.

Proof. Let $v = \begin{bmatrix} x_1 \\ \vdots \\ x_n \end{bmatrix}$ and $w = \begin{bmatrix} y_1 \\ \vdots \\ y_n \end{bmatrix}$ be in $\mathbb{C}^{(n)}$ and $A = (a_{rs})$ in

$M_n(\mathbb{C})$. Thus $Av = (a_{rs})\begin{bmatrix} x_1 \\ \vdots \\ x_n \end{bmatrix} = \begin{bmatrix} z_1 \\ \vdots \\ z_n \end{bmatrix}$, where for each r,

$z_r = \Sigma_{s=1}^n a_{rs} x_s$.

Therefore,

$$(Av, w) = \sum_{r=1}^n z_r \bar{y}_r = \sum_{r=1}^n \sum_{s=1}^n (a_{rs} x_s) \bar{y}_r = \sum_{r=1}^n \sum_{s=1}^n a_{rs} x_s \bar{y}_r.$$

On the other hand, $A^* = (b_{rs})$, where $b_{rs} = \bar{a}_{sr}$. So $A^* w =$

$(b_{rs})\begin{bmatrix} y_1 \\ \vdots \\ y_n \end{bmatrix} = \begin{bmatrix} t_1 \\ \vdots \\ t_n \end{bmatrix}$, where for each r, $t_r = \Sigma_{s=1}^n b_{rs} y_s = \Sigma_{s=1}^n \bar{a}_{sr} y_s$.

Therefore,

$$(v, A^*w) = \sum_{r=1}^n x_r \bar{t}_r = \sum_{r=1}^n \Sigma_{s=1}^n x_r \overline{(\bar{a}_{sr} y_s)} = \sum_{r=1}^n \sum_{s=1}^n x_r a_{sr} \bar{y}_s.$$

If we call r by the letter s and s by the letter r (don't forget, these are just dummy indices) we have

$$(v, A^*w) = \sum_{r=1}^n \sum_{s=1}^n x_r a_{sr} \bar{y}_s = \sum_{r=1}^n \sum_{s=1}^n a_{rs} x_s \bar{y}_r = (Av, w),$$

from above. This finishes the proof.   ■

From here on in—in this section—every result and every proof parrots virtually verbatim those in Section 1.12. We leave the proofs as exercises. Again, if you get stuck trying to do the exercises, look back at the relevant part in Chapter 1. But before doing this, see if you can come up with proofs of your own.

Theorem 3.5.4. If $A, B \in M_n(\mathbb{C})$ and $a \in \mathbb{C}$, then

1. $(A^*)^* = A^{**} = A$;
2. $(A + B)^* = A^* + B^*$;

3. $(aA)^* = \bar{a}A^*$;

4. $(AB) = B^*A^*$.

Although we did this Section 3.4 by direct computations, see if you can carry out the proofs using inner products and the important fact that $(Av, w) = (v, A^*w)$. [You will also need that $v = 0$ if $(v, w) = 0$ for all w.]

We go down the list of theorems.

Theorem 3.5.5. An element $A \in M_n(\mathbb{C})$ is such that $(Av, Aw) = (v, w)$ for all v, w in $\mathbb{C}^{(n)}$ if and only if $A^*A = I$.

Definition. A matrix A in $M_n(\mathbb{C})$ is called *unitary* if $A^*A = I$.

For instance, the matrix $A = (1/\sqrt{2})\begin{bmatrix} 1 & i \\ i & 1 \end{bmatrix}$ is unitary.

So a unitary matrix is one that does not disturb the inner product of two vectors in $\mathbb{C}^{(n)}$. Equivalently, a unitary matrix is an invertible matrix whose inverse is A^*. In the case of the unitary matrix $A = (1/\sqrt{2})\begin{bmatrix} 1 & i \\ i & 1 \end{bmatrix}$, its inverse is $A^* = (1/\sqrt{2})\begin{bmatrix} 1 & -i \\ -i & 1 \end{bmatrix}$.

Definition. If $v \in \mathbb{C}^{(n)}$, then the *length* of v, denoted by $\|v\|$, is defined by $\|v\| = \sqrt{(v, v)}$.

For example, the vector $\begin{bmatrix} 1 \\ i \end{bmatrix}$ has length 2 and both columns of the matrix $(1/\sqrt{2})\begin{bmatrix} 1 & -i \\ -i & 1 \end{bmatrix}$ have length 1.

Theorem 3.5.6. If $A \in M_n(\mathbb{C})$ is such that $(Aw, Aw) = (w, w)$ for all $w \in \mathbb{C}^{(n)}$, then A is unitary [hence $(Av, Aw) = (v, w)$ for all v and w in $\mathbb{C}^{(n)}$].

If $(Aw, Aw) = (w, w)$, then $\|Aw\| = \|w\|$. So unitary matrices preserve length and any matrix that preserves length must be unitary.

At the end of Section 3.4, the following occurred: If $A = A^*$ and for some $B \neq 0$, $AB = aB$, where $a \in \mathbb{C}$, then a is real. What this really says is that the characteristic roots of A are real. Do you recall from Chapter 1 what a characteristic root of A is? We define it anew.

Definition.   If $A \in M_n(\mathbb{C})$, then $a \in \mathbb{C}$ is a *characteristic root* of A if for some $v \neq 0$ in $\mathbb{C}^{(n)}$, $Av = av$. Such a vector v is then called a *characteristic vector* associated with a.

The same proof as that of Theorem 1.12.7 gives us the very important

Theorem 3.5.7.   If A is Hermitian, that is, if $A^* = A$, then the characteristic roots of A are real.

It is hard to exaggerate the importance of this theorem, not only in mathematics, but also in physics, chemistry, and statistics.
Finally, the exact analogue of Theorem 1.12.8 is

Theorem 3.5.8.   If $A \in M_n(\mathbb{C})$ is unitary then $|a| = 1$ for any characteristic root of A.

PROBLEMS

NUMERICAL PROBLEMS

1. Calculate the inner product (v, w) for the given v and w.

(a) $v = \begin{bmatrix} 1 \\ -3 \\ 2 \end{bmatrix}, \quad w = \begin{bmatrix} 0 \\ 1 \\ 5 \end{bmatrix}.$

(b) $v = \begin{bmatrix} i \\ 1 + i \\ 3i \\ 0 \end{bmatrix}, \quad w = \begin{bmatrix} 1 - i \\ 1 - i \\ 2 \\ 7 \end{bmatrix}.$

(c) $v = \begin{bmatrix} 1 \\ 0 \\ \vdots \\ 0 \end{bmatrix}, \quad w = \begin{bmatrix} 0 \\ 1 \\ \vdots \\ 0 \end{bmatrix}.$

(d) $v = \begin{bmatrix} i/2 \\ \pi^2 \\ 0 \\ -1/3 \end{bmatrix}, \quad w = \begin{bmatrix} \pi \\ \pi^2 \\ \pi^3 \\ i \end{bmatrix}.$

2. For the v and w in Problem 1, calculate (w, v) and show in each case that $(w, v) = \overline{(v, w)}$.

3. Find the length, $\sqrt{(v, v)}$, for the given v.

(a) $v = \begin{bmatrix} i \\ 2 \\ -\frac{1}{3} + i \\ 4i \end{bmatrix}$.

(b) $v = \begin{bmatrix} 1 \\ 2 \\ 3 \\ 4 \\ 5 \end{bmatrix}$.

(c) $v = \begin{bmatrix} i \\ 2i \\ 3i \\ 4i \\ 5i \end{bmatrix}$.

(d) $v = \begin{bmatrix} a \\ 0 \\ \vdots \\ 0 \end{bmatrix}$.

4. For the given matrices A and vectors v and w, calculate (Av, w), (v, A^*w) and verify that $(Av, w) = (v, A^*w)$.

(a) $A = \begin{bmatrix} 1 & i & 0 \\ 3-i & 2i & 7 \\ 6 & 5 & 0 \end{bmatrix}$, $v = \begin{bmatrix} 1 \\ 2 \\ 3 \end{bmatrix}$, $w = \begin{bmatrix} i \\ i^3 \\ 2 \end{bmatrix}$.

(b) $A = \begin{bmatrix} 1 & 2 & 3 & 4 \\ 2i & 0 & 1 & 2 \\ 3i & 0 & 0 & 1 \\ 4i & 1 & 2 & 0 \end{bmatrix}$, $v = \begin{bmatrix} i \\ -17i \\ 6 \\ 1 \end{bmatrix}$, $w = \begin{bmatrix} 7+i \\ 4 \\ 0 \\ 3i \end{bmatrix}$.

(c) $A = \begin{bmatrix} 0 & i & 0 & 0 \\ 0 & 0 & -i & 0 \\ 0 & 0 & 0 & 1 \\ 3i & 0 & 0 & 0 \end{bmatrix}$, $v = \begin{bmatrix} 1 \\ 2 \\ 3 \\ 4 \end{bmatrix}$, $w = \begin{bmatrix} 4i \\ -3i \\ 2i \\ i \end{bmatrix}$.

5. For the given v, find the form of all elements w in $v^{\perp}$.

(a) $v = \begin{bmatrix} 1 \\ -1 \\ 2 \\ 3 \end{bmatrix}$.

(b) $v = \begin{bmatrix} 1 \\ -i \\ 2 \\ 3i \end{bmatrix}$.

(c) $v = \begin{bmatrix} 1 \\ 0 \\ \vdots \\ 0 \end{bmatrix}$.

(d) $v = \begin{bmatrix} 0 \\ i \\ 0 \\ \vdots \\ 0 \end{bmatrix}$.

6. For the vectors $v = \begin{bmatrix} 1 \\ i \\ 2 \\ 0 \end{bmatrix}$, $w = \begin{bmatrix} 7 - i \\ 4 \\ .3i \\ 2 - i \end{bmatrix}$, recalling that the length of a vector u, $\|u\|$, is $\sqrt{(u, u)}$, show that $\|v + w\| \le \|v\| + \|w\|$.

7. Find all matrices A such that for $v = \begin{bmatrix} 1 \\ 2 \\ 3 \end{bmatrix}$ and $w = \begin{bmatrix} 3i \\ 2i \\ i \end{bmatrix}$, $(Av, Aw) = (v, w)$.

8. If $v, w \in \mathbb{C}^{(3)}$, compute (Av, w) and (v, A^*w), where $A = \begin{bmatrix} 0 & i & 0 \\ 2i & 0 & 0 \\ 0 & 3 & 3i \end{bmatrix}$, and show that $(Av, w) = (v, A^*w)$.

9. Prove that the following matrices are unitary by calculating A^*A.

(a) $A = \begin{bmatrix} 0 & 1 & 0 & 0 \\ 1 & 0 & 0 & 0 \\ 0 & 0 & 0 & 1 \\ 0 & 0 & 1 & 0 \end{bmatrix}$.

(b) $A = \begin{bmatrix} 1 - i & 0 \\ 0 & 2 + i \end{bmatrix} \begin{bmatrix} 1 + i & 0 \\ 0 & 2 - i \end{bmatrix}^{-1}$.

(c) $A = \begin{bmatrix} 0 & 0 & 0 & i \\ 0 & 0 & i & 0 \\ 0 & i & 0 & 0 \\ i & 0 & 0 & 0 \end{bmatrix}$.

MORE THEORETICAL PROBLEMS

Middle-Level Problems

10. Write out a complete proof of Theorem 3.5.4, following in the footsteps of what we did in Chapter 1.

11. Repeat Problem 10 for Theorem 3.5.5.

12. Repeat Problem 10 for Theorems 3.5.6, 3.5.7, and 3.5.8.

13. If $A \in M_n(\mathbb{R})$ is such that $(Av, Av) = (v, v)$ for all $v \in \mathbb{R}^{(n)}$, is A^*A necessarily equal to I? Either prove or produce a counterexample.

14. If A is both Hermitian and unitary, what possible values can the characteristic roots of A have?

15. If A is Hermitian and B is unitary, prove that BAB^{-1} is Hermitian.

16. If A is skew-Hermitian, show that the characteristic roots of A are pure imaginary.

17. If $A \in M_n(\mathbb{C})$ and $a \in \mathbb{C}$ is a characteristic root of A, prove that $aI - A$ is *not* invertible.

18. If $A \in M_n(\mathbb{C})$ and $a \in \mathbb{C}$ is a characteristic root of A, prove that for some matrix $B \neq 0$ in $M_n(\mathbb{C})$, $AB = aB$.

Harder Problems

19. If A is a skew-Hermitian, show that both $(I - A)$ and $(I + A)$ are invertible and that $C = (I - A)(I + A)^{-1}$ is unitary.

20. If A and B are unitary, prove that AB is unitary and A^{-1} is also unitary.

21. Show that the matrix $A = \begin{bmatrix} 0 & 0 & 1 & 0 \\ 1 & 0 & 0 & 0 \\ 0 & 1 & 0 & 0 \\ 0 & 0 & 0 & 1 \end{bmatrix}$ is unitary. Can you find

the characteristic roots of A?

22. Let the entries of $A = \begin{bmatrix} a & b & c \\ b & d & e \\ c & e & f \end{bmatrix}$ be real. Suppose that $g \in \mathbb{C}$

and there exists $v = \begin{bmatrix} x \\ y \\ z \end{bmatrix} \in M_3(\mathbb{C})$ such that $v \neq 0$ and $Av = gv$.

Show by direct computations that g is real and we can pick v so that its components are real.

3.6. BASES OF $F^{(n)}$

In studying $F^{(n)}$, where F is the reals, $\mathbb{R}$, or the complexes, $\mathbb{C}$, the set of elements

$$e_1 = \begin{bmatrix} 1 \\ 0 \\ \vdots \\ 0 \end{bmatrix}, e_2 = \begin{bmatrix} 0 \\ 1 \\ \vdots \\ 0 \end{bmatrix}, \ldots, e_n = \begin{bmatrix} 0 \\ 0 \\ \vdots \\ 1 \end{bmatrix},$$

where e_j is that element whose jth component is 1 and all of whose other components are 0, played a special role. Their most important property was that, given any $v \in F^{(n)}$, then

$$v = a_1 e_1 + a_2 e_2 + \cdots + a_n e_n,$$

where $a_1, a_2, \ldots, a_n$ are in F. Moreover, these a_r are unique in the sense that if

$$v = a_1 e_1 + a_2 e_2 + \cdots + a_n e_n = b_1 e_1 + b_2 e_2 + \cdots + b_n e_n,$$

where the b_j are in F, then

$$a_1 = b_1, a_2 = b_2, \ldots, a_n = b_n.$$

These properties are of great importance and we will abstract from them the notions of *linear independence*, *span*, and *basis*.

If $v_1, \ldots, v_r$ are elements of $F^{(n)}$ and $a_1, \ldots, a_r$ are elements of F, a vector $a_1 v_1 + \cdots + a_r v_r$ is called a *linear combination* of $v_1, \ldots, v_r$ *over* F. For example, any vector $v = \begin{bmatrix} a \\ b \\ c \end{bmatrix}$ of $F^{(3)}$ is a

linear combination $ae_1 + be_2 + ce_3$ of e_1, e_2, e_3 over F, and $F^{(3)}$ is the set of all such linear combinations. The only linear combination $ae_1 + be_2 + ce_3$ of e_1, e_2, e_3 which is 0 is the *trivial* linear combination $0e_1 + 0e_2 + 0e_3$, so that e_1, e_2, e_3 are linearly independent in the sense of

Definition.   The elements $v_1, v_2, \ldots, v_r$ in $F^{(n)}$ are said to be *linearly independent over* F if $a_1 v_1 + a_2 v_2 + \cdots + a_r v_r = 0$ only if each of $a_1, a_2, \ldots, a_r$ is 0.

If $v_1, v_2, \ldots, v_r \in F^{(n)}$ are linearly independent over F, suppose that

$$v = a_1 v_1 + \cdots + a_r v_r = b_1 v_1 + \cdots + b_r v_r,$$

where the a_j's and b_j's are in F. Then

$$(a_1 - b_1)v_1 + \cdots + (a_r - b_r)v_r = 0,$$

which, since the elements $v_1, \ldots, v_r$ are linearly independent over F forces $a_1 - b_1 = 0, \ldots, a_r - b_r = 0$, that is, $a_1 = b_1, a_2 = b_2, \ldots, a_r = b_r$. So the a_j are unique. This could serve as an alternative definition of linear independence.

EXAMPLE

If $F = \mathbb{R}$ and $z_1 = \begin{bmatrix} 1 \\ 2 \\ 3 \end{bmatrix}$, $z_2 = \begin{bmatrix} -1 \\ 3 \\ -3 \end{bmatrix}$, $z_3 = \begin{bmatrix} 0 \\ 1 \\ 1 \end{bmatrix}$ in $F^{(3)}$, are

they linearly independent over F? We must see if we can find a_1, a_2, a_3, not all 0, such that $a_1 z_1 + a_2 z_2 + a_3 z_3 = 0$. Since

$$a_1 z_1 + a_2 z_2 + a_3 z_3 = a_1 \begin{bmatrix} 1 \\ 2 \\ 3 \end{bmatrix} + a_2 \begin{bmatrix} -1 \\ 3 \\ -3 \end{bmatrix} + a_3 \begin{bmatrix} 0 \\ 1 \\ 1 \end{bmatrix}$$

$$= \begin{bmatrix} a_1 - a_2 \\ 2a_1 + 3a_2 + a_3 \\ 3a_1 - 3a_2 + a_3 \end{bmatrix},$$

for this to be 0 we must satisfy:

$$\begin{aligned} a_1 - a_2 &= 0 \\ 2a_1 + 3a_2 + a_3 &= 0 \\ 3a_1 - 3a_2 + a_3 &= 0. \end{aligned}$$

You can check that the only solution to these three equations is $a_1 = 0$, $a_2 = 0$, and $a_3 = 0$. Thus z_1, z_2, z_3 are linearly independent over $F = \mathbb{R}$.

Can we always determine easily whether vectors $v_1, \ldots, v_r$ are linearly independent? Write

$$v_1 = \begin{bmatrix} a_{11} \\ \vdots \\ a_{n1} \end{bmatrix}, \quad v_2 = \begin{bmatrix} a_{12} \\ \vdots \\ a_{n2} \end{bmatrix}, \quad \ldots, \quad v_r = \begin{bmatrix} a_{1r} \\ \vdots \\ a_{nr} \end{bmatrix},$$

so that the linear combination $x_1 v_1 + x_2 v_2 + \cdots + x_r v_r$ becomes

$$x_1 \begin{bmatrix} a_{11} \\ \vdots \\ a_{n1} \end{bmatrix} + x_2 \begin{bmatrix} a_{12} \\ \vdots \\ a_{n2} \end{bmatrix} + \cdots + x_r \begin{bmatrix} a_{1r} \\ \vdots \\ a_{nr} \end{bmatrix}$$

which in turn is

$$\begin{bmatrix} x_1 a_{11} + x_2 a_{12} + \cdots + x_r a_{1r} \\ \vdots \\ x_1 a_{n1} + x_2 a_{n2} + \cdots x_r a_{nr} \end{bmatrix} = \begin{bmatrix} a_{11} & \cdots & a_{1r} \\ \vdots & & \vdots \\ a_{n1} & \cdots & a_{nr} \end{bmatrix} \begin{bmatrix} x_1 \\ \vdots \\ x_r \end{bmatrix}$$

Then the v_j are the columns of the matrix $\begin{bmatrix} a_{11} & \cdots & a_{1r} \\ \vdots & & \vdots \\ a_{n1} & \cdots & a_{nr} \end{bmatrix}$ and they

are linearly independent if and only if $\begin{bmatrix} 0 \\ \vdots \\ 0 \end{bmatrix}$ is the *only* solution to the

equation $\begin{bmatrix} a_{11} & \cdots & a_{1r} \\ \vdots & & \vdots \\ a_{n1} & \cdots & a_{nr} \end{bmatrix}\begin{bmatrix} x_1 \\ \vdots \\ x_r \end{bmatrix} = \begin{bmatrix} 0 \\ \vdots \\ 0 \end{bmatrix}$. This, in turn, is equivalent to

the condition that an echelon matrix row equivalent $\begin{bmatrix} a_{11} & \cdots & a_{1r} \\ \vdots & & \vdots \\ a_{n1} & \cdots & a_{nr} \end{bmatrix}$

has r nonzero rows, by Corollary 2.4.2. So we have

Method to determine if $v_1, \ldots, v_r$ in $F^{(n)}$ are linearly independent

1. *Form the matrix $A = [v_1, \ldots, v_r]$ whose columns are the v_j and row reduce A to an echelon matrix B.*
2. *Then the vectors $v_1, \ldots, v_r$ are linearly independent if and only if B has r nonzero rows.*

EXAMPLE

Let's now use *this* method to determine whether the vectors

$$\begin{bmatrix} 1 \\ 2 \\ 3 \end{bmatrix}, \quad \begin{bmatrix} -1 \\ 3 \\ -3 \end{bmatrix}, \quad \begin{bmatrix} 0 \\ 1 \\ 1 \end{bmatrix}$$

considered in the preceding example are linearly independent. We

apply elementary row operations to the matrix $\begin{bmatrix} 1 & -1 & 0 \\ 2 & 3 & 1 \\ 3 & -3 & 1 \end{bmatrix}$, getting

$$\begin{bmatrix} 1 & -1 & 0 \\ 0 & 5 & 1 \\ 3 & -3 & 1 \end{bmatrix}, \quad \begin{bmatrix} 1 & -1 & 0 \\ 0 & 5 & 1 \\ 0 & 0 & 1 \end{bmatrix}, \quad \begin{bmatrix} 1 & -1 & 0 \\ 0 & 1 & \frac{1}{5} \\ 0 & 0 & 1 \end{bmatrix}.$$

Since the last matrix we get is an echelon matrix with three nonzero rows, the three given vectors are linearly independent.

If the elements $v_1, v_2, \ldots, v_r$ are not linearly independent over F they are called *linearly dependent over F*.

EXAMPLE

Let's also use our method to determine whether the vectors $\begin{bmatrix} 1 \\ 1 \\ 1 \end{bmatrix}$, $\begin{bmatrix} -i \\ 1-i \\ 0 \end{bmatrix}$, $\begin{bmatrix} 0 \\ i \\ -1 \end{bmatrix}$ are linearly independent over $\mathbb{C}$. We apply elementary row operations to the matrix $\begin{bmatrix} 1 & -i & 0 \\ 1 & 1-i & i \\ 1 & 0 & -1 \end{bmatrix}$,

getting $\begin{bmatrix} 1 & -i & 0 \\ 0 & 1 & i \\ 1 & 0 & -1 \end{bmatrix}$, $\begin{bmatrix} 1 & -i & 0 \\ 0 & 1 & i \\ 0 & i & -1 \end{bmatrix}$, and $\begin{bmatrix} 1 & -i & 0 \\ 0 & 1 & i \\ 0 & 0 & 0 \end{bmatrix}$. Since

the last matrix we get is an echelon matrix with only two nonzero rows, the three given vectors are linearly dependent over $\mathbb{C}$. If we so wish, we can easily get a nonzero linear combination by solving

the equation $\begin{bmatrix} 1 & -i & 0 \\ 0 & 1 & i \\ 0 & 0 & 0 \end{bmatrix} \begin{bmatrix} a \\ b \\ c \end{bmatrix} = 0$, getting a nonzero solution

$\begin{bmatrix} a \\ b \\ c \end{bmatrix} = \begin{bmatrix} 1 \\ -i \\ 1 \end{bmatrix}$. So we can use the values $1, -i, 1$ to express 0 as

a nontrivial linear combination of $\begin{bmatrix} 1 \\ 1 \\ 1 \end{bmatrix}$, $\begin{bmatrix} -i \\ 1-i \\ 0 \end{bmatrix}$, $\begin{bmatrix} 0 \\ i \\ -1 \end{bmatrix}$:

$$0 = 1\begin{bmatrix} 1 \\ 1 \\ 1 \end{bmatrix} - i\begin{bmatrix} -i \\ 1-i \\ 0 \end{bmatrix} + 1\begin{bmatrix} 0 \\ i \\ -1 \end{bmatrix}.$$

Since any vector $v = \begin{bmatrix} a \\ b \\ c \end{bmatrix}$ of $F^{(3)}$ is a linear combination

$ae_1 + be_2 + ce_3$ of e_1, e_2, e_3

over F and $F^{(3)}$ is the set of all such linear combinations,

$$\langle e_1, e_2, e_3 \rangle = F^{(3)}$$

and e_1, e_2, e_3 span $F^{(3)}$ in the sense of the

Definition.   The set of all linear combinations of $v_1, \ldots, v_r$ over F is denoted $\langle v_1, \ldots, v_r \rangle$ and is called the *span* of $v_1, \ldots, v_r$. If V is the set $\langle v_1, \ldots, v_r \rangle$, we say that *the vectors $v_1, \ldots, v_r$ span V.*

How do we know whether a vector v is in the span of given vectors

$$v_1 = \begin{bmatrix} a_{11} \\ \vdots \\ a_{n1} \end{bmatrix}, v_2 = \begin{bmatrix} a_{12} \\ \vdots \\ a_{n2} \end{bmatrix}, \ldots, v_r = \begin{bmatrix} a_{1r} \\ \vdots \\ a_{nr} \end{bmatrix}?$$

The linear combination $x_1 v_1 + x_2 v_2 + \cdots + x_r v_r$ is

$$\begin{bmatrix} a_{11} & \cdots & a_{1r} \\ \vdots & & \vdots \\ a_{n1} & \cdots & a_{nr} \end{bmatrix}\begin{bmatrix} x_1 \\ \vdots \\ x_r \end{bmatrix},$$

as we've just seen. Thus, expressing v as a linear combination of

$$\begin{bmatrix} a_{11} \\ \vdots \\ a_{n1} \end{bmatrix}, \ldots, \begin{bmatrix} a_{1r} \\ \vdots \\ a_{nr} \end{bmatrix}$$

is the same as finding a solution $\begin{bmatrix} x_1 \\ \vdots \\ x_r \end{bmatrix}$ to the equation

$$\begin{bmatrix} a_{11} & \cdots & a_{1r} \\ \vdots & & \vdots \\ a_{n1} & \cdots & a_{nr} \end{bmatrix}\begin{bmatrix} x_1 \\ \vdots \\ x_r \end{bmatrix} = v.$$

So, by Section 2.4, we have a straightforward way of settling whether or not v is a linear combination of $v_1, \ldots, v_r$.

Method to express $v \in F^{(n)}$ as a linear combination of $v_1, \ldots, v_r \in F^{(n)}$:

1. *Form the matrix $A = [v_1, \ldots, v_r]$ whose columns are the v_j.*
2. *Form the augmented matrix $[A, v]$ and row reduce it to an echelon matrix $[B, w]$.*
3. Then v is expressed as a linear combination $v = Ax$ of the v_j if and only if x is a solution to $Bx = w$.

EXAMPLE

Let's use this method to try to express the vectors $\begin{bmatrix} 1 \\ 2 \\ 6 \end{bmatrix}$ and

$\begin{bmatrix} 1 \\ 2 \\ 3 \end{bmatrix}$ as linear combinations of the vectors

$$\begin{bmatrix} 1 \\ 1 \\ 2 \end{bmatrix}, \begin{bmatrix} 2 \\ 3 \\ 5 \end{bmatrix}, \begin{bmatrix} 3 \\ 5 \\ 8 \end{bmatrix}, \begin{bmatrix} 4 \\ 5 \\ 9 \end{bmatrix}.$$

We first form the matrix $A = \begin{bmatrix} 1 & 2 & 3 & 4 \\ 1 & 3 & 5 & 5 \\ 2 & 5 & 8 & 9 \end{bmatrix}$, which we will use

for both vectors.

To try to express $\begin{bmatrix} 1 \\ 2 \\ 6 \end{bmatrix}$ as a linear combination of

$$\begin{bmatrix} 1 \\ 1 \\ 2 \end{bmatrix}, \begin{bmatrix} 2 \\ 3 \\ 5 \end{bmatrix}, \begin{bmatrix} 3 \\ 5 \\ 8 \end{bmatrix}, \begin{bmatrix} 4 \\ 5 \\ 9 \end{bmatrix},$$

we form the augmented matrix

$$\begin{bmatrix} A, & \begin{matrix} 1 \\ 2 \\ 6 \end{matrix} \end{bmatrix} = \begin{bmatrix} 1 & 2 & 3 & 4 & 1 \\ 1 & 3 & 5 & 5 & 2 \\ 2 & 5 & 8 & 9 & 6 \end{bmatrix}$$

and row reduce it to the echelon matrix

$$\begin{bmatrix} 1 & 2 & 3 & 4 & 1 \\ 0 & 1 & 2 & 1 & 1 \\ 0 & 0 & 0 & 0 & 1 \end{bmatrix} = \begin{bmatrix} B, & \begin{matrix} 1 \\ 1 \\ 1 \end{matrix} \end{bmatrix}.$$

Since $\begin{bmatrix} 1 & 2 & 3 & 4 \\ 0 & 1 & 2 & 1 \\ 0 & 0 & 0 & 0 \end{bmatrix} \begin{bmatrix} x_1 \\ x_2 \\ x_3 \end{bmatrix} = \begin{bmatrix} 1 \\ 1 \\ 1 \end{bmatrix}$ has no solution x, no linear

combination $\begin{bmatrix} 1 & 2 & 3 & 4 \\ 1 & 3 & 5 & 5 \\ 2 & 5 & 8 & 9 \end{bmatrix} \begin{bmatrix} x_1 \\ x_2 \\ x_3 \end{bmatrix}$ of

$$\begin{bmatrix} 1 \\ 1 \\ 2 \end{bmatrix}, \quad \begin{bmatrix} 2 \\ 3 \\ 5 \end{bmatrix}, \quad \begin{bmatrix} 3 \\ 5 \\ 8 \end{bmatrix}, \quad \begin{bmatrix} 4 \\ 5 \\ 9 \end{bmatrix}$$

equals $\begin{bmatrix} 1 \\ 2 \\ 6 \end{bmatrix}$.

To try to express $\begin{bmatrix} 1 \\ 2 \\ 3 \end{bmatrix}$ as a linear combination of

$$\begin{bmatrix} 1 \\ 1 \\ 2 \end{bmatrix}, \quad \begin{bmatrix} 2 \\ 3 \\ 5 \end{bmatrix}, \quad \begin{bmatrix} 3 \\ 5 \\ 8 \end{bmatrix}, \quad \begin{bmatrix} 4 \\ 5 \\ 9 \end{bmatrix},$$

we form the augmented matrix

$$\begin{bmatrix} A, & \begin{matrix} 1 \\ 2 \\ 3 \end{matrix} \end{bmatrix} = \begin{bmatrix} 1 & 2 & 3 & 4 & 1 \\ 1 & 3 & 5 & 5 & 2 \\ 2 & 5 & 8 & 9 & 3 \end{bmatrix}$$

and row reduce it to the echelon matrix

$$\begin{bmatrix} 1 & 2 & 3 & 4 & 1 \\ 0 & 1 & 2 & 1 & 1 \\ 0 & 0 & 0 & 0 & 0 \end{bmatrix} = \begin{bmatrix} B, & \begin{matrix} 1 \\ 1 \\ 0 \end{matrix} \end{bmatrix}.$$

Since $\begin{bmatrix} 1 & 2 & 3 & 4 \\ 0 & 1 & 2 & 1 \\ 0 & 0 & 0 & 0 \end{bmatrix} x = \begin{bmatrix} 1 \\ 1 \\ 0 \end{bmatrix}$ has the solution $x = \begin{bmatrix} -1 \\ 1 \\ 0 \\ 0 \end{bmatrix}$, the

linear combination

$$\begin{bmatrix} 1 & 2 & 3 & 4 \\ 1 & 3 & 5 & 5 \\ 2 & 5 & 8 & 9 \end{bmatrix} \begin{bmatrix} -1 \\ 1 \\ 0 \\ 0 \end{bmatrix} = -1 \begin{bmatrix} 1 \\ 1 \\ 2 \end{bmatrix} + 1 \begin{bmatrix} 2 \\ 3 \\ 5 \end{bmatrix} + 0 \begin{bmatrix} 3 \\ 5 \\ 8 \end{bmatrix} + 0 \begin{bmatrix} 4 \\ 5 \\ 9 \end{bmatrix}$$

equals $\begin{bmatrix} 1 \\ 2 \\ 3 \end{bmatrix}$.

We could handle both cases *at once* by reducing the *doubly* augmented matrix

$$\begin{bmatrix} 1 & 2 & 3 & 4 & 1 & 1 \\ 1 & 3 & 5 & 5 & 2 & 2 \\ 2 & 5 & 8 & 9 & 6 & 3 \end{bmatrix}$$

to the echelon matrix

$$\begin{bmatrix} 1 & 2 & 3 & 4 & 1 & 1 \\ 0 & 1 & 2 & 1 & 1 & 1 \\ 0 & 0 & 0 & 0 & 1 & 0 \end{bmatrix},$$

which shows at once that we must look for the solutions to

$$\begin{bmatrix} 1 & 2 & 3 & 4 \\ 0 & 1 & 2 & 1 \\ 0 & 0 & 0 & 0 \end{bmatrix} x = \begin{bmatrix} 1 \\ 1 \\ 1 \end{bmatrix} \text{ for the case } \begin{bmatrix} 1 \\ 2 \\ 6 \end{bmatrix} \text{ (there clearly are none)}$$

and to $\begin{bmatrix} 1 & 2 & 3 & 4 \\ 0 & 1 & 2 & 1 \\ 0 & 0 & 0 & 0 \end{bmatrix} x = \begin{bmatrix} 1 \\ 1 \\ 0 \end{bmatrix} \text{ for the case } \begin{bmatrix} 1 \\ 2 \\ 3 \end{bmatrix}$.

Now that we have defined linear independence and span, we can define the notion of basis of $F^{(n)}$ over F.

Definition. The elements $v_1, \ldots, v_r$ in $F^{(n)}$ form a *basis* of $F^{(n)}$ over F if:

    1. $v_1, \ldots, v_r$ are linearly independent over F;

    2. $v_1, \ldots, v_r$ span $F^{(n)}$.

From what we did above we see that if $v_1, v_2, \ldots, v_r$ is a basis of $F^{(n)}$ over F and if $v \in F^{(n)}$, then $v = a_1 v_1 + a_2 v_2 + \cdots + a_r v_r$, where the a_j are in F, *in one and only one way*. In fact, we could use this as the definition of a basis of $F^{(n)}$ over F.

Theorem 3.6.1. *If a finite set of vectors spans $F^{(n)}$, then it has a subset $v_1, \ldots, v_r$ which is a basis for $F^{(n)}$.*

Proof. Take $v_1, \ldots, v_r$ to be a subset with r as small as possible which spans $F^{(n)}$. To show that $v_1, \ldots, v_r$ is a basis for $F^{(n)}$ it suffices to show that $v_1, \ldots, v_r$ is linearly independent. Suppose, to the contrary, that there is a linear combination $x_1 v_1 + \cdots + x_r v_r = 0$ with $x_t \neq 0$. Thus, v_t equals

$$v_t = -\frac{1}{x_t}(x_1 v_1 + \cdots + x_{t-1} v_{t-1} + x_{t+1} v_{t+1} + \cdots + x_r v_r)$$

So $v_1, \ldots, v_{t-1}, v_{t+1}, \ldots, v_r$ span $F^{(n)}$ since v_t is expressed in terms of them. But this contradicts our choice of $v_1, \ldots, v_r$ as a spanning set with as few elements as possible. So there can be no such linear combination; that is, $v_1, \ldots, v_r$ are linearly independent. ∎

Of course, the columns $e_1, e_2, \ldots, e_n$ of the $n \times n$ identity matrix form a basis of $F^{(n)}$ over F, which we call the *canonical basis* of $F^{(n)}$ over F.

Why, in our definition of basis for $F^{(n)}$, did we not simply use n instead of r? Are there bases $v_1, \ldots, v_r$ for $F^{(n)}$ where r is not equal to n? The answer is "no"! Why, then, do we not use n instead of r? We will *prove* that r equals n. We then know that r and n are equal without having to *assume* so.

Theorem 3.6.2.   Let $v_1, \ldots, v_r$ be a linear independent set of vectors in $F^{(n)}$ and suppose that the vectors $v_1, \ldots, v_r$ are contained in $<w_1, \ldots, w_s>$. Then $r \leq s$.

Proof.   We can show instead that if $r > s$, then we can find a linear dependence $x_1 v_1 + x_2 v_2 + \cdots + x_r v_r = 0$ (where not all of the $x_1, \ldots, x_r$ are 0). To do this, write

$$v_1 = a_{11} w_1 + a_{21} w_2 + \cdots + a_{s1} w_s$$
$$\vdots \qquad \vdots$$
$$v_r = a_{1r} w_1 + a_{2r} w_2 + \cdots + a_{sr} w_s.$$

Then the equation $0 = x_1 v_1 + x_2 v_2 + \cdots + x_r v_r$ becomes

$$0 = x_1 a_{11} w_1 + x_1 a_{21} w_2 + \cdots + x_1 a_{s1} w_s +$$
$$\vdots \qquad \vdots$$
$$x_r a_{1r} w_1 + x_r a_{2r} w_2 + \cdots + x_r a_{sr} w_s$$
$$= (a_{11} x_1 + a_{12} x_2 + \cdots + a_{1r} x_r) w_1 +$$
$$\vdots \qquad \vdots$$
$$(a_{s1} x_1 + a_{s2} x_2 + \cdots + a_{sr} x_r) w_s.$$

Since $r > s$, the system of equations

$$a_{11}x_1 + a_{12}x_2 + \cdots + a_{1r}x_r = 0$$

$$\vdots \qquad\qquad\qquad \vdots$$

$$a_{s1}x_1 + a_{s2}x_2 + \cdots + a_{sr}x_r = 0$$

has a nonzero solution, by Corollary 2.4.3. Thus, we find $x_1v_1 + x_2v_2 + \cdots + x_rv_r = 0$. So they are linearly dependent over F. ∎

The following corollaries are straightforward and you may do them as easy exercises.

Corollary 3.6.3. Let $v_1, \ldots, v_r$ and $w_1, \ldots, w_s$ be two linear independent sets of vectors in $F^{(n)}$ such that the spans $<v_1, \ldots, v_r>$, $<w_1, \ldots, w_s>$ are equal. Then $r = s$.

Corollary 3.6.4. Any basis for $F^{(n)}$ has n elements.

Corollary 3.6.5. Any set of n linearly independent elements of $F^{(n)}$ is a basis of $F^{(n)}$.

Proof. If $v_1, \ldots, v_n$ are linearly independent elements of $F^{(n)}$ and if v is an arbitrary element of $F^{(n)}$, we must show that v is a linear combination of $v_1, \ldots, v_n$. Suppose not. Then the vectors

$$v_1, \ldots, v_n,$$

v are linearly dependent by Theorem 3.6.2. Letting

$$x_1v_1 + \cdots + x_nv_n + x_{n+1}v = 0$$

where not all of the x_s are 0, we must have $x_{n+1} \neq 0$. Why? If x_{n+1}, then

$$x_1v_1 + \cdots + x_nv_n = 0,$$

which, by the linear independence of $v_1, \ldots, v_n$, forces that

$$x_1 = x_2 = \cdots = x_n = 0.$$

In other words, all the coefficients are 0 which is not the case. But then

$$v = -\frac{1}{x_{n+1}}(x_1v_1 + \cdots + x_nv_n)$$

$$= \left(-\frac{x_1}{x_{n+1}}\right)v_1 + \cdots + \left(-\frac{x_n}{x_{n+1}}\right)v_n. \quad ∎$$

Corollary 3.6.6. Any set of n elements that spans $F^{(n)}$ is a basis of $F^{(n)}$.

Proof. By Theorem 3.6.1, any set of n elements which spans $F^{(n)}$ contains a basis for $F^{(n)}$ which, by Corollary 3.6.4, has n elements. So the set itself is a basis. ∎

PROBLEMS

NUMERICAL PROBLEMS

1. Determine if the following elements are linearly independent or linearly dependent.

   (a) $\begin{bmatrix} 1 \\ 1 \\ 0 \end{bmatrix}$, $\begin{bmatrix} 1 \\ 0 \\ 1 \end{bmatrix}$, $\begin{bmatrix} 0 \\ 1 \\ 1 \end{bmatrix}$ in $\mathbb{C}^{(3)}$.

   (b) $\begin{bmatrix} 1 \\ i \\ 0 \end{bmatrix}$, $\begin{bmatrix} i \\ 0 \\ 1 \end{bmatrix}$, $\begin{bmatrix} 0 \\ i \\ 1 \end{bmatrix}$, $\begin{bmatrix} i \\ i \\ i \end{bmatrix}$ in $\mathbb{C}^{(3)}$.

   (c) $\begin{bmatrix} 1 \\ 2 \end{bmatrix}$, $\begin{bmatrix} 2 \\ 1 \end{bmatrix}$ in $\mathbb{R}^{(2)}$.

   (d) $\begin{bmatrix} 1 \\ 2 \\ 3 \\ 4 \end{bmatrix}$, $\begin{bmatrix} i \\ -i \\ 1 \\ 4 \end{bmatrix}$, $\begin{bmatrix} 2i \\ 4i \\ 6i \\ 8i \end{bmatrix}$ in $\mathbb{C}^{(4)}$.

   (e) $\begin{bmatrix} 0 \\ 0 \\ 0 \\ 0 \end{bmatrix}$, $\begin{bmatrix} 1 \\ 1 \\ 0 \\ 0 \end{bmatrix}$, $\begin{bmatrix} 0 \\ 1 \\ 0 \\ 0 \end{bmatrix}$ in $\mathbb{R}^{(4)}$.

   (f) $\begin{bmatrix} 1 \\ 2 \\ 0 \end{bmatrix}$, $\begin{bmatrix} i \\ 0 \\ 1 + i \end{bmatrix}$, $\begin{bmatrix} 2 + i \\ 4 \\ 1 + i \end{bmatrix}$ in $\mathbb{C}^{(3)}$.

2. Which of sets of the elements in Problem 1 form a basis of their respective $F^{(n)}$'s?

3. In $F^{(3)}$ verify that Ae_1, Ae_2, and Ae_3 form a basis over F where
$$A = \begin{bmatrix} 1 & 2 & 3 \\ -1 & 0 & 0 \\ 0 & 0 & 2 \end{bmatrix}.$$

4. For what values of a in F do the elements Ae_1, Ae_2, Ae_3 *not* form a basis of $F^{(3)}$ over F if $A = \begin{bmatrix} 1 & 2 & 3 \\ -1 & 0 & 0 \\ 0 & 0 & a \end{bmatrix}$?

5. Let $A = \begin{bmatrix} 1 & 0 & 2 \\ 4 & -5 & 6 \\ 2 & 0 & 4 \end{bmatrix}$. Show that Ae_1, Ae_2, and Ae_3 do *not* form a basis of $F^{(3)}$ over F.

6. In $F^{(2)}$ show that any three elements must be linearly dependent over F.

7. In $F^{(3)}$ show that any four elements are linearly dependent over F.

8. If $\begin{bmatrix} 1 & 0 & a \\ 2 & 1 & b \\ 3 & 2 & c \end{bmatrix}$ is not invertible, show that the following elements

$$\begin{bmatrix} 1 \\ 2 \\ 3 \end{bmatrix}, \quad \begin{bmatrix} 0 \\ 1 \\ 2 \end{bmatrix}, \quad \begin{bmatrix} a \\ b \\ c \end{bmatrix} \text{ must be linearly dependent.}$$

9. Show that if the matrix in Problem 8 is invertible, then the vectors

$$\begin{bmatrix} 1 \\ 2 \\ 3 \end{bmatrix}, \quad \begin{bmatrix} 0 \\ 1 \\ 2 \end{bmatrix}, \quad \begin{bmatrix} a \\ b \\ c \end{bmatrix} \text{ are linearly independent.}$$

MORE THEORETICAL PROBLEMS

Easier Problems

10. Prove Corollaries 3.6.3 and 3.6.4.

3.7. CHANGE OF BASIS OF $F^{(n)}$

Starting with the canonical basis $e_1, \ldots, e_n$ as one basis for $F^{(n)}$ over F, let's also take $w_1, \ldots, w_n$ to be another basis of $F^{(n)}$ over F and consider the matrix C whose columns are $w_1, \ldots, w_n$.

Definition. We call C the *matrix of the change of basis* from the basis $e_1, \ldots, e_n$ to the basis $w_1, \ldots, w_n$.

EXAMPLE

Recall from an example in Section 3.6 with $F = \mathbb{R}$ and

$$z_1 = \begin{bmatrix} 1 \\ 2 \\ 3 \end{bmatrix}, z_2 = \begin{bmatrix} -1 \\ 3 \\ -3 \end{bmatrix}, z_3 = \begin{bmatrix} 0 \\ 1 \\ 1 \end{bmatrix}$$

that z_1, z_2, z_3 are linearly independent over $\mathbb{R}$. Thus z_1, z_2, z_3 forms a basis for $\mathbb{R}^{(3)}$. (Prove!) The corresponding matrix of the change of basis is

$$C = \begin{bmatrix} 1 & -1 & 0 \\ 2 & 3 & 1 \\ 3 & -3 & 1 \end{bmatrix}.$$

What properties does the matrix C of the change of basis enjoy? Can it be any old matrix? No! As we now show, C must be an *invertible matrix*.

Theorem 3.7.1.   If C is the matrix of the change of basis from the canonical basis $e_1, \ldots, e_n$ to the basis $w_1, \ldots, w_n$, then C is invertible.

Proof.   By the definition of C, $Ce_1 = w_1, \ldots, Ce_n = w_n$. Define the mapping $S: F^{(n)} \to F^{(n)}$ as follows:

Given $v \in F^{(n)}$, express v as $v = b_1 w_1 + \cdots + b_n w_n$ with the $b_j \in F$ and define S by $Sv = b_1 e_1 + \cdots + b_n e_n$.

Note that $Sw_r = e_r$. We leave it to the reader to show that S is also a linear transformation on $F^{(n)}$, and so is a matrix (s_{rs}). Since

$$(SC)e_r = S(Ce_r) = Sw_r = e_r,$$

we see that SC is the identity mapping as far as elements $e_1, \ldots, e_n$ are concerned. But this implies that SC is the identity mapping on $F^{(n)}$, since $e_1, \ldots, e_n$ form a basis of $F^{(n)}$ over F. (Prove!) But then we have that $(s_{rs})(c_{rs}) = I$. Similarly, we have $(c_{rs})(s_{rs})$, since $CSw_r = Ce_r = w_r$ and the w_r also form a basis. Since $SC = CS = I$, C is invertible with inverse S.   ∎

So given any basis $w_1, \ldots, w_n$ of $F^{(n)}$ over F, the matrix C whose columns are $w_1, \ldots, w_n$ is an invertible matrix. On the other hand, if B is an invertible matrix in $M_n(F)$ we claim that its columns $z_1 = Be_1, \ldots, z_n = Be_n$ form a basis of $F^{(n)}$ over F. Why? To begin with, since B is invertible, we know that, given $v \in F^{(n)}$, then

$$v = (BB^{-1})v = B(B^{-1}v),$$

so $v = Bw$, where $w = B^{-1}v$ is in $F^{(n)}$. Furthermore, we have $w = a_1e_1 + \cdots + a_ne_n$, where the a_r are in F and are unique. Thus

$$v = Bw = B(a_1e_1 + \cdots + a_ne_n) = a_1Be_1 + \cdots + a_nBe_n$$
$$= a_1z_1 + \cdots + a_nz_n.$$

So every element in $F^{(n)}$ is realizable in the form $a_1z_1 + \cdots + a_nz_n$, the second requisite condition defining a basis.

To prove that $z_1, \ldots, z_n$ is indeed a basis of $F^{(n)}$ over F, we must verify that

$$b_1z_1 + \cdots + b_nz_n = 0$$

forces

$$b_1 = b_2 = \cdots = b_n = 0.$$

But if $b_1z_1 + \cdots + b_nz_n = 0$, then $0 = A(b_1z_1 + \cdots + b_nz_n)$, where $A = B^{-1}$. Since $Be_r = z_r$, we have that $Az_r = e_r$ for every r. Therefore,

$$0 = A(b_1z_1 + \cdots + b_nz_n)$$
$$= b_1Az_1 + \cdots + b_nAz_n$$
$$= b_1e_1 + \cdots + b_ne_n.$$

Because $e_1, \ldots, e_n$ is a basis of $F^{(n)}$ over F we have that each $b_j = 0$. So $z_1, \ldots, z_n$ are linearly independent over F. So we see that $z_1 = Be_1, \ldots, Z_n = Be_n$ form a basis of $F^{(n)}$ over F.

Combining what we just did with Theorem 3.7.1 we get a description of *all possible bases* of $F^{(n)}$ over F, namely,

Theorem 3.7.2.   Given any invertible matrix A in $M_n(F)$, then $w_1 = Ae_1, \ldots, w_n = Ae_n$ form a basis of $F^{(n)}$ over F. Moreover, given any basis $z_1, \ldots, z_n$ of $F^{(n)}$ over F, then for some invertible matrix A in $M_n(F)$, $z_r = Ae_r$ for $1 \le r \le n$.

PROBLEMS

NUMERICAL PROBLEMS

1. Show that $\begin{bmatrix} 1 \\ 1 \\ 0 \\ 1 \end{bmatrix}$, $\begin{bmatrix} 1 \\ 0 \\ 1 \\ 1 \end{bmatrix}$, $\begin{bmatrix} 0 \\ 0 \\ 1 \\ 1 \end{bmatrix}$, $\begin{bmatrix} 0 \\ 1 \\ 0 \\ 2 \end{bmatrix}$ form a basis of $F^{(4)}$ over F.

2. Find the matrix of the change of basis from the basis $e_1, \ldots, e_n$ to that of Problem 1.

3. Show that we can make $\begin{bmatrix} 0 \\ 0 \\ 1 \\ 1 \end{bmatrix}$ the first column of some matrix A in $M_4(F)$ such that A is invertible.

4. Show that we can make $\begin{bmatrix} 1 \\ 1 \\ 0 \\ 1 \end{bmatrix}$, $\begin{bmatrix} 0 \\ 1 \\ 0 \\ 2 \end{bmatrix}$ the first and second columns of some matrix A in $M_4(F)$ such that A is invertible.

5. Show that we can make $\begin{bmatrix} 1 \\ 2 \\ 3 \\ 4 \end{bmatrix}$, $\begin{bmatrix} 0 \\ 0 \\ 1 \\ 1 \end{bmatrix}$, $\begin{bmatrix} 0 \\ 0 \\ 0 \\ 1 \end{bmatrix}$ the first, second, and fourth columns of some matrix A in $M_4(F)$ such that A is invertible.

6. Determine for which real numbers a, b, c, d the four vectors
$$a\begin{bmatrix} 1 \\ 1 \\ 0 \\ 1 \end{bmatrix} + b\begin{bmatrix} 1 \\ 0 \\ 1 \\ 1 \end{bmatrix} + c\begin{bmatrix} 0 \\ 0 \\ 1 \\ 1 \end{bmatrix} + d\begin{bmatrix} 0 \\ 1 \\ 0 \\ 2 \end{bmatrix}, \begin{bmatrix} 1 \\ 0 \\ 1 \\ 1 \end{bmatrix}, \begin{bmatrix} 0 \\ 0 \\ 1 \\ 1 \end{bmatrix}, \begin{bmatrix} 0 \\ 1 \\ 0 \\ 2 \end{bmatrix}$$ are linearly independent.

7. Determine for which real numbers a, b, c, d the four vectors
$$a\begin{bmatrix} 1 \\ 1 \\ 0 \\ 1 \end{bmatrix} + b\begin{bmatrix} 1 \\ 0 \\ 1 \\ 1 \end{bmatrix}, c\begin{bmatrix} 0 \\ 0 \\ 1 \\ 1 \end{bmatrix} + d\begin{bmatrix} 0 \\ 1 \\ 0 \\ 2 \end{bmatrix}, \begin{bmatrix} 1 \\ 0 \\ 1 \\ 1 \end{bmatrix}, \begin{bmatrix} 0 \\ 1 \\ 0 \\ 2 \end{bmatrix}$$ are linearly independent.

8. Compute the matrices $A = \begin{bmatrix} 0 & 1 & 0 & 0 \\ c & 0 & 0 & d \\ 1 & 0 & 0 & 0 \\ 0 & 0 & 1 & 0 \end{bmatrix}^2$ and

$B = \begin{bmatrix} 0 & 1 & 0 & 0 \\ c & 0 & 0 & d \\ 1 & 0 & 0 & 0 \\ 0 & 0 & 1 & 0 \end{bmatrix}^3$ and determine the values of c and d for which

their columns are bases of $F^{(4)}$.

9. For what values of c and d are the vectors

$$\begin{bmatrix} 0 & 1 & 0 & 0 \\ c & 0 & 0 & d \\ 1 & 0 & 0 & 0 \\ 0 & 0 & 1 & 0 \end{bmatrix}^2 \begin{bmatrix} 1 \\ 1 \\ 0 \\ 1 \end{bmatrix}, \quad \begin{bmatrix} 0 & 1 & 0 & 0 \\ c & 0 & 0 & d \\ 1 & 0 & 0 & 0 \\ 0 & 0 & 1 & 0 \end{bmatrix}^2 \begin{bmatrix} 1 \\ 0 \\ 1 \\ 1 \end{bmatrix},$$

$$\begin{bmatrix} 0 & 1 & 0 & 0 \\ c & 0 & 0 & d \\ 1 & 0 & 0 & 0 \\ 0 & 0 & 1 & 0 \end{bmatrix}^2 \begin{bmatrix} 0 \\ 0 \\ 1 \\ 1 \end{bmatrix}, \quad \begin{bmatrix} 0 & 1 & 0 & 0 \\ c & 0 & 0 & d \\ 1 & 0 & 0 & 0 \\ 0 & 0 & 1 & 0 \end{bmatrix}^2 \begin{bmatrix} 0 \\ 1 \\ 0 \\ 2 \end{bmatrix}$$

linearly independent?

10. Show by direct methods that the columns of the matrix
$\begin{bmatrix} 1 & 3 & 4 & e \\ 3 & 1 & 4 & d \\ 1 & 0 & 0 & 0 \\ 0 & 1 & 0 & 0 \end{bmatrix}$ are a basis for $F^{(4)}$ if and only if the columns of

the matrix $\begin{bmatrix} 1 & 3 & 1 & 0 \\ 3 & 1 & 0 & 1 \\ 4 & 4 & 0 & 0 \\ e & d & 0 & 0 \end{bmatrix}$ are a basis for $F^{(10)}$.

MORE THEORETICAL PROBLEMS

Easier Problems

11. Find the matrix of the change of basis from the basis $e_1, \ldots, e_n$ to the basis $w_1 = e_2, w_2 = e_3, \ldots, w_{n-1} = e_n, w_n = e_1$.

12. If $w \neq 0$ is in $F^{(3)}$, show that we can find elements w_2 and w_3 in $F^{(3)}$ so that w, w_2, w_3 form a basis of $F^{(3)}$ over F.

13. If $A = (a_{jk})$ is in $M_4(F)$, show that A is invertible if and only if

$$\begin{bmatrix} a_{11} \\ a_{21} \\ a_{31} \\ a_{41} \end{bmatrix}, \quad \begin{bmatrix} a_{12} \\ a_{22} \\ a_{32} \\ a_{42} \end{bmatrix}, \quad \begin{bmatrix} a_{13} \\ a_{23} \\ a_{33} \\ a_{43} \end{bmatrix}, \quad \text{and} \quad \begin{bmatrix} a_{14} \\ a_{24} \\ a_{34} \\ a_{44} \end{bmatrix} \quad \text{are linearly independent}$$

over F.

14. Given $w \neq 0$ in $F^{(4)}$, show that we can make w the first column of some matrix A in $M_4(F)$ such that A is invertible.

15. If A in Problem 14 is found, what can you say about the columns of A?

16. Do Problem 14 for the explicit vector $w = \begin{bmatrix} i \\ 0 \\ -i \\ 5 + i \end{bmatrix}$ in $\mathbb{C}^{(4)}$.

17. Find a basis of $\mathbb{C}^{(4)}$ over $\mathbb{C}$ in which the vector $w = \begin{bmatrix} i \\ 0 \\ -i \\ 5 + i \end{bmatrix}$ is the first basis element.

Middle-Level Problems

18. Given $w \neq 0$ in $F^{(n)}$, show that you can find elements $w_2, \ldots, w_n$ in $F^{(n)}$ such that $w, w_2, \ldots, w_n$ form a basis of $F^{(n)}$ over F.

19. In $F^{(4)}$ show that any 5 elements are linearly dependent over F.

Harder Problems

20. Generalize Problem 18 as follows: given $v_1, \ldots, v_m$ in $F^{(n)}$ which are linearly independent over F, and $m < n$, show that you can find elements $w_{m+1}, \ldots, w_n$ in $F^{(n)}$ such that $v_1, \ldots, v_m, w_{m+1}, \ldots, w_n$ form a basis of $F^{(n)}$ over F.

21. A matrix P in $M_n(F)$ is called a *permutation matrix* if its nonzero entries consist of exactly one 1 in each row and in each column. What does P do to the canonical basis of $F^{(n)}$?

22. Using the results of Problem 21, show that a permutation matrix is invertible. What is its inverse?

23. In $F^{(4)}$ if w_1, w_2, w_3 are nonzero elements, show that there is some element $v \in F^{(4)}$ such that v cannot be expressed as

$$v = a_1 w_1 + a_2 w_2 + a_3 w_3$$

for any choice of a_1, a_2, a_3 in F.

24. Show that if $w_1, \ldots, w_m$ are in $F^{(n)}$, where $m < n$, then there is an element v in $F^{(n)}$ which cannot be expressed as

$$v = a_1 w_1 + \cdots + a_m w_m$$

for any $a_1, \ldots, a_m$ in F.

25. If $w_1, \ldots, w_n$ in $F^{(n)}$ are such that *every* $v \in F^{(n)}$ can be expressed as $v = a_1 w_1 + \cdots + a_n w_n$, where $a_1, \ldots, a_n \in F$, show that $w_1, \ldots, w_n$ is a basis for $F^{(n)}$.

26. If P and Q are permutation matrices in $M_n(F)$, prove that PQ is also a permutation matrix.

27. How many permutation matrices are there in $M_n(F)$?

28. If P is a permutation matrix, prove that P' is actually equal to P^{-1}.

3.8. INVERTIBLE MATRICES

In Section 3.7 we showed that an $n \times n$ matrix A is invertible if and only if its columns form a basis for $F^{(n)}$. This gives us a description of all possible bases of $F^{(n)}$ over F in terms of invertible matrices. But given an $n \times n$ matrix A, how do we know whether it is invertible? We give some equivalent conditions in

Theorem 3.8.1. Let A be an $n \times n$ matrix and let $v_1, \ldots, v_n$ denote the columns of A. Then the following conditions are equivalent:

    1.   The only solution to $Ax = 0$ is $x = 0$;

    2. A is $1-1$;

    3.   The columns of A are linearly independent;

    4.   The columns of A form a basis;

    5. A is onto;

    6. A is invertible.

    Proof. The strategy of the proof is to show first that (1) implies (2), (2) implies (3), (3) implies (4), (4) implies (5), (5) implies (1). From this it will follow that the first five conditions are all equivalent. These conditions certainly imply that A is invertible, condition (6). Conversely, if A is invertible, then A is certainly $1-1$ so that A satisfies (2). But then it follows that all six conditions must be equivalent.

    So it suffices to show that the first five conditions are equivalent, which we now do.

1. *implies* (2): If the only solution to $Ax = 0$ is $x = 0$, and if $Au = Av$, then $A(u - v) = Au - Av = 0$ implies that $u - v = 0$, that is, $u = v$. So A is $1-1$ and hence (1) implies (2).

2. *implies* (3): The columns of A are $Ae_1, Ae_2, \ldots, Ae_n$. So if

$$x_1 Ae_1 + \cdots + x_n Ae_n = 0, \text{ then } A \begin{bmatrix} x_1 \\ \vdots \\ x_n \end{bmatrix} = 0, \text{ which implies that}$$

$\begin{bmatrix} x_1 \\ \vdots \\ x_n \end{bmatrix} = 0$ because A is $1-1$. So the columns $Ae_1, \ldots, Ae_n$ of A are

linearly independent and (2) implies (3).

3. *implies* (4): If the columns of A are linearly independent, they are a basis for $F^{(n)}$ by Corollary 3.6.5. So (3) implies (4).

4. *implies* (5): If the columns of A form a basis for $F^{(n)}$, any $y \in F^{(n)}$ can be expressed as a linear combination $x_1 Ae_1 + \cdots + x_n Ae_n$

of the columns of A, so $y = A \begin{bmatrix} x_1 \\ \vdots \\ x_n \end{bmatrix}$. So, (4) implies (5).

5. *implies* (1): If A is onto, its columns $Ae_1, \ldots, Ae_n$ span $F^{(n)}$ and so are a basis for $F^{(n)}$ by Corollary 3.6.6.   ■

We now get

Corollary 3.8.2.   If $A, B \in M_n(F)$ and AB is invertible, then A and B are invertible.

Proof.   Since AB is invertible, AB is $1-1$ and onto, by Theorem 3.8.1. Since AB is $1-1$, B is also $1-1$ (Prove!), so B is invertible by Theorem 3.8.1. Since AB is onto, A is also onto (Prove!), so A is invertible by Theorem 3.8.1.   ■

Since an $n \times n$ matrix A is invertible if and only if its columns are linearly independent, by Theorem 3.8.1, we can use the method of Section 3.6 for determining whether the columns of A are linearly independent as a

Method to determine whether an
$n \times n$ matrix is invertible

1. Row reduce A to an echelon matrix B.
2. Then A is invertible if and only if all diagonal entries of B are 1.

EXAMPLE

To see whether $A = \begin{bmatrix} 1 & 1 & 1 \\ 1 & 2 & 1 \\ 1 & 2 & 2 \end{bmatrix}$ is invertible, we row reduce

A to an echelon matrix $\begin{bmatrix} 1 & 1 & 1 \\ 0 & 1 & 0 \\ 0 & 0 & 1 \end{bmatrix}$. Since its diagonal entries all

equal 1, A is invertible.

Let A be an invertible matrix with columns $v_1, \ldots, v_n$. We know that the $v_1, \ldots, v_n$ form a basis, so we can find scalars c_{rs} such that $e_s = \sum_{r=1}^{n} c_{rs} v_r$ for all s. Then

$$ e_s = \sum_{r=1}^{n} c_{rs} v_r = \sum_{r=1}^{n} c_{rs} A_r = A \left(\sum_{r=1}^{n} c_{rs} e_r \right) $$

and

$$ A^{-1} e_s = \sum_{r=1}^{n} c_{rs} e_r. $$

This means that column s of A^{-1} is $\sum_{r=1}^{n} c_{rs} e_r$, that is, c_{rs} is the (r, s) entry of A^{-1}. Is this useful? Yes! It enables us to find the entries of A^{-1} by expressing the e_s as linear combinations of $v_1, \ldots, v_n$.

To be specific, let's use our method from Section 3.7 to express the vector e_s as linear combinations of $v_1, \ldots, v_n$. Following our method with $A = [v_1, \ldots, v_n]$, we form the augmented matrix $[A, e_s]$ and row reduce it to an echelon matrix $[B, z_s]$. Then B is an upper triangular matrix with diagonal entries all equal to 1. So we can further row reduce $[B, z_s]$ until B becomes the identity matrix and we get a matrix $[I, w_s]$. Then $Ax = e_s$ if and only if $Ix = w_s$, that is, if and only if $x = w_s$. So w_s is the solution to the equation $Ax = e_s$ for each s. It follows that w_s is $A^{-1} e_s$, that is, w_s is column s of A^{-1} for each s. So we have computed the columns of A^{-1} and we get A^{-1} as $[w_1, \ldots, w_n]$.

In practice, we compute all of the w_s at once. Instead of forming the *singly* augmented matrices $[A, e_s]$ and reducing each of them to $[I, w_s]$, we form the *n-fold* augmented matrix

$$[A, e_1, \ldots, e_n] = [A, I]$$

and now reduce it to the matrix

$$[I, w_1, \ldots, w_n] = [I, A^{-1}].$$

This gives the

Method to compute the inverse of an invertible *n* × *n* matrix *A*

1. *Form the* n × 2n *matrix* [A, I] *whose columns are the columns* v_j *of A followed by the columns* e_k *of I.*
2. *Row reduce* [A, I] *to the echelon matrix* [I, Z]. *(If you find that this is not possible because the first n entries of some row become 0, then A is not invertible.)*
3. *Then Z is the matrix* A^{-1}.

EXAMPLE

Suppose that we wish to determine whether the matrix

$$A = \begin{bmatrix} 1 & 2 & 3 \\ 1 & 3 & 5 \\ 2 & 5 & 9 \end{bmatrix}$$

is invertible. And if so, we want to compute A^{-1}. To try to compute the inverse of A, we form the matrix

$$[A, I] = \begin{bmatrix} 1 & 2 & 3 & 1 & 0 & 0 \\ 1 & 3 & 5 & 0 & 1 & 0 \\ 2 & 5 & 9 & 0 & 0 & 1 \end{bmatrix}$$

and row reduce it to the matrix

$$\begin{bmatrix} 1 & 2 & 3 & 1 & 0 & 0 \\ 0 & 1 & 2 & -1 & 1 & 0 \\ 0 & 1 & 3 & -2 & 0 & 1 \end{bmatrix},$$

then to the echelon matrix

$$\begin{bmatrix} 1 & 2 & 3 & 1 & 0 & 0 \\ 0 & 1 & 2 & -1 & 1 & 0 \\ 0 & 0 & 1 & -1 & -1 & 1 \end{bmatrix}.$$

Since the diagonal elements are all 1, the matrix is invertible and we continue until it is row reduced to

$$\begin{bmatrix} 1 & 2 & 0 & 4 & 3 & -3 \\ 0 & 1 & 0 & 1 & 3 & -2 \\ 0 & 0 & 1 & -1 & -1 & 1 \end{bmatrix}$$

and finally to

$$\begin{bmatrix} 1 & 0 & 0 & 2 & -3 & 1 \\ 0 & 1 & 0 & 1 & 3 & -2 \\ 0 & 0 & 1 & -1 & -1 & 1 \end{bmatrix} = [I, A^{-1}],$$

so

$$A^{-1} = \begin{bmatrix} 2 & -3 & 1 \\ 1 & 3 & -2 \\ -1 & -1 & 1 \end{bmatrix}.$$

PROBLEMS

NUMERICAL PROBLEMS

1. Determine those values for a for which the matrix $\begin{bmatrix} 1 & a & 0 \\ 2 & 1 & 0 \\ 1 & 3 & 1 \end{bmatrix}$ is not invertible.

2. Determine those values for a for which the matrix $\begin{bmatrix} 1 & a & 0 \\ 3 & 4 & 1 \\ 1 & 3 & 1 \end{bmatrix}$ is not invertible.

3. Determine those values for a for which the matrix $\begin{bmatrix} 2 & a & 1 \\ 3 & 4 & 1 \\ 1 & 3 & 1 \end{bmatrix}$ is not invertible.

4. Compute the inverse of $\begin{bmatrix} 2 & 3 & 1 \\ 3 & 4 & 1 \\ 1 & 3 & 1 \end{bmatrix}$ by row reduction of $[A, I]$.

MORE THEORETICAL PROBLEMS

Easier Problems

5. Let A and B be $n \times n$ matrices such that $AB = I$ and $Ax = 0$ only if $x = 0$. Without using the results of Sections 3.6 through 3.8, show that $A(BA - I) = 0$ and use this to show that $AB = BA$ and A is invertible.

6. Compute the inverse of the matrix $\begin{bmatrix} 1 & a & 0 \\ 1 & 1 & 1 \\ 0 & 0 & 1 \end{bmatrix}$ for all a for which it exists.

7. Show that if an $n \times n$ matrix A is invertible and B is row equivalent to A, then B is invertible.

8. Show that an $n \times n$ matrix A is row equivalent to I if and only if A is invertible.

Middle-Level Problems

9. Show that if A is row equivalent to B, then AC is row equivalent to BC for any $n \times n$ matrices A, B, C.

Harder Problems

10. Using Problems 8 and 9, show that two $n \times n$ matrices A and B are row equivalent if and only if $B = UA$ for some invertible $n \times n$ matrix U.

3.9. MATRICES AND BASES

Given $F^{(n)}$, it is not aware of what particular basis one uses for it. As far as it is concerned, one basis is just as good as any other. If T is a linear transformation on $F^{(n)}$, all $F^{(n)}$ cares about is that T satisfies $T(v + w) = T(v) + T(w)$ and $T(av) = aT(v)$ for all $v, w \in F^{(n)}$ and all $a \in F$. How we represent T—as a matrix or otherwise—is our business and not that of $F^{(n)}$. For us, the canonical basis is often best because its components are so simple—but $F^{(n)}$ could not care less.

Given a linear transformation T on $F^{(n)}$ we found, by using the canonical basis, that T is a matrix. However, the canonical basis is not holy. *We could talk in an analogous way about the matrix of T in any basis.*

To do this, let $v_1, \ldots, v_n$ be a basis of $F^{(n)}$ over F and let T be a linear transformation on $F^{(n)}$. If we knew $Tv_1, \ldots, Tv_n$, we would know how T acts on any element v in $F^{(n)}$. Why? Since $v \in F^{(n)}$ and $v_1, \ldots, v_n$ is a basis of $F^{(n)}$ over F, we know that

$$v = a_1 v_1, + \cdots + a_n v_n,$$

where the $a_1, \ldots, a_n$ are in F and are unique. So

$$Tv = T(a_1 v_1 + \cdots + a_n v_n) = a_1 Tv_1 + \cdots + a_n Tv_n.$$

So knowing each of $Tv_1, \ldots, Tv_n$ allows us to know *everything* about how T acts on arbitrary elements of $F^{(n)}$.

Because Tv_s is in $F^{(n)}$ for every $s = 1, \ldots, n$ and since $v_1, \ldots, v_n$ is a basis of $F^{(n)}$ over F, Tv_s is realizable in a unique way as

$$Tv_s = b_{1s} v_1 + b_{2s} v_2 + \cdots + b_{ns} v_n.$$

If we have the n^2 elements b_{rs} in hand, then we know T exactly how it acts on $F^{(n)}$.

Definition. We call the matrix (b_{rs}) such that

$$Tv_s = b_{1s} v_1 + b_{2s} v_2 + \cdots + b_{ns} v_n$$

for all s the *matrix of T in the basis $v_1, \ldots, v_n$.*

In summation notation, the condition on the entries b_{rs} of the matrix of T in the basis $v_1, \ldots, v_n$ is

$$Tv_s = \sum_{r=1}^{n} b_{rs} v_r.$$

When we showed that any linear transformation over $F^{(n)}$ was a matrix what we really showed was that its matrix in the canonical basis $e_1, \ldots, e_n$, namely, the matrix having the vectors $Te_1, \ldots, Te_n$ as its columns, also mapped e_r to Te_r for all r.

Before going on, let's look at an example. Let T be the matrix
$$\begin{bmatrix} 1 & 2 & 3 \\ -1 & 4 & 7 \\ 5 & 0 & 1 \end{bmatrix}$$ and regard T as a linear transformation of $F^{(3)}$. Then the

matrix of T (as linear transformation) in the canonical basis is

$$\begin{bmatrix} 1 & 2 & 3 \\ -1 & 4 & 7 \\ 5 & 0 & 1 \end{bmatrix}.$$ The vectors

$$v_1 = \begin{bmatrix} 0 \\ 1 \\ 1 \end{bmatrix}, \qquad v_2 = \begin{bmatrix} 1 \\ 0 \\ 1 \end{bmatrix}, \qquad v_3 = \begin{bmatrix} 1 \\ 1 \\ 0 \end{bmatrix}$$

can readily be shown to be a basis of $F^{(3)}$ over F. (Do it!) How do we go about finding the matrix of the T above in the basis v_1, v_2, v_3? To find the expression for

$$Tv_1 = \begin{bmatrix} 1 & 2 & 3 \\ -1 & 4 & 7 \\ 5 & 0 & 1 \end{bmatrix}\begin{bmatrix} 0 \\ 1 \\ 1 \end{bmatrix} = \begin{bmatrix} 5 \\ 11 \\ 1 \end{bmatrix}$$

in terms of v_1, v_2, v_3, we follow the method in Section 3.6 for expressing $\begin{bmatrix} 5 \\ 11 \\ 1 \end{bmatrix}$ as a linear combination of v_1, v_2, v_3. We form the augmented

matrix

$$[v_1, v_2, v_3, v] = \begin{bmatrix} 0 & 1 & 1 & 5 \\ 1 & 0 & 1 & 11 \\ 1 & 1 & 0 & 1 \end{bmatrix}$$

and row reduce it, getting

$$\begin{bmatrix} 1 & 0 & 1 & 11 \\ 0 & 1 & 1 & 5 \\ 0 & 0 & -2 & -15 \end{bmatrix}.$$

Solving

$$\begin{bmatrix} 1 & 0 & 1 \\ 0 & 1 & 1 \\ 0 & 0 & -2 \end{bmatrix} x = \begin{bmatrix} 11 \\ 5 \\ -15 \end{bmatrix}$$

for x by back substitution, we get $x = \begin{bmatrix} \frac{7}{2} \\ -\frac{5}{2} \\ \frac{15}{2} \end{bmatrix}$. So $x = \begin{bmatrix} \frac{7}{2} \\ -\frac{5}{2} \\ \frac{15}{2} \end{bmatrix}$ is also a solution to

$$[v_1, v_2, v_3]x = \begin{bmatrix} 0 & 1 & 1 \\ 1 & 0 & 1 \\ 1 & 1 & 0 \end{bmatrix} x = \begin{bmatrix} 5 \\ 11 \\ 1 \end{bmatrix}$$

and we get

$$Tv_1 = \begin{bmatrix} 5 \\ 11 \\ 1 \end{bmatrix} = [v_1, v_2, v_3] \begin{bmatrix} \frac{7}{2} \\ -\frac{5}{2} \\ \frac{15}{2} \end{bmatrix} = \tfrac{7}{2}v_1 - \tfrac{5}{2}v_2 + \tfrac{15}{2}v_3.$$

By the recipe given above for determining the matrix of T in the basis v_1, v_2, v_3, the first column of the matrix of T in the basis v_1, v_2, v_3 is

$$\begin{bmatrix} \frac{7}{2} \\ -\frac{5}{2} \\ \frac{15}{2} \end{bmatrix}.$$

Similarly,

$$Tv_2 = \begin{bmatrix} 1 & 2 & 3 \\ -1 & 4 & 7 \\ 5 & 0 & 1 \end{bmatrix} \begin{bmatrix} 1 \\ 0 \\ 1 \end{bmatrix} = \begin{bmatrix} 4 \\ 6 \\ 6 \end{bmatrix} = [v_1, v_2, v_3] \begin{bmatrix} 4 \\ 2 \\ 2 \end{bmatrix}$$

$$= 4v_1 + 2v_2 + 2v_3. \text{ (Verify!)}$$

So the second column of the matrix of T in the basis v_1, v_2, v_3 is

$$\begin{bmatrix} 4 \\ 2 \\ 2 \end{bmatrix}.$$

Finally,

$$Tv_3 = \begin{bmatrix} 1 & 2 & 3 \\ -1 & 4 & 7 \\ 5 & 0 & 1 \end{bmatrix} \begin{bmatrix} 1 \\ 1 \\ 0 \end{bmatrix} = \begin{bmatrix} 3 \\ -3 \\ 5 \end{bmatrix} = [v_1, v_2, v_3] \begin{bmatrix} \frac{5}{2} \\ \frac{5}{2} \\ \frac{5}{2} \end{bmatrix}$$

$$= (\tfrac{5}{2})v_1 + (\tfrac{5}{2})v_2 + (\tfrac{1}{2})v_3. \text{ (Verify!)}$$

so the third column of the matrix of T in the basis v_1, v_2, v_3, is

$$\begin{bmatrix} \frac{5}{2} \\ \frac{5}{2} \\ \frac{1}{2} \end{bmatrix}.$$

Therefore, the matrix of T in the basis v_1, v_2, v_3 is

$$\begin{bmatrix} \frac{7}{2} & 4 & \frac{5}{2} \\ -\frac{5}{2} & 2 & \frac{5}{2} \\ \frac{15}{2} & 2 & \frac{1}{2} \end{bmatrix}.$$

So *the two matrices* $\begin{bmatrix} 1 & 2 & 3 \\ -1 & 4 & 7 \\ 5 & 0 & 1 \end{bmatrix}$, $\begin{bmatrix} \frac{7}{2} & 4 & \frac{5}{2} \\ -\frac{5}{2} & 2 & \frac{5}{2} \\ \frac{15}{2} & 2 & \frac{1}{2} \end{bmatrix}$ *represent the same*

linear transformation T, but in different bases: the first in the canonical basis, the second in the basis

$$v_1 = \begin{bmatrix} 0 \\ 1 \\ 1 \end{bmatrix}, \qquad v_2 = \begin{bmatrix} 1 \\ 0 \\ 1 \end{bmatrix}, \qquad v_3 = \begin{bmatrix} 1 \\ 1 \\ 0 \end{bmatrix}.$$

Forget about linear transformations for a moment; given the two matrices

$$\begin{bmatrix} 1 & 2 & 3 \\ -1 & 4 & 7 \\ 5 & 0 & 1 \end{bmatrix}, \qquad \begin{bmatrix} \frac{7}{2} & 4 & \frac{5}{2} \\ -\frac{5}{2} & 2 & \frac{5}{2} \\ \frac{15}{2} & 2 & \frac{1}{2} \end{bmatrix},$$

and viewing them just as matrices, how on earth are they related? *Because we know that they came from the same linear transformation—in different bases—common sense tells us that they should be related.* But how? That is precisely what we are going to find out in the next theorem.

Theorem 3.9.1   Regard the $n \times n$ matrix T as a linear transformation of $F^{(n)}$. Then the matrix A of T in a basis $v_1, \ldots, v_n$ of $F^{(n)}$ is $A = C^{-1}TC$, where C is the matrix whose columns are $v_1, \ldots, v_n$.

Proof.   Letting $A = (a_{rs})$, we have $T(v_s) = \sum_{r=1}^{n} a_{rs}v_r$ for $s = 1, \ldots, n$. Therefore,

$$C^{-1}TC(e_s) = C^{-1}T(v_s) = C^{-1}\left(\sum_{r=1}^{n} a_{rs}v_r \right)$$

$$= \sum_{r=1}^{n} a_{rs}C^{-1}(v_r) = \sum_{r=1}^{n} a_{rs}e_r.$$

Thus $C^{-1}TC(e_s) = \sum_{r=1}^{n} a_{rs}e_r$ for all s and $C^{-1}TC = A$.   ∎

We first illustrate the theorem with a simple example. Let

$$v_1 = \begin{bmatrix} 0 \\ 1 \end{bmatrix}, v_2 = \begin{bmatrix} -1 \\ 0 \end{bmatrix}.$$

Then C is $C = \begin{bmatrix} 0 & -1 \\ 1 & 0 \end{bmatrix}$. Notice that $C^{-1} = \begin{bmatrix} 0 & 1 \\ -1 & 0 \end{bmatrix}$. Suppose that

$T = \begin{bmatrix} 3 & -1 \\ 1 & 5 \end{bmatrix}$. Since

$$T(v_1) = T(e_2) = \begin{bmatrix} -1 \\ 5 \end{bmatrix} = 5v_1 + 1v_2,$$

$$T(v_2) = T(-e_2) \begin{bmatrix} -3 \\ -1 \end{bmatrix} = 3v_2 - 1v_1,$$

the matrix of T in v_1, v_2 is

$$A = \begin{bmatrix} 5 & -1 \\ 1 & 3 \end{bmatrix}.$$

According to our theorem, A should equal

$$C^{-1}TC = \begin{bmatrix} 0 & 1 \\ -1 & 0 \end{bmatrix} \begin{bmatrix} 3 & -1 \\ 1 & 5 \end{bmatrix} \begin{bmatrix} 0 & -1 \\ 1 & 0 \end{bmatrix} = \begin{bmatrix} 5 & -1 \\ 1 & 3 \end{bmatrix},$$

as it is.

We return to the example preceding the theorem. Since

$$v_1 = \begin{bmatrix} 0 \\ 1 \\ 1 \end{bmatrix}, \quad v_2 = \begin{bmatrix} 1 \\ 0 \\ 1 \end{bmatrix}, \quad v_3 = \begin{bmatrix} 1 \\ 1 \\ 0 \end{bmatrix},$$

the C of the theorem is

$$\begin{bmatrix} 0 & 1 & 1 \\ 1 & 0 & 1 \\ 1 & 1 & 0 \end{bmatrix}.$$

Our matrix T was defined as

$$T = \begin{bmatrix} 1 & 2 & 3 \\ -1 & 4 & 7 \\ 5 & 0 & 1 \end{bmatrix}.$$

As we saw, the matrix of T in v_1, v_2, v_3 is

$$A = \begin{bmatrix} \frac{7}{2} & 4 & \frac{5}{2} \\ -\frac{5}{2} & 2 & \frac{5}{2} \\ \frac{15}{2} & 2 & \frac{1}{2} \end{bmatrix}.$$

We want to verify that $A = C^{-1}TC$. Since computing the inverse of C is a little messy, we will check the slightly weaker relation $CA = TC$. So, is the equation

$$\begin{bmatrix} 0 & 1 & 1 \\ 1 & 0 & 1 \\ 1 & 1 & 0 \end{bmatrix}\begin{bmatrix} \frac{7}{2} & 4 & \frac{5}{2} \\ -\frac{5}{2} & 2 & \frac{5}{2} \\ \frac{15}{2} & 2 & \frac{1}{2} \end{bmatrix} = \begin{bmatrix} 1 & 2 & 3 \\ -1 & 4 & 7 \\ 5 & 0 & 1 \end{bmatrix}\begin{bmatrix} 0 & 1 & 1 \\ 1 & 0 & 1 \\ 1 & 1 & 0 \end{bmatrix}$$

correct? Doing the matrix multiplication we see that both sides equal

$$\begin{bmatrix} 5 & 4 & 3 \\ 11 & 6 & 3 \\ 1 & 6 & 5 \end{bmatrix}.$$ So the theorem checks out—no surprise—in this example.

We can use our theorem to relate the matrix of a linear transformation T of $F^{(n)}$ in one basis to its matrix in another, as we observe in the

Corollary 3.9.2.   Let T be a linear transformation of $F^{(n)}$ and let $v_1, \ldots, v_n$ and $w_1, \ldots, w_n$ be two bases of $F^{(n)}$. Let A be the matrix of T in $v_1, \ldots, v_n$ and B the matrix of T in $w_1, \ldots, w_n$. Then $CAC^{-1} = DBD^{-1}$, where C is the matrix whose columns are $v_1, \ldots, v_n$ and D the matrix whose columns are $w_1, \ldots, w_n$.

Proof.   By Theorem 3.9.1, both CAC^{-1} and DBD^{-1} equal T when we regard the linear transformation T as a matrix.   ■

Corollary 3.9.2 shows us that the matrix B of T in $w_1, \ldots, w_n$ is related to the matrix A of T in $v_1, \ldots, v_n$ by $CAC^{-1} = DBD^{-1}$. Solving for B, we get

$$B = D^{-1}CAC^{-1}D = U^{-1}AU,$$

where $U = C^{-1}D$. Since $CU = D$, this implies that $w_s = \sum_{r=1}^{n} u_{rs}v_r$ for all s. (Prove!)

Definition.   We call U the *matrix of the change of basis* from the basis v_r to the basis w_s.

By these observations, we have proved the

Theorem 3.9.3.   Let T be a linear transformation of $F^{(n)}$ and let $v_1, \ldots, v_n$ and $w_1, \ldots, w_n$ be two bases for $F^{(n)}$. Then the matrix B of T in $w_1, \ldots, w_n$ is

$$B = U^{-1}AU,$$

where A is the matrix of T in $v_1, \ldots, v_n$ and $U = (u_{rs})$ is the matrix of the change of basis; that is, $w_s = \sum_{r=1}^{n} u_{rs}v_r$ for all s.

EXAMPLE

$$\text{Let } T = \begin{bmatrix} 1 & 0 & 1 \\ 1 & 0 & 0 \\ 2 & 0 & 0 \end{bmatrix} \text{ and } v_1 = \begin{bmatrix} 1 \\ 0 \\ 0 \end{bmatrix}, v_2 = \begin{bmatrix} 1 \\ 1 \\ 0 \end{bmatrix}, v_3 = \begin{bmatrix} 0 \\ 0 \\ 2 \end{bmatrix}.$$

According to Theorem 3.9.3, the matrix of T in v_1, v_2, v_3 is

$A = C^{-1}TC$, where $C = \begin{bmatrix} 1 & 1 & 0 \\ 0 & 1 & 0 \\ 0 & 0 & 2 \end{bmatrix}$. Following the method of

Section 3.8 for finding inverses, we find $C^{-1} = \begin{bmatrix} 1 & -1 & 0 \\ 0 & 1 & 0 \\ 0 & 0 & \frac{1}{2} \end{bmatrix}$,

so we get

$$A = \begin{bmatrix} 1 & -1 & 0 \\ 0 & 1 & 0 \\ 0 & 0 & \frac{1}{2} \end{bmatrix}\begin{bmatrix} 1 & 0 & 1 \\ 1 & 0 & 0 \\ 2 & 0 & 0 \end{bmatrix}\begin{bmatrix} 1 & 1 & 0 \\ 0 & 1 & 0 \\ 0 & 0 & 2 \end{bmatrix} = \begin{bmatrix} 0 & 0 & 2 \\ 1 & 1 & 0 \\ 1 & 1 & 0 \end{bmatrix}.$$

Similarly, for

$$w_1 = \begin{bmatrix} 0 \\ 1 \\ 0 \end{bmatrix}, \quad w_2 = \begin{bmatrix} 0 \\ 0 \\ 1 \end{bmatrix}, \quad w_3 = \begin{bmatrix} 1 \\ 0 \\ 0 \end{bmatrix},$$

we let $D = \begin{bmatrix} 0 & 0 & 1 \\ 1 & 0 & 0 \\ 0 & 1 & 0 \end{bmatrix}$, so $D^{-1} = \begin{bmatrix} 0 & 1 & 0 \\ 0 & 0 & 1 \\ 1 & 0 & 0 \end{bmatrix}$ and the matrix of

T in the basis w_1, w_2, w_3 is

$$B = D^{-1}TD$$

$$= \begin{bmatrix} 0 & 1 & 0 \\ 0 & 0 & 1 \\ 1 & 0 & 0 \end{bmatrix} \begin{bmatrix} 1 & 0 & 1 \\ 1 & 0 & 0 \\ 2 & 0 & 0 \end{bmatrix} \begin{bmatrix} 0 & 0 & 1 \\ 1 & 0 & 0 \\ 0 & 1 & 0 \end{bmatrix}$$

$$= \begin{bmatrix} 0 & 0 & 1 \\ 0 & 0 & 2 \\ 0 & 1 & 1 \end{bmatrix}.$$

The change of basis matrix from v_1, v_2, v_3 to w_1, w_2, w_3 is

$$U = C^{-1}D = \begin{bmatrix} 1 & -1 & 0 \\ 0 & 1 & 0 \\ 0 & 0 & \frac{1}{2} \end{bmatrix} \begin{bmatrix} 0 & 0 & 1 \\ 1 & 0 & 0 \\ 0 & 1 & 0 \end{bmatrix} = \begin{bmatrix} -1 & 0 & 1 \\ 1 & 0 & 0 \\ 0 & \frac{1}{2} & 0 \end{bmatrix},$$

which checks out since

$$\begin{bmatrix} 0 \\ 1 \\ 0 \end{bmatrix} = -1\begin{bmatrix} 1 \\ 0 \\ 0 \end{bmatrix} + 1\begin{bmatrix} 1 \\ 1 \\ 0 \end{bmatrix} + 0\begin{bmatrix} 0 \\ 0 \\ 2 \end{bmatrix}$$

$$\begin{bmatrix} 0 \\ 0 \\ 1 \end{bmatrix} = 0\begin{bmatrix} 1 \\ 0 \\ 0 \end{bmatrix} + 0\begin{bmatrix} 1 \\ 1 \\ 0 \end{bmatrix} + (\tfrac{1}{2})\begin{bmatrix} 0 \\ 0 \\ 2 \end{bmatrix}$$

$$\begin{bmatrix} 1 \\ 0 \\ 0 \end{bmatrix} = 1\begin{bmatrix} 1 \\ 0 \\ 0 \end{bmatrix} + 0\begin{bmatrix} 1 \\ 1 \\ 0 \end{bmatrix} + 0\begin{bmatrix} 0 \\ 0 \\ 2 \end{bmatrix}.$$

According to our theorem, the matrices $A = \begin{bmatrix} 0 & 0 & 2 \\ 1 & 1 & 0 \\ 1 & 1 & 0 \end{bmatrix}$ and

$B = \begin{bmatrix} 0 & 0 & 1 \\ 0 & 0 & 2 \\ 0 & 1 & 1 \end{bmatrix}$ of T in the two bases should be related by $B = U^{-1}AU$,

or by $UB = AU$. Checking, we find that this is in fact so:

$$\begin{bmatrix} -1 & 0 & 1 \\ 1 & 0 & 0 \\ 0 & \frac{1}{2} & 0 \end{bmatrix} \begin{bmatrix} 0 & 0 & 1 \\ 0 & 0 & 2 \\ 0 & 1 & 1 \end{bmatrix} = \begin{bmatrix} 0 & 1 & 0 \\ 0 & 0 & 1 \\ 0 & 0 & 1 \end{bmatrix}$$

$$= \begin{bmatrix} 0 & 0 & 2 \\ 1 & 1 & 0 \\ 1 & 1 & 0 \end{bmatrix} \begin{bmatrix} -1 & 0 & 1 \\ 1 & 0 & 0 \\ 0 & \frac{1}{2} & 0 \end{bmatrix}.$$

The relation expressed in Theorem 3.9.3, namely, $B = U^{-1}AU$ is an important one in matrix theory. It is called *similarity*.

Definition.   If $A, B \in M_n(F)$, then B is said to be *similar* to A if $B = C^{-1}AC$ for some invertible matrix C in $M_n(F)$. We denote this by $A \sim B$.

Similarity behaves very much like equality in that it obeys certain rules.

Theorem 3.9.4.   If A, B, C are in $M_n(F)$, then

1. $A \sim A$.
2. $A \sim B$ implies that $B \sim A$.
3. $A \sim B, B \sim C$ implies that $A \sim C$.

Proof.   To see that $A \sim A$ we need an invertible matrix M such that $A = M^{-1}AM$; well, $M = I$ certainly does the trick. So $A \sim A$.

If $A \sim B$, then $B = M^{-1}AM$, hence $A = (M^{-1})^{-1}B(M^{-1})$. Because M^{-1} is invertible we have that $B \sim A$.

Finally, if $A \sim B$ and $B \sim C$, then $A = M^{-1}BM$ and $B = N^{-1}CN$. Thus $A = M^{-1}(N^{-1}CN)M = (NM)^{-1}C(NM)$, and since NM is invertible, we get $A \sim C$.   ■

If $A \in M_n(F)$, then the set of all matrices B such that $A \sim B$ is a very important subset of $M_n(F)$.

Definition.   The *similarity class* of A, written as cl (A), is

$$\text{cl } (A) = \{B \in M_n(F) \mid A \sim B\}.$$

Note that A *is contained in its similarity class* cl (A). Moreover, A *is contained in no other similarity class* by

Theorem 3.9.5.   If A and B are in $M_n(F)$, then either their similarity classes are equal or they have no element in common.

Proof.   Suppose that the class of A and that of B have some matrix G in common. So $A \sim G$ and $B \sim G$. By property (b) in Theorem 3.9.4, $G \sim B$, hence, by property (c) in Theorem 3.9.4, $A \sim B$. So if $X \in$ cl (A), then $X \sim A$ and since $A \sim B$, we get that $X \sim B$, that is, $X \in$ cl (B). Therefore, cl (A) is contained in cl (B). But the argument works when we interchange the roles of A and B, so we get that cl (B) is contained in cl (A). Therefore, cl $(A) =$ cl (B).   ■

If B and A are similar, if we interpret A as a linear transformation in the canonical basis, then we can interpret B as the matrix of the same linear transformation in the basis $w_1, \ldots, w_n$, where $w_j = C(e_j)$ and where $B = C^{-1}AC$. This is exactly what Theorem 3.9.1 tells us. So in a certain sense, A and B can be viewed as coming from the same linear transformation.

Given a linear transformation T, there is nothing sacred impelling us to view T as a matrix in the canonical basis. We can view it as a matrix in any basis of our choice. *Why not pick a basis in which the matrix of T is the simplest looking?* Let's take an example. If T is the linear transformation whose matrix in the canonical basis is $\begin{bmatrix} 1 & 1 \\ 0 & 2 \end{bmatrix}$, what is the matrix in the basis $w_1 = e_1$, $w_2 = e_1 + e_2$? Since

$$T(w_1) = T(e_1) = e_1 = w_1$$

and

$$T(w_2) = T(e_1) + T(e_2) = e_1 + e_1 + 2e_2 = 2(e_1 + e_2) = 2w_2,$$

we see that the matrix of T in the basis w_1, w_2 is

$$\begin{bmatrix} 1 & 0 \\ 0 & 2 \end{bmatrix},$$

a diagonal matrix. In some sense this is a nicer looking matrix than

$$\begin{bmatrix} 1 & 1 \\ 0 & 2 \end{bmatrix}.$$

Since we are free to pick any basis we want, to get the matrix of a linear transformation, and any two bases give us matrices which are similar for this transformation, it is desirable to find nice matrices as road signs for our similarity classes. There are several different types of such road signs used in matrix theory. These are called *canonical forms*. To check that two matrices are similar then becomes checking if these two matrices have the same canonical form.

PROBLEMS

NUMERICAL PROBLEMS

1. Find the matrix B of T in the given basis, where T is the linear transformation whose matrix is given in the canonical basis.

(a) $T = \begin{bmatrix} 1 & 2 \\ 0 & 1 \end{bmatrix}$ in the basis $w_1 = e_2$, $w_2 = e_1$.

(b) $T = \begin{bmatrix} 1 & 1 & i \\ 0 & 0 & 1 \\ 2 & i & 0 \end{bmatrix}$ in the basis $w_1 = e_2$, $w_2 = e_3$, $w_3 = e_1$.

(c) $T = \begin{bmatrix} -1 & 0 & 1 \\ 0 & 1 & -1 \\ -4 & 1 & 0 \end{bmatrix}$ in the basis $w_1 = e_1 + e_2 + e_3$,

$w_2 = e_1 + e_2$, $w_3 = e_3$.

(d) $T = \begin{bmatrix} 1 & 2 & 0 & 1 \\ 0 & 1 & 2 & 0 \\ 0 & 0 & 1 & 2 \\ 0 & 0 & 0 & 1 \end{bmatrix}$ in the basis $w_1 = e_1$, $w_2 = e_1 + e_2$,

$w_3 = e_1 - e_2$, $w_4 = e_4$.

2. In each part of Problem 1 find an invertible matrix C such that $B = C^{-1}AC$, where A is the matrix T as given.

3. Find a basis w_1, w_2 of $\mathbb{C}^{(2)}$ such that the matrix of $\begin{bmatrix} 1 & i \\ 0 & 1 \end{bmatrix}$ in this

basis is $\begin{bmatrix} 1 & 1 \\ 0 & 1 \end{bmatrix}$.

4. Find a basis w_1, w_2, w_3 of $\mathbb{C}^{(3)}$ in which the matrix of

$\begin{bmatrix} 1 & 2 & 3 \\ 0 & i & 4 \\ 0 & 0 & -i \end{bmatrix}$ is diagonal.

5. Show that $\begin{bmatrix} 1 & 2 & 3 \\ 4 & 5 & 6 \\ 7 & 8 & 9 \end{bmatrix}$ and $\begin{bmatrix} 1 & 2 & 3 \\ 6 & 5 & 4 \\ 9 & 8 & 7 \end{bmatrix}$ are not similar.

(*Hint:* Use traces.)

6. Show that all matrices $\begin{bmatrix} 1 & a & b \\ 0 & 2 & c \\ 0 & 0 & 3 \end{bmatrix}$ are similar to $\begin{bmatrix} 1 & 0 & 0 \\ 0 & 2 & 0 \\ 0 & 0 & 3 \end{bmatrix}$.

7. Show that $\begin{bmatrix} 1 & 7 \\ -6 & -1 \end{bmatrix}$ and $\begin{bmatrix} i & 10 \\ -4 & -i \end{bmatrix}$ are similar in $M_2(\mathbb{C})$.

8. Show that $\begin{bmatrix} 1 & 0 & 1 \\ 0 & 1 & 0 \\ 0 & 0 & 1 \end{bmatrix}$ is *not* similar to I.

9. If A is a scalar matrix, find all elements in the similarity class of A.

MORE THEORETICAL PROBLEMS

Easier Problems

10. If you know the matrix of a linear transformation T in a basis $v_1, \ldots, v_n$ of $F^{(n)}$, what is the matrix of T in the basis $e_1, \ldots, e_n$ of $F^{(n)}$?

11. If A and B are the matrices of a linear transformation T in two different bases, show that tr $(A) = $ tr (B).

Middle-Level Problems

12. What is the matrix of $\begin{bmatrix} a_1 & & & 0 \\ & a_2 & & \\ & & \ddots & \\ 0 & & & a_n \end{bmatrix}$ in the basis $w_1 = e_{r_1}, \ldots, w_n = e_{r_n}$, where $e_{r_1}, \ldots, e_{r_n}$ are $e_1, \ldots, e_n$ in some order?

13. If $v_1, \ldots, v_m \in F^{(n)}$ are characteristic vectors associated with $a_1, \ldots, a_m$ which are *distinct* characteristic roots of $A \in M_n(F)$, show that $v_1, \ldots, v_m$ are linearly independent over F.

14. If the matrix $A \in M_n(F)$ has n distinct characteristic roots in F, show that $C^{-1}AC$ is a diagonal matrix for some invertible C in $M_n(F)$.

15. Show that there is no matrix C, invertible in $M_4(F)$, such that $C^{-1}\begin{bmatrix} 1 & 2 & 0 & 1 \\ 0 & 1 & 2 & 0 \\ 0 & 0 & 1 & 2 \\ 0 & 0 & 0 & 1 \end{bmatrix} C$ is a diagonal matrix.

16. Prove that if $A \in M_3(F)$ satisfies $A^3 = 0$, then for some invertible C in $M_3(F)$, $C^{-1}AC$ is an upper triangular matrix. What is the diagonal of $C^{-1}AC$?

17. If you know the matrix of a linear transformation T in the basis $v_1, \ldots, v_n$ of $F^{(n)}$, what is the matrix of T in terms of this in the basis $v_n, \ldots, v_1$ of $F^{(n)}$?

3.10. BASES AND INNER PRODUCTS

We saw in Section 3.9 that the nature of a given basis of $F^{(n)}$ over F has a strong influence on the form of a given linear transformation on $F^{(n)}$. If $F = \mathbb{C}$ (or $\mathbb{R}$) we saw earlier that if $v = \begin{bmatrix} x_1 \\ \vdots \\ x_n \end{bmatrix}$ and $w = \begin{bmatrix} y_1 \\ \vdots \\ y_n \end{bmatrix}$

are in $\mathbb{C}^{(n)}$, then their inner product $(v, w) = \sum_{r=1}^{n} x_r \bar{y}_r$ enjoys some very nice properties. How can we use the inner product on $\mathbb{C}^{(n)}$ or $\mathbb{R}^{(n)}$? This will be the major theme of this section.

Recall that two elements v and w of $\mathbb{C}^{(n)}$ are said to be orthogonal if $(v, w) = 0$. Suppose that $\mathbb{C}^{(n)}$ has a basis $v_1, \ldots, v_n$ where whenever $r \neq s$, $(v_r, v_s) = 0$. We give such a basis a name.

Definition. The basis $v_1, \ldots, v_n$ of $\mathbb{C}^{(n)}$ (or $\mathbb{R}^{(n)}$) is called an *orthogonal basis* if $(v_r, v_s) = 0$ for $r \neq s$.

What advantage does such a basis enjoy over just any old base? Well, for one thing, if we have a vector w in $\mathbb{C}^{(n)}$, say, we know that $w = a_1 v_1 + \cdots + a_n v_n$. Can we find a nice expression for these coefficients a_r? If we consider (w, v_r) we get

$$(w, v_r) = (a_1 v_1 + \cdots + a_n v_n, v_r)$$
$$= a_1 (v_1, v_r) + \cdots + a_n (v_n, v_r),$$

and since $(v_s, v_r) = 0$ if $s \neq r$, we end up with $a_r (v_r, v_r) = (w, v_r)$. Because $v_r \neq 0$, we have $(v_r, v_r) \neq 0$ and so $a_r = (w, v_r)/(v_r, v_r)$. So if we knew all the (v_r, v_r), we would have a nice, intrinsic expression for the a_r's in terms of w and the v_r's.

This expression would be even nicer if all the $(v_r, v_r) = 1$, that is, *if every v_j had length* 1. A vector of length 1 is called a *unit vector*.

Definition. An orthogonal basis $v_1, \ldots, v_n$ of $\mathbb{C}^{(n)}$ (or $\mathbb{R}^{(n)}$) is called an *orthonormal basis* of $\mathbb{C}^{(n)}$ (or $\mathbb{R}^{(n)}$) if each v_r is a unit vector.

So an orthonormal basis $v_1 \ldots, v_n$ is a basis of $\mathbb{C}^{(n)}$ (or $\mathbb{R}^{(n)}$) such that $(v_r, v_s) = 0$ if $r \neq s$, and $(v_r, v_r) = 1$ for all $r = 1, 2, \ldots, n$. Note that the canonical basis $e_1, \ldots, e_n$ is an orthonormal basis of $\mathbb{R}^{(n)}$ over $\mathbb{R}$ and of $\mathbb{C}^{(n)}$ over $\mathbb{C}$.

If $v_1, \ldots, v_n$ is an orthonormal basis of $\mathbb{C}^{(n)}$, say over $\mathbb{C}$, the computation above showed that if $w = a_1 v_1 + \cdots + a_n v_n$, then $a_r = (w, v_r)/(v_r, v_r) = (w, v_r)$ since $(v_r, v_r) = 1$. Plugging this in the expression for w, we obtain

Lemma 3.10.1. If $v_1, \ldots, v_n$ is an orthonormal basis of $\mathbb{C}^{(n)}$ (or $\mathbb{R}^{(n)}$), then given $w \in \mathbb{C}^{(n)}$ (or $\mathbb{R}^{(n)}$), $w = (w, v_1)v_1 + \cdots + (w, v_n)v_n$.

Suppose that $v_1, \ldots, v_n$ is an orthonormal basis of $F^{(n)}$. Given $u = \sum_{r=1}^{n} a_r v_r$ and $w = \sum_{s=1}^{n} b_s v_s$, what does their inner product (u, w) look like in terms of the a's and b's?

Lemma 3.10.2. If $v_1, \ldots, v_n$ is an orthonormal basis of $F^{(n)}$ and $u = \sum_{r=1}^{n} a_r v_r$, $w = \sum_{s=1}^{n} b_s v_s$ are in $F^{(n)}$, then $(u, w) = \sum_{s=1}^{n} a_s \bar{b}_s$.

Proof. Before going to the general result, let's look at a simple case. Suppose that $u = a_1 v_1 + a_2 v_2$ and $w = b_1 v_1 + b_2 v_2$. Then

$$
\begin{aligned}
(u, w) &= (a_1 v_1 + a_2 v_2, b_1 v_1 + b_2 v_2) \\
&= (a_1 v_1, b_1 v_1 + b_2 v_2) + (a_2 v_2, b_1 v_1 + b_2 v_2) \\
&= a_1(v_1, b_1 v_1 + b_2 v_2) + a_2(v_2, b_1 v_1 + b_2 v_2) \\
&= a_1(v_1, b_1 v_1) + a_1(v_1, b_2 v_2) + a_2(v_2, b_1 v_1) + a_2(v_2, b_2 v_2) \\
&= a_1 \bar{b}_1(v_1, v_1) + a_1 \bar{b}_2(v_1, v_2) + a_2 \bar{b}_1(v_2, v_1) + a_2 \bar{b}_2(v_2, v_2) \\
&= a_1 \bar{b}_1 + a_2 \bar{b}_2
\end{aligned}
$$

since $(v_1, v_1) = (v_2, v_2) = 1$ and $(v_1, v_2) = (v_2, v_1) = 0$. Notice that in carrying out this computation we have used the additive properties of the inner product and the rules for pulling a scalar out of an inner product.

The argument just given for the simple case is the basic clue of how the lemma should be proved. As we did above,

$$
(u, w) = \left(\sum_{r=1}^{n} a_r v_r, \sum_{s=1}^{n} b_s v_s \right) = \sum_{r=1}^{n} \sum_{s=1}^{n} a_r \bar{b}_s (v_r, v_s),
$$

by the defining properties of the inner product. In this double sum, the only nonzero contribution comes if $r = s$, for otherwise $(v_r, v_s) = 0$. So the double sum reduces to a single sum, namely,

$$
(u, w) = \sum_{s=1}^{n} a_s \bar{b}_s (v_s, v_s),
$$

and since $(v_s, v_s) = 1$ we get that

$$(u, w) = \sum_{s=1}^{n} a_s \overline{b}_s. \quad \blacksquare$$

Notice that if $v_1 \ldots, v_n$ is the canonical basis $e_1 \ldots, e_n$, then the result shows that if $u = \sum_{r=1}^{n} a_r e_r$ and $w = \sum_{s=1}^{n} b_s e_s$, then

$$(u, w) = \sum_{s=1}^{n} a_s \overline{b}_s.$$

This should come as no surprise because this is precisely the way in which we defined the inner product (u, w).

Suppose that $v_1, \ldots, v_n$ is an orthonormal basis of $F^{(n)}$. We saw in Theorem 3.8.1 that the linear transformation T defined by $T(e_r) = v_r$ for $r = 1, 2, \ldots, n$ is invertible. However, both $e_1 \ldots, e_n$ and $v_1, \ldots, v_n$ are not any old bases of $F^{(n)}$ but are, in fact, both orthonormal. Surely this should force some further condition on T other than it be invertible. Indeed, it does!

Theorem 3.10.3   If $v_1, \ldots, v_n$ is an orthonormal basis of $F^{(n)}$, then the linear transformation T, defined by $v_r = Te_r$ for $r = 1, 2, \ldots, n$ has the property that for all u, w in $F^{(n)}$, $(Tu, Tw) = (u, w)$.

Proof. If $u = \sum_{r=1}^{n} a_r e_r$ and $w = \sum_{r=1}^{n} b_r e_r$ are in $F^{(n)}$, then $(u, w) = \sum_{s=1}^{n} a_s \overline{b}_s$. Now, by the definition of T and the fact that it is a linear transformation on $F^{(n)}$,

$$Tu = T\left(\sum_{r=1}^{N} a_r e_r\right) = \sum_{r=1}^{n} T(a_r e_r) = \sum_{r=1}^{n} a_r Te_r = \sum_{r=1}^{n} a_r v_r.$$

Similarly, $Tw = \sum_{s=1}^{n} b_s v_s$. By the result of Lemma 3.10.2, we find that

$$(Tu, Tw) = \sum_{s=1}^{n} a_s \overline{b}_s = (u, w),$$

as asserted. $\blacksquare$

In the theorem, the condition $(Tu, Tw) = (u, w)$ for all $u, w \in F^{(n)}$ is equivalent to the condition $TT^* = I$, which we single out in the

Definition.   A linear transformation T of $F^{(n)}$ is *unitary* if $TT^* = I$.

We illustrate Theorem 3.10.3 with an example. If $w_1 = e_1 + e_2$ and $w_2 = e_1 - e_2$, what is (w_1, w_2)? Calculating,

$$
\begin{aligned}
(w_1, w_2) &= (e_1 + e_2, e_1 - e_2) \\
&= (e_1, e_1) - (e_1, e_2) + (e_2, e_1) - (e_2, e_2) \\
&= 1 - 0 + 0 - 1 = 0.
\end{aligned}
$$

So w_1 and w_2 are orthogonal. Also, $(w_1, w_1) = 2$ and $(w_2, w_2) = 2$ (verify!), so the length of w_1, that is, $\sqrt{(w_1, w_1)}$ is $\sqrt{2}$, as is that of w_2. So if we let $v_1 = (1/\sqrt{2})w_1$ and $v_2 = (1/\sqrt{2})w_2$, then

$$(v_1, v_1) = ([1/\sqrt{2}]w_1, [1/\sqrt{2}]w_1) = \tfrac{1}{2}(w_1, w_1) = \tfrac{1}{2}2 = 1.$$

Similarly, $(v_2, v_2) = 1$. So v_1 and v_2 form an orthonormal basis of $F^{(2)}$.

The change of basis linear transformation T is defined by $T(e_j) = v_j$ for $j = 1, 2$. So

$$T(e_1) = v_1 = (1/\sqrt{2})w_1 = (1/\sqrt{2})e_1 + (1/\sqrt{2})e_2,$$

and

$$T(e_2) = (1/\sqrt{2})e_1 - (1/\sqrt{2})e_2.$$

So the matrix of T in the basis e_1, e_2 is

$$
\begin{bmatrix}
1/\sqrt{2} & 1/\sqrt{2} \\
1/\sqrt{2} & -1/\sqrt{2}
\end{bmatrix}.
$$

Notice that

$$
TT^* =
\begin{bmatrix}
1/\sqrt{2} & 1/\sqrt{2} \\
1/\sqrt{2} & -1/\sqrt{2}
\end{bmatrix}
\begin{bmatrix}
1/\sqrt{2} & 1/\sqrt{2} \\
1/\sqrt{2} & -1/\sqrt{2}
\end{bmatrix}
=
\begin{bmatrix}
1 & 0 \\
0 & 1
\end{bmatrix};
$$

that is, T is unitary.

Theorem 3.10.3 can be generalized. We leave the proof of the next result to the reader.

Theorem 3.10.4. Let $v_1, \ldots, v_n$ and $w_1, \ldots, w_n$ be two orthonormal bases of $F^{(n)}$. Then the linear transformation T defined by $T(v_r) = w_r$ for $r = 1, 2, \ldots, n$ is unitary.

What Theorem 3.10.4 says is that changing bases from one orthonormal basis to another is carried out by a unitary transformation.

What do these results translate into matricially? By Theorem 3.9.3, if A is a matrix, then the matrix of A—as a linear transformation—in a basis $v_1, \ldots, v_n$ is $B = C^{-1}AC$, where C is the matrix defined by $v_r = C(e_r)$ for some $r = 1, 2, \ldots, n$. If, furthermore, $v_1, \ldots, v_n$ is an orthonormal basis, what is implied for C? We have an exercise that $CC^* = I$; that is, C is unitary. That is,

Theorem 3.10.5. If A is a matrix in $M_n(F)$, then A is transformed as a matrix, in the orthonormal basis $v_1, \ldots, v_n$, into the matrix $B = C^{-1}AC$, where $v_r = C(e_r)$ and C is unitary.

In closing this section, we mention a useful theorem on independence of orthogonal vectors.

Theorem 3.10.6. If $v_1, \ldots, v_r$ are nonzero mutually orthogonal vectors in $F^{(n)}$, then they are linearly independent.

Proof. If $a_1 v_1 + \cdots + a_r v_r = 0$ and $1 \leq s \leq r$, then $0 = (a_1 v_1 + \cdots + a_r v_r, v_s) = a_s(v_s, v_s)$. Since v_s is nonzero and $(v_s, v_s) \neq 0$, a_s must be 0. Since this is true for all s, the $v_1, \ldots, v_r$ are linearly independent. ∎

PROBLEMS

NUMERICAL PROBLEMS

1. Check which of the following are orthonormal bases of the given $F^{(n)}$.

(a) $\begin{bmatrix} \frac{1}{2} \\ 0 \\ 1 \end{bmatrix}$, $\begin{bmatrix} 0 \\ 1 \\ 0 \end{bmatrix}$, $\begin{bmatrix} -1 \\ 1 \\ 1 \end{bmatrix}$ in $F^{(3)}$.

(b) $\begin{bmatrix} i/\sqrt{2} \\ -i/\sqrt{2} \end{bmatrix}$, $\begin{bmatrix} -1/\sqrt{2} \\ 1/\sqrt{2} \end{bmatrix}$ in $\mathbb{C}^{(2)}$.

(c) $\begin{bmatrix} 1 \\ 0 \\ 0 \end{bmatrix}$, $\begin{bmatrix} 0 \\ 1 \\ 0 \end{bmatrix}$, $\begin{bmatrix} \frac{1}{2} \\ 0 \\ \sqrt{3}/2 \end{bmatrix}$ in $F^{(3)}$.

2. Find an element in $F^{(3)}$ orthogonal to both $\begin{bmatrix} 1 \\ 0 \\ 0 \end{bmatrix}$ and $\begin{bmatrix} 0 \\ 1 \\ 1 \end{bmatrix}$.

3. Find the matrix of the change of basis from $\begin{bmatrix} 1 \\ 0 \\ 0 \end{bmatrix}$, $\begin{bmatrix} 0 \\ 1 \\ 0 \end{bmatrix}$, $\begin{bmatrix} 0 \\ 0 \\ 1 \end{bmatrix}$

to $\begin{bmatrix} i/\sqrt{2} \\ 0 \\ -1/\sqrt{2} \end{bmatrix}$, $\begin{bmatrix} 0 \\ 1 \\ 0 \end{bmatrix}$, $\begin{bmatrix} 0 \\ 0 \\ i \end{bmatrix}$ in $\mathbb{C}^{(3)}$.

4. Verify that the matrix you obtain in Problem 3 is unitary.

5. For what value of a is the basis $\begin{bmatrix} a \\ -a \end{bmatrix}$ and $\begin{bmatrix} i/\sqrt{2} \\ i/\sqrt{2} \end{bmatrix}$ an orthonormal basis of $\mathbb{C}^{(2)}$?

MORE THEORETICAL PROBLEMS

Middle-level Problems

6. Give a complete proof of Theorem 3.10.4.

7. Give a complete proof of Theorem 3.10.5.

8. If $v \neq 0$ is in $F^{(n)}$, for what values of a in F is $\dfrac{v}{a}$ a unit vector?

9. If $v_1, \ldots, v_n$ is an orthogonal basis of $F^{(n)}$, construct from them an orthonormal basis of $F^{(n)}$.

10. If v_1, v_2, v_3 is a basis of $F^{(3)}$, use them to construct an orthogonal basis of $F^{(3)}$. How would you make this new basis into an orthonormal one?

11. If v_1, v_2, v_3 are in $F^{(4)}$, show that you can find a vector w in $F^{(4)}$ that is orthogonal to all of v_1, v_2, and v_3.

12. If T is a unitary linear transformation on $F^{(n)}$ and $T(v_1), \ldots, T(v_n)$ is an orthonormal basis of $F^{(n)}$, show that $v_1, \ldots, v_n$ is an orthonormal basis of $F^{(n)}$.

Harder Problems

13. If A is a symmetric matrix in $M_n(F)$ and if $v_1, \ldots, v_n$ in $F^{(n)}$ are nonzero and such that $Av_1 = a_1v_1$, $Av_2 = a_2v_2, \ldots, Av_n = a_nv_n$, where the a_j are *distinct* elements of F, show

(a) $v_1, \ldots, v_n$ is an orthogonal basis of $F^{(n)}$.

(b) We can construct from $v_1, \ldots, v_n$ an orthogonal basis of $F^{(n)}$, $w_1, \ldots, w_n$, such that $Aw_j = a_jw_j$ for $j = 1, 2, \ldots, n$.

(c) There exists a matrix C in $M_n(F)$ such that

$$C^{-1}AC = \begin{bmatrix} a_1 & & & 0 \\ & a_2 & & \\ & & \ddots & \\ 0 & & & a_n \end{bmatrix}.$$

14. In Part (c) of Problem 13 show that we can find a unitary C such that

$$C^{-1}AC = \begin{bmatrix} a_1 & & & 0 \\ & a_2 & & \\ & & \ddots & \\ 0 & & & a_n \end{bmatrix}.$$

4

MORE ON $n \times n$ MATRICES

4.1. SUBSPACES

When we first considered $F^{(n)}$ we observed that $F^{(n)}$ enjoys two fundamental properties, namely, if $v, w \in F^{(n)}$, then $v + w$ is also in $F^{(n)}$, and if $a \in F$, then av is again in $F^{(n)}$. It is quite possible that a nonempty subset of $F^{(n)}$ could imitate $F^{(n)}$ in this regard and also satisfy these two properties. Such subsets are thus distinguished among all other subsets of $F^{(n)}$. We single them out in

Definition. A nonempty subset W of $F^{(n)}$ is called a *subspace* of $F^{(n)}$ if:

1. $v, w \in W$ implies that $v + w \in W$.
2. $a \in F, v \in W$ implies that $av \in W$.

A few simple examples of subspaces come to mind immediately. Clearly, $F^{(n)}$ itself is a subspace of $F^{(n)}$. Also, the set consisting of the element 0 alone is a subspace of $F^{(n)}$. These are called *trivial subspaces*.

A subspace $W \neq F^{(n)}$ is called a *proper subspace* of $F^{(n)}$. What are

214

some proper subspaces of $F^{(n)}$? Let

$$W = \left\{ \begin{bmatrix} x \\ y \\ z \end{bmatrix} \in F^{(3)} \mid x + y + z = 0 \right\}.$$

Then W is a proper subspace of $F^{(3)}$. Why? If $\begin{bmatrix} a \\ b \\ c \end{bmatrix}$ and $\begin{bmatrix} u \\ v \\ w \end{bmatrix}$ are in W,

then $a + b + c = 0$ and $u + v + w = 0$, hence

$$(a + u) + (b + v) + (c + w) = 0.$$

Thus the element

$$\begin{bmatrix} a + u \\ b + v \\ c + w \end{bmatrix} = \begin{bmatrix} a \\ b \\ c \end{bmatrix} + \begin{bmatrix} u \\ v \\ w \end{bmatrix}$$

satisfies the requirements for membership in W, hence is in W. Similarly,

$t\begin{bmatrix} a \\ b \\ c \end{bmatrix} = \begin{bmatrix} ta \\ tb \\ tc \end{bmatrix}$ is such that $ta + tb + tc = 0$ if $a + b + c = 0$. Thus

$t\begin{bmatrix} a \\ b \\ c \end{bmatrix}$ is in W for $t \in F$ and $\begin{bmatrix} a \\ b \\ c \end{bmatrix}$ in W. Therefore, W is a subspace of

$F^{(3)}$. The element $\begin{bmatrix} 1 \\ 0 \\ 0 \end{bmatrix}$ is not in W since $1 + 0 + 0 \neq 0$. So, W is also

proper.

In fact, the vectors $\begin{bmatrix} u_1 \\ \vdots \\ u_n \end{bmatrix}$, where the elements $u_1, \ldots, u_n$ are the

solutions of a system of homogeneous linear equations, form a subspace of $F^{(n)}$. (Prove!) The case above where $x + y + z = 0$ in $F^{(3)}$ is merely a special case of this.

If T is a linear transformation on $F^{(n)}$ and if

$$W = \{ T(v) \mid v \in F^{(n)} \}$$

then W is a subspace of $F^{(n)}$. Why? If $w_1 = Tv_1$ and $w_2 = Tv_2$, then

$$w_1 + w_2 = Tv_1 + Tv_2 = T(v_1 + v_2),$$

so is in W. Similarly, if $a \in F$ and $w_1 = Tv_1$, then

$$aw_1 = aTv_1 = T(av_1),$$

so aw_1 is in W. This subspace W, which we may denote as $T(F^{(n)})$, is called the *image space* of T. As we shall see, $T(F^{(n)})$ is a *proper* subspace of $F^{(n)}$ if and only if T is *not* invertible. For example, the image space

of $T = \begin{bmatrix} 1 & 1 & 3 \\ 0 & 1 & 1 \\ 1 & 0 & 2 \end{bmatrix}$ is the plane $W = \left\{ g\begin{bmatrix} 1 \\ 0 \\ 1 \end{bmatrix} + h\begin{bmatrix} 1 \\ 1 \\ 0 \end{bmatrix} \mid g, h \in \mathbb{R} \right\}$

since the last column is 2 times the first plus 1 times the second, so that

$$\begin{bmatrix} 1 & 1 & 3 \\ 0 & 1 & 1 \\ 1 & 0 & 2 \end{bmatrix}\begin{bmatrix} a \\ b \\ c \end{bmatrix} = a\begin{bmatrix} 1 \\ 0 \\ 1 \end{bmatrix} + b\begin{bmatrix} 1 \\ 1 \\ 0 \end{bmatrix} + c\begin{bmatrix} 3 \\ 1 \\ 2 \end{bmatrix}$$

$$= (a + 2c)\begin{bmatrix} 1 \\ 0 \\ 1 \end{bmatrix} + (b + c)\begin{bmatrix} 1 \\ 1 \\ 0 \end{bmatrix}$$

$$= g\begin{bmatrix} 1 \\ 0 \\ 1 \end{bmatrix} + h\begin{bmatrix} 1 \\ 1 \\ 0 \end{bmatrix} = \begin{bmatrix} g + h \\ h \\ g \end{bmatrix}$$

for $g = a + 2c$ and $h = b + c$:

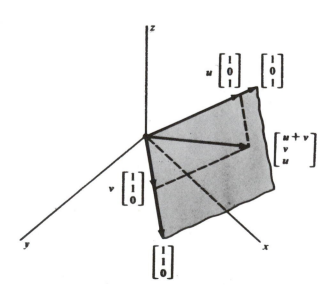

Finally, the span $<v_1, \ldots, v_r>$ over F of vectors $v_1, \ldots, v_r$ in $F^{(n)}$ is a subspace of $F^{(n)}$. Since we have tremendous flexibility in forming such a span (we simply choose a finite number of vectors any way we want to), you may wonder whether every subspace of $F^{(n)}$ can be gotten as such a span. The answer is "yes." In fact, every subspace has a basis in the sense of the following generalization of our definition for basis of $F^{(n)}$.

Definition. A *basis* for a subspace V of $F^{(n)}$ is a set $v_1, \ldots, v_r$ of vectors of $F^{(n)}$ such that

(1) The vectors $v_1, \ldots, v_r$ are linearly independent.

(2) $v_1 \ldots, v_r$ span V over F, that is $<v_1, \ldots, v_r> = V$.

How can we get a basis for a subspace V of $F^{(n)}$? Easily! We know that if $v_1, \ldots, v_r$ are linearly independent elements of $F^{(n)}$, $r \leq n$. Thus we can take a collection $v_1 \ldots, v_r$ of distinct linearly independent vectors of V over F which is *maximal* in the sense that any larger set $v_1, \ldots, v_r, v$ of vectors in V is linearly dependent over V. But then the vectors $v_1 \ldots, v_r$ form a basis. Why? We need only show that any vector v in V is a linear combination of $v_1, \ldots, v_r$. To show this, note that the set of vectors $v_1, \ldots, v_r, v$ must be linearly dependent by the maximality condition for $v_1, \ldots, v_r$. Letting $a_1, \ldots, a_r, a$ be elements of F (not all zero) such that

$$a_1 v_1 + \cdots + a_r v_r + av = 0.$$

We can't have $a = 0$ since the vectors $v_1 \ldots, v_r$ are linearly independent. After all, if $a = 0$, then $a_1 v_1 + \cdots + a_r v_r = 0$ and so $a_1 = \cdots a_r = 0$, since $v_1, \ldots, v_r$ are linearly independent. Hence we can solve for v as a linear combination

$$v = -\frac{1}{a}(a_1 v_1 + \cdots + a_r v_r) = -\frac{1}{a} a_1 v_1 - \cdots - \frac{1}{a} a_r v_r$$

of $v_1, \ldots, v_r$. Thus the linearly independent vectors $v_1, \ldots, v_r$ also span V, so form a basis for V over F. This proves the

Theorem 4.1.1. Every maximal linearly independent set $v_1, \ldots, v_r$ of vectors in a subspace V of $F^{(n)}$ is a basis for V over F.

By starting with any linearly independent set $v_1, \ldots, v_r$ of vectors in a subspace V of $F^{(n)}$ and enlarging it to a maximal linearly independent set $v_1, \ldots, v_s$ of vectors in V, we put the vectors $v_1, \ldots, v_r$ in a basis $v_1, \ldots, v_s$ for V over F, by Theorem 4.1.1. This procedure of *extending* the set $v_1, \ldots, v_r$ to a basis $v_1, \ldots, v_s$ is very useful, so we state our observations here in the

Corollary 4.1.2.   Every linearly independent set $v_1, \ldots, v_r$ of vectors in a subspace V of $F^{(n)}$ can be extended to a basis of V over F.

We saw in Corollary 3.6.3 that if $v_1, \ldots, v_r$ and $w_1, \ldots, w_s$ are two linear independent sets of vectors in $F^{(n)}$ such that $<v_1, \ldots, v_r>$ and $<w_1 \ldots, w_s>$ are equal, then $r = s$. This proves the important

Theorem 4.1.3.   Any two bases for a subspace V of $F^{(n)}$ over F have the same number of elements.

Thanks to Theorems 4.1.1 and 4.1.3, we now can associate a well-defined dimension to any subspace V of $F^{(n)}$.

Definition.   The *dimension* of a subspace V of $F^{(n)}$ over F is the number of elements of a basis for V over F. We denote it by dim (V).

We now proceed to investigate the properties of this dimension function, starting with

Theorem 4.1.4.   Let V and W be subspaces of $F^{(n)}$ and suppose that V is a *proper* subset of W. Then dim $(V) <$ dim (W).

Proof.   Take a basis $v_1, \ldots, v_r$ for V over F. Then extend the basis $v_1, \ldots, v_r$ of V to a basis $v_1, \ldots, v_s$ for W. Since V and W are not equal, r does not equal s. But then $r < s$.   ■

The case of Theorem 4.1.4 with $W = F^{(n)}$ is the

Corollary 4.1.5.   If V is a subspace of $F^{(n)}$ of dimension n, then $V = F^{(n)}$.

PROBLEMS

NUMERICAL PROBLEMS

1. In $F^{(3)}$ describe every element in $<w_1, \ w_2, \ w_3>$, where

$$w_1 = \begin{bmatrix} 1 \\ 0 \\ 0 \end{bmatrix}, \quad w_2 = \begin{bmatrix} 1 \\ 2 \\ 0 \end{bmatrix}, \quad w_3 = \begin{bmatrix} 1 \\ 2 \\ 3 \end{bmatrix}.$$

2. What is the dimension of $<w_1, w_2, w_3>$ in Problem 1?

3. In $\mathbb{C}^{(4)}$ find the form of every element in $<w_1, w_2, w_3>$, where

$$w_1 = \begin{bmatrix} 1 \\ 1 \\ 0 \\ 0 \end{bmatrix}, \quad w_2 = \begin{bmatrix} i \\ 0 \\ 1 \\ 1 \end{bmatrix}, \quad w_3 = \begin{bmatrix} i \\ i \\ 0 \\ 0 \end{bmatrix}.$$

4. What is the dimension of the $<w_1, w_2, w_3>$ in Problem 3 over $\mathbb{C}$?

5. If $A = \begin{bmatrix} 1 & 2 & 3 \\ 0 & 4 & 2 \\ 2 & 4 & 6 \end{bmatrix}$, find the dimension of $A(F^{(3)})$ over F. Describe

the general element in $A(F^{(3)})$.

MORE THEORETICAL PROBLEMS

Easier Problems

6. Prove Corollary 4.1.2.

7. Show that if $w_1, \ldots, w_r$ are in $F^{(n)}$, then $<w_1, \ldots, w_r>$ is a subspace of $F^{(n)}$.

8. Show that dim $(<w_1, \ldots, w_r>) \leq r$ and equals r if and only if $w_1, \ldots, w_r$ are linearly independent over F.

9. If $z_1, \ldots, z_m$ are in $<w_1, \ldots, w_r>$, prove that

$$<z_1, \ldots, z_m> \subset <w_1, \ldots, w_r>.$$

10. If V and W are subspaces of $F^{(n)}$, show that $V \cap W$ is a subspace of $F^{(n)}$.

11. If $W_1, \ldots, W_k$ are subspaces of $F^{(n)}$, show that

$$W_1 \cap W_2 \cap \cdots \cap W_k$$

is a subspace of $F^{(n)}$.

12. If $A \in M_n(F)$, let $W = \{v \in F^{(n)} \mid Av = 0\}$. Prove that W is a subspace of $F^{(n)}$.

13. If $A \in M_n(F)$, let V be the set

$$\{v \in F^{(n)} \mid A^k v = 0 \text{ for some positive integer } k \text{ depending on } v\}.$$

Show that V is a subspace of $F^{(n)}$.

14. If $A \in M_n(F)$, let $W_a = \{v \in F^{(n)} \mid Av = av\}$, where $a \in F$. Show that W_a is a subspace of $F^{(n)}$.

15. For $A \in M_n(F)$, let V_a be the set

$\{v \in F^{(n)} \mid (A - aI)^k v = 0$ for some positive integer k depending on $v\}$,

where $a \in F$. Prove that V_a is a subspace of $F^{(n)}$.

16. If $a \neq b$ are in F, show that $V_a \cap V_b = \{0\}$, where V_a, V_b are as defined in Problem 15.

17. If V and W are subspaces of $F^{(n)}$, define the *sum*, $V + W$, of V and W by $V + W = \{v + w \mid v \in V, w \in W\}$. Show that $V + W$ is a subspace of $F^{(n)}$.

18. If $V_1, \ldots, V_k$ are subspaces of $F^{(n)}$, how would you define the sum, $V_1 + V_2 + \cdots + V_k$, of $V_1, \ldots, V_k$? Show that what you get is a subspace of $F^{(n)}$.

Middle-Level Problems

19. If V and W are subspaces of $F^{(n)}$ such that $V \cap W = \{0\}$, show that every element in $V + W$ has a *unique* representation in the form $v + w$, where v is in V and w is in W. (In this case we call $V + W$ the *direct sum* of V and W and write it as $V \oplus W$.)

20. If V_1, V_2, V_3 are subspaces of $F^{(n)}$ such that every element of $V_1 + V_2 + V_3$ has a *unique* representation in the form $v_1 + v_2 + v_3$, where $v_1 \in V_1, v_2 \in V_2, v_3 \in V_3$, show that

$$V_1 \cap (V_2 + V_3) = V_2 \cap (V_1 + V_3) = V_3 \cap (V_1 + V_2) = \{0\}.$$

(In this case $V_1 + V_2 + V_3$ is called the *direct sum* of V_1, V_2, V_3 and is written as $V_1 \oplus V_2 \oplus V_3$.)

21. Generalize the result of Problem 20 to $V_1, \ldots, V_k$ subspaces of $F^{(n)}$.

Harder Problems

22. If $A \in M_n(F)$ and V is a subspace of $F^{(n)}$, we say that V is *invariant* under A if $Av \in V$ for all $v \in V$, that is, if $AV \subset V$. Prove that the subspace V_a of Problem 15 is invariant under A.

23. If T is a linear transformation on $F^{(n)}$ and V is a subspace of $F^{(n)}$ invariant under T (i.e., $TV \subset V$), show that $\tilde{T}$ defined on V by $\tilde{T}v = Tv$ is a mapping from V to V such that $\tilde{T}(v + w) = \tilde{T}v + \tilde{T}w$ and $\tilde{T}(cv) = c\tilde{T}(v)$ for all $c \in F$.

24. If V and W are subspaces of $F^{(n)}$ such that $F^{(n)} = V \oplus W$ (see Problem 19) and $v_1, \ldots, v_k$ is a basis of V over F and $w_1, \ldots,$

w_m is a basis of W over F, show that $v_1, \ldots, v_k, w_1, \ldots, w_m$ is a basis of $F^{(n)}$ over F.

25. If V, W, $v_1, \ldots, v_k, w_1, \ldots, w_m$ are as in Problem 24 and if T is a linear transformation on $F^{(n)}$ so that $T(V) \subset V$ and $T(W) \subset W$, show that the matrix of T in the bases $v_1, \ldots, v_k, w_1, \ldots, w_m$

    looks like $\left[\begin{array}{c|c} A & 0 \\ \hline 0 & B \end{array} \right]$, where A is a $k \times k$ matrix and B is an $m \times m$ matrix over F.

26. How could you describe A and B in Problem 25 in terms of the linear transformations $\tilde{T}v = Tv$ ($v \in V$) of V and $\tilde{T}w = Tw$ ($w \in W$) of W?

27. If $v \in F^{(n)}$, let $v^{\perp} = \{ w \in F^{(n)} \mid (v, w) = 0 \}$. Prove that $v^{\perp}$ is a subspace of $F^{(n)}$.

28. If W is a subspace of $F^{(n)}$ and $w_1, \ldots, w_m$ is a basis of W over F, suppose that $v \in F^{(n)}$ is such that $(w_j, v) = 0$ for $j = 1, 2, \ldots, m$. Show that $(w, v) = 0$ for all $w \in W$.

29. If W is a subspace of $F^{(n)}$, let

    $$W^{\perp} = \{ v \in F^{(n)} \mid (v, w) = 0 \text{ for all } w \in W \}.$$

    Show that $W^{\perp}$ is a subspace of $F^{(n)}$.

30. If W and $W^{\perp}$ are as in Problem 29, show that the sum $W + W^{\perp}$ is a direct sum.

NOTE TO THE READER. The problems from Section 4.1 on represent important ideas that we shall develop more formally in the next few sections. The results and techniques needed to solve them are not only important for matrices but will play a key role in discussing vector spaces. Don't give in to the temptation to peek ahead to solve them. Try to solve them on your own.

4.2. MORE ON SUBSPACES

We shall examine here some of the items that appeared as exercises in the problem set of Section 4.1. If you haven't tried those problems yet, don't read on. Go back and take a stab at the problems. Solving them on your own is n times as valuable for your understanding of the material than letting us solve them for you, with n a very large integer.

We begin with

Definition.   If V and W are subspaces of $F^{(n)}$, their *sum* $V + W$ is defined by $V + W = \{v + w \mid v \in V, w \in W\}$.

Of course, $V + W$ is not merely a subset of $F^{(n)}$; it inherits the basic properties of $F^{(n)}$. In short, $V + W$ is a subspace of $F^{(n)}$. We proceed to prove it now. The proof is easy but is the prototype of many proofs that we might need to verify that a given subset of $F^{(n)}$ is a subspace of $F^{(n)}$.

Lemma 4.2.1.   If V and W are subspaces of $F^{(n)}$, then $V + W$ is a subspace of $F^{(n)}$.

Proof.   What must we show in order to establish the lemma? We need only demonstrate two things: first, if z_1 and z_2 are in $V + W$, then $z_1 + z_2$ is also in $V + W$, and second, if a is any element of F and z is in $V + W$, then az is also in $V + W$.

Suppose then that z_1 and z_2 are in $V + W$. By the definition of $V + W$ we have that $z_1 = v_1 + w_1$ and $z_2 = v_2 + w_2$, where v_1, v_2 are in V and w_1, w_2 are in W. Therefore,

$$z_1 + z_2 = v_1 + w_1 + v_2 + w_2 = (v_1 + v_2) + (w_1 + w_2).$$

Because V and W are subspaces of $F^{(n)}$, we know that $v_1 + v_2$ is in V and $w_1 + w_2$ is in W. Hence $z_1 + z_2$ is in $V + W$.

If $a \in F$ and $z = v + w$ is in $V + W$, then $az = a(v + w) = av + aw$. However, V and W are subspaces of $F^{(n)}$, so $av \in V$ and $aw \in W$. Thus $az \in V + W$.

Having shown that $V + W$ satisfies the criteria for a subspace of $F^{(n)}$, we know that $V + W$ is a subspace of $F^{(n)}$.   ∎

Of course, having in hand the fact that $V + W$ is a subspace of $F^{(n)}$ allows us to prove that *the sum of any number of subspaces of $F^{(n)}$ is again a subspace of $F^{(n)}$*. We leave this as an exercise for the reader.

Some sums of subspaces are better than others. For instance, if

$$V_1 = \left\{ \begin{bmatrix} x \\ t \\ t \end{bmatrix} \mid x, t \in \mathbb{R} \right\} \text{ and } V_2 = \left\{ \begin{bmatrix} x \\ x \\ t \end{bmatrix} \mid x, t \in \mathbb{R} \right\}, V_1 + V_2 \text{ is } \mathbb{R}^{(3)},$$

but elements $v_1 + v_2$ from this sum can be expressed in more than one way, for example,

$$\begin{bmatrix} 1 \\ 2 \\ 2 \end{bmatrix} + \begin{bmatrix} 2 \\ 2 \\ 4 \end{bmatrix} = \begin{bmatrix} 3 \\ 4 \\ 6 \end{bmatrix} = \begin{bmatrix} 2 \\ 3 \\ 3 \end{bmatrix} + \begin{bmatrix} 1 \\ 1 \\ 3 \end{bmatrix}.$$

A better sum is $V_1 + V_3$, where $V_3 = \left\{ \begin{bmatrix} x \\ x \\ x \end{bmatrix} \mid x \in \mathbb{R} \right\}$. This sum is also

$\mathbb{R}^{(3)}$, but it also has the property that elements $v_1 + v_3$ (where $v_1 \in V_1$, $v_3 \in V_3$) can be expressed in only one way. This sum is an example of the special sums which we introduce in

Definition. If $V_1, \ldots, V_m$ is a set of subspaces of $F^{(n)}$, their sum $V_1 + V_2 + \cdots + V_m$ is called a *direct sum* if very element w in $V_1 + V_2 + \cdots + V_m$ has a *unique* representation as

$$w = v_1 + v_2 + \cdots + v_m,$$

where each v_j is in V_j for $j = 1, 2, \ldots, m$.

We shall denote the fact that the sum $V_1 + \cdots + V_m$ of the subspaces $V_1, \ldots, V_m$ is a direct sum by writing it as

$$V_1 \oplus V_2 \oplus \cdots \oplus V_m.$$

We consider an easier case first, that in which $m = 2$. When can we be sure that $V + W$ is a direct sum? We claim that all we need, in this special instance, is that $V \cap W = \{0\}$. Assuming that $V \cap W = \{0\}$, suppose $z \in V + W$ has two representations $z = v_1 + w_1$ and $z = v_2 + w_2$, where v_1, v_2 are in V and w_1, w_2 are in W. Therefore, $v_1 + w_1 = v_2 + w_2$, so that $v_1 - v_2 = w_2 - w_1$. But $v_1 - v_2$ is in V and $w_2 - w_1$ is in W, and since they are equal, we must have that $v_1 - v_2 \in V \cap W = \{0\}$ and $w_2 - w_1 \in V \cap W = \{0\}$. These yield that $v_1 = v_2$ and $w_1 = w_2$. So z only has one representation in the form $v + w$. Thus the sum of V and W is direct.

In the opposite direction, we claim that if $V + W = V \oplus W$, that is, the sum is direct, then $V \cap W = \{0\}$. Why? Suppose that u is in both V and W. Then certainly $u = u + 0$, where the u and 0 on the right-hand side are viewed as in V and W, respectively. However, we also know that $u = 0 + u$, where $0, u$ are in V and W, respectively. So $u + 0 = 0 + u$ in $V \oplus W$; by the directness of the sum we conclude that $u = 0$. Hence $V \cap W = \{0\}$.

We have proved

Theorem 4.2.2. If V and W are nonzero subspaces of $F^{(n)}$, their sum $V + W$ is a *direct sum* if and only if $V \cap W = \{0\}$.

Whereas the criterion that the sum of two subspaces be direct is quite simple, that for determining the directness of *more* than two subspaces is much more complicated.

It is no longer true that the sum $V_1 + V_2 + \cdots + V_m$ is a direct sum if and only if $V_j \cap V_k = \{0\}$ for all $j \neq k$.

It is true that if the sum $V_1 + \cdots + V_m$ is direct, then $V_j \cap V_k = \{0\}$ for $j \neq k$. The proof of this is a slight adaptation of the argument given above showing $V \cap W = 0$. *But the condition is not sufficient.* The best way of demonstrating this is by an example. In $F^{(3)}$ let

$$V_1 = \left\{ \begin{bmatrix} a \\ b \\ 0 \end{bmatrix} \,\middle|\, a, b \in F \right\}, \qquad V_2 = \left\{ \begin{bmatrix} 0 \\ 0 \\ c \end{bmatrix} \,\middle|\, c \in F \right\},$$

$$V_3 = \left\{ \begin{bmatrix} d \\ d \\ d \end{bmatrix} \,\middle|\, d \in F \right\}.$$

We easily see that

$$V_1 + V_2 = F^{(3)},$$
$$V_1 \cap V_2 = V_1 \cap V_3 = V_2 \cap V_3 = \{0\},$$

yet the sum $V_1 + V_2 + V_3$ is *not* direct. To see why, simply express some nonzero element $\begin{bmatrix} a \\ b \\ 0 \end{bmatrix}$ of V_1 in terms of elements $\begin{bmatrix} 0 \\ 0 \\ c \end{bmatrix}$ and $\begin{bmatrix} d \\ d \\ d \end{bmatrix}$.

We can do this by taking $a = b = d = 1$ and $c = -1$, getting the two representations

$$1\begin{bmatrix} 1 \\ 1 \\ 0 \end{bmatrix} + 0\begin{bmatrix} 0 \\ 0 \\ -1 \end{bmatrix} + 0\begin{bmatrix} 1 \\ 1 \\ 1 \end{bmatrix} = \begin{bmatrix} 1 \\ 1 \\ 0 \end{bmatrix} = 1\begin{bmatrix} 0 \\ 0 \\ 0 \end{bmatrix} + 1\begin{bmatrix} 0 \\ 0 \\ -1 \end{bmatrix} + 1\begin{bmatrix} 1 \\ 1 \\ 1 \end{bmatrix}.$$

The necessary and sufficient condition for the directness of a sum is given in

Theorem 4.2.3.   If $V_1 \ldots, V_m$ are nonzero subspaces of $F^{(n)}$, then their sum is a direct sum if and only if

$$V_j \cap (V_1 + V_2 + \cdots + V_{j-1} + V_{j+1} + \cdots + V_m) = \{0\}$$

for each $j = 1, 2, \ldots, m$.

     We leave the proof of Theorem 4.2.3 to the reader with the reminder that all one must show is that every element in $V_1 + \cdots + V_m$ has a unique representation of the form $v_1 + v_2 + \cdots + v_m$, where $v_j \in V_j$, if and only if the condition stated in the theorem holds.

     We give an instance of when a sum is direct. If $W \neq \{0\}$ is a subspace of $F^{(n)}$, let

$$W^\perp = \{v \in F^{(n)} \mid (v, w) = 0 \text{ for all } w \in W\}.$$

That $W^\perp$ is a subspace of $F^{(n)}$ we leave to the reader (see Problem 29 of Section 4.1). We claim that

$F^{(n)} = W \oplus W^\perp$ *if we happen to know that W has an*
orthonormal basis.

(As we shall soon see, this is no restriction on W; every subspace W will be shown to have an orthornormal basis.)

     We need the following remark to carry out the proof:

If $w_1, \ldots, w_k$ is a basis of W and $v \in F^{(n)}$ is such that
$(v, w_j) = 0$ for $j = 1, 2, \ldots, k$, then $v \in W^\perp$.

We leave this as an exercise.

Theorem 4.2.4.   Let $V_1, \ldots, V_r$ be mutually orthogonal subspaces of $F^{(n)}$. Then the sum $V_1 + \cdots + V_r$ is direct.

     Proof.   Suppose that $v \in V_1 \cap (V_2 + \cdots + V_r)$. Then v is orthogonal to v (prove!), which implies that $v = 0$. So,

$$V_1 \cap (V_2 + \cdots + V_r) = \{0\}.$$

Similarly, for $s > 1$

$$V_s \cap (V_{j_2} + \cdots + V_{j_r}) = \{0\},$$

where $j_2, \ldots, j_r$ are the integers from 1 to r leaving out s. So the sum $V_1 + \cdots + V_r$ is direct.   ■

Theorem 4.2.5. If $W \neq \{0\}$ is a subspace of $F^{(n)}$, and if W has an orthonormal basis $w_1, \ldots, w_k$, then $F^{(n)} = W \oplus W^\perp$.

Proof. The sum $W + W^\perp$ is direct by Theorem 4.2.4. So it remains only to show that $W + W^\perp = F^{(n)}$. For this, let v be any element of $F^{(n)}$ and consider the element

$$z = v - (v, w_1)w_1 - (v, w_2)w_2 - \cdots - (v, w_k)w_k.$$

Calculating (z, w_j) for $j = 1, 2, \ldots, k$, we get

$$(z, w_j) = (v - (v, w_1)w_1 - \cdots - (v, w_k)w_k, w_j)$$
$$= (v, w_j) - (v, w_1)(w_1, w_j) - \cdots - (v, w_k)(w_k, w_j).$$

Because $w_1, \ldots, w_k$ is an orthonormal basis of W, $(w_r, w_s) = 0$ if $r \neq s$ and $(w_r, w_r) = 1$ for all r, s. So the sum above only has nonzero contributions from the term (v, w_j) and $-(v, w_j)(w_j, w_j)$. In a word, the value we get is

$$(z, w_j) = (v, w_j) - (v, w_j) = 0.$$

Therefore, z is orthogonal to a basis of W; by the remark preceding the theorem, $z \in W^\perp$. But

$$z = v - (v, w_1)w_1 - \cdots - (v, w_k)w_k,$$

hence

$$v = z + ((v, w_1)w_1 - \cdots - (v, w_k)w_k),$$

where $z \in W^\perp$. So we have shown that every v in $F^{(n)}$ is in $W + W^\perp$. Thus $F^{(n)} = W \oplus W^\perp$.   ∎

We repeat what we said earlier. The condition "W has an orthonormal basis" is no restriction whatsoever on W. Every nonzero subspace has an orthonormal basis. To show this is our next objective.

PROBLEMS

NUMERICAL PROBLEMS

1. If $A = \begin{bmatrix} 1 & -1 & 0 \\ 0 & 2 & 3 \\ 0 & 0 & 0 \end{bmatrix}$ and $B = \begin{bmatrix} 0 & 0 & 0 \\ 0 & 1 & 0 \\ 1 & 0 & 1 \end{bmatrix}$ are in $M_3(F)$, then

(a) Find the form of the general element of $A(F^{(3)})$ and of $B(F^{(3)})$.

(b) Show that from Part (a), $A(F^{(3)})$ and $B(F^{(3)})$ are subspaces of $F^{(3)}$.

(c) Find the general form of the elements of $A(F^{(3)} + B(F^{(3)})$.

(d) From the result of Part (c), show that $A(F^{(3)}) + B(F^{(3)})$ is a subspace of $F^{(3)}$.

(e) Is $A(F^{(3)}) + B(F^{(3)}) = (A + B)(F^{(3)})$?

(f) What is dim $(A(F^{(3)}) + B(F^{(3)}))$?

2. If $A = \begin{bmatrix} 1 & 2 \\ 0 & 0 \end{bmatrix}$ and $B = \begin{bmatrix} 0 & 0 \\ 1 & a \end{bmatrix}$, find conditions on a so that

$A(F^{(2)}) + B(F^{(2)}) \neq (A + B)(F^{(2)})$.

3. In Problem 1 find the explicit form of each element of

$$A(F^{(3)}) \cap B(F^{(3)}).$$

What is its dimension?

4. In Problem 3, verify that dim $(A(F^{(3)}) + B(F^{(3)}))$ is

$$\text{dim } (A(F^{(3)})) + \text{dim } (B(F^{(3)})) - \text{dim } (A(F^{(3)}) \cap B(F^{(3)})).$$

5. If $v_1 = \begin{bmatrix} 0 \\ 1 \\ 1 \\ 1 \end{bmatrix}$, $v_2 = \begin{bmatrix} 0 \\ 1 \\ 1 \\ 0 \end{bmatrix}$, $v_3 = \begin{bmatrix} 0 \\ 1 \\ 0 \\ 0 \end{bmatrix}$, $w_1 = \begin{bmatrix} 1 \\ 0 \\ 1 \\ 0 \end{bmatrix}$, $w_2 = \begin{bmatrix} 0 \\ 0 \\ 1 \\ 1 \end{bmatrix}$,

find the form of the general elements in $<v_1, v_2, v_3>$ and $<w_1, w_2>$. From these find the general element in $<v_1, v_2, v_3> + <w_1 w_2>$.

6. What are dim $(<v_1, v_2, v_3>)$, dim $(<w_1, w_2>$? What is dim $(<v_1, v_2, v_3> + <w_1, w_2>)$?

7. Is the sum $<v_1, v_2, v_3> + <w_1, w_2>$ in Problem 6 direct?

8. If $v_1 = \begin{bmatrix} 0 \\ 1 \\ 1 \\ 1 \end{bmatrix}$, $v_2 = \begin{bmatrix} 0 \\ 0 \\ 1 \\ 1 \end{bmatrix}$, $v_3 = \begin{bmatrix} 0 \\ 0 \\ 0 \\ 1 \end{bmatrix}$, and $w_1 = \begin{bmatrix} 1 \\ 0 \\ 0 \\ 0 \end{bmatrix}$,

$w_2 = \begin{bmatrix} i \\ 0 \\ 0 \\ 0 \end{bmatrix}$ are in $\mathbb{C}^{(4)}$, find $V = <v_1, v_2, v_3>$, $W = <w_1, w_2>$,

and $V + W$.

9. Is the sum $<v_1, v_2, v_3>$ + $<w_1, w_2>$ in Problem 8 direct?

10. If v_1, v_2, v_3, w_1 are as in Problem 8, for what values of

$$w = \begin{bmatrix} a \\ b \\ c \\ d \end{bmatrix}$$ is the sum $V + W$, where $V = <v_1, v_2, v_3>$ and

$W = <w_1, w>$, direct?

11. Find $<v_1, v_2>^{\perp}$ if $v_1 = \begin{bmatrix} 1 \\ 2 \\ -1 \end{bmatrix}$, $v_2 = \begin{bmatrix} 0 \\ 1 \\ 1 \end{bmatrix}$ are in $F^{(3)}$.

12. If $W = <e_1, e_2, \ldots, e_{n-1}>$ in $F^{(n)}$, find $W^{\perp}$.

13. In Problem 11 show that $<v_1, v_2> + <v_1, v_2>^{\perp} = F^{(3)}$.

MORE THEORETICAL PROBLEMS

Easier Problems

14. If $W = <e_1, \ldots, e_m>$ in $F^{(n)}$, show that $(W^{\perp})^{\perp} = W$.

15. If $V_1, \ldots, V_m$ are subspaces of $F^{(n)}$, define $V_1 + V_2 + \cdots + V_m$ and show that it is a subspace of $F^{(n)}$.

16. If $V_1 + \cdots + V_m$ is a *direct* sum of the subspaces $V_1, \ldots, V_m$ of $F^{(n)}$, show that $V_j \cap V_k = \{0\}$ if $j \neq k$.

17. If $v \in F^{(n)}$ is such that $(v, w_j) = 0$ for all basis elements $w_1, \ldots, w_k$ of W, show that $v \in W^{\perp}$.

18. If $w_1, \ldots, w_k$ is a basis of orthogonal vectors of the subspace W [i.e., $(w_r, w_s) = 0$ if $r \neq s$], show how we can use $w_1, \ldots, w_k$ to get an orthonormal basis of W.

19. If v_1, v_2, v_3 in $F^{(n)}$ are linearly independent over F, construct an orthonormal basis for $<v_1, v_2, v_3>$.

Middle-Level Problems

20. If v_1, v_2, v_3, v_4 is a basis of $W \subset F^{(n)}$, construct an orthonormal basis for $<v_1, v_2, v_3, v_4> = W$.

21. If $W \neq F^{(5)}$ is a subspace of $F^{(5)}$, show that $W + W^{\perp} = F^{(5)}$.

22. If W is a subspace of $F^{(5)}$, show that $(W^{\perp})^{\perp} = W$.

23. If W is any subspace of $F^{(n)}$, show that $(W^{\perp})^{\perp} \supset W$.

24. If $A \in M_5(F)$ is Hermitian and if W is a subspace of $F^{(5)}$ such that $A(W) \subset W$, prove that $A(W^{\perp}) \subset W^{\perp}$.

4.3. GRAM–SCHMIDT ORTHOGONALIZATION PROCESS

In Theorem 4.2.5 we proved that if W is a subspace of $F^{(n)}$ *which has an orthonormal basis*, then $W \oplus W^{\perp} = F^{(n)}$. It would be aesthetically satisfying and highly useful if we could drop the assumption "W has an orthonormal basis" in this theorem and prove for *all* subspaces W of $F^{(n)}$, $F^{(n)} = W \oplus W^{\perp}$. One way of achieving this is to show that *any* subspace of $F^{(n)}$ has an orthonormal basis. Fortunately this is true, and we shall prove it in a short while. But first we need the procedure for constructing such an orthonormal basis. This is provided us by what is known as the *Gram–Schmidt Orthogonalization Process*.

Note first that if we could produce an orthogonal basis $w_1, \ldots, w_k$ for a subspace W of $F^{(n)}$, then it is extremely easy to adapt this basis to get an orthonormal one. If $a_j = \|w_j\|$, then for $v_j = w_j/a_j$ we have

$$\|v_j\| = \|w_j/a_j\| = (1/a_j)\|w_j\| = 1.$$

So $v_1, \ldots, v_k$ are unit vectors in W. They are also orthogonal since

$$(v_r, v_s) = \left(\frac{w_r}{a_r}, \frac{w_s}{a_s}\right) = (1/a_r a_s)(w_r, w_s) = 0 \text{ for } r \neq s. \text{ Thus } v_1, \ldots,$$

v_k is an orthonormal basis of W.

So, to get to where we want to get to, *we must merely produce an orthogonal basis* of W. Before jumping to the general case we try some simple (and sample) examples.

Suppose that v_1 and v_2 in $F^{(n)}$ are linearly independent over F and let, as usual, $\langle v_1, v_2 \rangle$ be the subspace of $F^{(n)}$ spanned by v_1 and v_2 over F. We want to find w_1 and w_2 in $\langle v_1, v_2 \rangle$ which span $\langle v_1, v_2 \rangle$ over F and, furthermore, satisfy $(w_1, w_2) = 0$. Our choice for w_1 is very easy; namely, we let

$$w_1 = v_1.$$

How to get w_2? Let $w_2 = v_2 + av_1 = v_2 + aw_1$, where a in F is to be chosen judiciously. We want $(w_2, w_1) = 0$, that is we want $(v_2 + aw_1, w_1) = 0$. But $(v_2 + aw_1, w_1) = (v_2, w_1) + a(v_1, w_1)$; hence this is 0 if $a = -\dfrac{(v_2, w_1)}{(w_1, w_1)}$. Therefore, our choice for w_2 is

$$w_2 = v_2 - \frac{(v_2, w_1)}{(w_1, w_1)} w_1,$$

and since $v_1 = w_1$ and v_2 are linearly independent over F, we know that $w_2 \neq 0$. Note that w_1 and w_2 are in $\langle v_1, v_2 \rangle$; and, moreover, since

$v_1 = w_1$, $v_2 = w_2 + \dfrac{(v_2, w_1)}{(w_1, w_1)} w_1$, the span of w_1 and w_2 contains the span of v_1 and v_2. Thus $<w_1, w_2> \supset <v_1, v_2>$, which together with $<w_1, w_2> \subset <v_1, v_2>$, yields $<v_1, v_2> = <w_1, w_2>$. Finally, $(w_1, w_2) = 0$ since this was forced by our choice of a.

Let us go on to the case of v_1, v_2, v_3 linearly independent over F. *We will use the result proved above for two vectors to settle this case.* Let

$$w_1 = v_1,$$

$$w_2 = v_2 - \frac{(v_2, w_1)}{(w_1, w_1)} w_1,$$

as above. So $w_1 \neq 0$, $w_2 \neq 0$, and $(w_1, w_2) = 0$. We seek an element $w_3 \neq 0$ in $<v_1, v_2, v_3>$ such that $(w_3, w_1) = (w_3, w_2) = 0$ and $<v_1, v_2, v_3> = <w_1, w_2, w_3>$. Let $w_3 = v_3 + aw_2 + bw_1$. Since $(w_2, w_1) = 0$, to get $(w_3, w_1) = 0$, that is, $(v_3 + aw_2 + bw_1, w_1) = 0$, we have to satisfy $(v_3, w_1) + b(w_1, w_1) = 0$. So $b = -\dfrac{(v_3, w_1)}{(w_1, w_1)}$. To have $(w_3, w_2) = 0$, we want $(v_3 + aw_2 + bw_1, w_2) = 0$, and so $(v_3, w_2) + a(w_2, w_2) = 0$. Thus we use $a = -\dfrac{(v_3, w_2)}{(w_2, w_2)}$. So our choice for w_3 is

$$w_3 = v_3 - \frac{(v_3, w_2)}{(w_2, w_2)} w_2 - \frac{(v_3, w_1)}{(w_1, w_1)} w_1.$$

Is $w_3 \neq 0$? Since w_1, w_2 only involve v_1 and v_2, and since v_1, v_2, v_3 are linearly independent over F, $w_3 \neq 0$. Finally, is $<w_1, w_2, w_3>$ all of $<v_1, v_2, v_3>$? From the construction in the paragraph above, v_1 and v_2 are linear combinations of w_1 and w_2. By construction

$$v_3 = w_3 + \frac{(v_3, w_2)}{(w_2, w_2)} w_2 + \frac{(v_3, w_1)}{(w_1, w_1)} w_1,$$

so v_3 is a linear combination of w_1, w_2, w_3. Therefore,

$$<v_1, v_2, v_3> \subset <w_1, w_2, w_3> \subset <v_1, v_2, v_3>,$$

yielding that $<v_1, v_2, v_3> = <w_1, w_2, w_3>$.

The discussion of the cases of two and three vectors has deliberately been long-winded and perhaps overly meticulous. However, its payoff will be a simple discussion of the general case. What we do is a bootstrap operation, going from one case to the next one involving one more vector.

Suppose we know that for some positive integer k for any set $v_1, \ldots, v_k$ of linearly independent elements of $F^{(n)}$ we can find *mutually orthogonal* nonzero elements $w_1, \ldots, w_k$ in $\langle v_1, \ldots, v_k \rangle$ such that

(1) $\langle v_1, \ldots, v_k \rangle = \langle w_1, \ldots, w_k \rangle$

(2)   for each $j = 1, 2, \ldots, k$, $w_j \in \langle v_1, v_2, \ldots, v_j \rangle$.   (*)

(We certainly did this explicitly for $k = 2$ and 3.) Given $v_1, v_2, \ldots, v_k, v_{k+1}$ in $F^{(n)}$, linearly independent over F, can we find $w_1, \ldots, w_{k+1}$ *mutually orthogonal* nonzero elements in $\langle v_1, \ldots, v_{k+1} \rangle$ such that

(1) $\langle v_1, \ldots, v_{k+1} \rangle = \langle w_1, \ldots, w_{k+1} \rangle$

(2)   for each $j = 1, 2, \ldots, k, k + 1$,
    $w_j \in \langle v_1, v_2, \ldots, v_j \rangle$?   (**)

By (*) we have $w_1, \ldots, w_k$ already satisfying our requirements. How do we pick w_{k+1}? Guided by the cases done for $k = 2$ and $k = 3$ we simply let w_{k+1} be

$$v_{k+1} - \frac{(v_{k+1}, w_k)}{(w_k, w_k)} w_k - \frac{(v_{k+1}, w_{k-1})}{(w_{k-1}, w_{k-1})} w_{k-1} - \cdots - \frac{(v_{k+1}, w_1)}{(w_1, w_1)} w_1.$$

We leave the verification that w_{k+1} is nonzero and that $w_1, \ldots, w_{k+1}$ satisfy the requirements in (**) to the reader.

Knowing the result for $k = 3$, by the argument above, we know it to be true for $k = 4$. Knowing it for $k = 4$ we get it for $k = 5$, and so on, to get that the result is true for every positive integer n. We have proved

Theorem 4.3.1. If $v_1, \ldots, v_m$ in $F^{(n)}$ are linearly independent over F, then the subspace $\langle v_1, \ldots, v_m \rangle$ of $F^{(n)}$ has a basis $w_1, \ldots, w_m$ such that $(w_j, w_k) = 0$ for all $j \neq k$. That is, $\langle v_1, \ldots, v_m \rangle$ has an orthogonal basis, hence an orthonormal basis.

Theorem 4.3.1 has an immediate consequence:

Theorem 4.3.2. If $W \neq 0$ is a subspace of $F^{(n)}$, then W has an orthonormal basis.

Proof. W has a basis $v_1, \ldots, v_m$, so $W = \langle v_1, \ldots, v_m \rangle$. Thus, by Theorem 4.3.1, W has an orthonormal basis. ∎

In Theorem 4.2.5 we proved that if $W \neq 0$ is a subspace of $F^{(n)}$ and if W *has an orthonormal basis*, then $F^{(n)} = W \oplus W^\perp$. By Theorem

4.3.2, *every* nonzero subspace W of $F^{(n)}$ has an orthonormal basis. Thus we get the definitive form of Theorem 4.2.5,

Theorem 4.3.3. Given any subspace W of $F^{(n)}$, then $F^{(n)} = W \oplus W^{\perp}$.

Corollary 4.3.4. If W is a subspace of $F^{(n)}$, then

$$\dim (W^{\perp}) = n - \dim (W).$$

Proof. Let $w_1, \ldots, w_k$ be a basis of W over F, and $v_1, \ldots, v_r$ a basis of $W^{\perp}$. Then $w_1, \ldots, w_k, v_1, \ldots, v_r$ is a basis for $F^{(n)}$, so $n = k + v$ or $r = n - k$. This translates into $\dim (W^{\perp}) = n - \dim (W)$. ∎

From Theorem 4.3.3 we get a nice result, namely, given any set of nonzero mutually orthogonal elements of $F^{(n)}$, we can fill it out to an orthogonal basis of $F^{(n)}$. More explicitly,

Corollary 4.3.5. Given any subspace $W \neq \{0\}$ of $F^{(n)}$ and any subspace V of $F^{(n)}$ containing W, $V = W \oplus (V \cap (W^{\perp}))$.

Proof. If $v \in V$, then $v = w + u$ with $w \in W$ and $u \in W^{\perp}$. But then $u = v - w \in V$, since $W \subset V$. So $V = W + V \cap (W^{\perp})$. Obviously $W \cap (V \cap (W^{\perp})) = \{0\}$. ∎

Theorem 4.3.6. If $v_1, \ldots, v_k$ in $F^{(n)}$ are mutually orthogonal over F, then we can find mutually orthogonal elements $u_1, \ldots, u_{n-k}$ in $F^{(n)}$ such that $v_1, \ldots, v_k, u_1, \ldots, u_{n-k}$ is an orthogonal basis of $F^{(n)}$ over F.

Proof. Let $W = \langle v_1, \ldots, v_k \rangle$; so $v_1, \ldots, v_k$ is an orthogonal basis of W, by Theorem 3.10.6. By Theorem 4.3.3, $F^{(n)} = W \oplus W^{\perp}$. Let $u_1, \ldots, u_r$ be an orthogonal basis of $W^{\perp}$ over F. Then

$$v_1, \ldots, v_k, u_1, \ldots, u_r$$

is an orthogonal basis of $F^{(n)}$ over F. (Prove!) Since $\dim (F^{(n)}) = n$, and since $v_1, \ldots, v_k, u_1, \ldots, u_r$ is a basis of $F^{(n)}$ over F, $n = k + r$, so $r = n - k$. The theorem is proved. ∎

Note that we also get from Theorem 4.3.4 the result that given any set of linearly independent elements of $F^{(n)}$, we can fill it out to a basis for $F^{(n)}$. This is proved along the same lines as is the proof of Theorem 4.3.6.

We close this section with

Theorem 4.3.7. If $W \neq 0$ is a subspace of $F^{(n)}$, then $(W^{\perp})^{\perp} = W$.

Proof. Let $V = W^{\perp}$. If $w \in W$ and $v \in V$, then $(v, w) = 0$. Hence $w \in V^{\perp} = (W^{\perp})^{\perp}$. This says that $W \subset V^{\perp} = (W^{\perp})^{\perp}$.

By the Corollary 4.3.4, dim $(W^{\perp}) = n - $ dim (W). Also, by Corollary 4.3.4,

$$\dim (W^{\perp})^{\perp} = n - \dim (W^{\perp}) = n - (n - \dim (W)) = \dim (W).$$

Therefore, W is a subspace of $(W^{\perp})^{\perp}$ and is of the same dimension as $(W^{\perp})^{\perp}$. Thus $W = (W^{\perp})^{\perp}$. ∎

PROBLEMS

NUMERICAL PROBLEMS

1. For the given $v_1, \ldots, v_k$ use the Gram–Schmidt process to get an orthonormal basis of $\langle v_1, \ldots, v_k \rangle$.

(a) $v_1 = \begin{bmatrix} 1 \\ 2 \\ 3 \end{bmatrix}$, $v_2 = \begin{bmatrix} 0 \\ 1 \\ 2 \end{bmatrix}$.

(b) $v_1 = \begin{bmatrix} 1 \\ i \\ 0 \\ 0 \end{bmatrix}$, $v_2 = \begin{bmatrix} 0 \\ i \\ 1 \\ 0 \end{bmatrix}$, $v_3 = \begin{bmatrix} 1 \\ i \\ 0 \\ 0 \end{bmatrix}$ in $\mathbb{C}^{(4)}$.

(c) $v_1 = \begin{bmatrix} i \\ 2 \\ 3 \\ 4 \\ 5 \end{bmatrix}$, $v_2 = \begin{bmatrix} 1 \\ 1 \\ 1 \\ 1 \\ 1 \end{bmatrix}$, $v_3 = \begin{bmatrix} 0 \\ 1 \\ 1 \\ i \\ 0 \end{bmatrix}$ in $\mathbb{C}^{(5)}$.

(d) $v_1 = \begin{bmatrix} i \\ -1 \\ 1 \\ -1 \\ 1 \\ -1 \end{bmatrix}$, $v_2 = \begin{bmatrix} 0 \\ 1 \\ 2 \\ 0 \\ 1 \\ 1 \end{bmatrix}$, $v_3 = \begin{bmatrix} 1 \\ 0 \\ 1 \\ 0 \\ 1 \\ 0 \end{bmatrix}$, $v_4 = \begin{bmatrix} i \\ 0 \\ 0 \\ 0 \\ 0 \\ 1 \end{bmatrix}$ in $\mathbb{C}^{(6)}$.

2. For each part of Problem 1 find $\langle v_1, \ldots, v_k \rangle^{\perp}$ explicitly.

3. For each part of Problem 1 verify that

$$\langle v_1, \ldots, v_k \rangle + \langle v_1, \ldots, v_k \rangle^{\perp} = F^{(n)}.$$

4. For each part of Problem 1, find $(<v_1, \ldots, v_k>^{\perp})^{\perp}$ and show that $(v_1, \ldots, v_k>^{\perp})^{\perp} = <v_1, \ldots, v_k>$.

5. For each part of Problem 1, fill out the given $v_1, \ldots, v_k$ to a basis for $F^{(n)}$.

MORE THEORETICAL PROBLEMS

Easier Problems

6. Fill in all the details in the proof of Theorem 4.3.1 to show that $w_1, \ldots, w_{k+1}$ satisfy (**).

7. What are $W^{\perp}$ if $W = 0$, and if $W = F^{(n)}$?

8. If $A \in M_n(\mathbb{C})$ is Hermitian and a is a real number, let $W = \{v \in \mathbb{C}^{(n)} \mid Av = av\}$. Prove that $(A - aI)(\mathbb{C}^{(n)}) \subset W^{\perp}$.

Harder Problems

9. If $A \in M_n(\mathbb{C})$ is Hermitian, and W is a subspace of $\mathbb{C}^{(n)}$ such that $A(W) \subset W$, prove that $AW^{\perp} \subset W^{\perp}$.

10. In Problem 8 show that $A(W) \subset W$.

11. If the W in Problem 8 is of dimension $k \geq 1$ (so that a is a characteristic root of A), show that we can find an orthonormal basis of $F^{(n)}$ in which the matrix of A is of the form $\begin{bmatrix} aI_k & 0 \\ 0 & B \end{bmatrix}$, where I_k is the $k \times k$ unit matrix and B is a Hermitian $(n - k) \times (n - k)$ matrix.

4.4. RANK AND NULLITY

Given a matrix A In $M_n(F)$ [or, if you want, a linear transformation on $F^{(n)}$,] we shall attach two integers to A which help to describe the nature of A. These integers are merely the dimensions of two subspaces of $F^{(n)}$ intimately related to A.

The set $N = \{v \in F^{(n)} \mid Av = 0\}$ is a subspace of $F^{(n)}$, because $Av = Aw = 0$ implies that $A(v + w) = Av + Aw = 0$ and $A(av) = aA(v) = 0$ for all $a \in F$. It is called the *nullspace* of A.

Definition.   The *nullity* of A, written $n(A)$, is the dimension of N over F.

Thus if the nullity of A is $n = \dim (F^{(n)})$, then $N = F^{(n)}$, so $Av = 0$ for all $v \in F^{(n)}$. In other words, $A = 0$. At the other end of the spectrum, if $n(A) = 0$, then $N = 0$, so $Av = 0$ forces $v = 0$. This tells us that A is invertible.

The set $A(F^{(n)}) = \{Av \mid v \in F^{(n)}\}$ is called the *range* of A. It is a subspace of $F^{(n)}$. The second integer we attach to A is the dimension of the range of A.

Definition. The *rank* of A, written $r(A)$, is the dimension of $A(F^{(n)})$ over F.

So, if the rank of A is 0, then $A(F^{(n)}) = 0$, hence $A = 0$. On the other hand, if $r(A) = n$, then $A(F^{(n)}) = F^{(n)}$. This tells us that A is onto and so is invertible [so $n(A) = 0$]. Notice that in both these cases $n(A) + r(A) = n$. In fact, as the next theorem shows, this is always the case for any matrix A in $F^{(n)}$.

Theorem 4.4.1. Given $A \in M_n(F)$, then $n(A) + r(A) = n$. That is,

rank + nullity = $\dim (F^{(n)})$.

Proof. Letting $N = \{v \in F^{(n)} \mid Av = 0\}$, N is a subspace of $F^{(n)}$ of dimension $n(A)$. By Theorem 4.3.3, $F^{(n)} = N \oplus N^{\perp}$. Therefore, since $A(N) = 0$, $A(F^{(n)}) = A(N \oplus N^{\perp}) = A(N^{\perp})$. Hence $r(A) = \dim (A(F^{(n)})) = \dim (A(N^{\perp}))$.

Let $v_1, \ldots, v_k$ be a basis of $N^{\perp}$. We assert that $Av_1, \ldots, Av_k$ are linearly independent over F. Why? If $a_1 A(v_1) + \cdots + a_k A(v_k) = 0$, where $a_1, \ldots, a_k$ are in F, then $A(a_1 v_1 + \cdots + a_k v_k) = 0$. This puts $a_1 v_1 + \cdots + a_k v_k$ in N. However, since $v_1, \ldots, v_k$ are in $N^{\perp}$, $a_1 v_1 + \cdots + a_k v_k$ is in $N^{\perp}$. Then because $N \cap N^{\perp} = 0$ and $a_1 v_1 + \cdots + a_k v_k \in N \cap N^{\perp} = 0$, we get that

$$a_1 v_1 + \cdots + a_k v_k = 0.$$

By the linear independence of $v_1, \ldots, v_k$ we conclude that $a_1 = a_2 = \cdots = a_k = 0$. Hence $Av_1, \ldots, Av_k$ are linearly independent over F.

Because $Av_1, \ldots, Av_k$ span $A(N^{\perp})$ and are linearly independent over F, they form a basis of $A(N^{\perp})$ over F. Therefore, $k = \dim (A(N^{\perp})) = \dim (A(F^{(n)})) = r(A)$.

By the Corollary 4.3.4, $n = \dim (N) + \dim (N^{\perp}) = n(A) + k = n(A) + r(A)$. This proves the theorem. ∎

PROBLEMS

NUMERICAL PROBLEMS

1. Find the rank of the following matrices, A, by finding $A(F^{(n)})$.

(a)
$$\begin{bmatrix} 1 & 2 & 3 & 4 \\ 0 & -1 & 6 & 5 \\ 0 & 0 & 0 & 1 \\ 0 & 0 & 0 & 0 \end{bmatrix}.$$

(b)
$$\begin{bmatrix} 1 & -1 & 0 \\ 6 & 6 & 1 \\ -.5 & .5 & 0 \end{bmatrix}.$$

(c)
$$\begin{bmatrix} \frac{1}{4} & \frac{1}{4} & \frac{1}{4} & \frac{1}{4} \\ \frac{1}{4} & \frac{1}{4} & \frac{1}{4} & \frac{1}{4} \\ 0 & 0 & 0 & 0 \\ 0 & 0 & 0 & 0 \end{bmatrix}.$$

(d)
$$\begin{bmatrix} 0 & 0 & 0 & 0 & 0 \\ 1 & 0 & 0 & 0 & 0 \\ 0 & 1 & 0 & 0 & 0 \\ 0 & 0 & 1 & 0 & 0 \\ 0 & 0 & 0 & 1 & 0 \end{bmatrix}.$$

(e)
$$\begin{bmatrix} 1 & 0 & 1 \\ 0 & 1 & 0 \\ 1 & 0 & 0 \end{bmatrix}.$$

(f)
$$\begin{bmatrix} 1 & 0 & 0 \\ 0 & 0 & 0 \\ 0 & 0 & 0 \end{bmatrix}.$$

2. Find the nullity of the matrices in Problem 1 by finding N and its dimension.

3. Verify for the matrices in Problem 1 that $r(A) + n(A) = n$.

4. Find the rank of $\begin{bmatrix} 1 & 2 & 3 \\ 4 & 5 & 6 \\ 7 & 8 & 9 \end{bmatrix}\begin{bmatrix} 1 & 0 & 0 \\ 0 & 0 & 0 \\ 0 & 0 & 0 \end{bmatrix}$. How does it compare with

the ranks of $\begin{bmatrix} 1 & 2 & 3 \\ 4 & 5 & 6 \\ 7 & 8 & 9 \end{bmatrix}$ and $\begin{bmatrix} 1 & 0 & 0 \\ 0 & 0 & 0 \\ 0 & 0 & 0 \end{bmatrix}$?

5. If $A = \begin{bmatrix} 1 & 2 & 0 \\ 0 & 1 & 0 \\ 0 & 0 & 0 \end{bmatrix} \begin{bmatrix} 1 & 0 & 0 \\ 0 & 0 & 0 \\ 0 & 0 & 1 \end{bmatrix}$, find the rank of A and those of

$\begin{bmatrix} 1 & 2 & 0 \\ 0 & 1 & 0 \\ 0 & 0 & 0 \end{bmatrix}$ and $\begin{bmatrix} 1 & 0 & 0 \\ 0 & 0 & 0 \\ 0 & 0 & 1 \end{bmatrix}$.

MORE THEORETICAL PROBLEMS

Easier Problems

6. If $A = BC$, how does $r(A)$ compare to $r(B)$ and $r(C)$?

7. If $A = B + C$, how does $r(A)$ compare to $r(B)$ and $r(C)$?

8. If $r(A) = 1$, prove that $A^2 = aA$ for some $a \in F$.

9. Show that any matrix can be written as the sum of matrices of rank 1.

10. If E_{jk} are the matrix units, what are the ranks $r(E_{jk})$?

Middle-Level Problems

11. If $E^2 = E \in M_n(F)$, give an explicit form for the typical element in $N = \{v \in F^{(n)} \mid Ev = 0\}$.

12. For Problem 11, show that $N^\perp = E(F^{(n)})$ if $E^* = E = E^2$.

13. Use Problem 11 to find a basis of $F^{(n)}$ in which the matrix of E is of the form $\begin{bmatrix} I_k & 0 \\ 0 & 0 \end{bmatrix}$, where I_k is the $k \times k$ unit matrix.

14. How would you describe the k in Problem 13?

Harder Problems

15. If $E^2 = E$, show that tr (E) is an integer.

4.5. CHARACTERISTIC ROOTS

Our intention now is to interrelate algebraic properties of a matrix A in $M_n(F)$ with the fact that $F^{(n)}$ is of dimension n over F.

If v is any vector in $F^{(n)}$, consider the vectors $v, Av, A^2v, \ldots, A^nv$. Here we have written down $n + 1$ vectors in $F^{(n)}$, which is of dimension n over F. By Theorem 3.6.2 these vectors must be linearly dependent over F. Thus there are elements $a_0, a_1, \ldots, a_n$ in F, not all

0, such that

$$a_0 A^n v + a_1 A^{n-1} v + \cdots + a_{n-1} A v + a_n v = 0$$

So if $p(x)$ is the polynomial

$$p(x) = a_0 x^n + a_1 x^{n-1} + \cdots + a_{n-1} x + a_n,$$

replacing x by the matrix A, we obtain

$$p(A) = a_0 A^n + a_1 A^{n-1} + \cdots + a_{n-1} A + a_n I$$

and the statement on linear dependence translates into $p(A)(v) = 0$. Note that this polynomial $p(x)$ is not 0 and depends on v. Also note that $p(x)$ is of degree at most n.

Let $e_1, e_2, \ldots, e_n$ be the canonical basis of $F^{(n)}$. By the above, there exist nonzero polynomials $p_1(x), p_2(x), \ldots, p_n(x)$ of degree at most n such that $p_j(A)e_j = 0$ for $j = 1, 2, \ldots, n$. Consider the polynomial $p(x) = p_1(x)p_2(x) \ldots p_n(x)$; it is nonzero and of degree at most n^2. What can we say about $p(A)$? To begin with, $p(A)e_n = p_1(A) \cdots p_n(A)e_n = 0$ since $p_n(A)e_n = 0$. In fact, since $p(x)$ can also be written as a product with $p_j(x)$ at the end, for any j, and since $p_j(A)e_j = 0$, we get that $p(A)e_j = 0$ for $j = 1, 2, \ldots, n$. Since $p(A)$ is a matrix which annihilates every element of the basis $e_1, e_2, \ldots, e_n$ of $F^{(n)}$ over F, $p(A)$ must annihilate *every* element v in $F^{(n)}$. But this says that $p(A) = 0$. We have proved

Theorem 4.5.1.   Given $A \in M_n(F)$, then there exists a nonzero polynomial $p(x)$, of degree at most n^2 and with coefficients in F, such that $p(A) = 0$.

We shall see later, in the famous Cayley–Hamilton Theorem, that there is a specific and easy-to-construct polynomial $f(A)$ *of degree n* with coefficients in F, such that $f(A) = 0$.

Since there is some nonzero polynomial $p(x)$ which has the property that $p(A) = 0$, there is a nonzero polynomial of lowest possible degree—call it $q_A(x)$—with the property that $q_A(A) = 0$. If we insist that the highest coefficient of q_A be 1—*which we do*—then this polynomial is unique. (Prove!) We call it the *minimal polynomial for A*.

Consider the matrix $A = \begin{bmatrix} 1 & 0 & 0 \\ 0 & 1 & 0 \\ 0 & 0 & 0 \end{bmatrix}$. Then as a calculation reveals,

$A^2 = A$. We quickly then see that $q_A(x) = x^2 - x$. If we go through the rigmarole preceding Theorem 4.5.1, what polynomial $p(x)$ do we get? Well,

$$Ae_1 = \begin{bmatrix} 1 & 0 & 0 \\ 0 & 1 & 0 \\ 0 & 0 & 0 \end{bmatrix} \begin{bmatrix} 1 \\ 0 \\ 0 \end{bmatrix} = \begin{bmatrix} 1 \\ 0 \\ 0 \end{bmatrix} = e_1,$$

$$Ae_2 = \begin{bmatrix} 1 & 0 & 0 \\ 0 & 1 & 0 \\ 0 & 0 & 0 \end{bmatrix} \begin{bmatrix} 0 \\ 1 \\ 0 \end{bmatrix} = \begin{bmatrix} 0 \\ 1 \\ 0 \end{bmatrix} = e_2,$$

and

$$Ae_3 = \begin{bmatrix} 1 & 0 & 0 \\ 0 & 1 & 0 \\ 0 & 0 & 0 \end{bmatrix} \begin{bmatrix} 0 \\ 0 \\ 1 \end{bmatrix} = \begin{bmatrix} 0 \\ 0 \\ 0 \end{bmatrix}.$$

So the polynomials $p_1(x), p_2(x), p_3(x)$ used in the proof of Theorem 4.5.1 are, respectively, $p_1(x) = x - 1$, $p_2(x) = x - 1$, $p_3(x) = x$, and the polynomial

$$p(x) = p_1(x)p_2(x)p_3(x) = x(x - 1)^2 = x^3 - 2x^2 + x.$$

Note that $A^3 - 2A^2 + A = 0$.

Let $q_A(x) = x^m + a_1 x^{m-1} + \cdots + a_{m-1}x + a_m$ be the minimum polynomial for A. We claim that A is invertible if and only if $a_m \neq 0$. To see this, note that since

$$0 = q_A(A) = A^m + a_1 A^{m-1} + \cdots + a_{m-1}A + a_m I,$$

if $a_m \neq 0$, then

$$\left(\frac{A^{m-1} + a_1 A^{m-2} + \cdots + a_{m-1}I}{-a_m} \right) A = I$$

and A is invertible.

On the other hand, if $a_m = 0$, then

$$(A^{m-1} + a_1 A^{m-2} + \cdots + a_{m-1}I)A = 0.$$

Because the polynomial

$$h(x) = x^{m-1} + a_1 x^{m-2} + \cdots + a_{m-1}$$

is of lower degree than $q_A(x)$, $h(A)$ *cannot* be 0. Now $h(A)A = 0$ and $h(A) \neq 0$. We leave it as an exercise to show that A cannot be invertible.

Note, too, that since $h(A) \neq 0$, there is some element v in $F^{(n)}$ such that $w = h(A)(v) \neq 0$. However, $Aw = Ah(A)v = q_A(A)v = 0$.

Summarizing this long discussion, we have

Theorem 4.5.2.   Given A in $M_n(F)$, then A is invertible if and only if the constant term of its minimum polynomial is nonzero. Moreover:

1. If A is invertible, the inverse of A is a polynomial expression over F in A.

2. If A is not invertible, then
    (a) $Aw = 0$ for some $w \neq 0$ in $F^{(n)}$;
    (b) $AF^{(n)} \neq F^{(n)}$.

Proof.   Part (1) is clear from our discussion above, as is Part (a) of (2). For Part (b) of (2), we've observed in our discussion above that $h(A)A = 0$ and $h(A)$ is not 0. Thus $AF^{(n)}$ cannot be all of $F^{(n)}$, since otherwise $h(A)F^{(n)} = h(A)AF^{(n)} = 0$.   ∎

We talked several times earlier about the characteristic roots of a matrix. One time we said that $a \in F$ is a characteristic root of $A \in M_n(F)$ if $A - aI$ is *not* invertible. Another time we said that a is a characteristic root of A if for some $v \neq 0$ in $F^{(n)}$, $Av = av$; and we called such a vector v a characteristic vector associated with a.

Note that both these descriptions—in light of Theorem 4.5.2—are the same. For if $Av = av$, then $(A - aI)v = 0$; hence $A - aI$ certainly cannot be invertible. In the other direction, if $B = A - aI$ is not invertible, by Part 2 of Theorem 4.5.2 there exists $w \neq 0$ in $F^{(n)}$ with $Bw = 0$, that is, $(A - aI)w = 0$. This yields $Aw = aw$. So, recalling that $a \in F$ is a characteristic root of $A \in M_n(F)$ if there exists $v \neq 0$ in $F^{(n)}$ such that $Av = av$, we have

Theorem 4.5.3.   The element a in F is a characteristic root of the matrix A in $M_n(F)$ if and only if $(A - aI)$ is not invertible.

If we consider $A = \begin{bmatrix} 0 & 1 \\ -1 & 0 \end{bmatrix}$ in $M_2(\mathbb{R})$, then $A^2 + I = 0$ and A has *no real characteristic roots*. However, viewing A as sitting in $M_2(\mathbb{C})$, we have that A has $+i$ and $-i$ as characteristic roots in $\mathbb{C}$.

Note that $A - iI = \begin{bmatrix} -i & 1-i \\ -i & \end{bmatrix}$ and $A - iI$ is not invertible.

(Prove.) Also

$$A\begin{bmatrix} 1 \\ i \end{bmatrix} = \begin{bmatrix} 0 & 1 \\ -1 & 0 \end{bmatrix}\begin{bmatrix} 1 \\ i \end{bmatrix} = \begin{bmatrix} i \\ -1 \end{bmatrix} = i\begin{bmatrix} 1 \\ i \end{bmatrix},$$

so $(A - iI)\begin{bmatrix} 1 \\ i \end{bmatrix} = 0.$

What the example above shows is that whether or not a given matrix has characteristic roots in F depends strongly on whether $F = \mathbb{R}$ or $F = \mathbb{C}$.

Until now—except for the rare occasion—our theorems were equally true for $M_n(\mathbb{R})$ and $M_n(\mathbb{C})$. However, in speaking about characteristic roots, a real difference between the two, $M_2(\mathbb{R})$ and $M_2(\mathbb{C})$, crops up.

We elaborate on this. What we need, and what we shall use— although we do not prove it—is a very important theorem due to Gauss which is known as the *Fundamental Theorem of Algebra*. This theorem can be stated in several ways. One way says that

> *given a polynomial p(x) of degree at least* 1, *having complex coefficients, then p(x) has a root in* $\mathbb{C}$; *that is, there is an a* $\in \mathbb{C}$ *such that p(a)* $= 0.$

Another—and equivalent—version states that

> *given a polynomial p(x) of degree n* ≥ 1, *having complex coefficients (and leading coefficient* 1), *then p(x) factors as p(x)* $= (x - a_1)(x - a_2) \cdots (x - a_n)$, *where* $a_1, \ldots, a_n \in \mathbb{C}$ *are the roots of p(x). (It may happen that a given root appears several times.)*

We shall use both versions of this important theorem of Gauss. Unfortunately, to prove it would require the development of a rather elaborate mathematical machinery.

Let $a \in \mathbb{C}$ be a characteristic root of $A \in M_n(\mathbb{C})$. So there is an element $v \neq 0$ in $\mathbb{C}^{(n)}$ such that $Av = av$. Therefore,

$$A^2v = A(Av) = A(av) = aAv = a(av) = a^2v.$$

Continuing in this way we get

$$A^kv = a^kv \qquad \text{for } k \geq 1,$$

so that if $f(x)$ is any polynomial having complex coefficients, then

$$f(A)v = f(a)(v).$$

(Prove!) Thus if $f(A) = 0$, then $f(a) = 0$, that is, if a is a characteristic root of A and if $f(x)$ is a polynomial such that $f(A) = 0$, then $f(a) = 0$, that is, a is a root of $f(x)$. So any characteristic root of A must be a root of $q_A(x)$, the minimal polynomial for A over $\mathbb{C}$. However, $q_A(x)$ has only a finite number of roots in $\mathbb{C}$—this finite number being at most the degree of $q_A(x)$. Therefore,

> A has only a finite number of characteristic roots and these must be the roots of $q_A(x)$.

We want to sharpen this last statement even further. This is

Theorem 4.5.4. If $A \in M_n(\mathbb{C})$, then every characteristic root of A is a root of $q_A(x)$, the minimal polynomial for A. Furthermore, any root of $q_A(x)$ is a characteristic root of A.

Proof. In the discussion preceding the statement of Theorem 4.5.4, we showed the first half, namely, that a characteristic root of A must be a root of $q_A(x)$.

So all we have to do is to prove the second half, that part beginning with "Furthermore." Let $a_1, \ldots, a_k$ be the roots of $q_A(x)$. Then, by Gauss's theorem,

$$q_A(x) = (x - a_1)(x - a_2) \cdots (x - a_k).$$

Therefore,

$$0 = q_A(A) = (A - a_1 I)(A - a_2 I) \cdots (A - a_k I).$$

Given j, let $h_j(x)$ be the product of all the $(x - a_t)$ for $t \neq j$. So $q_A(x) = h_j(x)(x - a_j)$. Because

$$\text{degree of } h_j(x) = \text{degree of } q_A(x) - 1,$$

we have $h_j(A) \neq 0$ by the minimality of the degree of $q_A(x)$. So for some $w \neq 0$ in $\mathbb{C}^{(n)}$, $v = h_j(A)w \neq 0$. However,

$$(A - a_j I)(v) = (A - a_j I)h_j(A)(w) = q_A(A)w = 0.$$

This says that $Av = a_j v$, with $v \neq 0$. Therefore, a_j is a characteristic root of A. ∎

Let's take a closer look at some of this in the special case where our $n \times n$ matrix A is upper triangular. Then any polynomial $f(A)$ in A is also an upper triangular matrix. (Prove!) So we get a sharp corollary to Theorem 4.5.2, namely

Corollary 4.5.5.   Given an invertible upper triangular A in $M_n(\mathbb{C})$, its diagonal entries are all nonzero and its inverse A^{-1} is upper triangular.

    Proof.   Since the inverse of A is a polynomial in A, by Theorem 4.5.2, it is also upper triangular. (Prove!) Letting the diagonal elements of A and A^{-1} be $a_1, \ldots, a_n$ and $a_1^*, \ldots, a_n^*$, respectively, the diagonal elements of the product AA^{-1} of these two upper triangular matrices are $a_1 a_1^*, \ldots, a_n a_n^*$. (Prove!) Since $I = AA^{-1}$, all these diagonal entries are 1, so that the diagonal entries a_r of A must be nonzero. ∎

    If A is a diagonal matrix with all entries on the diagonal nonzero, then A is invertible and its inverse is a diagonal matrix with diagonal entries inverses of those of A. (Prove!) If A is an upper triangular *unipotent* $n \times n$ matrix, that is, one all of whose diagonal entries are 1, then A can be expressed as $A = I - N$, where N is an upper triangular *nilpotent* $n \times n$ matrix, that is, one all of whose diagonal entries are 0. We call N nilpotent because $N^k = 0$ for some k; and we call A unipotent because $(A - I)^k = 0$ for some k. Thus, the matrices

$$\begin{bmatrix} 0 & 3 & 6 \\ 0 & 0 & 4 \\ 0 & 0 & 0 \end{bmatrix}, \quad \begin{bmatrix} 0 & 5 \\ 0 & 0 \end{bmatrix}$$

are nilpotent and the matrices

$$\begin{bmatrix} 1 & 3 & 6 \\ 0 & 1 & 4 \\ 0 & 0 & 1 \end{bmatrix}, \quad \begin{bmatrix} 1 & 5 \\ 0 & 1 \end{bmatrix}$$

are unipotent. *An upper triangular $n \times n$ matrix A with $a_{11} = 0, \ldots,$ $a_{nn} = 0$ is nilpotent* since, for $k \geq 1$, the entries

$$b_{1,1+t}, \, b_{2,2+t}, \, \ldots, \, b_{n-t,n}$$

of $B = A^k$ are 0 for $0 \leq t \leq k - 1$. For example, if $n = 5$ and $k = 2$, t can be 0 or 1 and the entries $b_{11}, \ldots, b_{55}$ and $b_{12}, \ldots, b_{45}$ are 0. Why is it always true? It is true if $k = 1$, by hypothesis. In general, we go by induction on k. Assuming that it is true for k, it is also true for $k + 1$, since A^{k+1} equals $AA^k = AB$ and the entries

$$c_{r,r+t} = \sum_{j=1}^{n} a_{rj} b_{j,r+t}$$

of A^{k+1} are 0 for $1 \leq r \leq n - t$ and $0 \leq t \leq k$. Thus it is true for all k. If A is an upper triangular unipotent matrix, then A is again invertible.

In fact, letting $N = I - A$, we get

$$A^{-1} = (I - N)^{-1} = I + N + N^2 + \cdots + N^{n-1},$$

as you see by showing that

$$(I - N)(I + N + N^2 + \cdots + N^{n-1}) = I.$$

(Do so!) The upshot of this is the

Theorem 4.5.6.   Let $A \in M_n(F)$ be upper triangular. Then

1. A is invertible if and only if all of its diagonal entries are non-zero;

2. The diagonal entries of A are its characteristic roots (with possible duplications).

Proof.   If A is invertible, we've seen above that its diagonal entries are all nonzero. Suppose, conversely, that all diagonal entries of A are nonzero and let D be the diagonal matrix with the same diagonal entries as A. Then $U = D^{-1}A$ is unipotent, so U is invertible by what we showed in our earlier discussion. But then $A = DU$ is the product of invertible matrices, so is itself invertible. This proves (1).

Now let's prove (2). Since the diagonal entries of A are just those scalars a such that $A - aI$ has 0 as a diagonal entry, the diagonal entries of A are just those scalars a such that $A - aI$ is not invertible, by (1). But then, by Theorem 4.5.3, the diagonal entries of A are just its characteristic roots.   ■

This section—so far—has been replete with important results, many of them not too easy. So we break the discussion here to give the reader a chance to digest the material. We pick up this theme again in the next section.

PROBLEMS

NUMERICAL PROBLEMS

1. For the given v and A find a polynomial $p(x)$ such that $p(A)(v) = 0$.

(a) $v = \begin{bmatrix} 1 \\ 1 \end{bmatrix}$, $A = \begin{bmatrix} 1 & 2 \\ 2 & 1 \end{bmatrix}$.

(b) $v = \begin{bmatrix} 1 \\ 1 \\ 0 \end{bmatrix}$, $A = \begin{bmatrix} 1 & 2 & 0 \\ 2 & 1 & 0 \\ 0 & 0 & 0 \end{bmatrix}$.

(c) $v = \begin{bmatrix} 0 \\ 1 \\ -1 \end{bmatrix}$, $A = \begin{bmatrix} 1 & 0 & 0 \\ 0 & 1 & 0 \\ 1 & 2 & 3 \end{bmatrix}$.

(d) $v = \begin{bmatrix} 0 \\ -1 \\ 1 \\ -1 \end{bmatrix}$, $A = \begin{bmatrix} 0 & 1 & 0 & 0 \\ 0 & 0 & 1 & 0 \\ 0 & 0 & 0 & 1 \\ 0 & 0 & 1 & 0 \end{bmatrix}$.

(e) $v = \begin{bmatrix} 1 \\ 2 \\ 1 \end{bmatrix}$, $A = \begin{bmatrix} 0 & 0 & 0 \\ 1 & 0 & 0 \\ 0 & 1 & 0 \end{bmatrix}$.

2. (a) For $A = \begin{bmatrix} 0 & 1 & 0 \\ 0 & 0 & 1 \\ 1 & 0 & 0 \end{bmatrix}$, find polynomials $p_1(x), p_2(x), p_3(x)$ such

that $p_j(A)(e_j) = 0$ for $j = 1, 2, 3$.

(b) By directly calculating $p(A)$, where $p(x) = p_1(x)p_2(x)p_3(x)$, show that $p(A) = 0$.

3. From the result of **Problem 2**, express A^{-1} as a polynomial expression in A.

4. For the A in Problem 2, find its minimal polynomial over $\mathbb{C}$.

5. For the A in Problem 2 and the minimal polynomial $q_A(x)$ found in Problem 4, show, by explicitly exhibiting $av \neq 0$, that if a is a root of $q_A(x)$, then a is a characteristic root of A.

6. In Problem 5, show directly that if a is a root of $q_A(x)$, then $(A - aI)$ is not invertible.

7. (a) If $A = \begin{bmatrix} 1 & 2 & 3 \\ 0 & 4 & 5 \\ 0 & 0 & 6 \end{bmatrix}$, show $(A - I)(A - 4I)(A - 6I) = 0$.

(b) What are the characteristic roots of A?

(c) What is $q_A(x)$?

8. (a) If $A = \begin{bmatrix} 1 & 0 & 0 \\ 0 & 1 & 2 \\ 0 & 2 & 4 \end{bmatrix}$, find $q_A(x)$.

(b) Show that A has three distinct, real, characteristic roots.

(c) If a_1, a_2, a_3 are the three characteristic roots of A, find $v_1 \neq 0$, $v_2 \neq 0$, $v_3 \neq 0$ such that $Av_j = a_j v_j$ for $j = 1, 2, 3$.

(d) Verify that $(v_j, v_k) = 0$, for $j \neq k$, for the v_1, v_2, v_3 found in Part (c).

9. (a) If $A = \begin{bmatrix} 0 & 0 & 0 \\ 0 & 0 & 1 \\ 0 & -1 & 0 \end{bmatrix}$, find $q_A(x)$.

(b) Find the roots of $q_A(x)$ in $\mathbb{C}$.

(c) Verify that each root of $q_A(x)$ found in Part (b) is a characteristic root of A.

(d) Find v_1, v_2, v_3 in $\mathbb{C}^{(3)}$ such that $Av_j = a_j v_j$, $j = 1, 2, 3$, where a_1, a_2, a_3 are the characteristic roots of A.

(e) Verify that $(v_j, v_k) = 0$, for $j \neq k$, for v_1, v_2, v_3 of Part (d).

(f) By a direct calculation show that $q_A(A) = 0$.

10. For the A in Problem 8, find an invertible matrix U such that
$$U^{-1}AU = \begin{bmatrix} a_1 & 0 & 0 \\ 0 & a_2 & 0 \\ 0 & 0 & a_3 \end{bmatrix}.$$

11. For the A in Problem 9, find an invertible matrix U such that
$$U^{-1}AU = \begin{bmatrix} a_1 & 0 & 0 \\ 0 & a_2 & 0 \\ 0 & 0 & a_3 \end{bmatrix}.$$

12. For the matrices A in Problems 8 and 9, can you exhibit *unitary* matrices such that $U^{-1}AU = \begin{bmatrix} a_1 & 0 & 0 \\ 0 & a_2 & 0 \\ 0 & 0 & a_3 \end{bmatrix}$?

MORE THEORETICAL PROBLEMS

Middle-Level Problems

13. If $p(x)$ is a polynomial of least positive degree such that $p(A) = 0$, for $A \in M_n(F)$, and if the highest coefficient of $p(x)$ is 1, show that $p(x)$ is unique. [That is, if $q(x)$ is possibly another polynomial of least positive degree with highest coefficient equal to 1 such that $q(A) = 0$, then $p(x) = q(x)$.]

14. If $Av = av$, show that for any polynomial $f(x)$ having coefficients in F, $f(A)(v) = f(a)v$.

15. Let $A = \begin{bmatrix} 0 & 1 & 0 & 0 \\ 0 & 0 & 1 & 0 \\ 0 & 0 & 0 & 1 \\ 1 & 0 & 0 & 0 \end{bmatrix}$. Show that A has four distinct character-

istic roots in $\mathbb{C}$, and find these roots.

16. What is $q_A(x)$ for the A in Problem 15?

17. If a_1, a_2, a_3, a_4 are the characteristic roots of the A in Problem 15, find $v_1 \neq 0$, $v_2 \neq 0$, $v_3 \neq 0$, $v_4 \neq 0$ in $\mathbb{C}^{(4)}$ such that $Av_j = a_j v_j$ for $j = 1, 2, 3, 4$.

18. Show that v_1, v_2, v_3, v_4 in Problem 17 form a orthogonal basis of $\mathbb{C}^{(4)}$.

19. Show that we can find a unitary matrix U such that $U^{-1}AU = \begin{bmatrix} a_1 & 0 & 0 & 0 \\ 0 & a_2 & 0 & 0 \\ 0 & 0 & a_3 & 0 \\ 0 & 0 & 0 & a_4 \end{bmatrix}$ for the A in Problem 15.

20. If $A \in M_n(\mathbb{C})$ and $B = A^*$, the Hermitian adjoint of A, express $q_B(x)$ in terms of $q_A(x)$.

21. Express the relationship between the characteristic roots of A and of A^*.

22. If $A \in M_n(\mathbb{C})$ and $f(A)v = 0$ for some $v \neq 0$ in $\mathbb{C}^{(n)}$, is every root of $f(x)$ a characteristic root of A?

Harder Problems

23. If $q_A(x)$ is the minimal polynomial for A and if $p(x)$ is such that $p(A) = 0$, show that $p(x) = q_A(x)t(x)$ for some polynomial $t(x)$.

24. Let $A \in M_n(\mathbb{C})$ and $v \neq 0 \in \mathbb{C}^{(n)}$. Suppose that $u(x)$ is the polynomial of least positive degree, with highest coefficient 1, such that $u(A)v = 0$. Prove

   (a) Degree $u(x) \leq$ degree $q_A(x)$.

   (b) $q_A(x) = u(x)s(x)$ for some polynomial $s(x)$ with complex coefficients.

   (c) Every root of $u(x)$ is a characteristic root of A.

Very Hard Problems

25. If A is *not* invertible, show that there is a nonzero matrix B such that $AB = BA = 0$.   [*Hint:* Look at $q_A(x)$.]

26. If $A = \begin{bmatrix} a_1 & & & 0 \\ & a_2 & & \\ & & \ddots & \\ * & & & a_n \end{bmatrix}$ is a lower triangular matrix, show that

A satisfies $p(x) = 0$, where

$$p(x) = (x - a_1)(x - a_2) \cdots (x - a_n).$$

27. If A is as in Problem 26, is the rank of A always equal to the number of elements $a_1, a_2, \ldots, a_n$ which are nonzero? Explain.

28. If $A \neq 0$ is not invertible, show that there is a matrix B such that $AB = 0$ but $BA \neq 0$.   (**Hint:** Multiply the B found in Problem 25 on the right by a suitable matrix.)

29. If A is a matrix such that $AC = I$ for some matrix C, prove that $CA = I$ (thus A is invertible).

30. If A is a Hermitian matrix and $A^2 v = 0$ for some $v \in \mathbb{C}^{(n)}$, prove that $Av = 0$.

31. Generalize Problem 30 as follows: If A is Hermitian and $A^k v = 0$ for $k \geq 1$, then $Av = 0$.

32. If A is Hermitian and b is a characteristic root of A, show that if $(A - bI)^k v = 0$ for some $k \geq 1$, then $Av = bv$.

4.6. HERMITIAN MATRICES

Although one could not tell from the section title, we continue here in the spirit of Section 4.5. Our objective is to study the relationship between characteristic roots and the property of being Hermitian. We had earlier obtained one piece of this relationship, namely, we showed in Theorem 3.5.7 that the characteristic roots of a Hermitian matrix must be *real*. We now fill out the picture. Some of the results which we shall obtain came up in the problems of the preceding section. But we want to do all this officially here.

We begin with the following property of Hermitian matrices, which, as you may recall, are the matrices A in $M_n(\mathbb{C})$ such that $A = A^*$.

Lemma 4.6.1. If A in $M_n(\mathbb{C})$ is Hermitian and v in $\mathbb{C}^{(n)}$ is such that $A^k v = 0$ for $k \geq 1$, then $Av = 0$.

Proof. Consider the easiest case first—that for $k = 2$, that is, the case

$A^2v = 0.$

But then $(A^2v, v) = 0$ and, since $A = A^*$, we get

$0 = (A^2v, v) = (Av, A^*v) = (Av, Av).$

Since $(Av, Av) = 0$, the defining relations for inner product allow us to conclude that

$Av = 0.$

For the case $k > 2$, note that

$A^2(A^{k-2}v) = 0$

implies that

$A(A^{k-2}v) = 0,$

that is,

$A^kv = 0$ *implies that* $A^{k-1}v = 0.$

So for $k \geq 2$ we can go from $A^kv = 0$ to $A^{k-1}v = 0$ and we can safely say "continue this procedure" to end up, eventually, with $Av = 0.$ ■

If A is Hermitian and a is a characteristic root of A, then because a is real, $A - aI$ is Hermitian [since $(A - aI)^* = A^* - \bar{a}I = A - aI$ from the reality of a]. So as an immediate consequence of the lemma, we get the

Corollary 4.6.2. If a is a characteristic root of the Hermitian matrix A and v is such that $(A - aI)^kv = 0$ (for $k \geq 1$), then $Av = av$.

Proof. Because $A - aI$ is Hermitian, our lemma assures us that $(A - aI)v = 0$, that is, $Av = av$. ■

A fundamental implication of this corollary is to specify quite sharply the nature of the minimal polynomial, $q_A(x)$, of the Hermitian matrix A. This is

Theorem 4.6.3. If A is Hermitian with minimal polynomial $q_A(x)$, then $q_A(x)$ is of the form $q_A(x) = (x_1 - a_1) \cdots (x - a_t)$, where the $a_1, \ldots, a_t$ are *real* and *distinct*.

Proof. By the Fundamental Theorem of Algebra, we can factor $q_A(x)$ as

$$q_A(x) = (x_1 - a_1)^{m_1} \cdots (x - a_t)^{m_t},$$

where the m_r are positive integers and the $a_1, \ldots, a_t$ are distinct. By Theorem 4.5.4, $a_1, \ldots, a_t$ are all of the distinct characteristic roots of A, and by Theorem 3.5.7 they are real. So our job is to show that m_r is precisely 1 for each r.

Since $q_A(x)$ is the minimal polynomial for A, we have

$$0 = q_A(A) = (A - a_1 I)^{m_1} \cdots (A - a_t I)^{m_t}.$$

Applying these mappings to any v in $\mathbb{C}^{(n)}$, we have

$$0 = q_A(A)v = (A - a_1 I)^{m_1} \cdots (A - a_t I)^{m_t} v$$
$$= (A - a_1 I)^{m_1}[(A - a_2 I)^{m_2} \cdots (A - a_t I)^{m_t} v].$$

Letting w denote $(A - a_2 I)^{m_2} \cdots (A - a_t I)^{m_t} v$, this says that

$$(A - a_1 I)^{m_1} w = 0.$$

By Corollary 4.6.2 we get that

$$Aw = a_1 w.$$

So

$$(A - a_1 I)[(A - a_2 I)^{m_2} \cdots (A - a_t I)^{m_t} v] = 0$$

for all v in $\mathbb{C}^{(n)}$, whence

$$(A - a_1 I)(A - a_2 I)^{m_2} \cdots (A - a_t I)^{m_t} = 0.$$

Now, rearranging the order of the terms to put the second one first, we can proceed to give the same treatment to m_2 as we gave to m_1. Continuing with the other terms in turn, we end up with

$$(A - a_1 I)(A - a_2 I) \cdots (A - a_t I) = 0,$$

as claimed.   ∎

Since the minimum polynomial of A has real roots if A is Hermitian, we can prove

Theorem 4.6.4.   Let F be $\mathbb{R}$ or $\mathbb{C}$ and let $A \in M_n(F)$ be Hermitian. If W is a nonzero subspace of $F^{(n)}$ and $AW \subset W$, then W contains a characteristic vector of A.

Proof.   Let the minimum polynomial $q_A(x)$ of A be

$$(x - a_1) \cdots (x - a_r)$$

and choose the first integer $r \geq 1$ such that

$$(A - a_1I) \cdots (A - a_rI)W = 0.$$

If $r = 1$, then any nonzero element of W is a characteristic vector of A. If $r > 1$, then any nonzero vector of

$$(A - a_1I) \cdots (A - a_{r-1}I)W$$

is a characteristic vector of A associated with the characteristic root a_r.   ■

The results we have already obtained allow us to pass to an extremely important theorem. It is difficult to overstate the importance of this result, not only in all of mathematics, but even more so, in physics. In physics, an infinite dimensional version of what we have proved and are about to prove plays a crucial role. There the characteristic roots of a certain Hermitian linear transformation called the *Hamiltonian* are the *energies* of the physical state. The fact that such a linear transformation has, in a change of basis, such a simple form is vital in the quantum mechanics. In addition to all of this, it really is a very pretty result.

First, some preliminary steps.

Lemma 4.6.5. If A is Hermitian and a_1, a_2 are distinct characteristic roots of A, then, if $Av = a_1v$ and $Aw = a_2w$, we have $(v, w) = 0$.

Proof. Consider (Av, w). On one hand, since $Av = a_1v$, we have

$$(Av, w) = a_1(v, w) = a_1(v, w).$$

On the other hand,

$$(Av, w) = (v, A*w) = (v, Aw) = (v, a_2w) = a_2(v, w),$$

since a_2 is real. Therefore,

$$a_1(v, w) = (Av, w) = a_2(v, w) \text{ and } (a_1 - a_2)(v, w) = 0.$$

Because a_1 and a_2 are distinct, $a_1 - a_2$ is nonzero. Canceling, the outcome is that $(v, w) = 0$, as claimed in the lemma.   ■

The next result is really a corollary to Lemma 4.6.4, and we leave its proof as an exercise for the reader.

Lemma 4.6.6. If A is a Hermitian matrix in $M_n(\mathbb{C})$ and if $a_1, \ldots, a_t$ are distinct characteristic roots of A with corresponding characteristic vectors $v_1, \ldots, v_t \neq 0$ (i.e., $Av_r = a_r v_r$ for $1 \leq r \leq t$), then the vectors $v_1, \ldots, v_t$ are pairwise orthogonal (and therefore linearly independent) over $\mathbb{C}$.

Since $\mathbb{C}^{(n)}$ has at most n linearly independent vectors, by the result of Lemma 4.6.6, we know that $t \leq n$. So A has *at most* n distinct characteristic roots. Together with Theorem 4.6.3, we then have that $q_A(x)$ is of *degree at most* n. So A satisfies a polynomial of degree n or less over $\mathbb{C}$. This is a weak version of the Cayley–Hamilton Theorem, which will be proved in due course. We summarize these remarks into

Lemma 4.6.7. If A in $M_n(\mathbb{C})$ is Hermitian, then A has at most n distinct characteristic roots, and $q_A(x)$ is of degree at most n.

Let A be a Hermitian matrix in $M_n(\mathbb{C})$ acting on the vector space $V = \mathbb{C}^{(n)}$. If $a_1, \ldots, a_k$ are distinct characteristic roots of A, then we know that the a_r are real and the minimal polynomial of A is

$$q_A(x) = (x - a_1) \cdots (x - a_k).$$

Let $V_{a_r} = \{v \in V \mid Av = a_r v\}$ for $r = 1, \ldots, k$. Clearly, $A(V_{a_r}) \subset V_{a_r}$ for all r; in fact, A acts as a_r times the identity map on $V_{a_r} \neq \{0\}$.

Theorem 4.6.8. $V = V_{a_1} \oplus \cdots \oplus V_{a_k}$.

Proof. Let $Z = V_{a_1} + \cdots + V_{a_k}$. The subspaces V_{a_r} are mutually orthogonal by Lemma 4.6.5. By Theorem 4.2.4, their sum is direct. So

$$Z = V_{a_1} \oplus \cdots \oplus V_{a_k}.$$

Also, because $A(V_{a_r}) \subset V_{a_r}$, we have $A(Z) \subset Z$.

Now it suffices to show that $V = Z$. For this, let $Z^{\perp}$ be the orthogonal complement of Z in V. We claim that $A(Z^{\perp}) \subset Z^{\perp}$. Why? Since $A^* = A$, $A(Z) \subset Z$ and $(Z, Z^{\perp}) = 0$, we have

$$0 = (A(Z), Z^{\perp}) = (Z, A^*(Z^{\perp})) = (Z, A(Z^{\perp})).$$

But this tells us that $A(Z^{\perp}) \subset Z^{\perp}$. So if $Z^{\perp} \neq 0$, by Theorem 4.6.4 there is a characteristic vector of A in $Z^{\perp}$. But this is impossible since all the characteristic roots are used up in Z and $Z^{\perp} \cap V_{a_s} = \{0\}$ for all s. It follows that $Z^{\perp} = \{0\}$, hence $Z = Z^{\perp\perp} = \{0\}^{\perp} = V$, by Theorem 4.3.7. This, in turn, says that

$$V = Z = V_{a_1} \oplus \cdots \oplus V_{a_k}. \quad \blacksquare$$

An immediate and important consequence of Theorem 4.6.7 is

Theorem 4.6.9. Let A be a Hermitian matrix in $M_n(\mathbb{C})$. Then there exists an orthonormal basis of $\mathbb{C}^{(n)}$ in which the matrix of A is diagonal. Equivalently, there is a unitary matrix U in $M_n(\mathbb{C})$ such that the matrix $U^{-1}AU$ is diagonal.

Proof. For each V_{a_r}, $r = 1, \ldots, k$, we can find an orthonormal basis. These, put together, form an orthonormal basis of V since

$$V = V_{a_1} \oplus \cdots \oplus V_{a_k}.$$

By the definition of V_{a_r}, the matrix A acts like multiplication by the scalar a_r on V_{a_r}. So, in this put-together basis of V, the matrix of A is the diagonal matrix

$$\begin{bmatrix} a_1 I_{m_1} & & & 0 \\ & a_2 I_{m_2} & & \\ & & \ddots & \\ 0 & & & a_k I_{m_k} \end{bmatrix},$$

where I_{mr} is the $m_r \times m_r$ identity matrix and $m_r = \dim(V_{a_r})$ for all r. Since a change of basis from one orthonormal basis to another is achieved by a unitary matrix U we have that

$$U^{-1}AU = \begin{bmatrix} a_1 I_{m_1} & & & 0 \\ & a_2 I_{m_2} & & \\ & & \ddots & \\ 0 & & & a_k I_{m_k} \end{bmatrix}$$

for such a U and the theorem is proved. ∎

This theorem is usually described as:

A Hermitian matrix can be brought to diagonal form, or can be diagonalized, by a unitary matrix.

It is not true that every matrix can be brought to diagonal form. We leave as an exercise that there is no invertible matrix $\begin{bmatrix} a & b \\ c & d \end{bmatrix}$ such that $\begin{bmatrix} a & b \\ c & d \end{bmatrix}^{-1} \begin{bmatrix} 0 & 1 \\ 0 & 0 \end{bmatrix} \begin{bmatrix} a & b \\ c & d \end{bmatrix}$ is diagonal. So, in this regard, Hermitian matrices are remarkable.

There is an easy passage from Hermitian to skew-Hermitian matrices, and vice versa. If A is Hermitian, then $B = iA$ is skew-Hermitian, since $B* = (iA)* = -iA = -B$. Similarly, if A is skew-Hermitian, then iA is Hermitian. Thus we can easily get analogues, for the theorems we proved for Hermitian matrices, for skew-Hermitian matrices.

PROBLEMS

NUMERICAL PROBLEMS

1. By a direct calculation show that if $A = \begin{bmatrix} a & b \\ b & c \end{bmatrix}$, $a, b, c \in \mathbb{R}$, is such that $A^2 = 0$ then $A = 0$.

2. **(a)** Find the characteristic roots of $A = \begin{bmatrix} 0 & 2 \\ 2 & 0 \end{bmatrix}$.

   (b) From Part (a), find $q_A(x)$.

   (c) For each characteristic root s of A, find a nonzero element v in $\mathbb{R}^{(2)}$ such that $Av = sv$.

   (d) Verify that if v, w are as in Part (c), for distinct characteristic roots of A, then $(v, w) = 0$.

3. Do Problem 2 for the matrix $A = \begin{bmatrix} 1 & 0 & 0 \\ 0 & 0 & 2 \\ 0 & 2 & 0 \end{bmatrix}$.

4. Let $A = \begin{bmatrix} 0 & 1 \\ -1 & 0 \end{bmatrix}$. Show that A is invertible and that $\begin{bmatrix} 0 & -1 \\ 1 & 0 \end{bmatrix} A^{-1}$ is unitary.

5. **(a)** Show that $A = \begin{bmatrix} i & 0 & 0 \\ 0 & 0 & 1 \\ 0 & -1 & 0 \end{bmatrix}$ in $M_3(\mathbb{C})$ is skew-Hermitian and find its characteristic roots.

   (b) Calculate A^2 and show that A^2 is Hermitian.

   (c) Find A^{-1} and show that it is skew-Hermitian.

MORE THEORETICAL PROBLEMS

Easier Problems

6. Prove Lemma 4.6.6.

7. If A is skew-Hermitian, show that A^k is Hermitian if k is even and is skew-Hermitian if k is odd.

8. If A is invertible, show that

  (a) If A is Hermitian, then A^{-1} is Hermitian.

  (b) If A is skew-Hermitian, then A^{-1} is skew-Hermitian.

  (c) A^* is invertible (find its inverse).

9. If A is such that $AA^* = A^*A$ and A is invertible, then $U = A^*A^{-1}$ is unitary.

Middle-Level Problems

10. If $A \in M_n(\mathbb{C})$ and $V = \{v \in \mathbb{C}^{(n)} \mid (A - aI)^k v = 0$ for some $k\}$, show that if $BA = AB$, then $B(V) \subset V$.

11. If A is Hermitian and has only one distinct characteristic root, show that A must be a scalar matrix.

12. If $E^2 = E$ is Hermitian, for what values of $a \in \mathbb{C}$ is $I - aE$ unitary?

4.7. TRIANGULARIZING MATRICES WITH COMPLEX ENTRIES

The subject matter we are about to take up is somewhat more subtle—consequently somewhat more difficult—than the material treated hitherto. This difficulty is not so much in the technical nature of the proofs, but rather, in the appreciation, the first time one sees these things, of the importance and role that the results play in matrix theory.

Until now, we have used mathematical induction only informally, contenting ourselves with phrases such as "and so on" and "repeat the procedure." Many readers will have had some exposure to mathematical induction. However, for those who have had little or none, we give a quick synopsis of that topic.

Mathematical induction is a technique for proving theorems about integers, or in situations where integers measure some phenomenon. We state the *principle of mathematical induction* now:

Suppose that P(n) is a proposition such that
    1. *P(1) is true.*
    2. *Whenever we know that P(k) is true for some integer*
     k ≥ 1 then we know that P(k + 1) is also true.
Then we know that P(n) is true for all positive integers n.

Another form—equivalent to the one above—runs as follows:

Suppose that P(1) is true and that, for any m > 1, the truth of
P(k) for all positive integers 1 ≤ k < m implies the truth of P(m).
Then P(n) is true for all positive integers n.

What is the rationale for this? Knowing $P(1)$ to be valid by 1 tells us, by 2, that $P(2)$ is valid. The validity of $P(2)$, again by 2, assures that of $P(3)$. That of $P(3)$ gives us $P(4)$, and so on.

We illustrate this technique with an example, one that is overworked in almost all discussions of induction. We want to prove that

$$1 + 2 + \cdots + n = \frac{n(n + 1)}{2}.$$

So let the proposition $P(n)$ be that

$$1 + 2 + \cdots + n = \frac{n(n + 1)}{2}.$$

Clearly, if $n = 1$, then, since $1 = \frac{1(1 + 1)}{2}$, $P(1)$ is true. Suppose that we happen to know that $P(k)$ is true for some integer $k \geq 1$, that is,

$$1 + 2 + \cdots + k = \frac{k(k + 1)}{2}.$$

We want to show from this that $P(k + 1)$ is true, that is, that

$$1 + 2 + \cdots + (k + 1) = \frac{(k + 1)((k + 1) + 1)}{2}.$$

Now, by our assumption,

$$1 + 2 + \cdots + k + (k + 1) = \frac{k(k + 1)}{2} + (k + 1)$$

$$= \frac{(k + 1)(k + 2)}{2}$$

$$= \frac{(k + 1)((k + 1) + 1)}{2}.$$

So the validity of $P(k)$ implies that of $P(k + 1)$. Therefore, by the principle of mathematical induction, $P(n)$ is true for all positive integers n; that is,

$$1 + 2 + \cdots + n = \frac{n(n + 1)}{2}$$

for all positive integers n.

In the proof we are about to give we will go by induction on the dimension of V. Precisely what does this mean? It means that we will show two things:

1. The theorem we want to prove is true if $\dim V = 1$.

2. The truth of the theorem for all subspaces W of $\mathbb{C}^{(n)}$ of the same kind for which $\dim W$ is less than $\dim V$ implies the truth of the theorem for V.

We shall describe this procedure by saying: *We go by induction on the dimension of V.*

Theorem 4.7.1. Let V be a subspace of $\mathbb{C}^{(n)}$, T and element of $M_n(\mathbb{C})$ such that $T(V) \subset V$. Then we can find a basis $v_1, \ldots, v_t$ of V such that $Tv_s = \Sigma_{r=1}^{s} a_{rs}v_r$, that is, Tv_s is a linear combination of v_s and its predecessors, for $1 \le s \le t$. In other words, we can find a basis of V in which the $t \times t$ matrix (a_{rs}) of T is upper triangular.

Proof. We go by induction on $\dim V$. If $\dim V = 1$, then $V = Fv$ and $Tv = av$ for some scalar a. Since v is a basis for V, it does the trick and we have proved the result for $\dim V = 1$.

Next, let V be a subspace of $\mathbb{C}^{(n)}$ and suppose that the theorem is correct for all proper subspaces W of V such that $T(W) \subset W$. So, for such a subspace W, we can find a basis $w_1, \ldots, w_u$ such that Tw_s is a linear combination of w_s and its predecessors for $s = 1, 2, \ldots, u$.

We can factor $q_T(x)$ as a product

$$q_T(x) = (x - b_1) \cdots (x - b_k),$$

where $b_1, \ldots, b_k$ are elements (not necessarily all different) of $\mathbb{C}$. Since

$$0 = q_T(T) = (T - b_1I) \cdots (T - b_kI),$$

we can't have $(T - b_pI)(V) = V$ for all p, since otherwise

$$V = (T - b_k I)(V) = (T - b_{k-1}I)(T - b_k I)(V) = \cdots$$
$$= (T - b_1 I) \cdots (T - b_k I)(V) = q_T(T)(V) = 0.$$

Choose p such that $(T - b_p I)(V) \neq V$ and let $W = (T - b_p I)(V)$; hence W is a proper subspace of V. Because

$$T(W) = T((T - b_p I)(V))$$
$$= (T - b_p I)(T(V)) \subset (T - b_p I)(V) = W$$

(using the fact that $T(V) \subset V$), we have $T(W) \subset W$. Furthermore, since W is a proper subspace of V, $\dim W < \dim V$. By induction, we therefore can find a basis $w_1, \ldots, w_u$ such that for each $s = 1, 2, \ldots, u$, $Tw_s = \Sigma_{r=1}^{s} a_{rs} w_r$.

By Corollary 4.1.2, we can find elements $w_{u+1}, \ldots, w_t$ such that $w_1, \ldots, w_u, w_{u+1}, \ldots, w_t$ form a basis of V. We know from the above that for $s \leq u$, $Tw_s = \Sigma_{r=1}^{s} a_{rs} w_r$. What is the action of T on $w_{u+1}, \ldots, w_t$? Because $W = (T - b_p I)(V)$, each of the vectors $(T - b_p I)w_s$, for $s > u$, is in W, *so is a linear combination of* $w_1, \ldots, w_u$. That is, we can express them as

$$(T - b_p I)w_s = \sum_{r=1}^{u} a_{rs} w_r$$

for $s > u$. But this implies that even for $s > u$, we can express them as

$$(T - b_p I)w_s = \sum_{r=1}^{s} a_{rs} w_r$$

(take the coefficients to be 0 for $u < r \leq s$). So here too, Tw_s is a linear combination of w_s and its predecessors. Thus this is true for every element in the basis we created, namely, $w_1, \ldots, w_u, w_{u+1}, \ldots, w_t$. In other words, the desired result holds for V. Since $a_{rs} = 0$ for $r > s$, the matrix A of T in this basis is upper triangular.   ■

We single out a special case of the theorem, which is of special interest.

Theorem 4.7.2.   Let $T \in M_n(\mathbb{C})$. Then there exists a basis $v_1, \ldots, v_n$ of $\mathbb{C}^{(n)}$ over $\mathbb{C}$ such that, for each $s = 1, 2, \ldots, n$, $Tv_s = \Sigma_{r=1}^{s} a_{rs} v_r$. In other words, there exists a basis for $\mathbb{C}^{(n)}$ in which the matrix A of T is upper triangular.

Proof.   After all, $\mathbb{C}^{(n)}$ is a subspace of itself, so by a direct application of Theorem 4.7.1, we obtain Theorem 4.7.2.   ■

Theorem 4.7.2 has important implications for matrices. We begin to catalog such implications now.

Theorem 4.7.3. If $T \in M_n(\mathbb{C})$, then there exists an invertible matrix C in $M_n(\mathbb{C})$ such that $C^{-1}TC$ is an upper triangular matrix

$$\begin{bmatrix} a_{11} & & * \\ & \ddots & \\ 0 & & a_{nn} \end{bmatrix}.$$

Proof. Let $v_1, \ldots, v_n$ be a basis of $\mathbb{C}^{(n)}$ over $\mathbb{C}$ in which the matrix A of T is upper triangular. If C is the matrix whose columns are the vectors $v_1, \ldots, v_n$, then C is invertible and $A = C^{-1}TC$ by Theorem 3.9.1. ∎

We should point out that Theorem 4.7.3 does not hold if we are merely working over $\mathbb{R}$. The following example of $T = \begin{bmatrix} 0 & 1 \\ -1 & 0 \end{bmatrix}$ reveals this. We claim that there is no invertible matrix V with real entries such that

$$V^{-1}TV = \begin{bmatrix} a & b \\ 0 & c \end{bmatrix},$$

that is, such that

$$\begin{bmatrix} 0 & 1 \\ -1 & 0 \end{bmatrix} V = V \begin{bmatrix} a & b \\ 0 & c \end{bmatrix}.$$

For suppose that $V = \begin{bmatrix} u & v \\ w & x \end{bmatrix}$. Then

$$\begin{bmatrix} 0 & 1 \\ -1 & 0 \end{bmatrix}\begin{bmatrix} u & v \\ w & x \end{bmatrix} = \begin{bmatrix} u & v \\ w & x \end{bmatrix}\begin{bmatrix} a & b \\ 0 & c \end{bmatrix}$$

yields $w = au$, $-u = aw$. But then $w \neq 0$ (otherwise, $u = w = 0$ and the first basis vector is 0) and $w = au = a(-aw) = -a^2w$ and so $(1 + a^2)w = 0$, which is impossible since a is real.

Theorem 4.7.3 is one of the key results in matrix theory. It is usually described as:

Any matrix in $M_n(\mathbb{C})$ can be brought to triangular form in $M_n(\mathbb{C})$.

Theorem 4.7.3 has many important consequences for us. We begin with an example with $n = 3$. Suppose that T is in $M_3(\mathbb{C})$ and that we take an invertible matrix C to triangularize the matrix A of T in C, that is,

$$A = C^{-1}TC = \begin{bmatrix} a & b & c \\ 0 & d & e \\ 0 & 0 & f \end{bmatrix}.$$

Then the columns v_1, v_2, v_3 of C form a basis of $\mathbb{C}^{(3)}$ such that

$$Av_1 = av_1, \; Av_2 = bv_1 + dv_2, \; Av_3 = cv_1 + ev_2 + fv_3.$$

Therefore,

$$(A - aI)v_1 = 0, \; (A - dI)v_2 = bv_1, \; (A - fI)v_3 = cv_1 + ev_2.$$

Thus we can knock out the basis vectors one at a time:

$$(A - aI)v_1 = 0,$$
$$(A - aI)(A - dI)v_2 = (A - aI)(bv_1) = 0,$$
$$(A - aI)(A - dI)(A - fI)v_3 = (A - aI)(A - dI)(cv_1 + ev_2)$$
$$= 0. \quad \text{(Why?)}$$

Letting $D = (A - aI)(A - dI)(A - fI)$, it follows that D knocks out all three basis vectors at the same time, so that $Dv = 0$ for all v in $\mathbb{C}^{(3)}$. (Prove!) In other words, $D = 0$ and $(A - aI)(A - dI)(A - fI) = 0$. Expressing A as $C^{-1}TC$, we have

$$(C^{-1}TC - aI)(C^{-1}TC - dI)(C^{-1}TC - fI) = 0.$$

But this implies that

$$C^{-1}(T - aI)(T - dI)(T - fI)C = 0.$$

(Prove!) Multiplying the equation on the left by C and on the right by C^{-1}, we see that

$$(T - aI)(T - dI)(T - fI) = 0.$$

What does the equation $(T - aI)(T - dI)(T - fI) = 0$ tell us? Several things:

1. Any matrix in $M_3(\mathbb{C})$ satisfies a polynomial of degree 3 over $\mathbb{C}$.

2. This polynomial is the product of terms of the form $T - uI$, where the u are all of the characteristic roots of T, in this case, a, b, and c. (Prove!) It may happen that these characteristic roots are not distinct, that is, that some are included more than once in this polynomial.

What we did for T in $M_3(\mathbb{C})$ we shall do—in exactly the same way—for matrices in $M_n(\mathbb{C})$, to obtain results like (1) and (2) above. This is why we did the 3×3 case in so painstaking a way.

Theorem 4.7.4.   Given T in $M_n(\mathbb{C})$, there is a polynomial, $p(x)$, of *degree* n with complex coefficients such that $p(T) = 0$. If T is an upper

triangular   matrix $\begin{bmatrix} a_1 & & * \\ & \ddots & \\ 0 & & a_n \end{bmatrix}$,   one   such   polynomial $p(x)$ is

$(x - a_1) \cdots (x - a_n)$, and the $a_1, \ldots, a_n$ are the characteristic roots of T (all of them with possible duplication).

Proof.   The scheme for proving the result is just what we did for the 3×3 case, with the obvious modification needed for the general situation.

By Theorem 4.7.2 we have a basis $v_1, \ldots, v_n$ of $\mathbb{C}^{(n)}$ such that for each s, $Tv_s = \sum_{r=1}^{s} a_{rs}v_r$. Take any such basis, let V be the invertible matrix whose columns are the v_r and write the entries a_{kk} as a_k. Thus

$C^{-1}TC$ is the upper triangular matrix $A = \begin{bmatrix} a_1 & & * \\ & \ddots & \\ 0 & & a_n \end{bmatrix}$. Note that

$(T - a_sI)v_s$ is a linear combination of $v_1, \ldots, v_{s-1}$ for all s. By an induction—or going step by step as for the 3×3 case—we have that

$$(T - a_1I)(T - a_2I) \cdots (T - a_{s-1}I)v_k = 0$$

for $k = 1, 2, \ldots, s - 1$. But then

$$(T - a_1I)(T - a_2I) \cdots (T - a_{s-1}I)(T - a_sI)v_k = 0$$

not only for $k = 1, 2, \ldots, s + 1$, but also for $k = s$. (Why?) The vector $(T - a_sI)v_k$ is a linear combination of $v_1, \ldots, v_{s-1}$. This is true for all s, therefore, and so, in particular,

$$(T - a_1I) \cdots (T - a_nI)v_k = 0$$

for $k = 1, 2, \ldots, n$. Because

$$(T - a_1I) \cdots (T - a_nI)$$

annihilates a basis of $\mathbb{C}^{(n)}$, it must annihilate all of $\mathbb{C}^{(n)}$, that is, $(T - a_1I) \cdots (T - a_nI)v = 0$ for all v in $\mathbb{C}^{(n)}$. In short, the linear transformation $(T - a_1I) \cdots (T - a_nI)$ is 0.

It follows that the minimum polynomial $q_T(x)$ divides the polynomial $(x - a_1) \cdots (x - a_n)$. Since the characteristic roots of T are just the roots of its minimum polynomial, it follows that every characteristic root occurs as some a_r. Conversely, for any $1 \le r \le n$, the matrix $A - a_r I$ is not invertible, by Theorem 4.5.3. So, $T - a_r I = C(A - a_r I)C^{-1}$ is also not invertible (prove!), so that a_r is a characteristic root of T.   ∎

This sharpens considerably our previous result, namely, that T satisfies a polynomial of degree n^2, for we have cut down from n^2 to n. We could call this theorem the *weak Cayley–Hamilton Theorem*. Why weak? Later, we prove this again and, at the same time, give Cayley–Hamilton's explicit description of this polynomial of degree n.

Suppose that T is a *nilpotent* matrix, that is, that $T^k = 0$ for some k. If $a \in \mathbb{C}$ is a characteristic root of T, then $Tv = av$ for some nonzero vector v in $\mathbb{C}^{(n)}$. Thus

$$T^2v = T(Tv) = T(av) = aT(v) = aav = a^2v,$$
$$T^3v = T^2(Tv) = T^2(av) = aT^2(v) = aa^2v = a^3v,$$

and so on. By induction we can show that $T^kv = a^kv$ for all positive integers k. (Do so!) Since $T^k = 0$ for some k and $v \ne 0$, and since $0 = T^kv = a^kv$, it follows that $a = 0$. Therefore, the only characteristic root of T is 0. By Theorem 4.7.3 it follows that there is an invertible matrix C such that $A = C^{-1}TC$ is an upper triangular matrix with 0's on the diagonal, proving

Theorem 4.7.5. If T in $M_n(\mathbb{C})$ is nilpotent, then there is an invertible C in $M_n(\mathbb{C})$ such that the matrix $C^{-1}TC$ is an upper triangular matrix
$$\begin{bmatrix} 0 & & * \\ & \ddots & \\ 0 & & 0 \end{bmatrix} \text{ with 0's on the diagonal.}$$

In particular, the trace of the matrix $C^{-1}TC = \begin{bmatrix} 0 & & * \\ & \ddots & \\ 0 & & 0 \end{bmatrix}$ of

the nilpotent T in Theorem 4.7.5 is 0. But then also $0 = \text{tr } (C^{-1}TC) = \text{tr } (T)$.

We can combine Theorems 4.7.3 and 4.7.4 to obtain an important property of the trace function.

Theorem 4.7.6. If $T \in M_n(\mathbb{C})$, then tr $(T) = a_1 + \cdots + a_n$, where $a_1, a_2, \ldots, a_n$ are the characteristic roots of T (with possible duplications).

Proof. By Theorem 4.7.3 we know that for some invertible C,

$$A = C^{-1}TC = \begin{bmatrix} a_1 & & * \\ & \ddots & \\ 0 & & a_n \end{bmatrix}.$$ By Theorem 4.7.4 the diagonal

entries

are the characteristic roots of T. Now

$$\text{tr } T = \text{tr } C^{-1}TC = \text{tr } A = a_1 + \cdots + a_n. \quad \blacksquare$$

Of course, the a_r need not be distinct. In the simplest possible case,

namely, $T = I = \begin{bmatrix} 1 & & 0 \\ & \ddots & \\ 0 & & 1 \end{bmatrix}$ in $M_n(\mathbb{C})$, T has trace n, $A = I$ and this

n is the sum of n duplicates of the characteristic root 1 appearing on the diagonal of A.

How many duplicates of the various characteristic roots of T are needed in the trace? At this point we can only say that every characteristic root is needed at least once and the *total number* needed is the number of characteristic roots on the diagonal of A, including duplicates, which of course is n. Later, however, we answer this question definitively by saying that

> *The characteristic roots (including duplicates) are the roots of the characteristic polynomial.*

The *characteristic polynomial* is a very important player in the world of linear algebra, whose identity we reveal in Chapter 5.

PROBLEMS

NUMERICAL PROBLEMS

1. For the given T, find a C in $M_n(\mathbb{C})$, invertible, such that $C^{-1}TC$ is triangular.

(a) $T = \begin{bmatrix} 1 & 2 \\ 3 & 4 \end{bmatrix}$.

(b) $T = \begin{bmatrix} 0 & 1 \\ 1 & 1 \end{bmatrix}$.

(c) $T = \begin{bmatrix} 0 & i \\ -i & 0 \end{bmatrix}$.

(d) $T = \begin{bmatrix} 1 & 0 & 0 \\ 0 & 0 & i \\ 0 & -i & 0 \end{bmatrix}$.

(e) $T = \begin{bmatrix} 1 & 1 & 1 \\ 1 & 1 & 1 \\ 0 & 0 & 0 \end{bmatrix}$.

(f) $T = \begin{bmatrix} 0 & 1 & 0 \\ 0 & 0 & 1 \\ 1 & 0 & 0 \end{bmatrix}$.

2. For the matrices in Problem 1, find bases $v_1, \ldots, v_n$ such that Tv_s is a linear combination of its predecessors, for $s = 1, 2, 3, \ldots, n$.

3. Let $t = 2$, $V = \left\{ \begin{bmatrix} a \\ 0 \\ b \end{bmatrix} \mid a, b \in \mathbb{C} \right\}$, and $T = \begin{bmatrix} 0 & 1 & 0 \\ 0 & 2 & 0 \\ 1 & 2 & 3 \end{bmatrix}$. Find a polynomial $P_{T,V}(x)$ of degree t such that $P_{T,V}(T)(V) = 0$.

4. Let $t = 2$, $V = \left\{ \begin{bmatrix} a \\ b \\ c \end{bmatrix} \mid a, b, c \in \mathbb{C}, a + b + c = 0 \right\}$, and $T = \begin{bmatrix} 1 & 0 & 0 \\ 0 & 0 & 1 \\ 0 & 1 & 0 \end{bmatrix}$. Find a polynomial $P_{T,V}(x)$ of degree 2 such that $P_{T,V}(T)(V) = 0$.

5. If $T = \begin{bmatrix} a & b \\ c & d \end{bmatrix}$ satisfies $T^2 = 0$, show that $a + d = 0$.

6. Find an invertible matrix C such that $C \begin{bmatrix} 0 & 1 \\ 0 & 0 \end{bmatrix} = \begin{bmatrix} 0 & 0 \\ 1 & 0 \end{bmatrix} C$.

MORE THEORETICAL PROBLEMS

Easier Problems

7. Suppose that $TT' = T$. Then show that $T^2 = T$.

8. Suppose that

$$(A - aI)v_1 = 0,$$
$$(A - dI)v_2 = bv_1,$$
$$(A - fI)v_3 = cv_1 + ev_2,$$

and let $D = (A - aI)(A - dI)(A - fI)$. Prove that

$$(A - aI)(A - dI)(A - fI)v = 0$$

for every linear combination v of v_1, v_2, v_3.

9. Prove that the expression $(C^{-1}TC - aI)(C^{-1}TC - dI)(C^{-1}TC\ fI)$ equals $C^{-1}(T - aI)(T - dI)(T - fI)C$.

10. If T in $M_n(\mathbb{C})$ is such that $T^2 = T$, find the characteristic roots of T.

11. If T in $M_n(\mathbb{C})$ is such that $T^2 = T$, prove that tr $(T) = $ rank T.

12. Show that there exists an invertible matrix C such that

$$C^{-1} \begin{bmatrix} 0 & 0 & 0 \\ 1 & 0 & 0 \\ 0 & 1 & 0 \end{bmatrix} C = \begin{bmatrix} 0 & 1 & 0 \\ 0 & 0 & 1 \\ 0 & 0 & 0 \end{bmatrix}.$$

Middle-Level Problems

13. If T in $M_n(\mathbb{C})$ is such that $T^2 = T$, show that for some invertible C in $M_n(\mathbb{C})$, $C^{-1}TC$ is a matrix of the form $\begin{bmatrix} I_k & 0 \\ 0 & 0 \end{bmatrix}$ (I_k is the $k \times k$ identity matrix and the 0's represent blocks of the right sizes all of whose entries are 0), where k is the rank of T.

14. If T in $M_3(\mathbb{C})$ is such that tr $(T) = $ tr $(T^2) = $ tr $(T^3) = 0$, prove that T cannot be invertible.

15. If T in $M_3(\mathbb{C})$ is such that tr $(T) = $ tr $(T^2) = $ tr $(T^3) = 0$, prove that $T^3 = 0$.

4.8. EXPONENTIALS

Insofar as possible we have tried to keep the exposition of the material in the book self-contained. One exception has been our use of the Fundamental Theorem of Algebra, which we stated but didn't prove. Another has been our use, at several places of the calculus; this was based on the assumption that almost all the readers will have had some exposure to the calculus.

Now, once again, we come to a point where we diverge from this policy of self-containment. For solving systems of differential equations, and for its own interest, we want to talk about power series in matrices. Most especially we would like to introduce the notion of the exponential, e^A, of a matrix A. We shall make some statements about these things which we don't verify with proofs.

Definition.   The power series $a_0 I + a_1 A + \cdots A_n A^n + \cdots$ in the matrix A (in $M_n(\mathbb{C})$ or $M_n(\mathbb{R})$) is said to *converge* to a matrix B if in *each entry* of "the matrix given by our power series" it converges.

For instance, let's consider the matrix $A = \begin{bmatrix} 1 & 0 \\ 0 & 2 \end{bmatrix}$ and the power series

$$I + \left(\frac{1}{3}\right)A + \left(\frac{1}{3^2}\right)A^2 + \cdots + \left(\frac{1}{3^n}\right)A^n + \cdots,$$

which looks like

$$\begin{bmatrix} 1 & 0 \\ 0 & 1 \end{bmatrix} + \frac{1}{3}\begin{bmatrix} 1 & 0 \\ 0 & 2 \end{bmatrix} + \frac{1}{3^2}\begin{bmatrix} 1 & 0 \\ 0 & 2^2 \end{bmatrix} + \cdots + \frac{1}{3^n}\begin{bmatrix} 1 & 0 \\ 0 & 2^n \end{bmatrix} + \cdots.$$

Then we get convergence in the $(1, 1)$ entry to

$$1 + \frac{1}{3} + \frac{1}{3^2} + \cdots + \frac{1}{3^n} + \cdots = \frac{3}{2},$$

in the $(2, 1)$ and $(1, 2)$ entries to 0, and in the $(2, 2)$ entry to

$$1 + \frac{2}{3} + \left(\frac{2}{3}\right)^2 + \cdots + \left(\frac{2}{3}\right)^n + \cdots = 3.$$

So we declare that the power series

$$I + \left(\frac{1}{3}\right)A + \left(\frac{1}{3^2}\right)A^2 + \cdots + \left(\frac{1}{3^n}\right)A^n + \cdots$$

equals

$$\begin{bmatrix} \frac{3}{2} & 0 \\ 0 & 3 \end{bmatrix}.$$

Of course, this example is a very easy one to compute. In general things don't go so easily.

Definition. If $A \in M_n(F)$, $F = \mathbb{C}$ or $\mathbb{R}$, then the *exponential*, e^A, of A is defined by

$$e^A = I + A + \frac{A^2}{2!} + \frac{A^3}{3!} + \cdots + \frac{A^m}{m!} + \cdots$$

The basic fact that we need is that the power series e^A converges, in the sense described above, for all matrices A in $M_n(F)$.

To prove this is not so hard, but would require a rather lengthy and distracting digression. So we wave our hands and make use of this fact without proof.

We know for real or complex numbers a and b that $e^{a+b} = e^a e^b$. This can be proved using the power series definition

$$e^x = 1 + x + \frac{x^2}{2!} + \frac{x^3}{3!} + \cdots + \frac{x^m}{m!} + \cdots$$

of e^x for real and complex numbers x.

For matrices, however, this is no longer true in general; that is, e^{A+B} is not, in general, equal to $e^A e^B$.

Can you give an example where these are not equal? The reason for this is that to rearrange terms in the product $e^A e^B$ we need to know that A and B commute, that is, $AB = BA$. Then it is possible rearrange the terms into the terms for e^{A+B}, that is, it is possible to prove:

If A and B commute, then $e^{A+B} = e^A e^B$.

In particular, if $A = aI$ is a scalar matrix, then

1. $e^{aI} = e^a I$.
2.   Since $AB = BA$ for all B, $e^{aI+B} = e^{aI} e^B = e^a e^B$.

We illustrate what we have discussed with a simple example. Let $A = \begin{bmatrix} 1 & 1 \\ 0 & 1 \end{bmatrix}$. Then what is e^A? Note that $A = I + N$, where $N = \begin{bmatrix} 0 & 1 \\ 0 & 0 \end{bmatrix}$ and N satisfies $N^2 = 0$. Thus

$$A = I + N$$
$$A^2 = (I + N)^2 = I + 2N + N^2 = I + 2N,$$
$$A^3 = (I + N)^3 = I + 3N,$$
$$\vdots$$
$$A^k = (I + N)^k = I + kN.$$

Thus

$$e^A = I + I + \frac{I}{2!} + \frac{I}{3!} + \cdots + \frac{I}{k!} + \cdots$$

$$+ N + \frac{2N}{2!} + \frac{3N}{3!} + \cdots + \frac{kN}{k!} + \cdots$$

$$= I\left(1 + 1 + \frac{1}{2!} + \frac{1}{3!} + \cdots + \frac{1}{k!} + \cdots\right)$$

$$+ N\left(0 + 1 + \frac{2}{2!} + \frac{3}{3!} + \cdots + \frac{k}{k!} + \cdots\right)$$

$$= e(I + N) = eA = e\begin{bmatrix} 1 & 1 \\ 0 & 1 \end{bmatrix}.$$

We can also compute e^A as

$$e^A = e^{I+N} = e^I e^N = ee^N = e(I + N) = e\begin{bmatrix} 1 & 1 \\ 0 & 1 \end{bmatrix},$$

ending up with the same answer.

There is one particular kind of matrix for which the power series really is a polynomial, namely a *nilpotent* matrix N, that is, a matrix $N \in M_n(F)$ such that $N^k = 0$ for some positive integer k. For such N and k, $N^s = 0$ for all $s \geq k$, so that

$$e^N = I + N + \frac{N^2}{2!} + \frac{N^3}{3!} + \cdots + \frac{N^{k-1}}{(k-1)!},$$

a polynomial expression in N. We state this in

Theorem 4.8.1. If $N \in M_n(F)$ and $N^k = 0$ for some k, then

$$e^N = I + N + \frac{N^2}{2!} + \frac{N^3}{3!} + \cdots + \frac{N^{k-1}}{(k-1)!}.$$

A very simple example of this is afforded us by the $N = \begin{bmatrix} 0 & 1 \\ 0 & 0 \end{bmatrix}$ in the example above, where $N^2 = 0$ and e^N is just $I + N$.

Let's now look at the *matrix function* e^{tT} of the scalar variable t for an $n \times n$ matrix T. In our example above for $A = \begin{bmatrix} 1 & 1 \\ 0 & 1 \end{bmatrix}$, we computed the exponential e^A as $e \begin{bmatrix} 1 & 1 \\ 0 & 1 \end{bmatrix}$. Following in the footsteps of what we did there, we can express the exponential function e^{tA} as

$$e^{tA} = e^{tI + tN} = e^{tI} e^{tN} = e^t e^{tN} = e^t(I + tN) = e^t \begin{bmatrix} 1 & t \\ 0 & 1 \end{bmatrix}.$$

Of course, we can even express e^{tT} for any $T = aI + N$, where N is nilpotent, by the formula

$$e^{t(aI + N)} = e^{taI + tN} = e^{taI} e^{tN} = e^{ta} e^{tN},$$

where e^{tN} is given by Theorem 4.8.1.

Using this exponential function e^{tT}, we can solve the homogeneous system

$$x_1'(t) = t_{11}x_1(t) + \cdots + t_{1n}x_n(t), \qquad x_1(0) = x_{01}$$
$$\vdots \qquad\qquad\qquad\qquad \vdots$$
$$x_n'(t) = t_{n1}x_1(t) + \cdots + t_{nn}x_n(t), \qquad x_n(0) = x_{0n}$$

of linear differential equations by representing the system by a *matrix differential equation*

$$x'(t) = Tx(t), \ x(0) = x_0,$$

where $x(t)$ is the vector function $x(t) = \begin{bmatrix} x_1(t) \\ \vdots \\ x_n(t) \end{bmatrix}$, x_0 is the vector of initial conditions x_{0r}, that is, $x_0 = \begin{bmatrix} x_{01} \\ \vdots \\ x_{0n} \end{bmatrix}$, and $x'(t)$ is the derivative

$$x'(t) = \begin{bmatrix} x_1'(t) \\ \vdots \\ x_n'(t) \end{bmatrix}$$

of $x(t)$. This *differentiation of vector functions* satisfies the properties

1.   The derivative of $ax(t) = \begin{bmatrix} ax_1(t) \\ \vdots \\ ax_n(t) \end{bmatrix}$ is $ax'(t)$, for $a \in \mathbb{C}$.

2.   The derivative of $x(t) + y(t) = \begin{bmatrix} x_1(t) + y_1(t) \\ \vdots \quad\quad \vdots \\ x_n(t) + y_n(t) \end{bmatrix}$ is $x'(t) + y'(t)$.

3.   The derivative of $f(t)y(t) = \begin{bmatrix} f(t)y_1(t) \\ \vdots \\ f(t)y_n(t) \end{bmatrix}$ is

$$f(t)y'(t) + f'(t)y(t) = \begin{bmatrix} f(t)y_1'(t) \\ \vdots \\ f(t)y_n'(t) \end{bmatrix} + \begin{bmatrix} f'(t)y_1(t) \\ \vdots \\ f'(t)y_n(t) \end{bmatrix}.$$

4.   If $T \in M_n(\mathbb{C})$, then the derivative of $e^{tT}x_0$ is $Te^{tT}x_0$.

We leave proving properties (1)–(3) as an exercise for the reader. Property (4) is easy to prove *if we assume that the series $e^{tT}x_0$ can be differentiated term by term*. We could prove this assumption here, but its proof, which requires background on power series, would take us far afield. So we omit its proof and assume it instead. Given this assumption, how do we prove property (4)? Easily. Letting $f(t) = e^{tT}x_0$, we differentiate the series

$$f(t) = x_0 + \left(\frac{tT}{1!}\right)x_0 + \left(\frac{t^2T^2}{2!}\right)x_0 + \cdots + \left(\frac{t^kT^k}{k!}\right)x_0 + \cdots$$

term by term, getting its derivative

$$f'(t) = 0 + tx_0 + \left(\frac{tT^2}{1!}\right)x_0 + \cdots + \left(\frac{t^{k-1}T^k}{(k-1)!}\right)x_0 + \cdots$$

$$= T\left(x_0 + \left(\frac{tT}{1!}\right)x_0 + \cdots + \left(\frac{t^{k-1}T^{k-1}}{(k-1)!}\right)x_0 + \cdots\right)$$

$$= Te^{tT}x_0.$$

Since $e^{tT}x_0$ has derivative $Te^{tT}x_0$ and since $e^{(t-t_0)T}x_0 = e^{tT}e^{-t_0T}x_0$, the derivative of the function $x(t) = e^{(t-t_0)T}x_0$ is $Tx(t)$. (Prove!) Since $x(t_0) = x_0$, this proves the *existence* part of

Theorem 4.8.2. For any $T \in M_n(\mathbb{C})$, $x_0 \in \mathbb{C}^{(n)}$ and real number t_0, $x(t) = e^{(t-t_0)T}x_0$ is a solution to the matrix differential equation $x'(t) = Tx(t)$ such that $x(t_0) = x_0$. Moreover, this is the only such solution.

Proof. we've just seen that $x(t) = e^{(t-t_0)T}x_0$ is a solution to the equation $x'(t) = Tx(t)$ such that $x(t_0) = x_0$. It now remains only to show that *this solution is the only one*. Letting $u(t)$ denote any other such solution, we define $v = u - e^{(t-t_0)T}x_0$. Then

$$v(t_0) = u(t_0) - e^{0T}x_0 = x_0 - x_0 = 0.$$

To prove that $u = e^{(t-t_0)T}x_0$, we now simply use the equation $v(t_0) = 0$ to show that $v = 0$. Suppose, to the contrary, that v is nonzero. Choosing $d > 0$ so that T^d is linearly dependent on $T^0, T^1, \ldots, T^{d-1}$, the vector function T^dv is a linear combination of $v, Tv, \ldots, T^{d-1}v$. It follows that the span V of $v, Tv, \ldots, T^{d-1}v$ over $\mathbb{C}$ is mapped into itself by T. Let w be a characteristic vector of T in V, so that $Tw = aw$ for some scalar a. Since $w \in V$, and since $v' = Tv$ and $v(t_0) = 0$, w satisfies the conditions $w' = Tw$ and $w(t_0) = 0$. So $w' = aw$. Since the solutions to the differential equations

$$w_r'(t) = aw_r(t)$$

are of the form $w_r(t) = w_r e^{a(t-t_0)}$ for some constants w_r, w is of the form

$$w = \begin{bmatrix} w_1 e^{a(t-t_0)} \\ \vdots \\ y_n e^{a(t-t_0)} \end{bmatrix}.$$

Since $0 = w(t_0) = \begin{bmatrix} w_1 \\ \vdots \\ n \end{bmatrix}$, it follows that $w = 0$. But this contradicts our choice of w as a characteristic (and, therefore, nonzero) vector. So our hypothesis that v be nonzero leads to a contradiction. We conclude that $v = 0$ and $u = e^{(t - t_0)T} x_0$.   ■

EXAMPLE

Suppose that predators A and their prey B, for example two kinds of fish, have populations $P_A(t)$, $P_B(t)$ at time t with growth rates

$$P_A'(t) = \quad 0 P_A(t) + 1 P_B(t)$$
$$P_B'(t) = -2 P_A(t) + 3 P_B(t).$$

These growth rates reflect how much the predators A depend on their current population and the availability $P_B(t)$ of the prey B for their population growth; and how much the growth in population of the prey B is impeded by the presence $P_A(t)$ of the predators B. Assuming that at time $t = 5$ we have initial populations $P_A(5) = 30,000$ and $P_B(5) = 40,000$, how do we find the population functions $P_A(t)$ and $P_B(t)$?

We represent the system

$$P_A'(t) = \quad 0 P_A(t) + 1 P_B(t)$$
$$P_B'(t) = -2 P_A(t) + 3 P_B(t)$$

of growth rate equations by the differential matrix equation

$$\begin{bmatrix} P_A'(t) \\ P_B'(t) \end{bmatrix} = \begin{bmatrix} 0 & 1 \\ -2 & 3 \end{bmatrix} \begin{bmatrix} P_A(t) \\ P_B(t) \end{bmatrix}.$$

The initial value $P(5)$ is then $\begin{bmatrix} P_A(5) \\ P_B(5) \end{bmatrix} = \begin{bmatrix} 30,000 \\ 40,000 \end{bmatrix}$. So the desired solution is

$$\begin{bmatrix} P_A(t) \\ P_B(t) \end{bmatrix} = e^{(t-5)\begin{bmatrix} 0 & 1 \\ -2 & 3 \end{bmatrix}} \begin{bmatrix} 30,000 \\ 40,000 \end{bmatrix}.$$

PROBLEMS

NUMERICAL PROBLEMS

1. Show that if $A = aI$, then $e^A = e^a I$.

2. If $N^2 = 0$, find the form of e^A where $A = aI + N$.

3. If $N^3 = 0$, find the form of e^A where $A = aI + N$.

4. Given $A = \begin{bmatrix} 1 & 1 \\ 0 & 0 \end{bmatrix}$ and $B = \begin{bmatrix} 0 & 0 \\ 1 & 1 \end{bmatrix}$, is $e^{A+B} = e^A e^B$?

5. Verify that the real part $\begin{bmatrix} x_{1r}(t) \\ x_{2r}(t) \end{bmatrix}$ of any solution $\begin{bmatrix} x_1(t) \\ x_2(t) \end{bmatrix}$ to

$$\begin{bmatrix} x_1'(t) \\ x_2'(t) \end{bmatrix} = \begin{bmatrix} 0 & 1 \\ -1 & 0 \end{bmatrix} \begin{bmatrix} x_1(t) \\ x_2(t) \end{bmatrix} \text{ satisfies}$$

$$\begin{bmatrix} x_{1r}'(t) \\ x_{2r}'(t) \end{bmatrix} = \begin{bmatrix} 0 & 1 \\ -1 & 0 \end{bmatrix} \begin{bmatrix} x_{1r}(t) \\ x_{2r}(t) \end{bmatrix}.$$

6. Find a solution to the equation $x'(t) = Tx(t)$ where $T = \begin{bmatrix} 1 & 1 & 0 \\ 0 & 1 & 0 \\ 0 & 0 & 2 \end{bmatrix}$.

MORE THEORETICAL PROBLEMS

Easier Problems

7. If $N^2 = 0$, find the form of e^{tA} where $A = aI + N$.
8. If $N^3 = 0$, find the form of e^{tA} where $A = aI + N$.

Middle-Level Problems

9. For real numbers a and b prove, using the power series definition of e^x, that $e^{a+b} = e^a e^b$.

10. Imitate the proof in Problem 7 to show that if $A, B \in M_n(F)$ and $AB = BA$, then $e^{A+B} = e^A e^B$.

11. Show that $e^A e^{-A} = I = e^{-A} e^A$, that is, e^A is invertible, for all A.

12. Define the power series

$$\sin A = A - \frac{A^3}{3!} + \frac{A^5}{5!} + \cdots$$

$$\cos A = I - \frac{A^2}{2!} + \frac{A^4}{4!} - \cdots.$$

and prove that $(\sin A)^2 + (\cos A)^2 = I$.

13. Evaluate $\sin (e^{aI})$ and $\cos (e^{aI})$.

14. Show that the power series

$$1 + x + x^2 + \cdots + x^k + \cdots$$

does *not* converge when we replace x by $A = \begin{bmatrix} 1 & 1 \\ 1 & 1 \end{bmatrix}$ nor when

we replace x by $I = \begin{bmatrix} 1 & 0 \\ 0 & 1 \end{bmatrix}$.

15. If $A = \begin{bmatrix} \frac{1}{2} & 0 & 0 \\ 0 & \frac{1}{3} & 0 \\ 0 & 0 & \frac{1}{4} \end{bmatrix}$, does the power series in Problem 14 converge

for A? If so, to what matrix does it converge?

16. Evaluate e^A if $A = \begin{bmatrix} 0 & 1 \\ 1 & 0 \end{bmatrix}$.

17. Evaluate e^A if $A = \begin{bmatrix} 0 & -1 \\ 1 & 0 \end{bmatrix}$.

18. Show that $e^{iA} = \cos A + i \sin A$, where $\sin A$ and $\cos A$ are as defined in Problem 12.

19. If $A^2 = A$, evaluate e^A, $\cos A$, and $\sin A$.

20. If C is invertible and $A^2 = A$, show that $e^{C^{-1}AC} = C^{-1}e^AC$.

21. If C is invertible and $A^k = 0$, show that $e^{C^{-1}AC} = C^{-1}e^AC$.

22. The *derivative* of a matrix function

$$f(t) = \begin{bmatrix} a(t) & b(t) \\ c(t) & d(t) \end{bmatrix}$$

is defined as

$$f'(t) = \begin{bmatrix} a'(t) & b'(t) \\ c'(t) & d'(t) \end{bmatrix}.$$

Using the definitions in Problem 12, show for $A = \begin{bmatrix} a & 0 \\ 0 & d \end{bmatrix}$ and

$tA = \begin{bmatrix} ta & 0 \\ 0 & td \end{bmatrix}$ that the derivative of $\sin tA$ is $A \cos tA$. What is

the derivative of $\cos tA$? Of e^{tA}?

Harder Problems

23. Find the derivative of e^{tA} directly from the definitions (see Problem 22) for the matrices

(a) $A = \begin{bmatrix} a & 0 \\ 0 & b \end{bmatrix}$.

(b) $A = \begin{bmatrix} a & 1 \\ 0 & a \end{bmatrix}$.

(c) $A = \begin{bmatrix} a & 1 & 0 \\ 0 & a & 0 \\ 0 & 0 & b \end{bmatrix}$.

(d) $A = \begin{bmatrix} a & 1 & 0 \\ 0 & a & 1 \\ 0 & 0 & a \end{bmatrix}$.

(e) $A = \begin{bmatrix} a & 1 & 0 & 0 \\ 0 & a & 1 & 0 \\ 0 & 0 & a & 1 \\ 0 & 0 & 0 & a \end{bmatrix}$.

24. For each A in Problem 23, verify directly from the definitions (see Problem 22) that the derivative of e^{tA} equals Ae^{tA}.

25. Prove that the real part $\begin{bmatrix} x_{1r}(t) \\ x_{2r}(t) \end{bmatrix}$ of any solution $\begin{bmatrix} x_1(t) \\ x_2(t) \end{bmatrix}$ to a differential equation $\begin{bmatrix} x_1'(t) \\ x_2'(t) \end{bmatrix} = \begin{bmatrix} a & b \\ c & d \end{bmatrix}\begin{bmatrix} x_1(t) \\ x_2(t) \end{bmatrix}$ with real coefficient matrix $\begin{bmatrix} a & b \\ c & d \end{bmatrix}$ satisfies $\begin{bmatrix} x_{1r}'(t) \\ x_{2r}'(t) \end{bmatrix} = \begin{bmatrix} a & b \\ c & d \end{bmatrix}\begin{bmatrix} x_{1r}(t) \\ x_{2r}(t) \end{bmatrix}$.

5

DETERMINANTS

5.1. INTRODUCTION

In our discussion of 2×2 matrices in Chapter 1, we introduced the notion of the determinant of the matrix $\begin{bmatrix} a & b \\ c & d \end{bmatrix}$ as $\begin{vmatrix} a & b \\ c & d \end{vmatrix} = ad - bc$. This notion turned out to have many uses and consequences, even in so easy a case as the 2×2 matrices. For instance, the determinant cropped up in solving two linear equations in two unknowns, in determining the invertibility of a matrix, in finding the characteristic roots of a matrix, and in other contexts.

We shall soon define the determinant of an $n \times n$ matrix for any $n \geq 2$. The procedure will be a step-by-step one. Knowing what a 2×2 determinant is will allow us to define what a 3×3 determinant is. Then, knowing what a 3×3 determinant is, we will be able to define a 4×4 one, and so on. In other words, we are defining the determinant *inductively*.

Whereas the proofs for the 2×2 case of all the properties of the determinant were very easy, merely involving rather simple computations, the proofs of these properties for the general case of n are much more subtle and difficult. Knowing the results for the 2×2 case will always be the starting point for the argument—an argument by induction—and then the $n \times n$ case will be handled by knowing it for the $(n - 1) \times (n - 1)$ case.

To try to clarify the arguments used, we shall use these arguments for the special case of 3×3 matrices. Here everything is quite visible and so should give the reader an inkling of what is going on in general.

Definition. If $A = [a_{rs}] = \begin{bmatrix} a_{11} & a_{12} & \cdots & a_{1n} \\ a_{21} & a_{22} & \cdots & a_{2n} \\ \vdots & \vdots & & \vdots \\ a_{n1} & a_{n2} & \cdots & a_{nn} \end{bmatrix} \in M_n(F)$, then the

(r, s) *minor submatrix* A_{rs} of A is that matrix which remains when we cross out row r and column s of A.

For example, if $A = \begin{bmatrix} 1 & 2 & 3 & 4 \\ 4 & 3 & 2 & 1 \\ 0 & 1 & 2 & 3 \\ -1 & 4 & 1 & 1 \end{bmatrix}$, then the $(3, 2)$ minor sub-

matrix A_{32} is given by $A_{32} = \begin{bmatrix} 1 & 3 & 4 \\ 4 & 2 & 1 \\ -1 & 1 & 1 \end{bmatrix}$, that matrix which remains

on eliminating row 3 and column 2 of A.

The matrices A_{rs} are, of course, $(n - 1) \times (n - 1)$ matrices, *so we proceed by assuming that what is meant by their determinant is known* (as we outlined a few paragraphs above).

Definition. The (r, s) *minor* of A is given by the determinant M_{rs} of the minor submatrix A_{rs}.

Since this notion is so important, to try to understand it completely, let's look at some minors for 3×3 matrices.

If $A = \begin{bmatrix} 1 & 2 & 3 \\ 4 & 5 & 6 \\ 7 & 8 & 9 \end{bmatrix}$, then

$$A_{11} = \begin{bmatrix} 5 & 6 \\ 8 & 9 \end{bmatrix}$$

and

$$M_{11} = \begin{vmatrix} 5 & 6 \\ 8 & 9 \end{vmatrix} = 5 \cdot 9 - 8 \cdot 6 = 45 - 48 = -3,$$

while

$$A_{32} = \begin{bmatrix} 1 & 3 \\ 4 & 6 \end{bmatrix}$$

and

$$M_{32} = \begin{vmatrix} 1 & 3 \\ 4 & 6 \end{vmatrix} = 1 \cdot 6 - 4 \cdot 3 = 6 - 12 = -6.$$

We are now ready to define the determinant of any $n \times n$ matrix.

Definition.   If $A = [a_{rs}] \in M_n(F)$, then the *determinant* of A, written as det (A), or det (a_{rs}), or

$$\begin{vmatrix} a_{11} & a_{12} & \cdots & a_{1n} \\ a_{21} & a_{22} & \cdots & a_{2n} \\ \vdots & \vdots & & \vdots \\ a_{n1} & a_{n2} & \cdots & a_{nn} \end{vmatrix},$$

is defined as

$$a_{11}M_{11} - a_{21}M_{21} + \cdots + (-1)^{r+1}a_{r1}M_{r1} + \cdots + (-1)^{n+1}a_{n1}M_{n1}.$$

So in summation notation,

$$\text{det } (A) = \sum_{r=1}^{n} (-1)^{r+1}a_{r1}M_{r1}.$$

Before seeing what this definition says for a 3×3 matrix, let's note some aspects built into the definition.

1.   The emphasis is given to the *first column* of A—an unsymmetrical fact that will cause us some annoyance later—since we are using the minors of the first column of A. For this reason we call this expansion the *expansion by the minors of the first column of A*.

2. The signs attached to the minors *alternate*, going as $+,\ -,\ +,\ -,$ and so on.

3. We are not restricting the rows of A in the definition. This will be utterly essential for us in the next section.

EXAMPLES

1. $$\begin{vmatrix} 1 & 2 & 3 \\ 3 & -1 & 4 \\ 0 & 1 & 1 \end{vmatrix} = 1\begin{vmatrix} -1 & 4 \\ 1 & 1 \end{vmatrix} - 3\begin{vmatrix} 2 & 3 \\ 1 & 1 \end{vmatrix} + 0\begin{vmatrix} 2 & 3 \\ -1 & 4 \end{vmatrix}$$

$$= -5 - 3(-1) + 0 = -2.$$

2. $$\begin{vmatrix} 1 & 2 & 3 & 4 \\ 0 & 1 & 2 & 3 \\ 0 & 0 & 1 & 2 \\ 0 & 0 & 0 & 1 \end{vmatrix} = 1\begin{vmatrix} 1 & 2 & 3 \\ 0 & 1 & 2 \\ 0 & 0 & 1 \end{vmatrix} - 0\begin{vmatrix} 2 & 3 & 4 \\ 0 & 1 & 2 \\ 0 & 0 & 1 \end{vmatrix}$$

$$+ 0\begin{vmatrix} 2 & 3 & 4 \\ 1 & 2 & 3 \\ 0 & 0 & 1 \end{vmatrix} - 0\begin{vmatrix} 2 & 3 & 4 \\ 1 & 2 & 3 \\ 0 & 1 & 2 \end{vmatrix} = 1.$$

3. $$\begin{vmatrix} 1 & 0 & 0 & 0 \\ 2 & 1 & 0 & 0 \\ 3 & 2 & 1 & 0 \\ 4 & 3 & 2 & 1 \end{vmatrix} = 1\begin{vmatrix} 1 & 0 & 0 \\ 2 & 1 & 0 \\ 3 & 2 & 1 \end{vmatrix} - 2\begin{vmatrix} 0 & 0 & 0 \\ 2 & 1 & 0 \\ 3 & 2 & 1 \end{vmatrix}$$

$$+ 3\begin{vmatrix} 0 & 0 & 0 \\ 1 & 0 & 0 \\ 3 & 2 & 1 \end{vmatrix} - 4\begin{vmatrix} 0 & 0 & 0 \\ 1 & 0 & 0 \\ 2 & 1 & 0 \end{vmatrix} = ?$$

(Fill in the "?".)

Notice in Examples 2 and 3, which involve 3×3 determinants, that the calculations can be made because we know (in terms of 2×2 determinants) what a 3×3 determinant is. We leave it to the reader to show that both determinants in Examples 2 and 3 are equal to 1.

Examples 2 and 3 are examples of what we called triangular matrices. Recall that an *upper triangular matrix* is one all of whose entries *below* the main diagonal are 0, and that a *lower triangular matrix* is one all of whose entries *above* the main diagonal are 0.

For triangular matrices (upper or lower) the determinant is extremely easy to evaluate. Notice that in Examples 2 and 3 the determinant is just the product of the diagonal entries. This is the story in general, as our next theorem shows. The proof will show how we exploit knowing the result for the $(n - 1) \times (n - 1)$ case.

Theorem 5.1.1. The determinant of a triangular (upper or lower) matrix A is the product of its diagonal entries.

Proof. We first do the easier case of the upper triangular matrix. Let

$$A = \begin{bmatrix} a_{11} & a_{12} & \cdots & a_{1n} \\ 0 & a_{22} & \cdots & a_{2n} \\ \vdots & \vdots & & \vdots \\ 0 & 0 & \cdots & a_{nn} \end{bmatrix}.$$

Since the entries $a_{21}, a_{31}, \ldots, a_{n1}$ of the first column of A are all 0, our definition of the determinant of A gives us

$$\begin{vmatrix} a_{11} & a_{12} & \cdots & a_{1n} \\ 0 & a_{22} & \cdots & a_{2n} \\ \vdots & \vdots & & \vdots \\ 0 & 0 & \cdots & a_{nn} \end{vmatrix} = a_{11}M_{11} = a_{11} \begin{vmatrix} a_{22} & a_{23} & \cdots & a_{2n} \\ 0 & a_{33} & \cdots & a_{3n} \\ \vdots & \vdots & & \vdots \\ 0 & 0 & \cdots & a_{nn} \end{vmatrix}.$$

The $(1, 1)$ minor submatrix of A is itself an upper triangular matrix, *but an $(n - 1) \times (n - 1)$ one*. Hence by our induction procedure,

$$\begin{vmatrix} a_{22} & a_{23} & \cdots & a_{2n} \\ 0 & a_{33} & \cdots & a_{3n} \\ \vdots & \vdots & & \vdots \\ 0 & 0 & \cdots & a_{nn} \end{vmatrix} = a_{22}a_{33} \ldots a_{nn},$$

the product of its diagonal entries. Therefore,

$$\det (A) = a_{11}M_{11} = a_{11}(a_{22}a_{33} \cdots a_{nn}) = a_{11}a_{22}a_{33} \cdots a_{nn},$$

the product of the diagonal entries of A.

Now to the case of a lower triangular matrix. Suppose that

$$A = \begin{bmatrix} a_{11} & 0 & \cdots & 0 \\ a_{21} & a_{22} & \cdots & 0 \\ \vdots & \vdots & & \vdots \\ a_{n1} & a_{n2} & \cdots & a_{nn} \end{bmatrix}.$$

Then its determinant

$$
\begin{vmatrix}
a_{11} & 0 & \cdots & 0 \\
a_{21} & a_{22} & \cdots & 0 \\
\vdots & \vdots & & \vdots \\
a_{n1} & a_{n2} & \cdots & a_{nn}
\end{vmatrix}
$$

is $\det(A) = a_{11}M_{11} - a_{21}M_{21} + a_{31}M_{31} - \cdots + (-1)^{n+1}M_{n1}$. Notice that in computing M_{r1} for $r > 1$ the $0 \ \ 0 \ \cdots \ 0$ part of the first row stays in M_{r1}. So all that is left to compute in $\det(A)$ is

$$
a_{11}M_{11} = a_{11}
\begin{vmatrix}
a_{22} & 0 & \cdots & 0 \\
a_{32} & a_{33} & \cdots & 0 \\
\vdots & \vdots & & \vdots \\
a_{n2} & a_{n2} & \cdots & a_{nn}
\end{vmatrix}.
$$

By our induction, the determinant M_{11} of this $(n-1) \times (n-1)$ matrix is $a_{22}a_{33} \cdots a_{nn}$, the product of its diagonal entries. So

$$
\det(A) = a_{11}M_{11} = a_{11}a_{22} \cdots a_{nn},
$$

the product of the diagonal entries of A. This completes the proof of Theorem 5.1.1. ∎

Since for $r > 1$, M_{r1} has a first row consisting of all zeroes, the $(n-1) \times (n-1)$ minor M_{r1} is 0. (Prove!)

PROBLEMS

NUMERICAL PROBLEMS

1. Compute the following determinants.

(a) $\begin{vmatrix} 1 & 3 & 4 \\ 2 & 0 & 2 \\ 3 & 0 & 3 \end{vmatrix}.$

(b) $\begin{vmatrix} 1 & 0 & 5 & 6 \\ 2 & 0 & 6 & 5 \\ 3 & 0 & 5 & 6 \\ 4 & 0 & 6 & 5 \end{vmatrix}.$

(c) $\begin{vmatrix} 9 & 0 & 0 \\ 2 & 5 & 6 \\ -1 & -2 & 0 \end{vmatrix}$.

(d) $\begin{vmatrix} 9 & 2 & -1 \\ 0 & 5 & -2 \\ 0 & 6 & 0 \end{vmatrix}$.

(e) $\begin{vmatrix} -1 & 2 & -1 & 3 \\ 2 & 3 & 2 & 1 \\ 3 & 4 & 3 & 0 \\ 0 & 1 & 0 & 1 \end{vmatrix}$.

(f) $\begin{vmatrix} 3 & 0 & 0 & 0 \\ 4 & 2 & 1 & 0 \\ 0 & 1 & 2 & 3 \\ -1 & -1 & 1 & 1 \end{vmatrix}$.

2. Compute the determinants of the transposes of the matrices in Problem 1.

3. Show that $\begin{vmatrix} 1 & 3 & 4 \\ 8 & x & 2 \\ 3 & r & 3 \end{vmatrix}$ and $\begin{vmatrix} 1+3 & 3 & 4 \\ 8+x & x & 2 \\ 3+r & r & 3 \end{vmatrix}$ are equal.

4. Evaluate $\begin{vmatrix} 2 & 4 & 6 & 8 \\ 0 & 1 & 2 & 3 \\ 0 & 2 & 3 & 1 \\ 0 & 5 & 6 & -1 \end{vmatrix}$ and compare it with

$\begin{vmatrix} 1 & 2 & 3 & 4 \\ 0 & 1 & 2 & 3 \\ 0 & 2 & 3 & 1 \\ 0 & 5 & 6 & -1 \end{vmatrix}$.

5. Show that $\begin{vmatrix} 4 & 5 & 1 \\ 2 & f & s \\ 3 & d & 6 \end{vmatrix} + \begin{vmatrix} 1 & 5 & 4 \\ s & f & 2 \\ 6 & d & 3 \end{vmatrix} = 0$.

6. Compute $\begin{vmatrix} 2 & 3 & 4 \\ r & r & 2+r \\ t & t & 3+t \end{vmatrix}$.

7. Compute $\begin{vmatrix} 1 & 1 & 1 \\ a & b & c \\ a^2 & b^2 & c^2 \end{vmatrix}$.

8. Compute $\begin{vmatrix} 1 & 1 & 1 & 1 \\ 3 & 0 & 1 & 2 \\ 0 & 1 & 2 & 3 \\ 1 & 5 & 6 & 7 \end{vmatrix}$ and $\begin{vmatrix} 0 & 1 & 2 & 3 \\ 3 & 0 & 1 & 2 \\ 1 & 1 & 1 & 1 \\ 1 & 5 & 6 & 7 \end{vmatrix}$ and compare

the answers.

MORE THEORETICAL PROBLEMS

Easier Problems

9. Show that $\begin{vmatrix} 1 & 1 & 1 & 1 \\ a & b & c & d \\ a^2 & b^2 & c^2 & d^2 \\ a^3 & b^3 & c^3 & d^3 \end{vmatrix}$ is 0 if and only if some two of a, b,

c, d are equal.

10. Show that $\begin{vmatrix} a+r & b+s & c+t \\ d & e & f \\ g & h & k \end{vmatrix} = \begin{vmatrix} a & b & c \\ d & e & f \\ g & h & k \end{vmatrix} + \begin{vmatrix} r & s & t \\ d & e & f \\ g & h & k \end{vmatrix}$.

Middle-Level Problems

11. Prove that $\begin{vmatrix} qa_{11} & a_{12} & \cdots & a_{1n} \\ qa_{21} & a_{22} & \cdots & a_{2n} \\ \vdots & \vdots & & \vdots \\ qa_{n1} & a_{n2} & \cdots & a_{nn} \end{vmatrix} = q \begin{vmatrix} a_{11} & a_{12} & \cdots & a_{1n} \\ a_{21} & a_{22} & \cdots & a_{2n} \\ \vdots & \vdots & & \vdots \\ a_{n1} & a_{n2} & \cdots & a_{nn} \end{vmatrix}$.

12. Prove that $\begin{vmatrix} qa_{11} & qa_{12} & \cdots & qa_{1n} \\ a_{21} & a_{22} & \cdots & a_{2n} \\ \vdots & \vdots & & \vdots \\ a_{n1} & a_{n2} & \cdots & a_{nn} \end{vmatrix} = q \begin{vmatrix} a_{11} & a_{12} & \cdots & a_{1n} \\ a_{21} & a_{22} & \cdots & a_{2n} \\ \vdots & \vdots & & \vdots \\ a_{n1} & a_{n2} & \cdots & a_{nn} \end{vmatrix}$.

13. If A and B are upper triangular $n \times n$ matrices are such that $\det(aA + bB) = a \det(A) + b \det(B)$ for all $a, b \in F$, show that $\det(A) = \det(B) = 0$.

14. If A and B are lower triangular $n \times n$ matrices such that

$$\det (aA + bB) = a \det (A) + b \det (B)$$

for all a, b, $\in F$, show that $\det (A) = \det (B) = 0$.

15. If A and B are upper triangular $n \times n$ matrices, show that

$$\det (A + xB)$$

is

$$\det (A) + b_1 x + b_2 x^2 + \cdots + b_{n-1} x^{n-1} + \det (B) x^n$$

for suitable b_r.

Harder Problems

16. Evaluate
$$\begin{vmatrix} 1 & 1 & \cdots & 1 \\ a_1 & a_1^2 & & a_1^n \\ a_2 & a_2^2 & \cdots & a_2^n \\ \vdots & \vdots & & \vdots \\ a_{n-1} & a_{n-1}^2 & \cdots & a_{n-1}^n \end{vmatrix}$$
and show

$$\begin{vmatrix} 1 & 1 & \cdots & 1 \\ a_1 & a_1^2 & & a_1^n \\ a_2 & a_2^2 & \cdots & a_2^n \\ \vdots & \vdots & & \vdots \\ a_{n-1} & a_{n-1}^2 & \cdots & a_{n-1}^n \end{vmatrix} = 0$$
if some two of $a_1, \ldots, a_{n-1}$ are equal.

17. If A, B are 3×3 matrices, show that $\det (AB) = \det (A) \det (B)$.

18. If A and B are 3×3 matrices and B is invertible, show that $\det (A + xB) = 0$ for some x in F.

19. Is the result of Problem 18 correct if $B \neq 0$ is not invertible? Either prove or give a counterexample.

20. If A is an invertible 3×3 matrix, show that $\det (A) \neq 0$ and
$$\det (A^{-1}) = \frac{1}{\det (A)}. \qquad (\textbf{\textit{Hint:}} \quad \text{Use Problem 17.})$$

21. For a 3×3 matrix A, show that $\det (A) = \det (A')$.

22. If A and C are 3×3 matrices and C is invertible, show that $\det (C^{-1}AC) = \det (A)$.

5.2. PROPERTIES OF DETERMINANTS: ROW OPERATIONS

In defining the determinants of a matrix we did it by the expansion by the minors of the first *column*. In this way we did not interface very much with the *rows*. Fortunately, this will allow us to show some things that can be done by playing around with the *rows* of a determinant. We begin with

Theorem 5.2.1.   If a row of an $n \times n$ matrix A consists only of zeros, then det $(A) = 0$.

   Proof.   We go by induction on n. The result is trivially true for $n = 2$. Next let $n > 2$ and suppose that row r of A consists of zeros. Then every minor M_{k1}, for $k \neq r$, retains a row of zeros. So by induction each $M_{k1} = 0$ if $k \neq r$. On the other hand, $a_{r1} = 0$, whence

$$a_{r1}M_{r1} = 0.$$

Thus

$$\det (A) = \sum_{k=1}^{n} (-1)^{k+1}a_{k1}M_{k1} = 0. \quad \blacksquare$$

   Another property that is fairly easy to establish, using our induction type of argument, is

Theorem 5.2.2.   Suppose that a matrix B is obtained from A by multiplying each entry of some row of A by a constant u. Then

$$\det (B) = u \det (A).$$

   Proof.   Before doing the general case, let us consider the story for the general 3×3 matrix

$$A = \begin{bmatrix} a_{11} & a_{12} & a_{13} \\ a_{21} & a_{22} & a_{23} \\ a_{31} & a_{32} & a_{33} \end{bmatrix}$$

and

$$B = \begin{bmatrix} a_{11} & a_{12} & a_{13} \\ ua_{21} & ua_{22} & ua_{23} \\ a_{31} & a_{32} & a_{33} \end{bmatrix}.$$

Let M_{r1} be the $(r, 1)$ minor of A and N_{r1} that of B. So

$$\det (B) = a_{11}N_{r1} - ua_{21}N_{21} + a_{31}N_{31}.$$

Since the minor N_{21} does not involve the second row, we see that $N_{21} = M_{21}$. The other two minors, N_{11} and N_{31}, which are 2×2, have a row multiplied by u. So we have $N_{11} = uM_{11}$, $N_{31} = uM_{31}$. Thus

$$\begin{aligned}
\det (B) &= a_{11}(uM_{11}) - ua_{21}M_{21} + a_{31}(uM_{31}) \\
&= u(a_{11}M_{11} - a_{21}M_{21} + a_{31}M_{31}) \\
&= u \det (A).
\end{aligned}$$

The general case goes in a similar fashion. Let B be obtained from A by multiplying every entry in row r of A by u. So if $B = (b_{st})$, then $b_{st} = a_{st}$ if $s \neq r$ and $b_{rt} = ua_{rt}$. If N_{k1} is the $(k, 1)$ minor of B, then

$$\det (B) = \sum_{k=1}^{n} (-1)^{k+1} b_{k1} N_{k1}.$$

But N_{k1}, for $k \neq r$, is the same as M_{k1}, the $(k, 1)$ minor of A, except that one of its rows is that of M_{k1} multiplied by u. Since N_{k1} is $(n - 1) \times (n - 1)$, by induction $N_{k1} = uM_{k1}$ for $k \neq r$. On the other hand, since in forming N_{r1} we strike out row r of A, we see that $N_{r1} = M_{r1}$. Thus for $k \neq r$, $b_{k1}N_{k1} = a_{k1}N_{k1}$ while $b_{r1}N_{r1} = (ua_{r1})M_{r1}$, hence

$$\det (B) = \sum_{k=1}^{n} (-1)^{k+1} b_{k1} N_{k1} = \sum_{k=1}^{n} (-1)^{k+1} ua_{k1} M_{k1}$$

$$= u \sum_{k=1}^{n} (-1)^{k+1} a_{k1} M_{k1} = u \det (A). \quad \blacksquare$$

We come to some theorems of key importance in the theory of determinants.

Theorem 5.2.3. If the matrix B is obtained from A by interchanging two *adjacent* rows of A, then $\det (B) = -\det (A)$.

Proof. Again we illustrate the proof for 3×3 matrices. Suppose that in

$$A = \begin{bmatrix} a_{11} & a_{12} & a_{13} \\ a_{21} & a_{22} & a_{23} \\ a_{31} & a_{32} & a_{33} \end{bmatrix}$$

we interchange rows 2 and 3 to obtain the matrix

$$
B = \begin{bmatrix} a_{11} & a_{12} & a_{13} \\ a_{31} & a_{32} & a_{33} \\ a_{21} & a_{22} & a_{23} \end{bmatrix} = \begin{bmatrix} b_{11} & b_{12} & b_{13} \\ b_{21} & b_{22} & b_{23} \\ b_{31} & b_{32} & b_{33} \end{bmatrix}.
$$

Then

$$
\det (B) = b_{11} \begin{vmatrix} b_{22} & b_{23} \\ b_{32} & b_{33} \end{vmatrix} - b_{21} \begin{vmatrix} b_{12} & b_{13} \\ b_{32} & b_{33} \end{vmatrix} + b_{31} \begin{vmatrix} b_{12} & b_{13} \\ b_{22} & b_{23} \end{vmatrix}
$$

$$
= a_{11} \begin{vmatrix} a_{32} & a_{33} \\ a_{22} & a_{23} \end{vmatrix} - a_{31} \begin{vmatrix} a_{12} & a_{13} \\ a_{22} & a_{23} \end{vmatrix} + a_{21} \begin{vmatrix} a_{12} & a_{13} \\ a_{32} & a_{33} \end{vmatrix}.
$$

Notice in the first minor $N_{11} = \begin{vmatrix} a_{32} & a_{33} \\ a_{22} & a_{23} \end{vmatrix}$ of B, the rows of the minor

$M_{11} = \begin{vmatrix} a_{22} & a_{23} \\ a_{32} & a_{33} \end{vmatrix}$ corresponding to the second and third rows of A are

interchanged. So by the 2×2 case, $N_{11} = -M_{11}$. Notice also that

$$
N_{21} = \begin{vmatrix} a_{12} & a_{13} \\ a_{22} & a_{23} \end{vmatrix} = M_{31}
$$

and

$$
N_{31} = \begin{vmatrix} a_{12} & a_{13} \\ a_{32} & a_{33} \end{vmatrix} = M_{21}.
$$

So

$$
\begin{aligned}
\det (B) &= b_{11}N_{11} - b_{21}N_{21} + b_{31}N_{31} \\
&= -a_{11}M_{11} - a_{31}M_{31} + a_{21}M_{21} \\
&= -(a_{11}M_{11} - a_{21}M_{21} + a_{31}M_{31}) \\
&= -\det (A).
\end{aligned}
$$

since $b_{31} = a_{21}$ and $b_{21} = a_{31}$.

Now to the general case. Suppose that we interchange rows r and $r + 1$ of A to get B. Thus if $B = (b_{uv})$, then for $u \neq r, r + 1$ we have $b_{uv} = a_{uv}$, while $b_{rv} = a_{r+1,v}$ and $b_{r+1,v} = a_{rv}$. So if N_{u1} is the $(u, 1)$ minor of B, then

$$
\det (B) = \sum_{u=1}^{n} (-1)^{u+1} b_{u1} N_{u1}.
$$

If $u \neq r$ or $r + 1$, then in obtaining the minor N_{u1} from M_{u1}, the $(u, 1)$ minor of A, we have interchanged two adjacent rows of M_{u1}. So, for these, by the $(n - 1) \times (n - 1)$ case, $N_{u1} = -M_{u1}$. Also, $N_{r1} = M_{r+1,1}$ and $N_{r+1,1} = M_{r1}$ (Why? Look at the 3×3 case!) So $b_{r1}N_{r1} = a_{r+1,1}M_{r+1,1}$ and $b_{r+1,1}N_{r+1,1} = a_{r1}M_{r1}$. Hence

$$\begin{aligned} \det(B) &= b_{11}N_{11} - b_{21}N_{21} + \cdots + (-1)^{r+1}b_{r1}N_{r1} \\ &\quad + (-1)^{r+2}b_{r+1,1}N_{r+1,1} + \cdots + (-1)^{n+1}b_{n1}N_{n1} \\ &= -a_{11}M_{11} + a_{21}M_{21} - \cdots + (-1)^{r+1}a_{r+1,1}M_{r+1,1} \\ &\quad + (-1)^{r+2}a_{r1}M_{r1} - \cdots - (-1)^{n+1}a_{n1}M_{n1} \\ &= -(a_{11}M_{11} - a_{21}M_{21} + \cdots + (-1)^{r+1}a_{r1}M_{r1} \\ &\quad + (-1)^{r+2}a_{r+1,1}M_{r+1,1} + \cdots + (-1)^{n+1}a_{n1}M_{n1}) \\ &= -\det(A). \end{aligned}$$

So $\det(B) = -\det(A)$, which proves the theorem. ∎

We realize that the proof just given may be somewhat intricate. We advise the reader to follow its steps for the 4×4 matrices. This should help clarify the argument.

The next result has nothing to do with matrices or determinants, but we shall make use of it in those contexts.

Suppose that n balls, numbered from the left as $1, 2, \ldots, n$, are lined up in a row. If $s > r$, we claim that we can interchange balls r and s by a succession of an *odd* number of interchanges of adjacent ones. How? First interchange ball s with ball $s - 1$, then the ball $s - 2$, and so on. To bring ball s to position r thus requires $s - r$ interchanges of adjacent balls. Also ball r is now in position $r + 1$, so to move it to position s requires $s - (r + 1)$ interchanges of adjacent ones. So, in all, to interchange balls r and s, we have made

$$(s - r) + (s - (r + 1)) = 2(s - r) - 1,$$

an *odd* number, of interchanges of adjacent ones.

Let's see this in the case of four balls, **(1) (2) (3) (4)**, where we want to interchange the first and fourth balls. The sequence pictorially is

(1) (2) (3) (4)   (The initial configuration)
(1) (2) (4) (3)   (After the first interchange)
(1) (4) (2) (3)   ·(After the second interchange)

(4) (1) (2) (3)　　　(After the third interchange)
(4) (2) (1) (3)　　　(After the fourth interchange)
(4) (2) (3) (1)　　　(After the fifth interchange)

So altogether it took $5 = 2(4 - 1) - 1$ adjacent interchanges to effect the interchange of balls **(1)** and **(4)**.

We note the result proved above, not for balls, but for rows of a matrix, as

Lemma 5.2.4.　We can interchange any two rows of a matrix by an *odd* number of interchanges of adjacent rows.

This lemma, together with Theorem 5.2.3, has the very important consequence,

Theorem 5.2.5.　If we interchange two rows of a determinant, the sign changes.

Proof.　According to Lemma 5.2.4, we can effect the interchange of these two rows by an odd number of interchanges of adjacent rows. By Theorem 5.2.3, each such change of adjacent rows changes the sign of the determinant. Since we do this an odd number of times, the sign of the determinant is changed in the interchange of two of its rows.　■

Suppose that two rows of the matrix A are the same. So if we interchange these two rows, we don't change it, hence we don't change det (A). But by Theorem 5.2.5, on interchanging these two rows of A, the determinant changes sign. This can happen only if det $(A) = 0$. Hence

Theorem 5.2.6.　If two rows of A are equal, then det $(A) = 0$.

Before going formally to the next result, we should like to motivate the theorem we shall prove with an example.
Consider the 3×3 matrix

$$C = \begin{bmatrix} a_{11} & a_{12} & a_{13} \\ a_{21} + b_{21} & a_{22} + b_{22} & a_{23} + b_{23} \\ a_{31} & a_{32} & a_{33} \end{bmatrix}.$$

We should like to interrelate det (C) with two other determinants, det (A) and det (B), where

$$A = \begin{bmatrix} a_{11} & a_{12} & a_{13} \\ a_{21} & a_{22} & a_{23} \\ a_{31} & a_{32} & a_{33} \end{bmatrix}$$

and

$$B = \begin{bmatrix} a_{11} & a_{12} & a_{13} \\ b_{21} & b_{22} & b_{23} \\ a_{31} & a_{32} & a_{33} \end{bmatrix}.$$

Of course, since 3 is not a very large integer, we could expand det (A), det (B), det (C) by brute force and compare the answers. But this would not be very instructive. Accordingly, we'll carry out the discussion along the lines that we shall follow in the general case.

What do we hope to show? Our objective is to prove that det (C) = det (A) + det (B). Now

$$\det (C) = a_{11} \begin{vmatrix} a_{22} + b_{22} & a_{23} + b_{23} \\ a_{32} & a_{33} \end{vmatrix}$$

$$- (a_{21} + b_{21}) \begin{vmatrix} a_{12} & a_{13} \\ a_{32} & a_{33} \end{vmatrix}$$

$$+ a_{31} \begin{vmatrix} a_{12} & a_{13} \\ a_{22} + b_{22} & a_{23} + b_{23} \end{vmatrix}.$$

For the 2×2 case we readily have that

$$\begin{vmatrix} a_{22} + b_{22} & a_{23} + b_{23} \\ a_{32} & a_{33} \end{vmatrix} = \begin{vmatrix} a_{22} & a_{23} \\ a_{32} & a_{33} \end{vmatrix} + \begin{vmatrix} b_{22} & b_{23} \\ a_{32} & a_{33} \end{vmatrix}.$$

and

$$\begin{vmatrix} a_{12} & a_{13} \\ a_{22} + b_{22} & a_{23} + b_{23} \end{vmatrix} = \begin{vmatrix} a_{12} & a_{13} \\ a_{22} & a_{23} \end{vmatrix} + \begin{vmatrix} a_{12} & a_{13} \\ b_{22} & b_{23} \end{vmatrix}.$$

Substituting these values in the expression for det (C), we obtain

$$\det (C) = a_{11} \left(\begin{vmatrix} a_{22} & a_{23} \\ a_{32} & a_{33} \end{vmatrix} + \begin{vmatrix} b_{22} & b_{23} \\ a_{32} & a_{33} \end{vmatrix} \right)$$

$$- (a_{21} + b_{21}) \begin{vmatrix} a_{12} & a_{13} \\ a_{32} & a_{33} \end{vmatrix}$$

$$+ a_{31} \left(\begin{vmatrix} a_{12} & a_{13} \\ a_{22} & a_{23} \end{vmatrix} + \begin{vmatrix} a_{12} & a_{13} \\ b_{22} & b_{23} \end{vmatrix} \right).$$

So regrouping terms we get

$$\det (C) = \left(a_{11} \begin{vmatrix} a_{22} & a_{23} \\ a_{32} & a_{33} \end{vmatrix} - a_{21} \begin{vmatrix} a_{12} & a_{13} \\ a_{32} & a_{33} \end{vmatrix} + a_{31} \begin{vmatrix} a_{12} & a_{13} \\ a_{22} & a_{23} \end{vmatrix} \right)$$

$$+ \left(a_{11} \begin{vmatrix} b_{22} & b_{23} \\ a_{32} & a_{33} \end{vmatrix} - b_{21} \begin{vmatrix} a_{12} & a_{13} \\ a_{32} & a_{33} \end{vmatrix} + a_{31} \begin{vmatrix} a_{12} & a_{13} \\ b_{22} & b_{23} \end{vmatrix} \right).$$

We recognize the first expression

$$a_{11} \begin{vmatrix} a_{22} & a_{23} \\ a_{32} & a_{33} \end{vmatrix} - a_{21} \begin{vmatrix} a_{12} & a_{13} \\ a_{32} & a_{33} \end{vmatrix} + a_{31} \begin{vmatrix} a_{12} & a_{13} \\ a_{22} & a_{23} \end{vmatrix}$$

as det (A) and the second expression

$$a_{11} \begin{vmatrix} b_{22} & b_{23} \\ a_{32} & a_{33} \end{vmatrix} - b_{21} \begin{vmatrix} a_{12} & a_{13} \\ a_{32} & a_{33} \end{vmatrix} + a_{31} \begin{vmatrix} a_{12} & a_{13} \\ b_{22} & b_{23} \end{vmatrix}$$

as det (B). Consequently, det (C) = det (A) + det (B).

The pattern just exhibited will be that used in the proof of

Theorem 5.2.7.   If the matrix $C = (c_{uv})$ is such that for $u \neq r$, $c_{uv} = a_{uv}$ but for $u = r$, $c_{rv} = a_{rv} + b_{rv}$ then

$$\det (C) = \det (A) + \det (B),$$

where

$$A = \begin{bmatrix} a_{11} & a_{12} & \cdots & a_{1n} \\ \vdots & \vdots & & \vdots \\ a_{r-1,1} & a_{r-1,2} & \cdots & a_{r-1,n} \\ a_{r1} & a_{r2} & \cdots & a_{rn} \\ a_{r+1,1} & a_{r+1,2} & \cdots & a_{r1,n} \\ \vdots & \vdots & & \vdots \\ a_{n1} & a_{n2} & \cdots & a_{nn} \end{bmatrix}$$

and

$$B = \begin{bmatrix} a_{11} & a_{12} & \cdots & a_{1n} \\ \vdots & \vdots & & \vdots \\ a_{r-1,1} & a_{r-1,2} & \cdots & a_{r-1,n} \\ b_{r1} & b_{r2} & \cdots & b_{rn} \\ a_{r+1,1} & a_{r+1,2} & \cdots & a_{r+1,n} \\ \vdots & \vdots & & \vdots \\ a_{n1} & a_{n2} & \cdots & a_{nn} \end{bmatrix}.$$

Proof. The matrix C is such that in all the rows, except for row r, the entries are a_{uv} but in row r the entries are $a_{rv} + b_{rv}$. Let M_{k1}, N_{k1}, Q_{k1} denote the $(k, 1)$ minors of A, B, and C respectively. Thus

$$\det(C) = \sum_{u=1}^{n} (-1)^{k+1} c_{k1} Q_{k1}.$$

Note that the minor Q_{r1} is exactly the same as M_{r1} and N_{r1}, since in finding these minors row r has been crossed out, so the a_{rs} and b_{rs} do not enter the picture. So $Q_{r1} = M_{r1} = N_{r1}$. What about the other minors Q_{k1}, where $k \neq r$? Since row r is retained in this minor, one of its rows has the entries $a_{rv} + b_{rv}$. Because Q_{k1} is $(n - 1) \times (n - 1)$, by induction we have that $Q_{k1} = M_{k1} + N_{k1}$. So the formula for $\det(C)$ becomes

$$a_{11}(M_{11} + N_{11}) - a_{21}(M_{21} + N_{21}) + \cdots + (-1)^r a_{r-1,1}(M_{r-1,1} + N_{r-1,1})$$
$$+ (-1)^{r+1}(a_{r1} + b_{r1})Q_{r1}$$
$$+ (-1)^{r+2} a_{r+1,1}(M_{r+1,1} + N_{r+1,1}) + \cdots + (-1)^{n+1} a_{n1}(M_{n1} + N_{n1}).$$

Using that $Q_{r1} = M_{r1} = N_{r1}$, we can replace the term

$$(-1)^{r+1}(a_{r1} + b_{r1})Q_{r1}$$

by

$$(-1)^{r+1} a_{r1} M_{r1} + (-1)^{r+1} b_{r1} N_{r1}.$$

Regrouping then gives us the expression

$$(a_{11}M_{11} - a_{21}M_{21} + \cdots + (-1)^{r+1} a_{r1} M_{r1} + \cdots + (-1)^{n+1} a_{n1} M_{n1})$$
$$+ (a_{11}N_{11} - a_{21}N_{21} + \cdots + (-1)^{r+1} b_{r1} N_{r1} + \cdots + (-1)^{n+1} a_{n1} N_{n1})$$

for $\det(C)$. We recognize the first expression

$$(a_{11}M_{11} - a_{21}M_{21} + \cdots + (-1)^{r+1} a_{r1} M_{r1} + \cdots + (-1)^{n+1} a_{n1} M_{n1})$$

as det (A) and the second expression

$$(a_{11}M_{11} - a_{21}M_{21} + \cdots + (-1)^{r+1}a_{r1}M_{r1} + \cdots + (-1)^{n+1}a_{n1}M_{n1})$$

as det (B). In consequence, det (C) = det $(A$ + det (B), as claimed in the theorem. ■

For instance, from the theorem, we know that

$$\begin{vmatrix} 1 & 2 & 3 & 4 \\ 1 & 1 & 1 & 1 \\ 2 & 5 & 6 & 2 \\ 0 & 1 & 2 & 3 \end{vmatrix} = \begin{vmatrix} 1 & 2 & 3 & 4 \\ 1 & 1 & 1 & 1 \\ 1 & 4 & 5 & 1 \\ 0 & 1 & 2 & 3 \end{vmatrix} + \begin{vmatrix} 1 & 2 & 3 & 4 \\ 1 & 1 & 1 & 1 \\ 1 & 1 & 1 & 1 \\ 0 & 1 & 2 & 3 \end{vmatrix},$$

and the second determinant is 0 by Theorem 5.2.6, since two of its rows are equal. So

$$\begin{vmatrix} 1 & 2 & 3 & 4 \\ 1 & 1 & 1 & 1 \\ 2 & 5 & 6 & 2 \\ 0 & 1 & 2 & 3 \end{vmatrix} = \begin{vmatrix} 1 & 2 & 3 & 4 \\ 1 & 1 & 1 & 1 \\ 1 & 4 & 5 & 1 \\ 0 & 1 & 2 & 3 \end{vmatrix}.$$

We could proceed with this to get that

$$\begin{vmatrix} 1 & 2 & 3 & 4 \\ 1 & 1 & 1 & 1 \\ 2 & 5 & 6 & 2 \\ 0 & 1 & 2 & 3 \end{vmatrix} = \begin{vmatrix} 1 & 2 & 3 & 4 \\ 1 & 1 & 1 & 1 \\ 0 & 3 & 4 & 0 \\ 0 & 1 & 2 & 3 \end{vmatrix}$$

and even to get

$$\begin{vmatrix} 1 & 2 & 3 & 4 \\ 1 & 1 & 1 & 1 \\ 2 & 5 & 6 & 2 \\ 0 & 1 & 2 & 3 \end{vmatrix} = \begin{vmatrix} 0 & 1 & 2 & 3 \\ 1 & 1 & 1 & 1 \\ 0 & 3 & 4 & 0 \\ 0 & 1 & 2 & 3 \end{vmatrix}.$$

Can you see how this comes about? Notice that the last determinant has all but one entry equal to 0 in its first column, so it is relatively easy to evaluate, for we need but calculate the (2, 1) minor. By the way, what

are the exact values of $\begin{vmatrix} 0 & 1 & 2 & 3 \\ 1 & 1 & 1 & 1 \\ 0 & 3 & 4 & 0 \\ 0 & 1 & 2 & 3 \end{vmatrix}$ and of $\begin{vmatrix} 1 & 2 & 3 & 4 \\ 1 & 1 & 1 & 1 \\ 2 & 5 & 6 & 2 \\ 0 & 1 & 2 & 3 \end{vmatrix}$? Evaluate

by expanding as is given in the definition and compare the results. If you make no computational mistake you should find that they are equal.

The next result is an important consequence of Theorem 5.2.7. Before doing it we again look at a 3×3 example. Let

$$A = \begin{bmatrix} a_{11} & a_{12} & a_{13} \\ a_{21} & a_{22} & a_{23} \\ a_{31} & a_{32} & a_{33} \end{bmatrix}$$

and

$$B = \begin{bmatrix} a_{11} & a_{12} & a_{13} \\ a_{21} + qa_{31} & a_{22} + qa_{32} & a_{23} + qa_{33} \\ a_{31} & a_{32} & a_{33} \end{bmatrix}.$$

By Theorem 5.2.7,

$$\det (B) = \det (A) + \begin{vmatrix} a_{11} & a_{12} & a_{13} \\ qa_{31} & qa_{32} & qa_{33} \\ a_{31} & a_{32} & a_{33} \end{vmatrix}.$$

Moreover, by Theorem 5.2.2 we have

$$\begin{vmatrix} a_{11} & a_{12} & a_{13} \\ qa_{31} & qa_{32} & qa_{33} \\ a_{31} & a_{32} & a_{33} \end{vmatrix} = q \begin{vmatrix} a_{11} & a_{12} & a_{13} \\ a_{31} & a_{32} & a_{33} \\ a_{31} & a_{32} & a_{33} \end{vmatrix}$$

and then by Theorem 5.2.6 we have

$$\begin{vmatrix} a_{11} & a_{12} & a_{13} \\ a_{31} & a_{32} & a_{33} \\ a_{31} & a_{32} & a_{33} \end{vmatrix} = 0$$

since the last two rows of the latter determinant are equal. So we find that

$$\det (B) = \det (A) + \begin{vmatrix} a_{11} & a_{12} & a_{13} \\ qa_{31} & qa_{32} & qa_{33} \\ a_{31} & a_{32} & a_{33} \end{vmatrix} = \det (A) + 0$$

and, wonder of wonders, $\det (B) = \det (A)$.

The argument given is really the general argument but it never hurts to repeat, or almost repeat.

Theorem 5.2.8.   If the matrix B is obtained from A by adding a constant times one row of A to another row of A, then $\det (B) = \det (A)$.

Proof.   Suppose that we add q times row s of A to row r of A to get B. Then all the entries of B, except those in row r, are the same as the corresponding entries of A. But the (r, v) entry of B is $a_{rv} + qa_{sv}$. Thus by Theorem 5.2.7,

$$\det (B) = \det (A) + \begin{vmatrix} a_{11} & a_{12} & \cdots & a_{1n} \\ \vdots & \vdots & & \vdots \\ qa_{s1} & qa_{s2} & \cdots & qa_{sn} \\ \vdots & \vdots & & \vdots \\ a_{s1} & a_{s2} & \cdots & a_{sn} \\ \vdots & \vdots & & \vdots \\ a_{n1} & a_{n2} & \cdots & a_{nn} \end{vmatrix} \begin{array}{l} (\text{row } r) \\ \\ (\text{row } s) \end{array} .$$

By Theorem 5.2.2 we have

$$\begin{vmatrix} a_{11} & a_{12} & \cdots & a_{1n} \\ \vdots & \vdots & & \vdots \\ qa_{s1} & qa_{s2} & \cdots & qa_{sn} \\ \vdots & \vdots & & \vdots \\ a_{s1} & a_{s2} & \cdots & a_{sn} \\ \vdots & \vdots & & \vdots \\ a_{n1} & a_{n2} & \cdots & a_{nn} \end{vmatrix} = q \begin{vmatrix} a_{11} & a_{12} & \cdots & a_{1n} \\ \vdots & \vdots & & \vdots \\ a_{s1} & a_{s2} & \cdots & a_{sn} \\ \vdots & \vdots & & \vdots \\ a_{s1} & a_{s2} & \cdots & a_{sn} \\ \vdots & \vdots & & \vdots \\ a_{n1} & a_{n2} & \cdots & a_{nn} \end{vmatrix},$$

and by Theorem 5.2.7 we have

$$\begin{vmatrix} a_{11} & a_{12} & \cdots & a_{1n} \\ \vdots & \vdots & & \vdots \\ a_{s1} & a_{s2} & \cdots & a_{sn} \\ \vdots & \vdots & & \vdots \\ a_{s1} & a_{s2} & \cdots & a_{sn} \\ \vdots & \vdots & & \vdots \\ a_{n1} & a_{n2} & \cdots & a_{nn} \end{vmatrix} = 0,$$

since two rows of the latter determinant are equal. So we have

$$\det (B) = \det (A) + \begin{vmatrix} a_{11} & a_{12} & \cdots & a_{1n} \\ \vdots & \vdots & & \vdots \\ qa_{s1} & qa_{s2} & \cdots & qa_{sn} \\ \vdots & \vdots & & \vdots \\ a_{s1} & a_{s2} & \cdots & a_{sn} \\ \vdots & \vdots & & \vdots \\ a_{n1} & a_{n2} & \cdots & a_{nn} \end{vmatrix} = \det (A) + 0$$

and $\det (B) = \det (A)$, which proves the theorem. ∎

Theorem 5.2.8 is very useful in computing determinants. We illustrate its use with a 4×4 example. Suppose that we want to compute

the determinant $\begin{vmatrix} 5 & -1 & 6 & 7 \\ 1 & 3 & -1 & 2 \\ 4 & 5 & 0 & 1 \\ -1 & 6 & 2 & 2 \end{vmatrix}$. If we add -5 time the second

row to the first row, we get

$$\begin{vmatrix} 0 & -16 & 11 & -3 \\ 1 & 3 & -1 & 2 \\ 4 & 5 & 0 & 1 \\ -1 & 6 & 2 & 2 \end{vmatrix}.$$

Now add the second row to the fourth, which yields

$$\begin{vmatrix} 0 & -16 & 11 & -3 \\ 1 & 3 & -1 & 2 \\ 4 & 5 & 0 & 1 \\ 0 & 9 & 1 & 4 \end{vmatrix}.$$

Finally, adding -4 times the second row to the third results in

$$\begin{vmatrix} 0 & -16 & 11 & -3 \\ 1 & 3 & -1 & 2 \\ 0 & -7 & 4 & -7 \\ 0 & 9 & 1 & 4 \end{vmatrix}.$$

All these determinants are equal by Theorem 5.2.8. But what have we obtained by all this jockeying around? Because all the entries, except for the $(2, 1)$ one, in the first column are 0, the determinant merely becomes

$$(-1)^{2+1} \begin{vmatrix} -16 & 11 & -3 \\ -7 & 4 & -7 \\ 9 & 1 & 4 \end{vmatrix} = - \begin{vmatrix} -16 & 11 & -3 \\ -7 & 4 & -7 \\ 9 & 1 & 4 \end{vmatrix}.$$

In this way we have reduced the evaluation of a 4×4 determinant to that of a 3×3 one. We could carry out a similar game for the 3×3

matrix $\begin{vmatrix} -16 & 11 & -3 \\ -7 & 4 & -7 \\ 9 & 1 & 4 \end{vmatrix}$ to obtain zeros in its first column. For instance,

adding $16/9$ times row 3 to row 1 gives us a 0 in the corner:

$$\begin{vmatrix} 0 & 11 + \frac{16}{9} & -3 + \frac{64}{9} \\ -7 & 4 & -7 \\ 9 & 1 & 4 \end{vmatrix} = \begin{vmatrix} 0 & \frac{115}{9} & \frac{37}{9} \\ -7 & 4 & -7 \\ 9 & 1 & 4 \end{vmatrix}.$$

Then adding $7/9$ times row 3 to row 2 gives us another 0 in the first column:

$$\begin{vmatrix} 0 & \frac{115}{9} & \frac{37}{9} \\ 0 & 4 + \frac{7}{9} & -7 + \frac{28}{9} \\ 9 & 1 & 4 \end{vmatrix} = \begin{vmatrix} 0 & \frac{115}{9} & \frac{37}{9} \\ 0 & \frac{43}{9} & -\frac{35}{9} \\ 9 & 1 & 4 \end{vmatrix}.$$

Again, all these determinants are equal by Theorem 5.2.8. And again, since all the entries, except for the $(3, 1)$ one, in the first column are 0, the determinant merely becomes

$$(-1)^{3+1} 9 \begin{vmatrix} \frac{115}{9} & \frac{37}{9} \\ \frac{43}{9} & -\frac{35}{9} \end{vmatrix} = 9((\tfrac{115}{9})(-\tfrac{35}{9}) - (\tfrac{37}{9})(\tfrac{43}{9})) = -\frac{5616}{9}.$$

Of course, this playing around with the rows can almost be done visually, so it is a rather fast way of evaluating determinants. For a computer this is relatively child's play, and easy to program.

The material in this section has not been easy. If you don't understand it all the first time around, go back to it and try out what is being said for specific examples.

PROBLEMS

NUMERICAL PROBLEMS

1. Evaluate the determinants directly from the definition, and then evaluate them using the row operations.

(a) $\begin{vmatrix} 1 & 4 & -1 \\ 7 & 6 & 2 \\ \pi & -\pi & 1 \end{vmatrix}.$

(b) $\begin{vmatrix} 1+i & i & -i & 0 \\ 1 & 1 & 1 & 1 \\ i & i-1 & -i-1 & -1 \\ 5 & \pi & \pi^2 & 17 \end{vmatrix}.$

(c) $\begin{vmatrix} 5 & -1 & 6 & 7 \\ 1 & 3 & -1 & 2 \\ 4 & 5 & 0 & 1 \\ -1 & 6 & 2 & 2 \end{vmatrix}.$

(d)
$$\begin{vmatrix} 5 & -1 & 6 & 7 \\ 1 & 3 & -1 & 2 \\ 3 & 4 & 1 & -1 \\ 0 & 9 & 1 & 4 \end{vmatrix}.$$

2. Verify by a direct calculation or by row operations that

(a)
$$\begin{vmatrix} 1 & -1 & 6 \\ 4 & 3 & -11 \\ 6 & 2 & -1 \end{vmatrix} = \begin{vmatrix} 1 & -1 & 6 \\ 3 & 2 & -12 \\ 6 & 2 & -1 \end{vmatrix} + \begin{vmatrix} 1 & -1 & 6 \\ 1 & 1 & 1 \\ 6 & 2 & -1 \end{vmatrix}.$$

(b)
$$\begin{vmatrix} 1 & 2 & 4 \\ 6 & 7 & 9 \\ -11 & 5 & 10 \end{vmatrix} + \begin{vmatrix} 2 & 1 & 4 \\ 7 & 6 & 9 \\ 5 & -11 & 10 \end{vmatrix} = 0.$$

(c)
$$\begin{vmatrix} 1 & 2 & 4 \\ 6 & 7 & 9 \\ -11 & 5 & 10 \end{vmatrix} = \begin{vmatrix} 2 & 4 & 1 \\ 7 & 9 & 6 \\ 5 & 10 & -11 \end{vmatrix}.$$

(d)
$$\begin{vmatrix} 0 & 5 & 0 & 0 \\ 1 & 2 & -1 & 0 \\ 3 & 4 & 1 & 6 \\ 2 & 1 & 5 & 1 \end{vmatrix} = -5 \begin{vmatrix} 1 & -1 & 1 \\ 3 & 1 & 6 \\ 2 & 5 & 1 \end{vmatrix}.$$

3. Compute all the minors of

(a)
$$\begin{vmatrix} 1 & 3 & 0 \\ 3 & 4 & 5 \\ 0 & 0 & 1 \end{vmatrix}.$$

(b)
$$\begin{vmatrix} 1 & 2 & 0 \\ 0 & 1 & 1 \\ 6 & 1 & 2 \end{vmatrix}.$$

(c)
$$\begin{vmatrix} 1 & 2 & 3 & 4 \\ 0 & 3 & 0 & 4 \\ 1 & 0 & 1 & 0 \\ 0 & 1 & 0 & 1 \end{vmatrix}.$$

MORE THEORETICAL PROBLEMS

Easier Problems

4. If A is a triangular matrix, what are the necessary and sufficient conditions on A that det $(A) \neq 0$?

5. If A and B are both upper triangular matrices, show that det (AB) = det (A) det (B).

6. If A is an upper triangular matrix, prove that det $(A) = $ det (A').

Middle Level Problems

7. If A is a 3×3 matrix, show that A is invertible if det $(A) \neq 0$.

8. If A is a 3×3 matrix and A is invertible, show that det $(A) \neq 0$.

Harder Problems

9. In Problem 8 show that det $(A^{-1}) = \dfrac{1}{\det (A)}$.

Very Hard Problems

10. If A is an upper triangular matrix, show that for all B, det $(AB) = $ det (A) det (B), using row operations.

11. If A is a lower triangular matrix, show that for all B, det $(AB) = $ det (A) det (B).

5.3. PROPERTIES OF DETERMINANTS: COLUMN OPERATIONS

In Section 5.2 we saw what operations can be carried out on the rows of a given determinant. We now want to establish the analogous results for the columns of a given determinant. The style of many of the proofs will be similar to those given in Section 5.2. But there is a difference. Because the determinant was defined in terms of the expansion by minors of the first column, we shall often divide the argument into two parts: one, in which the first column enters in the argument, and two, in which the first column is not involved. For the results about the rows we didn't need such a case division, since no row was singled out among all the other rows.

With the experience in the manner of proof used in Section 5.2 under our belts, we can afford, at times, to be a little sketchier in our arguments than we were in Section 5.2. As before, the proof will be by induction, assuming the results to be correct for $(n - 1) \times (n - 1)$ matrices, and passing from this to the validity of the results for the $n \times n$ ones. We shall also try to illustrate what is going on by resorting to 3×3 examples.

We begin the discussion with

Theorem 5.3.1. If a column of A consists of zeros, then det $(A) = 0$.

Proof. If the column of zeros is the first column, then the proof is very easy, for det $(A) = \sum_{r=1}^{n} (-1)^{r+1} a_{r1} M_{r1}$ and since each $a_{r1} = 0$ we get that det $(A) = 0$.

What happens if this column of zeros is not the first one? In that case, since in expressing M_{r1} for $r > 1$, we see that each M_{r1} has a column of zeros. Because M_{r1} is an $(n - 1) \times (n - 1)$ determinant, we know that $M_{r1} = 0$. Hence det $(A) = 0$ by the formula

$$\det (A) = \sum_{r=1}^{n} (-1)^{r+1} a_{r1} M_{r1}. \quad \blacksquare$$

In parallel with what we did in Section 5.2 we now prove

Theorem 5.3.2. If the matrix B is obtained from A by multiplying each entry of some column of A by a constant, u, then det $(B) = u$ det (A).

Proof. Let's see the situation first for the 3×3 case. If B is obtained from A by multiplying the first column of A by u, then

$$B = \begin{bmatrix} ua_{11} & a_{12} & a_{13} \\ ua_{21} & a_{22} & a_{23} \\ ua_{31} & a_{32} & a_{33} \end{bmatrix},$$

where

$$A = \begin{bmatrix} a_{11} & a_{12} & a_{13} \\ a_{21} & a_{22} & a_{23} \\ a_{31} & a_{32} & a_{33} \end{bmatrix}.$$

Thus

$$\det (B) = ua_{11} \begin{vmatrix} a_{22} & a_{23} \\ a_{32} & a_{33} \end{vmatrix} - ua_{21} \begin{vmatrix} a_{12} & a_{13} \\ a_{32} & a_{33} \end{vmatrix} + ua_{31} \begin{vmatrix} a_{12} & a_{13} \\ a_{22} & a_{23} \end{vmatrix}$$

$$= u \left(a_{11} \begin{vmatrix} a_{22} & a_{23} \\ a_{32} & a_{33} \end{vmatrix} - a_{21} \begin{vmatrix} a_{12} & a_{13} \\ a_{32} & a_{33} \end{vmatrix} + a_{31} \begin{vmatrix} a_{12} & a_{13} \\ a_{22} & a_{23} \end{vmatrix} \right)$$

$$= u \det (A),$$

the desired result.

If a column other than the first is multiplied by u, then in each minor N_{k1} of B a column in N_{k1} is multiplied by u. Hence $N_{k1} = uM_{k1}$, by the 2×2 case, where M_{k1} is the $(k, 1)$ minor of det (A). So

$$\begin{aligned}
\det(B) &= a_{11}N_{11} - a_{21}N_{21} + a_{31}N_{31} \\
&= ua_{11}M_{11} - ua_{21}M_{21} + ua_{31}M_{31} \\
&= u(a_{11}M_{11} - a_{21}M_{21} + a_{31}M_{31}) \\
&= u \det(A).
\end{aligned}$$

We leave the details of the proof for the case of $n \times n$ matrices to the readers. In that way they can check if they have this technique of proof under control. The proof runs along pretty well as it did above for 3×3 matrices.   ∎

At this point the parallelism with Section 5.2 breaks. (It will be picked up later.) The result we now prove depends heavily on the results for row operations, and is a little tricky. So before embarking on the proof of the general result let's try the argument to be given on a 3×3 matrix.

Consider the 3×3 matrix

$$A = \begin{bmatrix} a & d & a \\ b & e & b \\ c & f & c \end{bmatrix}.$$

Notice that the first and third columns of A are equal. Suppose, for the sake of argument, that $a \neq 0$ (this really is not essential). By Theorem 5.2.8 we can add any multiple of the first row to any other row without changing the determinant. Add $-\dfrac{b}{a}$ times the first row to the second one and $-\dfrac{c}{a}$ times the first row to the third one. The resulting matrix B is

$$\begin{bmatrix} a & d & a \\ 0 & u & 0 \\ 0 & v & 0 \end{bmatrix},$$

where $u = e - \dfrac{b}{a}d$ and $v = f - \dfrac{c}{a}d$ (and are of no importance).

Thus

$$\det(A) = \det(B) = a \begin{vmatrix} u & 0 \\ v & 0 \end{vmatrix} = 0$$

since $\begin{vmatrix} u & 0 \\ v & 0 \end{vmatrix}$ has a column of zeros.

This simple example suggests

Theorem 5.3.3.   If two columns of A are equal, then det $(A) = 0$.

Proof.   If neither column in question is the first then for each minor M_{k1} we have two columns equal. So by the $(n - 1) \times (n - 1)$ case, $M_{k1} = 0$. Therefore, det $(A) = 0$.

Suppose then that the two columns involved are the first and the rth. If the first column of A consists only of zeros, then det $(A) = 0$ by Theorem 5.3.1, and we are done. So we may assume that $a_{s1} \neq 0$ for some s. If we interchange row s with the first row we get a matrix B whose $(1, 1)$ entry is nonzero. Moreover, by Theorem 5.2.5, doing this interchange merely changes the sign of the determinant. So if the determinant of B is 0, since det $(B) = -\det (A)$ we get the desired result, det $(A) = 0$.

So it is enough to carry out the argument for B. In other words, we may assume that the $(1, 1)$ entry a_{11} of A is nonzero.

By our assumption

$$A = \begin{bmatrix} a_{11} & & \overset{(r)}{a_{11}} & \\ a_{21} & & a_{21} & \\ \vdots & * & \vdots & * \\ a_{n1} & & a_{n1} & \end{bmatrix}$$

where the $*$'s indicate entries in which we have no interest, and where

the second $\begin{bmatrix} a_{11} \\ a_{21} \\ \vdots \\ a_{n1} \end{bmatrix}$ occurs in column r.

By Theorem 5.2.8 we can add any multiple of the first row to any other row without changing the determinant. Since $a_{11} \neq 0$, if we add $-\dfrac{a_{k1}}{a_{11}}$ times the first row to row k, for $k = 2, \ldots, n$, the resulting matrix C looks like

$$\begin{bmatrix} a_{11} & & a_{11} & \\ 0 & & 0 & \\ \vdots & * & \vdots & * \\ 0 & & 0 & \end{bmatrix},$$

and det $(C) = $ det (A). But what is det (C)? Since all the entries in the first column are 0 except for a_{11}, det $(C) = a_{11}$ times the $(1, 1)$ minor of C. But look at this minor! Its column $r - 1$, which comes from column r of A, consists only of zeros. Thus this minor is 0! Therefore, det $(C) = 0$, and since det $(C) = $ det (A) we get the desired result that det $(A) = 0$.   ■

We should like to prove that interchanging two columns of a determinant merely changes its sign. This is the counterpart of Theorem 5.2.5. We are almost there, but we first need

Theorem 5.3.4.   If $C = (c_{uv})$ is a matrix such that for $v \neq r$, $c_{uv} = a_{uv}$ and for $v = r$, $c_{ur} = a_{ur} + b_{ur}$ then det $(C) = $ det $(A) + $ det (B), where

$$A = \begin{bmatrix} a_{11} & a_{12} & \cdots & a_{1r} & \cdots & a_{1n} \\ a_{21} & a_{22} & \cdots & a_{2r} & \cdots & a_{2n} \\ \vdots & \vdots & & \vdots & & \vdots \\ a_{n1} & a_{n2} & \cdots & a_{nr} & \cdots & a_{nn} \end{bmatrix}$$

and

$$B = \begin{bmatrix} a_{11} & a_{12} & \cdots & b_{1r} & \cdots & a_{1n} \\ a_{21} & a_{22} & \cdots & b_{2r} & \cdots & a_{2n} \\ \vdots & \vdots & & \vdots & & \vdots \\ a_{n1} & a_{n2} & \cdots & b_{nr} & \cdots & a_{nn} \end{bmatrix}.$$

$\left(B \text{ is the same as } A \text{ except that its column } r \text{ is replaced by } \begin{bmatrix} b_{1r} \\ b_{2r} \\ \vdots \\ b_{nr} \end{bmatrix} . \right)$

Proof.   If $r = 1$, then the result is easy. Why? Because

$$\text{det } (C) = \sum_{k=1}^{n} (-1)^{k+1} c_{k1} Q_{k1}$$

$$= \sum_{k=1}^{n} (-1)^{k+1} (a_{k1} + b_{k1}) Q_{k1},$$

where Q_{k1} is the $(k, 1)$ minor of det (C). But what is Q_{k1}? Since the columns, other than the first, do not involve the b's, we have that

$Q_{k1} = M_{k1}$, the $(k, 1)$ minor of det (A). Thus

$$\det (C) = \sum_{k=1}^{n} (-1)^{k+1} a_{k1} M_{k1} + \sum_{k=1}^{n} (-1)^{k+1} b_{k1} M_{k1}.$$

We recognize the first sum as det (A) and the second one as det (B). Therefore, det $(C) = $ det $(A) + $ det (B).

Suppose, then, that $r > 1$. Then we have that in column $r - 1$ of Q_{k1} that each entry is of the form $c_{ur} = a_{ur} + b_{ur}$, while all the other entries come from A. Since Q_{k1} is an $(n - 1) \times (n - 1)$ matrix we have, by induction, that $Q_{k1} = M_{k1} + N_{k1}$ where N_{k1} is the $(k, 1)$ minor of det (B). Thus

$$\det (C) = \sum_{k=1}^{n} (-1)^{k+1} c_{k1} Q_{k1} = \sum_{k=1}^{n} (-1)^{k+1} a_{k1}(M_{k1} + N_{k1})$$

$$= \sum_{k=1}^{N} (-1)^{k+1} a_{k1} M_{k1} + \sum_{k=1}^{n} (-1)^{k+1} a_{k1} N_{k1}$$

$$= \det (A) + \det (B). \quad \blacksquare$$

If all the Σ's in the proof confuse you, try the argument out for the 3×3 situation.

Let's see that the result is correct for a specific 3×3 matrix. Let

$$A = \begin{bmatrix} 1 & 2 & 3 \\ 0 & 4 & 5 \\ 0 & 1 & 3 \end{bmatrix} = \begin{bmatrix} 1 & 2 & 1+2 \\ 0 & 4 & 1+4 \\ 0 & 1 & 2+1 \end{bmatrix}.$$

According to the theorem,

$$\det \begin{bmatrix} 1 & 2 & 3 \\ 0 & 4 & 5 \\ 0 & 1 & 3 \end{bmatrix} = \det \begin{bmatrix} 1 & 2 & 1 \\ 0 & 4 & 1 \\ 0 & 1 & 2 \end{bmatrix} + \det \begin{bmatrix} 1 & 2 & 2 \\ 0 & 4 & 4 \\ 0 & 1 & 1 \end{bmatrix}$$

$$= \det \begin{bmatrix} 1 & 2 & 1 \\ 0 & 4 & 1 \\ 0 & 1 & 2 \end{bmatrix} + 0$$

$$= \det \begin{bmatrix} 1 & 2 & 1 \\ 0 & 4 & 1 \\ 0 & 1 & 2 \end{bmatrix}.$$

Is that the case? Well,

$$\det \begin{bmatrix} 1 & 2 & 3 \\ 0 & 4 & 5 \\ 0 & 1 & 3 \end{bmatrix} = 1 \begin{vmatrix} 4 & 5 \\ 1 & 3 \end{vmatrix} = 12 - 5 = 7$$

and

$$\det \begin{bmatrix} 1 & 2 & 1 \\ 0 & 4 & 1 \\ 0 & 1 & 2 \end{bmatrix} = 1 \begin{vmatrix} 4 & 1 \\ 1 & 2 \end{vmatrix} = 8 - 1 = 7.$$

So they are indeed equal. No surprise!

We now can prove that interchanging two columns of det (A) changes the sign of det (A).

We do a specific case first. Let

$$A = \begin{bmatrix} a_{11} & a_{12} & a_{13} \\ a_{21} & a_{22} & a_{23} \\ a_{31} & a_{32} & a_{33} \end{bmatrix},$$

our old friend, and

$$B = \begin{bmatrix} a_{12} & a_{11} & a_{13} \\ a_{22} & a_{21} & a_{23} \\ a_{32} & a_{31} & a_{33} \end{bmatrix},$$

the matrix obtained from A by interchanging the first two columns of A. Let

$$C = \begin{bmatrix} a_{11} + a_{12} & a_{11} + a_{12} & a_{13} \\ a_{21} + a_{22} & a_{21} + a_{22} & a_{23} \\ a_{31} + a_{32} & a_{31} + a_{32} & a_{33} \end{bmatrix}.$$

Since the first two columns of C are equal, det $(C) = 0$ by Theorem 5.3.3. By Theorem 5.3.4 used several times, this determinant can also be written as

$$0 = \begin{vmatrix} a_{11} + a_{12} & a_{11} + a_{12} & a_{13} \\ a_{21} + a_{22} & a_{21} + a_{22} & a_{23} \\ a_{31} + a_{32} & a_{31} + a_{32} & a_{33} \end{vmatrix}$$

$$= \begin{vmatrix} a_{11} & a_{11} + a_{12} & a_{13} \\ a_{21} & a_{21} + a_{22} & a_{23} \\ a_{31} & a_{31} + a_{32} & a_{33} \end{vmatrix} + \begin{vmatrix} a_{12} & a_{11} + a_{12} & a_{13} \\ a_{22} & a_{21} + a_{22} & a_{23} \\ a_{32} & a_{31} + a_{32} & a_{33} \end{vmatrix}$$

$$= \begin{vmatrix} a_{11} & a_{11} & a_{13} \\ a_{21} & a_{21} & a_{23} \\ a_{31} & a_{31} & a_{33} \end{vmatrix} + \begin{vmatrix} a_{11} & a_{12} & a_{13} \\ a_{21} & a_{22} & a_{23} \\ a_{31} & a_{32} & a_{33} \end{vmatrix}$$

$$+ \begin{vmatrix} a_{12} & a_{11} & a_{13} \\ a_{22} & a_{21} & a_{23} \\ a_{32} & a_{31} & a_{33} \end{vmatrix} + \begin{vmatrix} a_{12} & a_{12} & a_{13} \\ a_{22} & a_{22} & a_{23} \\ a_{32} & a_{32} & a_{33} \end{vmatrix}.$$

Notice that the first and last of these four determinants have two equal columns, so are 0 by Theorem 5.3.3. Thus we are left with

$$0 = \begin{vmatrix} a_{11} & a_{12} & a_{13} \\ a_{21} & a_{22} & a_{23} \\ a_{31} & a_{33} & a_{33} \end{vmatrix} + \begin{vmatrix} a_{12} & a_{11} & a_{13} \\ a_{22} & a_{21} & a_{23} \\ a_{32} & a_{31} & a_{33} \end{vmatrix}.$$

Hence det $(B) = -\det(A)$.

The proof for the general case will be along the lines of the proof for the 3×3 matrix.

Theorem 5.3.5. If B is obtained from A by the interchange of two columns of A, then det $(B) = -\det(A)$.

Proof. Suppose that $A = (a_{rs})$ and that B is obtained from A by interchanging columns r and s. Consider the matrix $C = (c_{uv})$, where $c_{uv} = a_{uv}$ if $v \neq r$ or $v \neq s$ and where $c_{ur} = a_{ur} + a_{us}$ and $c_{us} = a_{ur} + a_{us}$. Then columns r and s of C are equal, therefore, by Theorem 5.3.3, det $(C) = 0$. Now, by several uses of Theorem 5.3.4, we get that

$$0 = \det(C) = \begin{vmatrix} a_{11} & \cdots & \overset{(\text{col. } r)}{a_{1r} + a_{1s}} & \cdots & \overset{(\text{col. } s)}{a_{1r} + a_{1s}} & \cdots & a_{11} \\ a_{21} & \cdots & a_{2r} + a_{2s} & \cdots & a_{2r} + a_{2s} & \cdots & a_{21} \\ \vdots & & \vdots & & \vdots & & \vdots \\ a_{n1} & \cdots & a_{nr} + a_{ns} & \cdots & a_{nr} + a_{ns} & \cdots & a_{n1} \end{vmatrix}$$

$$= \begin{vmatrix} a_{11} & \cdots & \overset{(r)}{a_{1r}} & \cdots & \overset{(\text{col. } s)}{a_{1r} + a_{1s}} & \cdots & a_{11} \\ a_{21} & \cdots & a_{2r} & \cdots & a_{2r} + a_{2s} & \cdots & a_{21} \\ \vdots & & \vdots & & \vdots & & \vdots \\ a_{n1} & \cdots & a_{nr} & \cdots & a_{nr} + a_{ns} & \cdots & a_{n1} \end{vmatrix}$$

$$+ \begin{vmatrix} a_{11} & \cdots & \overset{(r)}{a_{1s}} & \cdots & \overset{(\text{col. } s)}{a_{1r} + a_{1s}} & \cdots & a_{11} \\ a_{21} & \cdots & a_{2s} & \cdots & a_{2r} + a_{2s} & \cdots & a_{21} \\ \vdots & & \vdots & & \vdots & \vdots & \vdots \\ a_{n1} & \cdots & a_{ns} & \cdots & a_{nr} + a_{ns} & \cdots & a_{n1} \end{vmatrix}$$

$$= \det(A) + \begin{vmatrix} a_{11} & \cdots & \overset{(r)}{a_{1r}} & \cdots & \overset{(s)}{a_{1r}} & \cdots & a_{11} \\ a_{21} & \cdots & a_{2r} & \cdots & a_{2r} & \cdots & a_{21} \\ \vdots & & \vdots & & \vdots & & \vdots \\ a_{n1} & \cdots & a_{nr} & \cdots & a_{nr} & \cdots & a_{n1} \end{vmatrix}$$

$$+ \det(B) + \begin{vmatrix} a_{11} & \cdots & \overset{(r)}{a_{1s}} & \cdots & \overset{(s)}{a_{1s}} & \cdots & a_{11} \\ a_{21} & \cdots & a_{2s} & \cdots & a_{2s} & \cdots & a_{21} \\ \vdots & & \vdots & & \vdots & & \vdots \\ a_{n1} & \cdots & a_{ns} & \cdots & a_{ns} & \cdots & a_{n1} \end{vmatrix}$$

Since the second and last determinants on the right-hand side above have two equal columns, by Theorem 5.3.3 they are both 0. This leaves us with $0 = \det(A) + \det(B)$. Thus $\det(B) = -\det(A)$ and the theorem is proved. ∎

We close this section with the column analogue of Theorem 5.2.8.

If D is an $n \times n$ matrix, let's use the shorthand $D = (d_1, d_2, \ldots, d_n)$ to represent D, where the vector d_k is column k of D.

Theorem 5.3.6.   If a multiple of one column is added to another in $\det(A)$, the determinant does not change.

Proof.   Let $A = (a_1, \ldots, a_n)$ and suppose that we add q times column s to column r. The resulting matrix is $C = (a_1, \ldots, a_r + qa_s, \ldots, a_n)$, where only column r has been changed, as indicated.

Thus, by Theorem 5.3.4, we have that

$$\det(C) = \det(a_1, \ldots, a_r, \ldots, a_s, \ldots, a_n)$$
$$+ \det(a_1, \ldots, qa_s, \ldots, a_s, \ldots, a_n).$$

By Theorem 5.3.2, det $(a_1, \ldots, qa_s, \ldots, a_s, \ldots, a_n)$ equals

q det $(a_1, \ldots, a_s, \ldots, a_s, \ldots, a_n)$,

and since two of the columns of $(a_1, \ldots, a_s, \ldots, a_s, \ldots, a_n)$ are equal, we get

det $(a_1, \ldots, a_s, \ldots, a_s, \ldots, a_n) = 0$.

Thus we get that

det $(C) =$ det $(a_1, \ldots, a_r, \ldots, a_s, \ldots, a_n) =$ det (A).   ■

We do the proof just done in detail for 3×3 matrices. Suppose that

$$A = \begin{bmatrix} a_{11} & a_{12} & a_{13} \\ a_{21} & a_{22} & a_{23} \\ a_{31} & a_{32} & a_{33} \end{bmatrix}.$$

Then $C = (a_1, a_2 + qa_3, a_3)$ is the matrix

$$C = \begin{bmatrix} a_{11} & a_{12} + qa_{13} & a_{13} \\ a_{21} & a_{22} + qa_{23} & a_{23} \\ a_{31} & a_{32} + qa_{33} & a_{33} \end{bmatrix}$$

and using Theorems 5.3.4 and 5.3.2 as in the proof above, we have

$$\det (C) = \begin{vmatrix} a_{11} & a_{12} + qa_{13} & a_{13} \\ a_{21} & a_{22} + qa_{23} & a_{23} \\ a_{31} & a_{32} + qa_{33} & a_{33} \end{vmatrix}$$

$$= \begin{vmatrix} a_{11} & a_{12} & a_{13} \\ a_{21} & a_{22} & a_{23} \\ a_{31} & a_{32} & a_{33} \end{vmatrix} + \begin{vmatrix} a_{11} & qa_{13} & a_{13} \\ a_{21} & qa_{23} & a_{23} \\ a_{31} & qa_{33} & a_{33} \end{vmatrix}$$

$$= \det (A) + q \begin{vmatrix} a_{11} & a_{13} & a_{13} \\ a_{21} & a_{23} & a_{23} \\ a_{31} & a_{33} & a_{33} \end{vmatrix}$$

$$= \det (A) + 0$$

$$= \det (A).$$

PROBLEMS

NUMERICAL PROBLEMS

1. Verify by a direct computation that

(a) $\begin{vmatrix} 1 & -1 & 6 \\ 4 & \frac{1}{2} & 0 \\ 3 & -1 & 2 \end{vmatrix} = - \begin{vmatrix} 6 & -1 & 1 \\ 0 & \frac{1}{2} & 4 \\ 2 & -1 & 3 \end{vmatrix}$.

(b) $\begin{vmatrix} 1 & -1 & 6 \\ 4 & \frac{1}{2} & 0 \\ 3 & -1 & 2 \end{vmatrix} = \begin{vmatrix} 1 & 0 & 0 \\ 4 & 4 + \frac{1}{2} & -24 \\ 3 & 2 & -16 \end{vmatrix}$.

(c) $\begin{vmatrix} 1 & 2 & 3 & 4 \\ 4 & 3 & 2 & 1 \\ 3 & 2 & 1 & 4 \\ 5 & 10 & 15 & 20 \end{vmatrix} = \begin{vmatrix} 1 & 3 & 4 & 5 \\ 4 & 7 & 6 & 5 \\ 3 & 5 & 4 & 7 \\ 5 & 15 & 20 & 25 \end{vmatrix}$.

(d) $\begin{vmatrix} 1 & -2 & 6 \\ 4 & \frac{1}{2} & 0 \\ 3 & -1 & 2 \end{vmatrix} = \begin{vmatrix} 1 & 4 & 3 \\ -2 & \frac{1}{2} & -1 \\ 6 & 0 & 2 \end{vmatrix}$.

(e) $\begin{vmatrix} 1 & 2 & 3 \\ 4 & 5 & 6 \\ 7 & 8 & 9 \end{vmatrix} = \begin{vmatrix} 3 & 2 & 3 \\ 6 & 5 & 6 \\ 9 & 8 & 9 \end{vmatrix} + \begin{vmatrix} -2 & 2 & 3 \\ -2 & 5 & 6 \\ -2 & 8 & 9 \end{vmatrix}$.

2. Evaluate $\begin{vmatrix} 1 & a & b & c \\ 1 & a^2 & b^2 & c^2 \\ 1 & a^3 & b^3 & c^3 \\ 1 & a^4 & b^4 & c^4 \end{vmatrix}$.

3. When is the determinant in Problem 2 equal to 0?

MORE THEORETICAL PROBLEMS

Easier Problems

4. Complete the proof of Theorem 5.3.2 for $n \times n$ matrices.

5. Prove that if you carry out the operation on A of putting its first row into the second, the second row into the third, and the third row into the first, then the determinant remains unchanged.

Middle-Level Problems

6. Show that if the columns of A are linearly independent as vectors, then det $(A) = 0$.

7. Show that if the columns of A are linearly independent, then A is invertible.

8. Let $V = F^{(n)}$ and $W = F^{(m)}$. Suppose that f is a function of two variables such that

(1) $f(x, y) \in W$ for all $x, y \in V$;
(2) $f(ax, y) = f(x, ay) = af(x, y)$ for all $a \in F$, $x, y \in V$;
(3) $f(x + x', y) = f(x, y) + f(x', y)$ for all $x, x', y \in V$;
(4) $f(x, y + y') = f(x, y) + f(x, y')$ for all $x, y, y' \in V$.
(5) $f(x, x) = 0$ for all $x \in V$.

Show that $f(x, y) = -f(y, x)$ for all $x, y \in V$.

5.4. CRAMER'S RULE

The rules for operations with columns and rows of a determinant allow us to make use of determinants in solving systems of linear equations. Consider the system of equations

$$
\begin{aligned}
a_{11}x_n + \cdots + a_{1n}x_n &= y_1 \\
a_{21}x_n + \cdots + a_{2n}x_n &= y_2 \\
\vdots \qquad \cdots \qquad \vdots \quad &\quad \vdots \\
a_{n1}x_n + \cdots + a_{nn}x_n &= y_n
\end{aligned}
\tag{1}
$$

We can represent this system as a matrix-vector equation $Ax = y$, where A is the $n \times n$ matrix (a_{rs}) and $x, y \in F^{(n)}$, that is,

$$
\begin{bmatrix} a_{11} & \cdots & a_{1n} \\ \vdots & & \vdots \\ a_{n1} & \cdots & a_{nn} \end{bmatrix}
\begin{bmatrix} x_1 \\ \vdots \\ x_n \end{bmatrix}
=
\begin{bmatrix} y_1 \\ \vdots \\ y_n \end{bmatrix}.
$$

Consider x_r det (A). By Theorem 5.3.2 we can absorb this x_r in column r, that is,

$$
x_r \det (A) = \begin{vmatrix} a_{11} & a_{12} & \cdots & x_r a_{1r} & \cdots & a_{1n} \\ \vdots & \vdots & & \cdot & & \vdots \\ \vdots & \vdots & & \cdot & & \vdots \\ a_{n1} & a_{n2} & \cdots & x_r a_{nr} & \cdots & a_{nn} \end{vmatrix}.
$$

If we add x_1 times the first column, x_2 times the second, . . . , x_k times column k to column r of this determinant, for $k \neq r$, by Theorem 5.3.6 we do not change the determinant. Thus

$$x_r \det (A) = \begin{vmatrix} a_{11} & a_{12} & \cdots & a_{11}x_1 + \cdots + a_{1n}x_n & \cdots & a_{1n} \\ \vdots & \vdots & & \vdots & \vdots & \vdots \\ a_{n1} & a_{n2} & \cdots & a_{n1}x_1 + \cdots + a_{nn}x_n & \cdots & a_{nn} \end{vmatrix}.$$

$$\overset{(\text{column } r)}{}$$

Hence if $x_1, \ldots, x_n$ is a solution to the system (1,) we have

$$x_r \det (A) = \begin{vmatrix} a_{11} & a_{12} & \cdots & y_1 & \cdots & a_{1n} \\ \vdots & \vdots & & \vdots & & \vdots \\ a_{n1} & a_{n2} & \cdots & y_n & \cdots & a_{nn} \end{vmatrix} = \det (A_r),$$

$$\overset{(r)}{}$$

where the matrix A_r is obtained from A by replacing column r of A by

the vector $\begin{bmatrix} y_1 \\ \vdots \\ y_n \end{bmatrix}$, that is, the vector of the values on the right-hand side

of (1).

In particular, if $\det (A) \neq 0$, we have

Theorem 5.4.1 (**Cramer's Rule**). If $\det (A) \neq 0$, the solution to the system (1) of linear equations is given by

$$x_r = \frac{\det (A_r)}{\det (A)} \qquad \text{for } r = 1, 2, \ldots, n,$$

where A_r is as described above.

To illustrate the result, let's use it to solve the three linear equations in three unknowns:

$$\begin{array}{rcrcrcr} x_1 & + & x_2 & + & x_3 & = & 1 \\ 2x_1 & - & x_2 & + & 2x_3 & = & 2 \\ & & 3x_2 & - & 4x_3 & = & 3. \end{array}$$

Thus here the matrix A is

$$A = \begin{bmatrix} 1 & 1 & 1 \\ 2 & -1 & 2 \\ 0 & 3 & -4 \end{bmatrix}$$

and

$$A_1 = \begin{bmatrix} 1 & 1 & 1 \\ 2 & -1 & 2 \\ 3 & 3 & -4 \end{bmatrix}, \quad A_2 = \begin{bmatrix} 1 & 1 & 1 \\ 2 & 2 & 2 \\ 0 & 3 & -4 \end{bmatrix}, \quad A_3 = \begin{bmatrix} 1 & 1 & 1 \\ 2 & -1 & 2 \\ 0 & 3 & 3 \end{bmatrix}.$$

Evaluating all these determinants, we obtain

$$\det (A) = 12,$$

and

$$\det (A_1) = 21, \quad \det (A_2) = 0, \quad \det (A_3) = -9.$$

Consequently, by Cramer's rule,

$$x_1 = \tfrac{21}{12} = \tfrac{7}{4}, \quad x_2 = \tfrac{0}{12} = 0, \quad x_3 = -\tfrac{9}{12} = -\tfrac{3}{4}.$$

Checking, we see that we have indeed solved the system:

$$\begin{aligned} (\tfrac{7}{4}) + (0) + (-\tfrac{3}{4}) &= 1 \\ 2(\tfrac{7}{4}) - (0) + 2(-\tfrac{3}{4}) &= 2 \\ 3(0) - 4(-\tfrac{3}{4}) &= 3. \end{aligned}$$

Aside from its role in solving systems of n linear equations in n unknowns, Cramer's rule has a very important consequence interrelating the behavior of a matrix and its determinant. In the next result we get a very powerful and useful criterion for the invertibility of a matrix.

Theorem 5.4.2.   If $\det (A) \neq 0$, then A is invertible.

Proof.   If $\det (A) \neq 0$, then, by Cramer's rule, we can solve the

matrix-vector equation $Ax = v$ for any vector $v = \begin{bmatrix} y_1 \\ \vdots \\ y_n \end{bmatrix}$. In particular,

if $e_1, e_2, \ldots, e_n$ is the canonical basis of $F^{(n)}$, we can find vectors

$X_1 = \begin{bmatrix} x_{11} \\ \vdots \\ x_{n1} \end{bmatrix}, X_2 = \begin{bmatrix} x_{12} \\ \vdots \\ x_{n2} \end{bmatrix}, \ldots, X_n = \begin{bmatrix} x_{1n} \\ \vdots \\ x_{nn} \end{bmatrix}$ such that for each $r =$

$1, 2, \ldots, n$, $AX_r = e_r$. Thus if $B = \begin{bmatrix} x_{11} & & x_{1n} \\ \vdots & \cdots & \vdots \\ x_{n1} & & x_{nn} \end{bmatrix}$, from $AX_r =$

e_r we obtain

$$AB = A(X_1, \ldots, X_n) = (e_1, \ldots, e_n) = \begin{bmatrix} 1 & 0 & \cdots & 0 \\ 0 & & & \vdots \\ \vdots & & \ddots & 0 \\ 0 & \cdots & 0 & 1 \end{bmatrix} = I.$$

In short, A is invertible. ■

We shall later see that the converse of Theorem 5.4.2 is true, namely, that if A is invertible, then det $(A) \neq 0$.

PROBLEMS

NUMERICAL PROBLEMS

Use Cramer's rule.

1. Solve for x_1, x_2, x_3, x_4 in the system

$$
\begin{aligned}
x_1 + 2x_2 + 3x_3 + 4x_4 &= 1 \\
x_1 + 2x_2 + 3x_3 + 3x_4 &= 2 \\
x_1 + 2x_2 + 2x_3 + 2x_4 &= 3 \\
x_1 + x_2 + x_3 + x_4 &= 4.
\end{aligned}
$$

2. Find the solution $\begin{bmatrix} a \\ b \\ c \\ d \end{bmatrix}$ to $\begin{bmatrix} 5 & 5 & 5 & 5 \\ 4 & 4 & 5 & 5 \\ 4 & 5 & 4 & 5 \\ 4 & 5 & 0 & 0 \end{bmatrix} \begin{bmatrix} a \\ b \\ c \\ d \end{bmatrix} = \begin{bmatrix} 2 \\ \pi \\ 1 \\ 7 \end{bmatrix}.$

3. Solve for x_1, x_2, x_3 in the system

$$
\begin{aligned}
3x_1 + 2x_2 - 3x_3 &= 4 \\
3x_1 + 8x_2 + x_3 &= 9 \\
21x_1 + 22x_2 + x_3 &= -1.
\end{aligned}
$$

4. Find, for the vector $\begin{bmatrix} a \\ b \\ c \end{bmatrix}$, a vector $\begin{bmatrix} u \\ v \\ w \end{bmatrix}$ such that

$$\begin{bmatrix} 1 & 5 & -1 \\ 2 & 0 & 1 \\ 3 & 1 & 1 \end{bmatrix} \begin{bmatrix} u \\ v \\ w \end{bmatrix} = \begin{bmatrix} a \\ b \\ c \end{bmatrix}.$$

Harder Problems

5. If $A = (a_{rs})$ is an $n \times n$ matrix such that the vectors

$$v_1 = \begin{bmatrix} a_{11} \\ \vdots \\ a_{n1} \end{bmatrix}, \quad v_2 = \begin{bmatrix} a_{12} \\ \vdots \\ x_{n2} \end{bmatrix}, \quad \cdots \quad v_n = \begin{bmatrix} a_{1n} \\ \vdots \\ a_{nn} \end{bmatrix}$$

are linearly independent, using the column operations of Section 5.3, show that det $(A) \neq 0$.

5.5. PROPERTIES OF DETERMINANTS: OTHER EXPANSIONS

After the short respite from row and column operations, which was afforded us in Section 5.4, we return to a further examination of such operations.

In our initial definition of the determinant of A, the determinant was defined in terms of the expansion by minors of the first column. In this way we favored the first column over all the others. Is this really necessary? Can we define the determinant in terms of the expansion by the minors of any column? The answer to this is "yes," which we shall soon demonstrate.

To get away from the annoying need to alternate the signs of the minors, we introduce a new notion, closely allied to that of a minor.

Definition. If A is an $n \times n$ matrix, then the (r, s) *cofactor* A_{rs} of det (A) is defined by $A_{rs} = (-1)^{r+s} M_{rs}$, where M_{rs} is the (r, s) minor of det (A). (We used A_{rs} in another context in Section 5.1. It is not the same as the A_{rs} we now use to denote a cofactor.)

So, for example,

$$A_{57} = (-1)^{5+7} M_{57} = M_{57},$$

while

$$A_{34} = (-1)^{3+4} M_{34} = -M_{34}.$$

Note that in terms of cofactors we have that

$$\det (A) = \sum_{r=1}^{n} a_{r1} A_{r1}.$$

As we so often have done, let us look at a particular matrix, say,

$$A = \begin{bmatrix} 1 & 2 & 3 \\ 4 & 5 & 6 \\ 7 & 8 & 9 \end{bmatrix}.$$

What are the cofactors of the entries of the second column of A? They are

$$A_{12} = (-1)^3 M_{12} = - \begin{vmatrix} 4 & 6 \\ 7 & 9 \end{vmatrix} = 6$$

$$A_{22} = (-1)^4 M_{22} = + \begin{vmatrix} 1 & 3 \\ 7 & 9 \end{vmatrix} = -12$$

$$A_{32} = (-1)^5 M_{32} = - \begin{vmatrix} 1 & 3 \\ 4 & 6 \end{vmatrix} = 6.$$

So

$$a_{12}A_{12} + a_{22}A_{22} + a_{32}A_{32} = 2(6) + 5(-12) + 8(6) = 0.$$

If we expand det (A) by the first column, we get det $(A) = 0$. So what we might call the "expansion by the second column of A," $\sum_{r=1}^{3} a_{r2}A_{r2}$, in this particular instance turns out to be the same as det (A). Do you think this is happenstance?

The next theorem shows that the expansion by *any* column of A—using cofactors in this expansion—gives us det (A). In other words, *the first column is no longer favored over its colleagues.*

Theorem 5.5.1. For any s, det $(A) = \sum_{r=1}^{n} a_{rs}A_{rs}$.

Proof. Let $A = (a_{rs})$ and let B be the matrix

$$B = \begin{bmatrix} a_{1s} & a_{11} & a_{12} & \cdots & a_{1n} \\ a_{2s} & a_{21} & a_{22} & \cdots & a_{2n} \\ \vdots & \vdots & \vdots & & \vdots \\ a_{ns} & a_{n1} & a_{n2} & \cdots & a_{nn} \end{bmatrix}$$

obtained from A by moving column s to the first column. Notice that this is accomplished by $s - 1$ interchanges of adjacent columns, that is, by moving column across column $s - 1$, then across column $s - 2$, and so on. Since each such interchange changes the sign of the determinant (Theorem 5.2.5), we have that det $(B) = (-1)^{s-1}$ det (A).

But what is det (B)? Using the expansion by the first column of B, we calculate $(-1)^{s-1} \det (A) = \det (B)$ as

$$a_{1s}N_{11} - a_{2s}N_{21} + a_{31}N_{31} + \cdots + (-1)^{n+1}a_{ns}N_{n1}$$

where N_{r1} is the $(r, 1)$ minor of $\det (B)$. However, we claim that $N_{r1} = M_{rs}$, the (r, s) minor of $\det (A)$. Why? To construct N_{r1} we eliminate the first column of B and its row r. What is left is precisely what we get by eliminating column s and row r of A. So $N_{r1} = M_{rs}$. Thus our expression for $(-1)^{s-1} \det (A) = \det (B)$ becomes

$$a_{1s}M_{1s} - a_{2s}M_{2s} + a_{31}M_{3s} + \cdots + (-1)^{n+1}a_{ns}M_{ns}.$$

Because $A_{1s} = (-1)^{1+s}M_{1s}$, $A_{2s} = (-1)^{2+s}M_{2s}$ and so on, we get

$$(-1)^{s-1} \det (A) = \det (B) = (-1)^{1+s}a_{1s}A_{1s} - (-1)^{2+s}a_{2s}A_{2s}$$
$$+ (-1)^{3+s}a_{31}A_{3s} + \cdots + (-1)^{n+s}a_{ns}A_{ns}.$$

To get $\det (A)$, we multiply through by $(-1)^{s-1}$, ending up with

$$\det (A) = (-1)^{2s}a_{1s}A_{1s} - (-1)^{1+2s}a_{2s}A_{2s} + (-1)^{2+2s}a_{31}A_{3s}$$
$$+ \cdots + (-1)^{2s+2n}a_{ns}A_{ns}$$
$$= (-1)^{2s}a_{1s}A_{1s} + (-1)^{2+2s}a_{2s}A_{2s} + (-1)^{2+2s}a_{31}A_{3s}$$
$$+ \cdots + (-1)^{2s+2n}a_{ns}A_{ns}$$
$$= a_{1s}A_{1s} + a_{2s}A_{2s} + a_{31}A_{3s} + \cdots + a_{ns}A_{ns},$$

since the powers of -1 in the expression for $\det (A)$ above are *even*. This completes the proof. ∎

Theorem 5.5.1 allows us to say something about the expansion of a determinant of A by the minors or cofactors of the first row of A. Consider the particular matrix

$$A = \begin{bmatrix} 0 & a & 0 \\ b & c & d \\ e & f & g \end{bmatrix}.$$

By Theorem 5.5.1 we can expand A by the cofactors of the second column. Thus

$$\det (A) = a_{12}A_{12} + a_{22}A_{22} + a_{32}A_{32}$$
$$= (-1)^3 a_{12}M_{12} + (-1)^4 a_{22}M_{22} + (-1)^5 a_{32}M_{32}$$
$$= -a \begin{vmatrix} b & d \\ e & g \end{vmatrix} + c \begin{vmatrix} 0 & 0 \\ e & g \end{vmatrix} - f \begin{vmatrix} 0 & 0 \\ b & d \end{vmatrix} = -a \begin{vmatrix} b & d \\ e & g \end{vmatrix}$$
$$= -aM_{12},$$

since in the last two minors we have a row of zeros.

The argument just given is the model for that needed in the proof of

Theorem 5.5.2

$$
\begin{vmatrix}
0 & 0 & \cdots & 0 & a_{1s} & 0 & \cdots & 0 \\
a_{21} & a_{22} & \cdots & a_{2s-1} & a_{2s} & a_{2s+1} & \cdots & a_{2n} \\
\vdots & \vdots & & \vdots & \vdots & \vdots & & \vdots \\
a_{n1} & a_{n2} & \cdots & a_{ns-1} & a_{ns} & a_{ns-1} & \cdots & a_{nn}
\end{vmatrix} = a_{1s}A_{1s}.
$$

Proof. By Theorem 5.5.1 we may expand the determinant by cofactors of columns s. We thus get

$$
\begin{vmatrix}
0 & 0 & \cdots & 0 & a_{1s} & 0 & \cdots & 0 \\
a_{21} & a_{22} & \cdots & a_{2s-1} & a_{2s} & a_{2s+1} & \cdots & a_{2n} \\
\vdots & \vdots & & \vdots & \vdots & \vdots & & \vdots \\
a_{n1} & a_{n2} & \cdots & a_{ns-1} & a_{ns} & a_{ns-1} & \cdots & a_{nn}
\end{vmatrix}
$$

$$
= a_{1s}A_{1s} + a_{2s}A_{2s} + \cdots + a_{ns}A_{ns}.
$$

Notice, however, that in each of the cofactors $A_{2s}, \ldots, A_{ns}$ we have a row of zeros. Hence $A_{2s} = \cdots = A_{ns} = 0$. This leaves us with

$$
\begin{vmatrix}
0 & 0 & \cdots & 0 & a_{1s} & 0 & \cdots & 0 \\
a_{21} & a_{22} & \cdots & a_{2s-1} & a_{2s} & a_{2s+1} & \cdots & a_{2n} \\
\vdots & \vdots & & \vdots & \vdots & \vdots & & \vdots \\
a_{n1} & a_{n2} & \cdots & a_{ns-1} & a_{ns} & a_{ns-1} & \cdots & a_{nn}
\end{vmatrix} = a_{1s}A_{1s}. \quad \blacksquare
$$

Given the matrix $A = (a_{rs})$, then

$$
\det (A) =
\begin{vmatrix}
a_{11} & a_{12} & \cdots & a_{1n} \\
a_{21} & a_{22} & \cdots & a_{2n} \\
\vdots & \vdots & & \vdots \\
a_{n1} & a_{n2} & \cdots & a_{nn}
\end{vmatrix}
=
\begin{vmatrix}
a_{11} & 0 & \cdots & 0 \\
a_{21} & a_{22} & \cdots & a_{2n} \\
\vdots & \vdots & & \vdots \\
a_{n1} & a_{n2} & \cdots & a_{nn}
\end{vmatrix}
$$

$$
+
\begin{vmatrix}
0 & a_{12} & \cdots & 0 \\
a_{21} & a_{22} & \cdots & a_{2n} \\
\vdots & \vdots & & \vdots \\
a_{n1} & a_{n2} & \cdots & a_{nn}
\end{vmatrix}
$$

$$
\vdots
$$

$$+ \begin{vmatrix} 0 & 0 & \cdots & a_{1n} \\ a_{21} & a_{22} & \cdots & a_{2n} \\ \vdots & \vdots & & \vdots \\ a_{n1} & a_{n2} & \cdots & a_{nn} \end{vmatrix}$$

by Theorem 5.3.4. But by Theorem 5.5.2, we have

$$\begin{vmatrix} 0 & 0 & \cdots & a_{1s} & \cdots & 0 \\ a_{21} & a_{22} & \cdots & a_{2s} & \cdots & a_{2n} \\ \vdots & \vdots & & \vdots & & \vdots \\ a_{n1} & a_{n2} & \cdots & a_{ns} & \cdots & a_{nn} \end{vmatrix} = a_{1s}A_{1s}$$

for all $s = 1, \ldots, n$. Thus

$$\det (A) = a_{11}A_{11} + a_{12}A_{12} + \cdots + a_{1s}A_{1s} + \cdots + a_{1n}A_{1n}.$$

In other words, $\det (A)$ *can be evaluated by expanding it by the cofactors of the first row.*

In summation notation, this result is $\det (A) = \Sigma_{s=1}^{n} a_{1s}A_{1s}$. Using the definition $A_{1s} = (-1)^{1+s}M_{1s}$, it translates to *the expansion*

$$\det (A) = \sum_{s=1}^{n} (-1)^{1+s}a_{1s}M_{1s}$$

of $\det (A)$ *by the minors of the first row.*

We record this very important result as

Theorem 5.5.3. $\det (A) = \Sigma_{s=1}^{n} (-1)^{1+s}a_{1s}M_{1s} = \Sigma_{s=1}^{n} a_{1s}A_{1s}$, that is, we can evaluate $\det (A)$ by an expansion by the minors or cofactors of the *first row* of A.

An example is now in order. Let $A = \begin{bmatrix} 1 & 2 & 3 \\ -1 & 1 & -1 \\ 2 & 4 & 1 \end{bmatrix}$. The

minors of the first row are

$$\begin{vmatrix} 1 & -1 \\ 4 & 1 \end{vmatrix} = 5, \qquad \begin{vmatrix} -1 & -1 \\ 2 & 1 \end{vmatrix} = 1, \qquad \begin{vmatrix} -1 & 1 \\ 2 & 4 \end{vmatrix} = -6,$$

so that

$$\begin{vmatrix} 1 & 2 & 3 \\ -1 & 1 & -1 \\ 2 & 4 & 1 \end{vmatrix} = 1 \cdot 5 - 2 \cdot 1 + 3(-6) = -15.$$

Evaluate det (A) by the minors of the first column and you will get the same answer, namely, -15.

Theorem 5.5.3 itself has a very important consequence. Let A be a matrix on A' its transpose. Then

Theorem 5.5.4. det (A) = det (A').

Proof. Once again we look at the situation of a 3×3 matrix. Let

$$A = \begin{bmatrix} a_{11} & a_{12} & a_{13} \\ a_{21} & a_{22} & a_{23} \\ a_{31} & a_{32} & a_{33} \end{bmatrix}, \text{ hence } A' = \begin{bmatrix} a_{11} & a_{21} & a_{31} \\ a_{12} & a_{22} & a_{32} \\ a_{13} & a_{23} & a_{33} \end{bmatrix}.$$

Therefore, expanding det (A') by the minors of its first row, we have

$$\det (A') = a_{11} \begin{vmatrix} a_{22} & a_{32} \\ a_{23} & a_{33} \end{vmatrix} - a_{21} \begin{vmatrix} a_{12} & a_{32} \\ a_{13} & a_{33} \end{vmatrix} + a_{31} \begin{vmatrix} a_{12} & a_{22} \\ a_{13} & a_{23} \end{vmatrix}.$$

Expanding det (A) by the minors of its first column, we get

$$\det (A) = a_{11} \begin{vmatrix} a_{22} & a_{23} \\ a_{32} & a_{33} \end{vmatrix} - a_{21} \begin{vmatrix} a_{12} & a_{13} \\ a_{32} & a_{33} \end{vmatrix} + a_{31} \begin{vmatrix} a_{12} & a_{13} \\ a_{22} & a_{23} \end{vmatrix}.$$

Notice that each (r, s) minor in the expansion of det (A') is the 2×2 determinant of the transpose of the (s, r) minor submatrix of A (see Section 5.1). By the result for 2×2 matrices we then know that these minors are equal. This gives us that det (A') = det (A).

Now to the general case, assuming that the result is known for $(n - 1) \times (n - 1)$ matrices. The first row of A' is the first column of A, from the very definition of transpose. If U_{rs} is the (r, s) minor submatrix of A and V_{rs} the (r, s) minor submatrix of A', then $V_{rs} = U'_{sr}$. (Prove!) By induction, N_{rs}, the (r, s) minor of A', is given by

$$N_{rs} = \det (V_{rs}) = \det (U'_{sr}) = \det (U_{sr}) = M_{sr},$$

where M_{sr} is the (s, r) minor of A.

If b_{1s} is the $(1, s)$ entry of A' then $b_{1s} = a_{s1}$. Also, by Theorem 5.5.3,

$$\det (A') = \sum_{s=1}^{n} (-1)^{s+1} b_{1s} N_{1s} = \sum_{s=1}^{n} (-1)^{s+1} a_{s1} M_{s1} = \det (A).$$

This proves the theorem. ∎

For the columns we showed in Theorem 5.5.1 that we can calculate det (A) by the cofactors of any column of A. We would hope that a similar result holds for the rows of A. In fact, this is true; it is

Theorem 5.5.5   For any $r \geq 1$, det (A) = $\sum_{s=1}^{n} a_{rs}A_{rs}$, that is, we can expand det (A) by the cofactors of any row of A.

Proof.   We leave the proof to the reader, but we do provide a few hints of how to go about it. By Theorem 5.5.4, $|A| = |A'|$ and, by Theorem 5.5.1, $|A'|$ can be expanded by the cofactors of its rth column. Since the (s, r) cofactor of A' equals the (r, s) cofactor of A (Prove!), it follows that the expansion of $|A'|$ by the cofactors of its rth column equals the expansion of $|A|$ by the cofactors of its rth row.   ∎

PROBLEMS

NUMERICAL PROBLEMS

1. Verify that the expansion of the determinant of

$$A = \begin{bmatrix} 1 & 2 & 3 \\ 7 & -1 & 0 \\ 6 & 1 & 2 \end{bmatrix}$$

by the cofactors of the second column equals det (A).

2. Verify by a direct computation that $\begin{vmatrix} 1 & 0 & -5 \\ 6 & 2 & -1 \\ 5 & -11 & 6 \end{vmatrix}$, as evaluated

by expansion by the minors of the first row and as evaluated by expansion by the minors of the first column are equal.

3. Evaluate by the expansion of the minors of the first row:

(a) $\begin{vmatrix} 1 & 2 & 3 \\ -3 & 2 & 1 \\ 3 & -1 & 2 \end{vmatrix} + \begin{vmatrix} 2 & 1 & 3 \\ -3 & 2 & 1 \\ 2 & -1 & 2 \end{vmatrix}.$

(b) $\begin{vmatrix} 1 & 2 & 3 \\ -3 & 2 & 1 \\ 3 & 0 & 4 \end{vmatrix} + \begin{vmatrix} 2 & 1 & 3 \\ -3 & 2 & 1 \\ 6 & -5 & 7 \end{vmatrix}.$

(c) $\begin{vmatrix} 1 & 1 & 1 & 1 \\ 2 & 2 & 2 & 2 \\ 3 & 3 & 3 & 3 \\ 4 & 4 & 4 & 4 \end{vmatrix}$.

(d) $\begin{vmatrix} 0 & 1 & 2 & 3 \\ 4 & 5 & 6 & 7 \\ 0 & 8 & 9 & 10 \\ 0 & 1 & 1 & 1 \end{vmatrix}$.

(e) Compare the result of Part (d) with the result obtained using the expansion by minors of the first column.

MORE THEORETICAL PROBLEMS

Easier Problems

4. Show that if A has two proportional columns, then det $(A) = 0$.

5. For what value of a is $\begin{vmatrix} 1 & 5 & 9 & 0 \\ 2 & 6 & 10 & 0 \\ 3 & 7 & 0 & a \\ 4 & 8 & 0 & a \end{vmatrix} = 0$?

Middle-Level Problems

6. Completely prove Theorem 5.5.5.

5.6. ELEMENTARY MATRICES

We have seen in the preceding sections that there are certain operations which we can carry out on the row, or on the columns, of a matrix A which result in no change, or in a change of a very easy and specific sort, in det (A). Let's recall which they were. We list them for the rows, but remember, *the same thing holds for the columns*.

1. If we add the multiple of one row of A to another row of A we do not change the determinant.

2. If we interchange two rows of A, then the determinant changes sign.

3. If we multiply every entry in a given row of A by a constant q, then the determinant is merely multiplied by q.

Since we are dealing with matrices it is natural for us to ask whether these three operations can be achieved by matrix multiplication of A by three specific types of matrices. The answer is indeed "yes"! Before going to the general situation note, for the 3×3 matrices that:

1. $\begin{bmatrix} 1 & 0 & 0 \\ 0 & 1 & q \\ 0 & 0 & 1 \end{bmatrix} \begin{bmatrix} a_{11} & a_{12} & a_{13} \\ a_{21} & a_{22} & a_{23} \\ a_{31} & a_{32} & a_{33} \end{bmatrix} =$

$\begin{bmatrix} a_{11} & a_{12} & a_{13} \\ a_{21} + qa_{31} & a_{22} + qa_{32} & a_{23} + qa_{33} \\ a_{31} & a_{32} & a_{33} \end{bmatrix}$, so this multiplication

adds q times the third row to the second.

2. $\begin{bmatrix} 1 & 0 & 0 \\ 0 & 0 & 1 \\ 0 & 1 & 0 \end{bmatrix} \begin{bmatrix} a_{11} & a_{12} & a_{13} \\ a_{21} & a_{22} & a_{23} \\ a_{31} & a_{32} & a_{33} \end{bmatrix} = \begin{bmatrix} a_{11} & a_{12} & a_{13} \\ a_{31} & a_{32} & a_{33} \\ a_{21} & a_{22} & a_{23} \end{bmatrix}$, so this multipli-

cation interchanges rows 2 and 3.

3. $\begin{bmatrix} 1 & 0 & 0 \\ 0 & q & 0 \\ 0 & 0 & 1 \end{bmatrix} \begin{bmatrix} a_{11} & a_{12} & a_{13} \\ a_{21} & a_{22} & a_{23} \\ a_{31} & a_{32} & a_{33} \end{bmatrix} = \begin{bmatrix} a_{11} & a_{12} & a_{13} \\ qa_{21} & qa_{22} & qa_{23} \\ a_{31} & a_{32} & a_{33} \end{bmatrix}$, so this mul-

tiplication results in multiplying the second row by q.

As you can readily verify, *multiplying A on the right by*

$\begin{bmatrix} 1 & 0 & 0 \\ 0 & 1 & q \\ 0 & 0 & 1 \end{bmatrix}$, $\begin{bmatrix} 1 & 0 & 0 \\ 0 & 0 & 1 \\ 0 & 1 & 0 \end{bmatrix}$, $\begin{bmatrix} 1 & 0 & 0 \\ 0 & q & 0 \\ 0 & 0 & 1 \end{bmatrix}$

gives us the same story for the columns, except that the role of the indices is reversed namely

adds q time the *second* column to the *third*;

interchanges columns 3 and 2; and

multiplies column 2 by q, respectively.

So in the particular instance for 3×3 matrices, multiplying A on the *left* by the three matrices above achieves the three operations (1), (2), (3) above, and doing it on the *right* achieves the corresponding thing for the columns.

Fortunately, the story is equally simple for the $n \times n$ matrices. But first recall that the matrices E_{rs} introduced earlier in the book were defined by: E_{rs} is the matrix whose (r, s) entry is 1 and all of whose other entries are 0. The basic facts about these matrices were

1. If $A = (a_{rs})$, then $A = \sum_{r=1}^{n} \sum_{s=1}^{n} a_{rs} E_{rs}$.
2. If $s \neq t$, then $E_{rs} E_{tu} = 0$.
3. $E_{rs} E_{st} = E_{rt}$.

Note that in the example above, $\begin{bmatrix} 1 & 0 & 0 \\ 0 & 1 & q \\ 0 & 0 & 1 \end{bmatrix} = I + qE_{23}$,

$\begin{bmatrix} 1 & 0 & 0 \\ 0 & q & 0 \\ 0 & 0 & 1 \end{bmatrix} = I + (q - 1)E_{22}$, and $\begin{bmatrix} 1 & 0 & 0 \\ 0 & 0 & 1 \\ 0 & 1 & 0 \end{bmatrix}$, which is obtained

by interchanging the second and third columns of matrix I, can be written
in the (awkward) form $I + E_{23} + E_{32} - E_{22} - E_{33}$.

We are now ready to pass to the general context.

Definition. We define three types of matrices:

1. $A(r, s; q) = I + qE_{rs}$ for $r \neq s$.

2. $M(r; q) = I + (q - 1)E_{rs}$ for $q \neq 0$.

3. $I(r, s)$, the matrix obtained from the unit matrix I by interchanging columns r and s of I. [We then know that

$$I(r, s) = I + E_{rs} + E_{sr} - E_{rr} - E_{ss}$$

(Prove!), but we will not use this clumsy form of representing $I(r, s)$].

We call these three types of matrices the *elementary matrices* and
shall often represent them by the letter E or by E with *one* subscript.

What are the basic properties of these elementary matrices? We
leave the proof of the next theorem to the reader.

Theorem 5.6.1. For any matrix B,

1. $A(r, s; q)B$, for $r \neq s$, is that matrix obtained from B by adding
   q times row s to row r;

1'. $BA(r, s; q)$, for $r \neq s$, is that matrix obtained from B by adding
   q times column r to column s.

2. $M(r; q)B$ is that matrix obtained from B by multiplying each
   entry in row r by q.

2'. $BM(r; q)$ is that matrix obtained from B by multiplying each entry in column r by q.

3. $I(r, s)B$ is that matrix obtained from B by interchanging rows r and s of B.

3'. $BI(r, s)$ is that matrix obtained from B by interchanging columns s and r of B.

Using the fact that $A(r, s; q)$, for $r \neq s$, is triangular, and that $M(r; q)$ is diagonal, with 1's on the diagonal except for the (r, r) entry which is q, we have:

$$\det (A(r, s; q)) = 1 \text{ for } r \neq s \text{ and } \det (M(r; q)) = q.$$

Finally, by Theorem 5.2.5, since we obtain $I(r, s)$, for $r \neq s$, by interchanging two columns of I,

$$\det (I(r, s)) = -\det (I) = -1.$$

Consider the effect on the determinant of multiplying a given matrix B by an elementary matrix E.

1.   If $E = A(r, s; q)$, then $A(r, s; q)B$ is obtained from B by adding q times row s to row r. By Theorem 5.2.8 we have that

$$\det (A(r, s; q)B) = \det (B) = \det (A(r, s; q)) \det (B).$$

And similarly, $BA(r, s; q)$ is obtained from B by adding q times column r to column s. By Theorem 5.3.6 we get that

$$\det (BA(r, s; q)) = \det (B) = \det (B) \det (A(r, s; q)).$$

2.   If $E = M(r; q)$, then by Theorems 5.3.2 and 5.2.2 we know that

$$\det (BM(r; q)) = q \det (B) = \det (B) \det (M(r; q)),$$
$$\det (M(r; q)B) = q \det (B) = \det (M(r; q)) \det (B).$$

3.   Finally, if $E = I(r, s)$, by Theorems 5.2.5 and 5.3.5 we know that

$$\det (I(r, s)B) = -\det (B) = \det (I(r, s)) \det (B),$$
$$\det (BI(r, s)) = -\det (B) \det (I(r, s)).$$

We summarize this longish paragraph in

Theorem 5.6.2.   For any matrix B and any elementary matrix E,

$$\det (EB) = \det (E) \det (B)$$

and

$$\det (BE) = \det (B) \det (E).$$

By iteration we can extend Theorem 5.6.2 tremendously. Take, for example, two elementary matrices E_1 and E_2, and consider both E_1E_2B and E_1BE_2. What are the determinants of these matrices? Now $E_1E_2B = E_1(E_2B)$, so by Theorem 5.6.2,

$$\det (E_1E_2B) = \det (E_1) \det (E_2B) = \det (E_1) \det (E_2) \det (B).$$

Similarly,

$$\det (E_1BE_2) = \det (E_1) \det (B) \det (E_2).$$

We can continue this game and prove

Theorem 5.6.3. If $E_1, E_2, \ldots, E_m, E_{m+1}, \ldots, E_k$ are elementary matrices, then

1. $\det (E_1E_2 \cdots E_mE_{m+1} \cdots E_kB) =$

    $\det (E_1) \det (E_2) \cdots \det (E_m) \det (E_{m+1}) \cdots \det (E_k) \det (B)$;

2. $\det (E_1E_2 \cdots E_mBE_{m+1} \cdots E_k) =$

    $\det (E_1) \det (E_2) \cdots \det (E_m) \det (B) \det (E_{m+1}) \cdots \det (E_k)$.

Proof. Either go back to the paragraph before the statement of the theorem and say: "continuing in this way," or better still (and more formally), prove the result by induction. ■

A simple but important corollary to Theorem 5.6.3 is

Theorem 5.6.4. If $E_1, \ldots, E_k$ are elementary matrices, then

$$\det (E_1E_2 \cdots E_k) = \det (E_1) \det (E_2) \cdots \det (E_k).$$

Proof. In Part (1) of Theorem 5.6.3, put $B = I$. Then

$$\det (E_1E_2 \cdots E_kI) = \det (E_1) \det (E_2) \cdots \det (E_k) \det (I)$$
$$= \det (E_1) \det (E_2) \cdots \det (E_k). \quad ■$$

Theorem 5.6.4 will be of paramount importance to us. It is the key to proving the most basic property of the determinant. That property is that

$$\det (AB) = \det (A) \det (B)$$

for all $n \times n$ matrices A and B. This will be shown in the next section.

We close this section with an easy remark about the elementary matrices. The remark is checked by an easy multiplication.

Theorem 5.6.5. If E is an elementary matrix, and then E is invertible and E^{-1} is an elementary matrix.

Proof. We run through the three types of elementary matrices.

1. If $E = A(r, s; q)$, with $r \neq s$, then $E = I + qE_{rs}$. So

$$E(I - qE_{rs}) = (I + qE_{rs})(I - qE_{rs})$$
$$= I + qE_{rs} - qE_{rs} + q^2E_{rs}^2 = I$$

since $E_{rs}^2 = 0$ (because $r \neq s$). So $E^{-1} = I - qE_{rs} = A(r, s; -q)$.

2. If $E = M(r; q) = \begin{bmatrix} 1 & & & & 0 \\ & \ddots & & & \\ & & q & & \\ & & & \ddots & \\ 0 & & & & 1 \end{bmatrix}$ with $q \neq 0$ then

$$E^{-1} = \begin{bmatrix} 1 & & & & 0 \\ & \ddots & & & \\ & & q^{-1} & & \\ & & & \ddots & \\ 0 & & & & 1 \end{bmatrix} = M(r; q^{-1}).$$

3. If $E = I(r, s)$, with $r \neq s$, then interchanging columns r and s twice brings us back to the original. That is, $E^2 = I$. Hence $E^{-1} = E = I(r, s)$. ∎

PROBLEMS

NUMERICAL PROBLEMS

1. Compute the product of the 3×3 elementary matrices $A(2, 3; 1)$, $A(2, 3; 2)$, $A(2, 3; 3)$, $A(2, 3; 4)$, $A(2, 3; 5)$ and show that the order of these factors does not affect your answer.

2. Compute the product of the 3×3 elementary matrices $A(2, 3; 1)$, $A(2, 3; 2)$, $A(1, 3; 3)$ in every possible order.

3. Compute the product

$$B = I(3, 2)A(1, 2; 5)\begin{bmatrix} 5 & 0 & 0 \\ 0 & 3 & 0 \\ 0 & 0 & 2 \end{bmatrix}A(1, 2; -5)I(2, 3).$$

Then show that the determinant of B is 30, that is, the determinants

of $\begin{bmatrix} 5 & 0 & 0 \\ 0 & 3 & 0 \\ 0 & 0 & 2 \end{bmatrix}$ and B are equal.

4. Compute the inverses of the matrices

$$B, I(3, 2)A(1, 2; 5), \quad \begin{bmatrix} 5 & 0 & 0 \\ 0 & 3 & 0 \\ 0 & 0 & 2 \end{bmatrix}, \quad A(1, 2; -5), \quad I(2, 3)$$

in Problem 3 and show that the inverse of B equals

$$I(3, 2)A(1, 2; 5)\begin{bmatrix} 5 & 0 & 0 \\ 0 & 3 & 0 \\ 0 & 0 & 2 \end{bmatrix}^{-1} A(1, 2; -5)I(2, 3).$$

5. Compute $A(1, 3; 5)A(1, 3; 8)$.

6. Compute $A(1, 3; -5)A(1, 3; 8)A(1, 3; 5)$.

7. Compute $I(1, 3)A(1, 2; 3)$ and $A(1, 2; 3)I(1, 3)$.

8. Compute $I(1, 3)A(1, 2; 3)I(1, 3)$ and $A(1, 2; -3)I(1, 3)A(1, 4; 3)$.

9. Compute $I(1, 3)M(3; 3)I(1, 3)$ and $M(3; \frac{1}{3})I(1, 3)M(3; 3)$.

MORE THEORETICAL PROBLEMS

Easier Problems

10. Verify that multiplying B by $A(r, s; q)$ on the left adds q times row s to row r.

11. Verify that multiplying B by $A(r, s; q)$ on the right adds q times column r to column s.

12. Give a formula for the powers E^k of the elementary matrices E in the following cases:

(a) $E = I(a, b)$.

(b) $E = M(a; u)$.

13. Describe the entries of the matrix

$$J = A(a, b; u)I(a, b)A(a, b; -u) \quad \text{(with } a \text{ not equal to } b\text{)}$$

and verify that $J^2 = I$ (the identity matrix).

14. Show that transposes of elementary matrices are also elementary matrices by describing them explicitly in each case.

Middle-Level Problems

15. Compute $A(a, b; u)A(a, b; v)$ for a not equal to b for any v.

16. Compute $A(a, b; u)^k$ for a not equal to b for any k.

17. Using that $I(r, s) = I + E_{rs} + E_{sr} - E_{rr} - E_{ss}$, for $r \neq s$, prove, using the multiplication rules of the E_{uv}'s, that $I(r, s)^2 = I$.

5.7. THE DETERMINANT OF THE PRODUCT

Several results will appear in this section. One theorem stands out far and above all of the other results. We show that

$$\det (AB) = \det (A) \det (B)$$

for all matrices A and B in $M_n(F)$. With this result in hand we shall be able to do many things. For example, given a matrix A, using determinants we shall construct a polynomial, $P_A(x)$, whose roots are the characteristic roots of A. We'll be able to prove the important Cayley–Hamilton Theorem and a host of other nice theorems.

Our first objective is to show that an invertible matrix is the product of elementary matrices, that is, those introduced in Section 5.6.

Theorem 5.7.1. If B is an invertible matrix, then B is the product of elementary matrices.

Proof. The proof will be a little long, but it gives you an actual algorithm for factoring B as a product of elementary matrices.

Since B is invertible, it has no column of zeros, by Theorem 5.3.1. Hence the first column cannot consist only of zeros; in other words, $b_{r1} \neq 0$ for some s. If $r = 1$, that is, $b_{11} \neq 0$, fine. If $b_{11} = 0$, consider the matrix

$$B^{(1)} = I(r, 1)B.$$

In it the $(1, 1)$ entry is b_{r1} since we obtain $B^{(1)}$ from B by interchanging row r with the first row. We want to show that $B^{(1)}$ is the product of elementary matrices. If so then

$$I(r, 1)B = E_1 \cdots E_k,$$

where $E_1, \ldots, E_k$ are elementary matrices and

$$B = I(r, 1)E_1 \cdots E_k$$

would be a product of elementary matrices as well. We therefore proceed assuming that $b_{11} \neq 0$ [in other words, we are really carrying out the argument for $B^{(1)}$].

Consider

$$A\left(s, 1; -\frac{b_{s1}}{b_{11}}\right)B$$

for $s > 1$. By the property of $A(s, 1; -\frac{b_{s1}}{b_{11}})$, this matrix $A(s, 1; -\frac{b_{s1}}{b_{11}})B$

is obtained from B by adding $-\frac{b_{s1}}{b_{11}}$ times the first row of B to row s of

B. Consequently, the $(s, 1)$ entry of $A(s, 1; -\frac{b_{s1}}{b_{11}})B$ is

$$b_{s1} + \left(-\frac{b_{s1}}{b_{11}}\right)b_{11} = 0.$$

So if we do this for $s = 2, 3, \ldots, n$—that is, we act on B by the product of

$$A\left(n, 1; -\frac{b_{n1}}{b_{11}}\right), \ldots, A\left(2, 1; -\frac{b_{21}}{b_{11}}\right)$$

—we arrive at a matrix all of whose entries in the first column are 0 except for the $(1, 1)$ entry. If we let $E_s = A(s, 1; -\frac{b_{s1}}{b_{11}})$, this last statement becomes

$$C = E_n E_{n-1} \cdots E_2 B = \begin{bmatrix} b_{11} & b_{12} & b_{13} & \cdots & b_{1n} \\ 0 & c_{22} & c_{23} & \cdots & c_{2n} \\ 0 & c_{32} & c_{33} & \cdots & c_{3n} \\ \vdots & \vdots & \vdots & & \vdots \\ 0 & c_{n2} & c_{n3} & \cdots & c_{nn} \end{bmatrix}.$$

Now consider what happens to C when we multiply it on the right by the elementary matrix $F_s = A(1, s; -\frac{b_{1s}}{b_{11}})$. Since CF_s is obtained

from C by adding $-\frac{b_{1s}}{b_{11}}$ times the first column of C to column s of C, the $(1, s)$ entry of CF_s, for $s > 1$, is $b_{1s} + \left(-\dfrac{b_{1s}}{b_{11}}\right)b_{11} = 0$. So

$$D = E_n \cdots E_2 B F_2 \cdots F_n$$

$$= CF_2 \cdots F_n = \begin{bmatrix} b_{11} & 0 & 0 & \cdots & 0 \\ 0 & c_{22} & c_{23} & \cdots & c_{2n} \\ 0 & c_{32} & c_{33} & \cdots & c_{3n} \\ \vdots & \vdots & \vdots & & \vdots \\ 0 & c_{n2} & c_{n3} & \cdots & c_{nn} \end{bmatrix}.$$

Now D is invertible, since the E_s, F_r, and B are invertible; in fact,

$$D^{-1} = F_n^{-1} \cdots F_2^{-1} B^{-1} E_2^{-1} \cdots E_n^{-1}.$$

It follows from the invertibility of D that b_{11} is nonzero, that the $(n-1) \times (n-1)$ matrix

$$G = \begin{bmatrix} c_{22} & c_{23} & \cdots & c_{2n} \\ c_{32} & c_{33} & \cdots & c_{3n} \\ \vdots & \vdots & & \vdots \\ c_{n2} & c_{n3} & \cdots & c_{nn} \end{bmatrix}$$

is invertible and that the inverse of

$$D = \left[\begin{array}{c|c} b_{11} & 0 \cdots 0 \\ \hline 0 & \\ \vdots & G \\ 0 & \end{array}\right]$$

is the matrix

$$D^{-1} = \left[\begin{array}{c|c} b_{11}^{-1} & 0 \cdots 0 \\ \hline 0 & \\ \vdots & G^{-1} \\ 0 & \end{array}\right].$$

(Prove!) Moreover, we can express D as

$$D = M(1; b_{11}) \left[\begin{array}{c|c} 1 & 0 \cdots 0 \\ \hline 0 & \\ \vdots & G \\ 0 & \end{array}\right].$$

By induction, we know that G is a product $G = G_1 \cdots G_k$ of elementary $(n - 1) \times (n - 1)$ matrices $G_1, \ldots, G_k$. For each G_j of these, we define

$$\bar{G}_j = \begin{bmatrix} 1 & 0 \cdots 0 \\ \hline 0 & \\ \vdots & G_j \\ 0 & \end{bmatrix},$$

which is an $n \times n$ elementary matrix. We leave it as an exercise for the reader to show that $G = G_1 \cdots G_k$ implies that

$$\begin{bmatrix} 1 & 0 \cdots 0 \\ \hline 0 & \\ \vdots & G \\ 0 & \end{bmatrix} = \begin{bmatrix} 1 & 0 \cdots 0 \\ \hline 0 & \\ \vdots & G_1 \\ 0 & \end{bmatrix} \cdots \begin{bmatrix} 1 & 0 \cdots 0 \\ \hline 0 & \\ \vdots & G_k \\ 0 & \end{bmatrix}.$$

Multiplying by $M(1; b_{11})$, the left-hand side becomes D and the right-hand side becomes $M(1; b_{11})\bar{G}_1 \cdots \bar{G}_k$. So, this equation becomes

$$D = M(1; b_{11})\bar{G}_1 \cdots \bar{G}_k.$$

Since $D = E_n \cdots E_2 B F_2 \cdots F_n$, we can now express B as

$$\begin{aligned} B &= E_2^{-1} \cdots E_n^{-1} D F_n^{-1} \cdots F_2^{-1} \\ &= E_2^{-1} \cdots E_n^{-1} M(1; b_{11})\bar{G}_1 \cdots \bar{G}_k F_n^{-1} \cdots F_2^{-1}, \end{aligned}$$

which is a product of elementary matrices. This proves the theorem. ∎

If the proof just given strikes you as long and hard, try it out for 3×3 and 4×4 matrices, following the proof we just gave step by step. You will see that each step is fairly easy. What may confuse you in the general proof was the need for a certain amount of notation.

This last theorem gives us readily

Theorem 5.7.2.   If A and B are invertible $n \times n$ matrices, then $\det(AB) = \det(A)\det(B)$.

Proof.   By Theorem 5.7.1 we can write

$$A = E_1 \cdots E_k \quad \text{and} \quad B = F_1 \cdots F_m,$$

where the E_r and F_s are elementary matrices. So, by Theorem 5.6.4,

$$\begin{aligned} \det(AB) &= \det(E_1 \cdots E_k F_1 \cdots F_m) \\ &= \det(E_1) \cdots \det(E_k)\det(F_1) \cdots \det(F_m) \end{aligned}$$

and, again by Theorem 5.6.4,

$$\det (E_1) \cdots \det (E_k) = \det (E_1, \cdots E_k) = \det (A)$$

and

$$\det (F_1) \cdots \det (F_m) = \det (F_1 \cdots F_m) = \det (B).$$

Putting this all together gives us $\det (AB) = \det (A) \det (B)$.   ∎

From Theorem 5.7.2 we get the immediate consequence

Theorem 5.7.3. *A is invertible if and only if $\det (A) \neq 0$. Moreover, if A is invertible then $\det (A^{-1}) = \det (A)^{-1}$.*

Proof. If $\det (A) \neq 0$, we have that A is invertible; this is merely Theorem 5.4.2. On the other hand, if A is invertible, then from $AA^{-1} = I$, using Theorem 5.7.2, we get

$$1 = \det (I) = \det (AA^{-1}) = \det (A) \det (A^{-1}).$$

This give us at the same time that $\det (A)$ is nonzero and that $\det (A^{-1}) = \det (A)^{-1}$.   ∎

As a consequence of Theorem 5.7.3 we have that if A is not invertible, then $\det (A) = 0$. But if A is not invertible, then AB is not invertible for any B by Corollary 3.8.2; so $\det (AB) = 0 = \det (A) \det (B)$. Therefore, Theorem 5.7.2 can be sharpened to

Theorem 5.7.4. *If A and B are in $M_n(F)$, then*

$$\det (AB) = \det (A) \det (B).$$

Proof. If both A and B are invertible, this result is just Theorem 5.7.2. If one of A or B is not invertible then AB is not invertible, so $\det (AB) = 0 = \det (A) \det (B)$ since one of $\det (A)$ or $\det (B)$ is 0. ∎

An important corollary to these last few theorems is

Theorem 5.7.5. *If C is an invertible matrix, then $\det (C^{-1}AC) = \det (A)$ for all A.*

Proof. By Theorem 5.7.4, $\det (C^{-1}AC) = \det ((C^{-1}A)C) = \det (C(C^{-1}A)) = \det (CC^{-1}A) = \det (A)$.   ∎

With this, we close this section. We cannot exaggerate the importance of Theorems 5.7.4 and 5.7.5. As you will see, we shall use them to good effect in what is to come.

PROBLEMS

NUMERICAL PROBLEMS

1. Evaluate $\det\left(\begin{bmatrix} 1 & 0 & 0 \\ 1 & 2 & 0 \\ 1 & 2 & 3 \end{bmatrix}\begin{bmatrix} 1 & 4 & 4 \\ 0 & 2 & 1 \\ 0 & 0 & 3 \end{bmatrix}\right)$ by multiplying the matrices out and computing the determinant.

2. Do Problem 1 by making use of Theorem 5.7.4.

3. Express the following as products of elementary matrices:

   (a) $\begin{bmatrix} 1 & 2 \\ 3 & 4 \end{bmatrix}$.

   (b) $\begin{bmatrix} 0 & 2 & 3 \\ 1 & -1 & 6 \\ 5 & 3 & 1 \end{bmatrix}$.

   (c) $\begin{bmatrix} 1 & 0 & 0 & 0 \\ 2 & 3 & 0 & 0 \\ 4 & 5 & 6 & 0 \\ 7 & 8 & 9 & 10 \end{bmatrix}$.

   (*Hint:* Follow the steps of the proof of Theorem 5.7.1.)

MORE THEORETICAL PROBLEMS

4. Prove that for any matrix A, $\det(A^k) = (\det(A))^k$ for all $k \geq 1$.

5. Prove that $\det(ABA^{-1}B^{-1}) = 1$ if A and B are invertible $n \times n$ matrices.

6. If A is a skew-symmetric matrix in $M_n(\mathbb{R})$, prove that if n is odd, then $\det(A) = 0$.

5.8. THE CHARACTERISTIC POLYNOMIAL

Given a matrix A we shall now construct a polynomial $p_A(x)$ whose roots are the characteristic roots of A. We do this immediately in the

Definition. The *characteristic polynomial*, $p_A(x)$, of A is defined by $p_A(x) = \det (xI - A)$.

So, for example, if $A = \begin{bmatrix} 1 & 2 & 3 \\ 4 & 5 & 6 \\ 7 & 8 & 9 \end{bmatrix}$, then

$$p_A(x) = \det (xI - A) = \begin{vmatrix} x - 1 & -2 & -3 \\ -4 & x - 5 & -6 \\ -7 & -8 & x - 9 \end{vmatrix}.$$

$$= x^3 - 15x^2 - 18x. \text{ (Verify!)}$$

The key property that the characteristic polynomial enjoys is

Theorem 5.8.1. The complex number a is a characteristic root of A if and only if a is a root of $p_A(x)$.

Proof. Recall that a number a is a characteristic root of A in $M_n (\mathbb{C})$ if and only if $aI - A$ is not invertible or, equivalently, if $Av = av$ for some $v \neq 0$ in $\mathbb{C}^{(n)}$. But by Theorem 5.7.3, $aI - A$ is not invertible if and only if $\det (aI - A) = 0$. Since

$$0 = \det (aI - A) = p_A(a),$$

we have proven the theorem. ∎

Given $A, C \in M_n(F)$ and C invertible, then

$$C^{-1}(xI - A)C = C^{-1}(xI)C - C^{-1}AC = xI - C^{-1}AC.$$

Therefore, by Theorem 5.7.5,

$$\det (xI - A) = \det (C^{-1}(xI - A)C) = \det (xI - C^{-1}AC).$$

What this says is precisely

Theorem 5.8.2. If A, C are in $M_n(F)$ and C is invertible, then

$$p_A(x) = p_{C^{-1}AC}(x);$$

that is, A and $C^{-1}AC$ have the same characteristic polynomial. Thus A and $C^{-1}AC$ have the same characteristic roots.

Proof. By definition,

$$p_A(x) = \det (xI - A)$$

and

$$p_{C^{-1}AC}(x) = \det(xI - C^{-1}AC).$$

Since

$$\det(xI - A) = \det(C^{-1}(xI - A)C) = \det(xI - C^{-1}AC),$$

as we saw above, it follows that

$$p_A(x) = p_{C^{-1}AC}(x)$$

and A and $C^{-1}AC$ have the same characteristic polynomials. But then, by Theorem 5.8.1, A and $C^{-1}AC$ also have the same characteristic roots. ∎

Going back to the example of $A = \begin{bmatrix} 1 & 2 & 3 \\ 4 & 5 & 6 \\ 7 & 8 & 9 \end{bmatrix}$, for which we saw

that

$$p_{C^{-1}AC}(x) = x^3 - 15x^2 - 18x,$$

we have that the characteristic roots of A, namely the roots of $p_{C^{-1}AC}(x) = x^3 - 15x^2 - 18x$, are 0 and the roots of $x^2 - 15x - 18$. The roots of this quadratic polynomial are

$$\frac{15 + 3\sqrt{33}}{2}, \quad \frac{15 - 3\sqrt{33}}{2}.$$

So the characteristic roots of A are

$$0, \quad \frac{15 + 3\sqrt{33}}{2}, \quad \frac{15 - 3\sqrt{33}}{2}.$$

Note one further property of $p_A(x)$. If A' denotes the transpose of A, then $(xI - A)' = xI - A'$. But we know by Theorem 5.5.4 that the determinant of a matrix equals that of its transpose. This allows us to prove

Theorem 5.8.3 For any $A \in M_n(F)$, $p_A(x) = p_{A'}(x)$, that is, A and A' have the same characteristic polynomial.

Proof. By definition,

$$p_{A'}(x) = \det(xI - A') = \det((xI - A)')$$
$$= \det(xI - A) = p_A(x). \quad ∎$$

Given A in $M_n(F)$ and $p_A(x)$, we know that the roots of $p_A(x)$ all lie in $\mathbb{C}$ and

$$p_A(x) = (x - r_1) \cdots (x - r_n)$$

with $r_1, \ldots, r_n \in \mathbb{C}$, where the $r_1, \ldots, r_n$ are the roots of $p_A(x)$, and the characteristic roots of A (the roots need not be distinct).

We ask: What is $p_A(0)$? Setting $x = 0$ in the equation above, we get

$$p_A(0) = (0 - r_1) \cdots (0 - r_n) = (-1)^n r_1 \cdots r_n.$$

On the other hand, $p_A(x) = \det(xI - A)$, hence

$$p_A(0) = \det(-A) = \det((-I)A) = \det(-I)\det(A)$$
$$= (-1)^n \det(A).$$

Comparing these two evaluations of $p_A(0)$, we come up with

$$(-1)^n r_1 \cdots r_n = (-1)^n \det(A).$$

Thus we get the

Theorem 5.8.4.   If $A \in M_n(F)$, then $\det(A)$ is the product of the characteristic roots of A.

PROBLEMS

NUMERICAL PROBLEMS

1. Compute $p_A(x)$ for $A = 0$ and for $A = I$.

2. Compute $p_A(x)$ for A a lower triangular matrix.

3. Compute $p_A(x)$ for

(a) $A = \begin{bmatrix} 1 & 0 & 1 \\ 2 & 0 & 2 \\ 5 & 6 & -1 \end{bmatrix}.$

(b) $A = \begin{bmatrix} 1 & 0 & 0 \\ 0 & a & b \\ 0 & c & d \end{bmatrix}.$

(c) $A = \begin{bmatrix} 0 & 1 & 0 \\ 1 & 0 & 0 \\ 0 & 0 & 1 \end{bmatrix}$.

(d) $A = \begin{bmatrix} 1 & 2 & 0 & 0 \\ 3 & 4 & 0 & 0 \\ 0 & 0 & 5 & 6 \\ 0 & 0 & 7 & 8 \end{bmatrix}$.

4. Find the characteristic roots for the matrices in Problem 3, and show directly that det (A) is the product of the roots of the characteristic polynomial of A.

MORE THEORETICAL PROBLEMS

Middle-Level Problems

5. If at least one of A or B is invertible, prove that $p_{AB}(x) = p_{BA}(x)$.

6. Find $p_A(x)$ for $A = \begin{bmatrix} 0 & 0 & 0 & -c_0 \\ 1 & 0 & 0 & -c_1 \\ 0 & 1 & 0 & -c_2 \\ 0 & 0 & 1 & -c_3 \end{bmatrix}$.

7. If $A \in M_n(F)$ is nilpotent $(A^k = 0$ for some positive integer $k)$, show that $p_A(x) = x^n$.

8. If $A = I + N$, where N is nilpotent, find $p_A(x)$.

9. If A^* is the Hermitian adjoint of A, express $p_{A^*}(x)$ in terms of $p_A(x)$.

10. If A is invertible express $p_{A^{-1}}(x)$ in terms of $p_A(x)$ if

$$p_A(x) = x^n + a_1 x^{n-1} + \cdots + a_n.$$

11. If A and B in $M_n(F)$ are upper triangular, prove that $p_{AB}(x) = p_{BA}(x)$.

NOTE: It is a fact that for all A, B, in $M_n(F)$, $p_{AB}(x) = p_{BA}(x)$. However, if neither A nor B is invertible, it is difficult to prove. So although we alert you to this fact, we do not prove it for you.

12. Show that if an invertible $n \times n$ matrix D is of the form

$$D = \begin{bmatrix} b_{11} & 0 \cdots 0 \\ \hline 0 & \\ \vdots & G \\ 0 & \end{bmatrix},$$

then b_{11} is nonzero, G is invertible and the inverse of D is

$$D^{-1} = \begin{bmatrix} b_{11}^{-1} & 0 \cdots 0 \\ \hline 0 & \\ \vdots & G^{-1} \\ 0 & \end{bmatrix}.$$

13. For any elementary $(n-1) \times (n-1)$ matrix G, we define $\overline{G}$ to be the matrix

$$\overline{G} = \begin{bmatrix} 1 & 0 \cdots 0 \\ \hline 0 & \\ \vdots & G \\ 0 & \end{bmatrix}$$

obtained by adorning G with one more row and column as displayed. Show that $\overline{G}$ is an elementary matrix. Then prove the following product rule:

$$\begin{bmatrix} 1 & 0 \cdots 0 \\ \hline 0 & \\ \vdots & G \\ 0 & \end{bmatrix} \begin{bmatrix} 1 & 0 \cdots 0 \\ \hline 0 & \\ \vdots & H \\ 0 & \end{bmatrix} = \begin{bmatrix} 1 & 0 \cdots 0 \\ \hline 0 & \\ \vdots & GH \\ 0 & \end{bmatrix}.$$

5.9. THE CAYLEY–HAMILTON THEOREM

We saw earlier that if A is in $M_n(F)$ then A *satisfies* a polynomial $p(x)$ of degree at most n^2 whose coefficients are in F, that is,

$$p(A) = 0.$$

What we shall show here is that A *actually satisfies a specific polynomial of degree n* having coefficients in F. In fact, this polynomial will turn out to be the characteristic polynomial of A. This is a famous theorem in the subject and is known as the *Cayley–Hamilton Theorem*.

Before proving the Cayley–Hamilton Theorem in its full generality we prove it for the special case in which A is an upper triangular matrix. From there, using a result on triangularization that we proved earlier (Theorem 4.7.3), we shall be able to establish the full Cayley–Hamilton Theorem.

Theorem 5.9.1   If A is an upper triangular matrix, then A satisfies $p_A(x)$.

Proof.   If

$$A = \begin{bmatrix} a_{11} & a_{12} & a_{13} & \cdots & a_{1n} \\ 0 & a_{22} & a_{23} & \cdots & a_{2n} \\ 0 & 0 & a_{33} & \cdots & a_{3n} \\ \vdots & \vdots & & \ddots & \vdots \\ 0 & 0 & \cdots & \cdots 0 & a_{nn} \end{bmatrix},$$

then

$$xI - A = \begin{bmatrix} x - a_{11} & -a_{12} & -a_{13} & \cdots & -a_{1n} \\ 0 & x - a_{22} & -a_{23} & \cdots & -a_{2n} \\ 0 & 0 & x - a_{33} & \cdots & -a_{3n} \\ \vdots & \vdots & & \ddots & \vdots \\ 0 & 0 & \cdots & & 0 \; x - a_{nn} \end{bmatrix},$$

so

$$p_A(x) = \det(xI - A) = (x - a_{11}) \cdots (x - a_{nn}).$$

If you look back at Theorem 4.7.4, we proved there that A satisfies

$$(x - a_{11}) \cdots (x - a_{nn}).$$

So we get that A satisfies $p_A(x) = (x - a_{11}) \cdots (x - a_{nn}).$ ∎

In Theorem 4.7.3 it was shown that if $A \in M_n \; (\mathbb{C})$, then for some invertible matrix, C, $C^{-1}AC$ is upper triangular. Hence $C^{-1}AC$ satisfies $p_{C^{-1}AC}(x)$. By Theorem 5.8.2, $P_{C^{-1}AC}(x) = p_A(x)$. So $C^{-1}AC$ satisfies $p_A(x)$. If

$$p_A(x) = x^n + a_1 x^{n-1} + \cdots + a_n,$$

this means that

$$0 = p_A(C^{-1}AC) = (C^{-1}AC)^n + a_1(C^{-1}AC)^{n-1} + \cdots + a_n I.$$

However,

$$(C^{-1}AC)^k = C^{-1}A^k C$$

for all $k \geq 1$. Using this, it follows from the preceding equation that

$$0 = C^{-1}A^n C + a_1 C^{-1}A^{n-1} C + \cdots + a_n I$$
$$= C^{-1}(A^n + a_1 A^{n-1} + \cdots + a_n I)C.$$

Multiplying from the left by C and from the right by C^{-1}, we then obtain

$$0 = A^n + a_1 A^{n-1} + \cdots + a_n I = p_A(A),$$

which proves

Theorem 5.9.2 (Cayley–Hamilton). Every $A \in M_n(F)$ satisfies its characteristic polynomial.

PROBLEMS

NUMERICAL PROBLEMS

1. Prove that the following matrices A satisfy $p_A(x)$ by directly computing $p_A(A)$.

(a) $A = \begin{bmatrix} 1 & 2 & 3 \\ 4 & 5 & 6 \\ 7 & 8 & 9 \end{bmatrix}$.

(b) $A = \begin{bmatrix} 1 & 0 & 0 & 0 \\ 0 & 1 & 2 & 3 \\ 0 & 4 & 5 & 6 \\ 0 & 7 & 8 & 9 \end{bmatrix}$.

(c) $A = \begin{bmatrix} 1 & 1 & 0 & 0 \\ 2 & 3 & 0 & 0 \\ 0 & 0 & 4 & 5 \\ 0 & 0 & 6 & 7 \end{bmatrix}$.

2. If $C = \begin{bmatrix} 1 & 0 & 0 \\ 0 & 2 & 0 \\ 0 & 1 & 3 \end{bmatrix}$, find C^{-1} and by a direct computation show

that $P_{C^{-1}AC}(x) = p_A(x)$ for $A = \begin{bmatrix} 1 & 2 & 3 \\ 4 & 5 & 6 \\ 7 & 8 & 9 \end{bmatrix}$.

MORE THEORETICAL PROBLEMS

Harder Problems

3. If $A = \begin{bmatrix} a & 0 \cdots 0 \\ 0 & \\ \vdots & B \\ 0 & \end{bmatrix}$, where B is an $(n-1) \times (n-1)$ matrix,

   prove that $p_A(x) = (x - a)p_B(x)$.

4. In Problem 3 prove that $p_A(A) = 0$ if $p_B(B) = 0$.

5. If $A = \begin{bmatrix} a & b & 0 & \cdots & 0 \\ c & d & 0 & \cdots & 0 \\ 0 & 0 & & & \\ \vdots & \vdots & & B & \\ 0 & 0 & & & \end{bmatrix}$, where B is an $(n-2) \times (n-2)$

   matrix, prove that $p_A(x) = p_C(x)p_B(x)$, where $C = \begin{bmatrix} a & b \\ c & d \end{bmatrix}$.

6

RECTANGULAR MATRICES. MORE ON SYSTEMS OF LINEAR EQUATIONS

6.1. RECTANGULAR MATRICES

An $m \times n$ *matrix* is an array $A = (a_{rs})$ of scalars having m rows and n columns. As for square matrices, a_{rs} denotes the (r, s) *entry* of A, that is, the scalar located in row r and column s. A *rectangular matrix* is just an $m \times n$ matrix, where m and n are positive integers. Thus every square matrix is a rectangular matrix. Row vectors are $1 \times n$ matrices, column vectors are $m \times 1$ matrices, and scalars are 1×1 matrices so that all these are rectangular matrices as well.

We can generalize our operations on square matrices to operations on rectangular matrices. Whenever operations were defined before (e.g., product of an $n \times n$ matrix and an $n \times 1$ column vector, product of two 1×1 scalars, sum of two $n \times 1$ column vectors), you should verify that they are special cases of the operations that we are about to define. We sometimes refer to $m \times n$ as the *shape* of an $m \times n$ matrix.

So we will define addition for the matrices of the same shape and multiplication UC for a matrix U of shape $b \times m$ and a matrix C of shape $m \times n$.

Definition. The *product* of the $b \times m$ matrix (u_{rs}) and the $m \times n$ matrix (v_{rs}) is the $b \times n$ matrix (w_{rs}) whose (r, s) entry w_{rs} is given by the expression $w_{rs} = \sum_{t=1}^{m} u_{rt}v_{ts}$.

Definition. Let A and B be the $m \times n$ matrices (a_{rs}), (b_{rs}) and let c be any scalar. Then $A + B$ is the $m \times n$ matrix $(a_{rs} + b_{rs})$, cA is the matrix (ca_{rs}), $-A$ is the matrix $(-a_{rs})$, $A - B$ is the matrix $A + (-B) = (a_{rs} - b_{rs})$, 0 denotes the $m \times n$ matrix all of whose entries are 0, I or I_m denote the $m \times m$ identity matrix.

Whenever a concept for square matrices makes just as good sense in the case of rectangular matrices, we use it freely without formally introducing it. So, for instance, the transpose of the matrix $\begin{bmatrix} 1 & 2 \\ 3 & 5 \\ 6 & 7 \end{bmatrix}$ is

the matrix $\begin{bmatrix} 1 & 3 & 6 \\ 2 & 5 & 7 \end{bmatrix}$.

Some properties of operations on square matrices carry over routinely to rectangular matrices, as long as their shapes are compatible; that is,

> *you can always multiply a $b \times m$ matrix by an $m \times n$ matrix for any n, and you can only add two matrices if they both have the same shape.*

We list some of these properties in the following theorem.

Theorem 6.1.1. Rectangular $m \times m$ matrices satisfy the following properties when their shapes are compatible:

1. $(A + B) + C = A + (B + C)$;
2. $\quad\quad A + 0 = A$;
3. $\quad A + (-A) = 0$;
4. $\quad\quad A + B = B + A$;
5. $\quad\quad (AB)C = A(BC)$;
6. $\quad\quad I_m A = A$ and $AI_n = A$;

7. $(A + B)C = AC + BC$;

8. $C(A + B) = CA + CB$;

9. $(a + b)C = aC + bC$;

10. $(ab)C = a(bC)$;

11. $1C = C$.

PROBLEMS

NUMERICAL PROBLEMS

1. Compute the product $\begin{bmatrix} 1 & 2 & 3 & 2 & 1 \\ 2 & 3 & 4 & 3 & 2 \\ 3 & 4 & 5 & 4 & 3 \end{bmatrix} \begin{bmatrix} 1 & 1 \\ 2 & 1 \\ 3 & 2 \\ 4 & 1 \\ 5 & 1 \end{bmatrix}$.

2. Find matrices A, B, C, D such that the C is 6×6, the product $ABCD$ makes sense, and $ABCD = \begin{bmatrix} 1 & 2 & 3 \\ 3 & 2 & 1 \end{bmatrix}$.

3. Compute the determinants of $\begin{bmatrix} 1 & 1 & 1 \\ 3 & 2 & 1 \end{bmatrix} \begin{bmatrix} 1 & 3 \\ 1 & 2 \\ 1 & 1 \end{bmatrix}$ and

$\begin{bmatrix} 1 & 3 \\ 1 & 2 \\ 1 & 1 \end{bmatrix} \begin{bmatrix} 1 & 1 & 1 \\ 3 & 2 & 1 \end{bmatrix}$.

MORE THEORETICAL PROBLEMS

Easier Problems

4. Give an example of an $m \times n$ matrix A and $n \times m$ matrix B such that $AB = 0$ and the rows of A and the columns of B are linearly independent.

Middle-Level Problems

5. Show that if $m < n$ and A is an $m \times n$ matrix, the determinant of $A'A$ is 0.

Harder Problems

6. Show that the determinant of $A'A$ is nonzero for any real $m \times n$ matrix A whose columns are linearly independent.

6.2. LINEAR TRANSFORMATIONS FROM $F^{(n)}$ TO $F^{(m)}$

In Chapter 2 we talked about systems of linear equations and how to solve them. Now we return to this topic and discuss it from a somewhat different perspective.

As in Chapter 2, we represent a system

$$a_{11}x_1 + \cdots + a_{1n}x_n = y_1$$
$$\vdots \qquad\qquad \vdots \quad \vdots$$
$$a_{m1}x_1 + \cdots + a_{mn}x_n = y_m$$

of m linear equations in n variables by the matrix equation $Ax = y$,

where A is the $m \times n$ matrix $\begin{bmatrix} a_{11} & \cdots & a_{1n} \\ \vdots & & \vdots \\ a_{m1} & \cdots & a_{mn} \end{bmatrix}$ and x and y are the

column vectors $x = \begin{bmatrix} x_1 \\ \vdots \\ x_n \end{bmatrix}$ and $y = \begin{bmatrix} y_1 \\ \vdots \\ y_m \end{bmatrix}$. Now, as in the case $m = n$

in Chapter 3, we regard an $m \times n$ matrix A as a mapping from $F^{(n)}$ to $F^{(m)}$ sending $x \in F^{(n)}$ to $Ax \in F^{(m)}$, where Ax denotes the product of A and x and the system of equations above describes the mapping sending x to Ax in terms of coordinates. From the properties of matrix addition, multiplication, and scalar multiplication given in Theorem 6.1.1, *these mappings Ax are linear in the sense of*

Definition.   A *linear mapping* from $F^{(n)}$ to $F^{(m)}$ is a function f from $F^{(n)}$ to $F^{(m)}$ such that

$$f(u + v) = f(u) + f(v)$$

and

$$f(cv) = cf(v)$$

for all $u, v \in F^{(n)}$ and $c \in F$.

The properties $f(u + v) = f(u) + f(v)$ and $f(cv) = cf(v)$ are the *linearity properties* of f.

EXAMPLE

The mapping $f\left(\begin{bmatrix} a \\ b \\ d \end{bmatrix}\right) = \begin{bmatrix} 3a + d \\ b - a \end{bmatrix}$ is a linear mapping.

We can see this either by direct verification of the linearity properties for f or, as we now do, by finding a 2×3 matrix A such that

$f\left(\begin{bmatrix} a \\ b \\ d \end{bmatrix}\right) = A\begin{bmatrix} a \\ b \\ d \end{bmatrix}$. If we let $A = \begin{bmatrix} 3 & 0 & 1 \\ -1 & 1 & 0 \end{bmatrix}$, then

$$A\begin{bmatrix} a \\ b \\ d \end{bmatrix} = \begin{bmatrix} 3 & 0 & 1 \\ -1 & 1 & 0 \end{bmatrix}\begin{bmatrix} a \\ b \\ d \end{bmatrix} = \begin{bmatrix} 3a + d \\ b - a \end{bmatrix} = f\left(\begin{bmatrix} a \\ b \\ d \end{bmatrix}\right).$$

So since Ax is a linear mapping and $Ax = f(x)$ for all x, f is a linear mapping.

We have the following theorem, whose proof is exactly the same as the proof of Theorem 3.2.2 for the $n \times n$ case.

Theorem 6.2.1.   The linear mappings from $F^{(n)}$ to $F^{(m)}$ are just the $m \times n$ matrices.

In Chapter 2, given an $m \times n$ matrix A and vector $y \in F^{(n)}$, we asked: What is the set of all solutions to the equation $Ax = y$? From the point of view that A is a linear transformation, we are asking: What is the set of elements $x \in F^{(n)}$ that are mapped by A into y? Of course, this set of elements is nonempty if and only if y is the image Ax of some $x \in F^{(n)}$.

When $y = 0$, 0 is the image $A0$ of 0, so the set of elements $x \in F^{(n)}$ that are mapped into 0 is always nonempty. In fact, this set

$N = \{x \in F^{(n)}| \ Ax = 0\}$ is a subspace of $F^{(n)}$ (prove!) which we call the *nullspace* of A. This is of great importance since, as we are about to see, this implies that the set of solutions to $Ax = y$ is either empty or a *translation* $v + N = \{v + u \mid u \in N\}$ of a subspace N of $F^{(n)}$.

Of course, the equation $Ax = y$ may have no solution x or it may have exactly one. Otherwise, there are at least two different solutions u and v, in which case the homogeneous equation $Ax = 0$ has infinitely many solutions $c(u - v)$, c being any scalar. Why? Since A is a linear transformation, we can use its linearity properties to compute

$$A(c(u - v)) = c(A(u - v))$$
$$= c(Au - Av) = c(y - y) = c(0) = 0.$$

So the equation $Ax = 0$ has infinitely many different solutions, in fact, one for each scalar value c. But then it follows that the equation $Ax = y$ also has infinitely many solutions $v + c(u - v)$, c being any scalar. Why? Again, simply compute

$$A(v + c(u - v)) = Av + A(c(u - v))$$
$$= Av + 0 = y + 0 = y.$$

These simple observations prove

Theorem 6.2.2.   For any $m \times n$ matrix A and any $y \in F^{(n)}$, one of the following is true:

1. $Ax = y$ has no solution x;
2. $Ax = y$ has exactly one solution x;
3. $Ax = y$ has infinitely many solutions x.

More important than this theorem itself is the realization we get from its proof that any two solutions u and v to the equation $Ax = y$ give us a solution $u - v$ to the homogeneous equation $Ax = 0$:

$$A(u - v) = 0.$$

Why is this important? Letting N be the nullspace

$$N = \{w \in F^{(n)} \mid Aw = 0\},$$

we have $u - v \in N$ and $u = v + w$ with $w \in N$ for any two solutions u, v of $Ax = y$. Conversely, if $Av = y$ and $u = v + w$ with $w \in N$, then $Au = y$. (Prove!) So, we get

Theorem 6.2.3.   Let A be an $m \times n$ matrix and let $y \in F^{(m)}$. If $Av = y$, then the set of solutions x to the equation $Ax = y$ is

$$v + N = \{v + w \mid w \in N\},$$

where N is the nullspace of A.

By this theorem, we can split up the job of finding the set $v + N$ of all the solutions to $Ax = 0$ to finding one v such that $Av = y$ and finding the nullspace N of A. So, we are now confronted with the question: What is the nullspace N of A, and how do we get one solution $x = v$ to the equation $Ax = y$, if it exists?

In Chapter 2, given an $m \times n$ matrix A and $y \in F^{(m)}$, we gave methods to determine whether the matrix equation $Ax = y$ has a solution x and, if so, to find the solutions:

1. *Form the augmented matrix $[A, y]$ and reduce it to a row equivalent echelon matrix $[B, z]$.*
2. *Then $Ax = y$ has a solution if and only if $z_r = 0$ whenever row r of B is 0.*
3. *All solutions x can be found by solving $Bx = z$ by back substitution.*

Now, we can get one solution v to $Bx = z$ and express the solutions to $Ax = y$ as $v + w$, where w is in the nullspace of B, so we have

Method to solve $Ax = y$ for x

1. *Form the augmented matrix $[A, y]$ and reduce it to a row equivalent echelon matrix $[B, z]$.*
2. *If a solution exists, get one solution-v to $Bv = z$ by setting all independent variables equal to zero and finding the corresponding solution v by back substitution.*
3. *Find the nullspace N of B, that is, the set of all solutions w to $Bw = 0$, by back substitution.*
4. *If a solution exists, the set of solutions x to $Ax = y$ is then $v + N = \{v + w \mid Bw = 0\}$.*

EXAMPLE

Let's return to the problem in Section 2.2 of solving the matrix equation

$$\begin{bmatrix} 2 & 3 & 1 \\ -2 & 1 & 3 \end{bmatrix} \begin{bmatrix} p \\ q \\ r \end{bmatrix} = \begin{bmatrix} 6000 \\ 2000 \end{bmatrix}.$$

We reduce $\begin{bmatrix} 2 & 3 & 1 & 6000 \\ -2 & 1 & 3 & 2000 \end{bmatrix}$ successively to

$$\begin{bmatrix} 2 & 3 & 1 & 6000 \\ 0 & 4 & 4 & 8000 \end{bmatrix}, \quad \begin{bmatrix} 2 & 3 & 1 & 6000 \\ 0 & 1 & 1 & 2000 \end{bmatrix}, \quad \begin{bmatrix} 1 & \frac{3}{2} & \frac{1}{2} & 3000 \\ 0 & 1 & 1 & 2000 \end{bmatrix}.$$

To get one solution v of the equivalent matrix equation

$$\begin{bmatrix} 1 & \frac{3}{2} & \frac{1}{2} \\ 0 & 1 & 1 \end{bmatrix} \begin{bmatrix} p \\ q \\ r \end{bmatrix} = \begin{bmatrix} 3000 \\ 2000 \end{bmatrix},$$

we set the independent variable r to the value $r = 0$ and solve for

$q = 2000$ and $p = 0$, getting $v = \begin{bmatrix} 0 \\ 2000 \\ 0 \end{bmatrix}$. To get the elements

of the nullspace of $\begin{bmatrix} 1 & \frac{3}{2} & \frac{1}{2} \\ 0 & 1 & 1 \end{bmatrix}$, we solve

$$\begin{bmatrix} 1 & \frac{3}{2} & \frac{1}{2} \\ 0 & 1 & 1 \end{bmatrix} \begin{bmatrix} p \\ q \\ r \end{bmatrix} = 0$$

by back substitution, finding $q = -r$ and $p = r$ and then getting

$w = \begin{bmatrix} r \\ -r \\ r \end{bmatrix}$ as the general element. So the general solution to the

equation $\begin{bmatrix} 2 & 3 & 1 \\ -2 & 1 & 3 \end{bmatrix} \begin{bmatrix} p \\ q \end{bmatrix} = \begin{bmatrix} 6000 \\ 2000 \end{bmatrix}$ is

$$v + w = \begin{bmatrix} 0 \\ 2000 \\ 0 \end{bmatrix} + \begin{bmatrix} r \\ -r \\ r \end{bmatrix} = \begin{bmatrix} r \\ 2000 - r \\ r \end{bmatrix},$$

which agrees with the solution which we got in Section 2.2.

The principal players here are the subspace

$$AF^{(n)} = \{Ax \mid x \in F^{(n)}\}$$

of images Ax $(x \in F^{(n)})$ of A and the nullspace

$$N = \{x \in F^{(n)} \mid Ax = 0\}$$

of A. Since the image Ax of x under A is also the linear combination $x_1 v_1 + \cdots + x_n v_n$ of the columns $v_1, \ldots, v_n$ of A, $AF^{(n)}$ is also the span of the columns of A, that is, the *column space* of A. In terms of the nullspace and column space of A, the answer to the question

"What is the set of all solutions of the equation $Ax = y$?"

is

1. $Ax = y$ has a solution x if and only if y is in the column space $AF^{(n)}$ of A.

2. For $y \in AF^{(n)}$, $Av = y$ implies that the set of solutions to $Ax = y$ is $v + N$ where N is the nullspace of A, and conversely.

We discuss the nullspace and column space of A in the next section, where an important relationship between them is determined.

PROBLEMS

NUMERICAL PROBLEMS

1. Show that the mapping $f\left(\begin{bmatrix} q \\ w \\ u \end{bmatrix}\right) = \begin{bmatrix} 2u - q - w \\ w - 2q - u \end{bmatrix}$ is a linear

    mapping from $F^{(3)}$ to $F^{(2)}$.

2. Describe the mapping f in Problem 1 as a 2×3 matrix.

3. Show that $\begin{bmatrix} 1 & 2 & 3 & 4 \\ 1 & 3 & 4 & 7 \\ 1 & 2 & 6 & 12 \end{bmatrix} \begin{bmatrix} a \\ s \\ d \\ c \end{bmatrix} = \begin{bmatrix} 0 \\ 0 \\ 3 \end{bmatrix}$ has infinitely many solu-

    tions $\begin{bmatrix} a \\ s \\ d \\ c \end{bmatrix}$.

4. Find a basis for the nullspace N of $\begin{bmatrix} 1 & 2 & 3 & 4 \\ 1 & 3 & 3 & 3 \\ 1 & 2 & 6 & 12 \end{bmatrix}$ and express

the set of solutions to $\begin{bmatrix} 1 & 2 & 3 & 4 \\ 1 & 3 & 3 & 3 \\ 1 & 2 & 6 & 12 \end{bmatrix} \begin{bmatrix} a \\ b \\ c \\ d \end{bmatrix} = \begin{bmatrix} 3 \\ 3 \\ 8 \end{bmatrix}$ in the form

$\begin{bmatrix} a \\ b \\ c \\ d \end{bmatrix} + N$ for some specific $\begin{bmatrix} a \\ b \\ c \\ d \end{bmatrix} \in F^{(4)}$.

5. If $F = \mathbb{R}$, show that there is a unique solution $\begin{bmatrix} a \\ b \\ c \\ d \end{bmatrix}$ of shortest

length to the equation in Problem 4.

MORE THEORETICAL PROBLEMS

Easier Problems

6. Let N be any subspace of $\mathbb{R}^{(n)}$. Show that N is the nullspace of some $n \times n$ matrix A.

Middle-Level Problems

7. let N be any subspace of $\mathbb{R}^{(n)}$ and let v be any vector in $\mathbb{R}^{(n)}$. For any $m \geq n - \dim(N)$, show that there is an $m \times n$ matrix A such that $v + N$ is the set of solutions x to $Ax = y$ for some $y \in F^{(m)}$.

8. Show that if $A \in M_n(\mathbb{R})$ and Ax is the projection of y on the column space of A, then $A'Ax = A'y$.

6.3. THE NULLSPACE AND COLUMN SPACE OF AN $m \times n$ MATRIX

In Section 6.2 we found in discussing the set of solutions to an equation $Ax = y$ that the nullspace and column space of A played central roles. Here, you will recall, the column space of an $m \times n$ matrix A is the set

$AF^{(n)} = \{Ax \mid x \in F^{(n)}\}$ of linear combinations Ax of the columns of A and the nullspace of A, denoted Nullspace (A), is $\{x \in F^{(n)} \mid Ax = 0\}$. The dimension of the column space of A is called the *rank* of A, denoted by $r(A)$, whereas the dimension of Nullspace (A) is called the *nullity* of A, denoted by $n(A)$.

Since $Ax = 0$ and $Bx = 0$ have the same solutions if A and B are row equivalent $m \times n$ matrices, by Theorem 2.3.1, we have

Theorem 6.3.1. The nullspaces of any two row equivalent $m \times n$ matrices are equal.

From this theorem it follows that the nullity of two row equivalent matrices are the same. Are their ranks the same too? If $w_1, \ldots, w_r$ is a basis for the column space of an $m \times n$ matrix A and if U is an invertible $m \times m$ matrix, then $Uw_1, \ldots, Uw_r$ is a basis for the column space of UA. Why? Surely, they span the column space of UA. And if

$$c_1 Uw_1 + \cdots + c_r Uw_r = 0,$$

then $U(c_1 w_1 + \cdots c_r w_r) = 0$. Since U is invertible, it follows that

$$c_1 w_1 + \cdots + c_r w_r = 0,$$

so the c_s are 0. It follows from all this that $Uw_1, \ldots, Uw_r$ is a basis for the column space of UA. Consequently, A and UA have the same rank. This proves

Theorem 6.3.2. If A and B are row equivalent $m \times n$ matrices, then A and B have the same rank and nullity.

Using this theorem, we now proceed to show that the rank plus the nullity of any $m \times n$ matrix A is n. In Section 4.4 we showed this to be true for square matrices. For $m \times n$ matrices we really do not need to prove very much, since this result is simply an enlightened version of Theorem 2.4.1. Why is it an enlightened version of Theorem 2.4.1? The rank and nullity of A appear in disguise in Theorem 2.4.1 as the values $n - m'$ (the nullity of A) and m' (the rank of A), which are defined by taking m' to be the number of nonzero rows of any echelon matrix B which is row equivalent to A. What do we really need to prove? We just need to prove that m' is the rank of A (see Theorem 6.3.4) and $n - m'$ is the nullity of A (an easy exercise).

To keep our discussion self-contained, however, we do not use Theorem 2.4.1 in showing that the rank plus nullity is n. And we don't use the swift approach proof using inner products given in Chapter 4 in the case $m = n$, since it does not easily generalize. Instead, our approach is to replace the matrix A by a matrix row equivalent to it, which is a reduced echelon matrix in the sense of the following definition.

Definition.   An $m \times n$ matrix A is a *reduced echelon matrix* if A is an echelon matrix such that the leading entry in any nonzero row is the only nonzero entry in its column.

Theorem 6.3.3.   Any $m \times n$ matrix is row equivalent to a reduced echelon matrix.

Proof.   We know from Chapter 2 that any $m \times n$ matrix A is row equivalent to an echelon matrix. But any echelon matrix U can be reduced to a reduced echelon matrix using elementary row operations to get rid of nonzero entries above any given leading entry. How? If the leading entry of column s is the (r, s) entry, we simply apply the operation Add $(t, r; -u_{ts})$ for $t = 1, 2, \ldots, r - 1$. Doing this for column s for $s = 1, 2, \ldots, n$ gives us the reduction.   ■

The key to finding the rank of A is

Theorem 6.3.4.   The rank $r(A)$ of an $m \times n$ matrix A is the number of nonzero rows in any echelon matrix B row equivalent to A.

Proof.   The rank of A and B are equal, by Theorem 6.3.2. If we further reduce B to a reduced echelon matrix, using elementary row operations, the rank is not changed and the number of nonzero rows is not changed. So we may as well assume that B itself is a *reduced* echelon matrix. Then each nonzero row gives us a leading entry 1, which in turn gives us the corresponding column in the $m \times m$ identity matrix. So if there are r nonzero rows, they correspond to the first r columns of the $m \times m$ identity matrix. In turn, these r columns of the $m \times m$ identity matrix form a basis for the column space of B, since all columns have all entries equal to 0 in row k for all $k > r$. This means that $r(A) = r$.   ■

Let's look at this in a specific example.

EXAMPLE

The rank of the echelon matrix $A = \begin{bmatrix} 1 & 3 & 0 & 0 \\ 0 & 0 & 1 & 0 \\ 0 & 0 & 0 & 1 \\ 0 & 0 & 0 & 0 \end{bmatrix}$ is 3 since

A has three nonzero rows. The three nonzero rows correspond to the columns 1, 3, 4, which are the first three columns of the 4×4 identity matrix. On the other hand, the rank of the nonechelon

matrix $B = \begin{bmatrix} 1 & 3 & 0 & 0 \\ 0 & 0 & 1 & 1 \\ 0 & 0 & 1 & 1 \\ 0 & 0 & 0 & 0 \end{bmatrix}$ is 2 even though it also has three non-

zero rows.

The *row space* of an $m \times n$ matrix is the span of its rows over F. Since the row space does not change when an elementary row operation is performed (Prove!), the row spaces of any two row equivalent $m \times n$ matrices are the same. This is used in the proof of

Corollary 6.3.5.   The row and column spaces of an $m \times n$ matrix A have the same dimension, namely $r(A)$.

Proof.   The dimension of the row space of any $m \times n$ matrix A equals the dimension of the row space of any reduced echelon matrix row equivalent to it. So we may as well assume that the matrix A itself is a row reduced echelon matrix. Then its nonzero rows are certainly linearly independent. Since the number of nonzero rows is $r(A)$, $r(A)$ is the dimension of the row space of A.   ■

EXAMPLE

To find the rank (dimension of the column space) of

$$\begin{bmatrix} 1 & 2 & 3 & 4 & 4 & 3 & 6 \\ 2 & 1 & 1 & 1 & 1 & 1 & 1 \\ 3 & 3 & 3 & 3 & 3 & 3 & 3 \end{bmatrix},$$

calculate the dimension of the row space, which is 3 since the rows are linearly independent. (Verify!)

We now show that as for square matrices, the sum of the rank and nullity of an $m \times n$ matrix is n. Since the rank $r(A)$ and nullity $n(A)$ of an $m \times n$ matrix A do not change if an elementary row or column operation is applied to A (Prove!), we can use both elementary row and column operations in our proof. We use a column version of row equivalence, where we say that two $m \times n$ matrices A and B are *column equivalent* if there exists an invertible $n \times n$ matrix U such that $B = AU$.

Theorem 6.3.6.   Let A be an $m \times n$ matrix. Then the sum of the rank and nullity of A is n.

Proof.   Since two row or column equivalent matrices have the same rank and nullity, and since A is row equivalent to a reduced echelon matrix, we can assume that A itself is a reduced echelon matrix. Then, using elementary column operations, we can further reduce A so that it has at most one nonzero entry in any row or column. Using column interchanges if necessary, we can even assume that the entries

$$a_{11}, \ldots, a_{rr} \text{ (where } r \text{ is the rank of } A)$$

are nonzero, and therefore equal to 1, and all other entries of A are 0. But then

$$Ax = \begin{bmatrix} 1 & & & & & & 0 \\ & \ddots & & & & & \\ & & 1 & & & & \\ & & & 0 & & & \\ & & & & \ddots & & \\ 0 & & & & & & 0 \end{bmatrix} \begin{bmatrix} x_1 \\ \vdots \\ x_r \\ x_{r+1} \\ \vdots \\ x_r \end{bmatrix} = \begin{bmatrix} x_1 \\ \vdots \\ x_r \\ 0 \\ \vdots \\ 0 \end{bmatrix},$$

so $Ax = 0$ if and only if $x_1 = \cdots = x_r = 0$, which implies that the dimension of Nullspace (A) is $n - r$.   ∎

EXAMPLE

   To find the nullity of

$$\begin{bmatrix} 1 & 2 & 3 & 4 & 4 & 3 & 6 \\ 2 & 1 & 1 & 1 & 1 & 1 & 1 \\ 3 & 3 & 3 & 3 & 3 & 3 & 3 \end{bmatrix},$$

find its rank r. The nullity is then $n - r$. We saw that the rank is 3 in the example above, so the nullity is $7 - 3 = 4$.

This theorem enables us to prove

Corollary 6.3.7. Let A be an $m \times n$ matrix A such that the equation $Ax = y$ has one and only one solution $x \in F^{(n)}$ for every $y \in F^{(m)}$. Then m equals n and A is invertible.

    Proof. The nullity of A is 0, since $Ax = 0$ has only one solution. So the rank of A is $n - 0 = n$. Since $Ax = y$ has a solution for every $y \in F^{(m)}$, the column space of A is $F^{(m)}$. Since the dimension of $F^{(n)}$ is n, we get that $F^{(m)}$ has dimension equal to the rank of A, that is, n. So $m = n$. Since A has rank n, it follows that A is invertible. ■

PROBLEMS

NUMERICAL PROBLEMS

1. Find the dimension of the column space of

$$\begin{bmatrix} 1 & 2 & 3 & 4 & 4 & 3 & 6 \\ 2 & 1 & 1 & 1 & 1 & 1 & 1 \\ 3 & 3 & 4 & 5 & 5 & 4 & 7 \end{bmatrix}$$ by finding the dimension of its row

space.

2. Find the nullity of $\begin{bmatrix} 1 & 2 & 3 & 4 & 4 & 3 & 6 \\ 2 & 1 & 1 & 1 & 1 & 1 & 1 \\ 3 & 3 & 4 & 5 & 5 & 4 & 7 \end{bmatrix}$ using Problem 1 and

Theorem 6.3.4.

3. Find a reduced echelon matrix row equivalent to the matrix

$$\begin{bmatrix} 1 & 2 & 3 & 4 & 4 & 3 & 6 \\ 2 & 1 & 1 & 1 & 1 & 1 & 1 \\ 3 & 3 & 4 & 5 & 5 & 4 & 7 \end{bmatrix}$$ and use it to compute the nullspace of

this matrix.

4. Find the nullspace of the transpose $\begin{bmatrix} 1 & 2 & 3 & 4 & 4 & 3 & 6 \\ 2 & 1 & 1 & 1 & 1 & 1 & 1 \\ 3 & 3 & 5 & 5 & 5 & 4 & 7 \end{bmatrix}$ of

the matrix of Problem 3 by any method.

MORE THEORETICAL PROBLEMS

Easier Problems

5. Show that the rank and nullity of an $m \times n$ matrix A do not change if an elementary row or column operation is applied to A.

6. If A is an $m \times n$ matrix, show that the nullity of A equals the nullity of the transpose A' of A if and only if $m = n$.

7. If the set of solutions to $Ax = y$ is a subspace, show that $y = 0$.

Middle-Level Problems

8. Let A be a 3×2 matrix with real entries. Show that the equation $A'Ax = A'y$ has a solution x for all y in $\mathbb{R}^{(3)}$.

9. Show that if two reduced $m \times n$ echelon matrices are row equivalent, they are equal.

Harder Problems

10. Show that if an $m \times n$ matrix is row equivalent to two reduced echelon matrices C and D, then $C = D$.

11. Let A be an $m \times n$ matrix with real entries. Show that the equation $A'Ax = A'y$ has a solution x for all y in $\mathbb{R}^{(m)}$.

C H A P T E R

7

ABSTRACT VECTOR SPACES

7.1. INTRODUCTION, DEFINITIONS, AND EXAMPLES

Before getting down to the formal development of the mathematics involved, a few short words on what we'll be about. What will be done is from a kind of turned-about point of view. At first, most of the things that will come up will be things already encountered by the reader in $F^{(n)}$. It may seem as a massive *deja-vu*. That's not surprising: it *is* a massive *deja-vu,* but with a twist.

Everything we did in Chapter 3 will make an appearance in due course, but in a new, abstract setting. In addition, some fundamental notions—notions that we have not seen to date, which are universal in all of mathematics—will show up.

Let F be $\mathbb{R}$ or $\mathbb{C}$, the real or complex numbers, respectively. For much of the development, whether F is $\mathbb{R}$ or $\mathbb{C}$ will have no relevance. As in our discussion of $F^{(n)}$, at some crucial point, where we shall require roots of polynomials, the use of $F = \mathbb{C}$ will become essential.

In Chapter 3 we saw that in $F^{(n)}$, the set of all column vectors

$$\begin{bmatrix} a_1 \\ a_2 \\ \vdots \\ a_n \end{bmatrix}$$ over F, there are certain natural ways of combining elements—

addition and multiplication by a scalar—which satisfied some nice, formal, computational rules. These properties, which are concrete and specific in $F^{(n)}$, serve to guide us to one of the most basic, abstract structures in all of mathematics, known as a *vector space*. the notion is closely modeled on the concrete $F^{(n)}$.

Definition. A set V is called a *vector space over F* if on V we have two operations, called *addition* and *multiplication by a scalar*, such that the following hold:

(A) *Rules for Addition*
1. If $v, w \in V$, then $v + w \in V$.
2. $v + w = w + v$ for all $v, w \in V$.
3. $v + (w + z) = (v + w) + z$ for all v, w, z in V.
4. There exists an element 0 in V such that $v + 0 = v$ for every $v \in V$.
5. Given $v \in V$ there exists a $w \in V$ such that $v + w = 0$.

(B) *Rules for Multiplication by a Scalar*
6. If $a \in F$ and $v \in V$, then $av \in V$.
7. $a(v + w) = av + aw$ for all $a \in F, v, w \in V$.
8. $(a + b)v = av + bv$ for all $a, b \in F, v \in V$.
9. $a(bv) = (ab)v$ for all $a, b \in F, v \in V$.
10. $1v = v$ for all $v \in V$, where 1 is the unit element of F.

These rules governing the operations in V seem very formal and formidable. They probably are. But what they really say is something quite simple:

Go ahead and calculate in V as we have been used to doing in $F^{(n)}$.

The rules just give us the enabling framework for carrying out such calculations.

Note that property 4 guarantees us that V has at least one element, hence V is *nonempty*.

Before doing anything with vector spaces it is essential that we see a large cross section of examples. In this way we can get the feeling of how wide and encompassing the concept really is.

In the examples to follow, we shall define the operations and verify, at most, that $v + w$ and av are in V for $v, w \in V$ and $a \in F$. The verification that each example, is indeed, an example of a vector space is quite straightforward, and is left to the reader.

EXAMPLES

1. Let $V = F^{(n)}$, or any subspace of $F^{(n)}$, with the operations of those of Chapter 3. Since these examples are the prime motivator for the general concept of vector space, it is not surprising that they are vector spaces over F. If $n = 1$, then $V = F^{(1)} = F$, showing how $V = F$ is itself regarded as a vector space over F.

2. As a variation of the theme of Example 1, let $F = \mathbb{R}$ and $V = \mathbb{C}^{(n)}$ with the usual addition and multiplication of elements in $\mathbb{C}^{(n)}$. It is immediate that $V = \mathbb{C}^{(n)}$ is a vector space over $F = \mathbb{R}$ with the usual product by a scalar between elements of F and of $\mathbb{C}$. If $n = 1$, then $V = \mathbb{C}^{(1)} = \mathbb{C}$, showing how $V = \mathbb{C}$ is regarded as a vector space over $\mathbb{R}$.

3. Let V consist of all the functions $a \cos(x) + b \sin(x)$, where $a, b \in \mathbb{R}$, with the usual operations used for functions. If $v = a \cos(x) + b \sin(x)$ and $w = a_1 \cos(x) + b_1 \sin(x)$, then $v + w = (a + a_1) \cos(x) + (b + b_1) \sin(x)$, so $v + w$ is in V by the very criterion for membership in V. Similarly, $cv \in V$ if $c \in \mathbb{R}$, $v \in V$. So, V is a vector space over $\mathbb{R}$.

4. Let V consist of all real-valued solutions of the differential equations

$$\frac{d^2 f(x)}{dx^2} + f(x) = 0.$$

From the theory of differential equations we know that V consists precisely of the elements $a \cos(x) + b \sin(x)$, with $a, b \in \mathbb{R}$. In other words, this example coincides with Example 3. For the sake of variety, we verify that $f + g$ is in V if f, g are in V by another method. Since $f, g \in V$,

$$\frac{d^2 f(x)}{dx^2} + f(x) = 0 \quad \text{and} \quad \frac{d^2 g(x)}{dx^2} + g(x) = 0,$$

whence

$$\frac{d^2}{dx^2}(f + g) + (f + g) = \left(\frac{d^2 f}{dx^2} + f\right) + \left(\frac{d^2 f}{dx^2} + g\right) = 0.$$

Thus $f + g$ is in V. Similarly, cf is in V if $c \in \mathbb{R}$ and $f \in V$.

5. Let V be the set of all real-valued solutions $f(x)$ of the differential equation $\dfrac{d^2 f}{dx^2} + f(x) = 0$ such that $f(0) = 0$. It is clear that V is contained in the vector space of Example 4, and is itself a vector space over $\mathbb{R}$. So it is appropriate to call it a *subspace* (you can guess from Chapter 3 what this should mean) of Example 4. What is the form of the typical element of V?

6. Let V be the set of all polynomials $f(x) = a_0 x^n + \cdots + a_n$ where the a_i are in $\mathbb{C}$ and n is any nonnegative integer. We add and multiply polynomials by constants in the usual way. With these operations V becomes a vector space over $\mathbb{C}$.

7. Let W be the set of all polynomials over $\mathbb{C}$ of degree 4 or less. That is,

$$W = \{a_0 x^4 + a_1 x^3 + a_2 x^2 + a_3 x + a_4 \mid a_0, \ldots, a_4 \in \mathbb{C}\},$$

with the operations those of Example 6. It is immediate that W is a vector space over $\mathbb{C}$ and is a subspace of Example 6.

8. Let V be the set, $M_n(F)$, of $n \times n$ matrices over F with the matrix addition and scalar multiplication used in Chapter 3. V is a vector space over F. If we view an $n \times n$ matrix (a_{rs}) as a strung-out n^2-tuple, then $M_n(F)$ resembles $F^{(n^2)}$ as a vector space. Thus if

$$\begin{bmatrix} a_{11} & a_{12} \\ a_{21} & a_{22} \end{bmatrix} \in M_2(F),$$

we "identify" it with the column vector

$$\begin{bmatrix} a_{11} \\ a_{21} \\ a_{12} \\ a_{22} \end{bmatrix} \in F^{(4)}$$

and we get the "resemblance" between $M_2(F)$ and $F^{(4)}$ mentioned above.

9. Let V be any vector space over F and let $v_1, \ldots, v_n$ be n

given elements in V. An element

$$x = a_1v_1 + \cdots + a_nv_n,$$

where the a_i are in F, is called a *linear combination* of $v_1, \ldots, v_n$ *over* F. Let

$$<v_1, \ldots, v_n>$$

denote the set of *all linear combinations of* $v_1, \ldots, v_n$ *over* F. We claim that $<v_1, \ldots, v_n>$ is a vector space over F; it is also a subspace of V over F. We leave these to the reader.

We call $<v_1, \ldots, v_n>$ *the subspace of V spanned over F by* $v_1, \ldots, v_n$.

This is a very general construction of producing new vector spaces from old. We saw how important $<v_1, \ldots, v_n>$ was in the study of $F^{(n)}$. It will be equally important here.

Let's look at an example for some given V of $<v_1, \ldots, v_n>$. Let V be the set of all polynomials in x over $\mathbb{C}$ and let

$$v_1 = x, \ v_2 = 1 - x, \ v_3 = 1 + x + x^2, \ v_4 = x^3.$$

What is $<v_1, \ldots, v_n>$? It consists of all

$$a_1v_1 + \cdots + a_4v_4,$$

where $a_1, \ldots, a_4$ are in $\mathbb{C}$. In other words, a typical element of $<v_1, \ldots, v_n>$ looks like

$$a_1x + a_2(1 - x) + a_3(1 + x + x^2) + a_4x^3,$$

that is,

$$(a_2 + a_3) + (a_1 - a_2 + a_3)x + a_3x^2 + a_4x^3.$$

We leave it to the reader to verify that *every* polynomial of degree 3 or less so realizable. Thus, in this case, $<v_1, \ldots, v_4>$ consists of all polynomials over $\mathbb{C}$ of degree 3 or less. Put more tersely, we are saying that

$$<x, \ 1 - x, \ 1 + x + x^2, \ x^3> = <1, x, x^2, x^3> = V.$$

10. Let $V = M_n(F)$, viewed as a vector space over F. Let

$$W = \{(a_{rs}) \in M_n(F) \mid \text{tr} \ (a_{rs}) = 0\}.$$

Since tr $(A + B) = $ tr $(A) + $ tr (B) and tr $(aA) = a$ tr (A), we get that W is a vector space over F. Of course, W is a subspace of V.

11. Let $V = M_n(\mathbb{R})$ be viewed as a vector space over $\mathbb{R}$ and let

$$W = \{A \in M_n(\mathbb{R}) \mid A = A'\}.$$

Thus W is the set of symmetric matrices in $M_n(\mathbb{R})$. Because $(A + B)' = A' + B'$ and $(cA)' = cA'$, for $A, B \in M_n(\mathbb{R})$ and $c \in \mathbb{R}$, we see that W is a subspace of V over $\mathbb{R}$.

12. Let V be the set of all real-valued functions on the closed unit interval $[0, 1]$. For the operations in V we use the usual operations of functions. Then V is a vector space over $\mathbb{R}$. Let

$$W = \{f(x) \in V \mid f(\tfrac{1}{2}) = 0\}.$$

Thus if $f, g \in W$, then $f(\tfrac{1}{2}) = g(\tfrac{1}{2}) = 0$, hence

$$(f + g)(\tfrac{1}{2}) = f(\tfrac{1}{2}) + g(\tfrac{1}{2}) = 0,$$

and

$$af(\tfrac{1}{2}) = 0$$

for $a \in \mathbb{R}$. So W is a vector space over $\mathbb{R}$ and is a subspace of V.

13. Let V be the set of all *continuous* real-valued functions on the closed unit interval. Since the sum of continuous functions is a continuous function and the product of a continuous function by a constant is continuous, we get that V is a vector space over $\mathbb{R}$. If W is the set of real-valued functions differentiable on the closed unit interval then $W \subset V$, and by the properties of differentiable functions, W is a vector space over $\mathbb{R}$ and a subspace of V.

The reader easily perceives that many of our examples come from the calculus and real variable theory. This is not just because these provide us with natural examples. We deliberately made the choice of these to stress to the reader that the concept of vector space is not just an algebraic concept. It turns up all over the place.

In Chapter 4, in discussing subspaces, we gave many diverse examples of subspaces of $F^{(n)}$. Each such, of course, provides us with an example of a vector space over F. We would suggest to readers that they go back to Chapter 4 and give those old examples a new glance.

7.2. SUBSPACES

With what we did in Chapter 4 on subspaces of $F^{(n)}$ there should be few surprises in store for us in the material to come. It might be advisable to look back and give oneself a quick review of what is done there.

Before getting to the heart of the matter we must dispose of a few small, technical items. These merely tell us it is all right to go ahead and compute in a natural way.

To begin with, we assume that in a vector space V the sum of any two elements is again in V. By combining elements we immediately get from this that the sum of *any finite number of elements of V is again in V*. Also, the associative law

$$v + (w + z) = (v + w) + z$$

can be extended to any finite sum and *any* bracketing. So we can write such sums without the use of parentheses. To prove this is not hard but is tedious and noninstructive, so we omit it and go ahead using the general associative law without proof.

Possibly more to the point are some little computational details. We go through them in turn.

1. *If $v + w = v + z$, then $w = z$.*

There exists an element $u \in V$ such that $u + v = 0$; hence $u + (v + w) = u + (v + z)$. In other words, $w = 0 + w = (u + v) + w = u + (v + z) = (u + v) + z = 0 + z = z$, resulting in $w = z$. This little rule allows us to cancel in sums.

2. *If $0 \in F$ and $v \in V$, then $0v = 0$, the zero element of V.*

Since $0v + 0v = (0 + 0)v = 0v = 0v + 0$, we get $0v + 0v = 0v + 0$, so that $0v = 0$ by (1) above.

3. $v + (-1)v = 0$.

For $0 = (1 + (-1))v = v + (-1)v$. So $(-1)v$ acts as the negative for v in V. We denote it $(-1)v = -v$.

4. *If $a \neq 0$ is in F and $av = 0$ then $v = 0$.*

For $0 = av$ gives us $0 = a^{-1}0 = a^{-1}(av) = (a^{-1}a)v = 1v = v$.

5. *If $a \neq 0$ is in F and $av + a_2v_2 + \cdots + a_kv_k = 0$, then $v = (-a_2/a)v_2 + \cdots + (-a_k/a)v_k$.*

We leave the proof of this to the reader. Again we stress that a result such as (5) is a license to carry on, in our usual way, in computing in V.

With these technical points out of the way we can get down to the business at hand.

Definition. Let V be a vector space over F. Then a nonempty subset W of V is called a *subspace* of V (over F) if, relative to the operations of V, W forms a vector space over F.

Two subspaces of V come immediately to mind. $W = \{0\}$ and $W = V$. We leave it to the reader to verify that these two are indeed subspaces of V. By a *proper* subspace W of V we shall mean a subspace of V such that $W \neq V$.

How do we recognize whether or not a given nonempty subset W of V is a subspace of V? Easily. All we have to do is to verify that $u, w \in W$ implies that $u + w \in W$ and $au \in W$ for all $a \in F$, $u \in W$. Why do we have to verify only these two properties? Because most of the other rules defining a vector space hold in W because they already hold in the larger set V. What we really need is that $0 \in W$ and, for every $w \in W$, $-w$ is in W. Now, since W is nonempty, let $z \in W$. Then $0z = 0$ is in W, hence $0 \in W$. Similarly, for w in W, $(-1)w = -w$ is in W. So W is a subspace of V.

We record this as

Lemma 7.2.1. A nonempty subset W of V is a subspace of V if and only if:

1. $u, w \in W$ implies that $u + w \in W$;
2. $a \in F$, $w \in W$ implies that $aw \in W$.

If we look back at the examples in Section 7.1, we note that we have singled out many of the examples as subspaces of some others. But just for the sake of practice we verify in a specific instance that something is a subspace.

Let V be the set of all real-valued functions defined on the closed unit interval. Then V is a vector space over $\mathbb{R}$. Let W be the set of all continuous functions in V. We assert that W is a subspace of V. Why? If $f, g \in W$, then each of them is continuous, but the sum of continuous functions is continuous. Hence $f + g$ is continuous, and as such, it enjoys the condition for membership in W. In short, $f + g$ is in W. Similarly, $af \in W$ if $a \in F$ and $f \in W$.

Let U be the set of all differentiable functions in V. Because a differentiable function is automatically continuous, we get that $U \subset W$. Furthermore, if f, g are differentiable, so are $f + g$ and af, with $a \in F$, differentiable. Thus U is a subspace of V. Since U lies in W, U is also a subspace of W.

Of paramount importance to us is Example 9 of Section 7.1. It gives us a ready way of producing subspaces of a given vector space, at will. Recall that given $v_1, \ldots, v_n \in V$, then $\langle v_1, \ldots, v_n \rangle$ is the set of all linear combinations $a_1 v_1 + \cdots + a_n v_n$ of $v_1, \ldots, v_n$ over F. Let's verify that $\langle v_1, \ldots, v_n \rangle$ is really a subspace of V over F. If

$$u = a_1 v_1 + \cdots + a_n v_n$$

and

$$z = b_1 v_1 + \cdots + b_n v_n,$$

where the a_r and b_r are in F, then

$$
\begin{aligned}
u + z &= (a_1 v_1 + \cdots + a_n v_n) + (b_1 v_1 + \cdots + b_n v_n) \\
&= (a_1 + b_1) v_1 + \cdots + (a_n + b_n) v_n.
\end{aligned}
$$

(Verify!) So $u + z$ is a linear combination of $v_1, \ldots, v_n$ over F, hence lies in $\langle v_1, \ldots, v_n \rangle$. Similarly, if $c \in F$, then

$$c(a_1 v_1 + \cdots a_n v_n) = (ca_1) v_1 + \cdots + (ca_n) v_n,$$

so is a linear combination of $v_1, \ldots, v_n$ over F. As such it lies in $\langle v_1, \ldots, v_n \rangle$. We have checked out that $\langle v_1, \ldots, v_n \rangle$ is a subspace of V over F. We call it the subspace of V *spanned* by $v_1, \ldots, v_n$ over F.

In $F^{(3)}$, if we let

$$
v_1 = \begin{bmatrix} 1 \\ 1 \\ 1 \end{bmatrix}, \quad
v_2 = \begin{bmatrix} 1 \\ 0 \\ 1 \end{bmatrix}, \quad
v_3 = \begin{bmatrix} 1 \\ 1 \\ 0 \end{bmatrix}, \quad
v_4 = \begin{bmatrix} 1 \\ 2 \\ 3 \end{bmatrix},
$$

then $\langle v_1, v_2, v_3, v_4 \rangle$ is the set of all

$$
a_1 v_1 + a_2 v_2 + a_3 v_3 + a_4 v_4 = \begin{bmatrix} a_1 + a_2 + a_3 + a_4 \\ a_1 + a_3 + 2a_4 \\ a_1 + a_2 + 3a_4 \end{bmatrix}.
$$

We leave it to the reader that *any* vector $\begin{bmatrix} b_1 \\ b_2 \\ b_3 \end{bmatrix}$ in $F^{(3)}$ is so realizable.

Hence $\langle v_1, v_2, v_3, v_4 \rangle = F^{(3)}$.

This last example is quite illuminating. In it the given vectors $v_1, \ldots, v_4$ spanned all of $V = F^{(3)}$ over F. This type of situation is the one that will be of greatest interest to us.

Definition. The vector space V over F is said to be *finite-dimensional over F* if $V = \langle v_1, \ldots, v_n \rangle$ for some vectors $v_1, \ldots, v_n$ in V.

In other words, V is finite-dimensional over F if there is a *finite* set $v_1, \ldots, v_n$ in V such that, given *any* $u \in V$, $u = a_1 v_1 + \cdots + a_n v_n$ for some appropriate $a_1, \ldots, a_n$ in F.

Which of the examples in Section 7.1 are finite-dimensional over F? We give the answer as "yes" or "no" to each in turn. The reader should verify that these answers are the right ones:

Example 1: Yes. Example 2: Yes. Example 3: Yes.
Example 4: Yes. Example 5: Yes. Example 6: No.
Example 7: Yes. Example 8: Yes. Example 9: Yes.
Example 10: Yes. Example 11: Yes. Example 12: No.
Example 13: No.

By the very definition of finite-dimensionality we know that the span $W = \langle v_1, \ldots, v_n \rangle$ of elements $v_1, \ldots, v_n$ of a vector space V is not only a subspace of V but is also finite-dimensional over F.

For almost everything that we shall do in the rest of the book the vector space V will be finite-dimensional over F. Therefore, you may assume that V is finite dimensional unless we indicate that it is not.

PROBLEMS

NUMERICAL PROBLEMS

1. Show that $W = \{0\}$ and $W = V$ are subspaces of V.

2. Verify that Examples 1 through 5 and 7 through 11 are finite-dimensional over F and explain why Examples 6, 12, and 13 are not.

3. Show that $v_1 = \begin{bmatrix} 1 \\ 1 \\ 1 \end{bmatrix}$, $v_2 = \begin{bmatrix} 1 \\ 0 \\ 1 \end{bmatrix}$, $v_3 = \begin{bmatrix} 1 \\ 1 \\ 0 \end{bmatrix}$, $v_4 = \begin{bmatrix} 1 \\ 2 \\ 3 \end{bmatrix}$ span $F^{(3)}$ over F.

4. Let V be the vector space of polynomials in x having real coefficients. If $v_1 = x$, $v_2 = 2x$, $v_3 = x^2$, $v_4 = x - 6x^2$, find the span of v_1, v_2, v_3, v_4 over $\mathbb{R}$.

5. In Problem 4 show that $<v_1, v_2, v_3, v_4> = <v_1, v_4>$.

MORE THEORETICAL PROBLEMS

Easier Problems

6. If $v_1, \ldots, v_n$ span V over F, show that $v_1, \ldots, v_n, w_1, \ldots, w_m$ also span V over F, for any $w_1, w_2, \ldots, w_m$ in V.

7. If $v_1, \ldots, v_n$ span V over F and $a_1 v_1 + a_2 v_2 + \cdots + a_n v_n = 0$, where $a_1 \neq 0$, show that $v_2, \ldots, v_n$ already span V over F.

8. If $w_1, \ldots, w_k$ are in $<v_1, \ldots, v_n>$, show that $<w_1, \ldots, w_k>$ is contained in $<v_1, \ldots, v_n>$.

9. Let V be a vector space over $\mathbb{C}$. As such we consider V also as a vector space over $\mathbb{R}$, by using as scalars only the elements of $\mathbb{R}$. Show that if V is finite-dimensional over $\mathbb{C}$, then V is finite-dimensional over $\mathbb{R}$.

10. In Problem 9 show that if V is finite-dimensional over $\mathbb{R}$, then V is also finite-dimensional over $\mathbb{C}$.

11. If U and W are subspaces of V, show that $U \cap W$ is a subspace of V.

12. If U and W are subspaces of V, let $U + W$ be defined by

$$U + W = \{u + w \mid u \in U, w \in W\}.$$

Show that $U + W$ is a subspace of V over F.

13. Let V and W be two vector spaces over F. Let

$$V \oplus W = \{(v, w) \mid v \in V, w \in W\},$$

where we define $(v_1, w_1) + (v_2, w_2) = (v_1 + v_2, w_1 + w_2)$ and $a(v_1, w_1) = (av_1, aw_1)$ for all $v_1, v_2 \in V$ and $w_1, w_2 \in W$ and $a \in F$. Prove that $V \oplus W$ is a vector space over F. ($V \oplus W$ is called the *direct sum* of V and W.)

Middle-Level Problems

14. If, in Problem 12, U and W are finite-dimensional over F, show that $U + W$ is finite-dimensional over F.

15. If V and W are finite-dimensional over F, show that $V \oplus W$ is also finite-dimensional over F.

16. Let $\tilde{V} = \{(v, 0) \mid v \in V\}$ and $\tilde{W} = \{(0, w) \mid w \in W\}$. Show that $\tilde{V}$ and $\tilde{W}$ are subspaces of $V \oplus W$ such that $V \oplus W = \tilde{V} + \tilde{W}$ and $\tilde{V} \cap \tilde{W} = \{0\}$.

17. If $V = <v_1, \ldots, v_m>$ and $W = <w_1, \ldots, w_m>$, show that $V \oplus W = <\tilde{v}_1, \ldots, \tilde{v}_m, \tilde{w}_1, \ldots, \tilde{w}_n>$, where $\tilde{v}_i = (v_i, 0)$ and $\tilde{w}_j = (0, w_j)$.

18. Show that in the vector space $V \oplus V$ over F, the set $V = \{(v, v)\}$ is a subspace of $V \oplus V$ over F.

19. If $v \in V$, let $Fv = \{av \mid a \in F\}$. Show that Fv is a subspace of V.

20. If $v_1, \ldots, v_n$ are in V, show that their span $<v_1, \ldots, v_n>$ is

$$Fv_1 + \cdots + Fv_n.$$

(See Problem 12.)

Harder Problems

21. Let V be a vector space over F and U, W subspaces of V. Define $f: U \oplus W \longrightarrow V$ by $f(u, w) = u + w$. Then show that f maps $U \oplus W$ *onto* $U + W$. Furthermore, show that if $x, y \in U \oplus W$, then $f(x + y) = f(x) + f(y)$ and $f(ax) = af(x)$ for all $a \in F$.

22. In Problem 21, let $Z = \{(u, w) \mid f(u, w) = 0\}$. Show that Z is a subspace of $U \oplus W$ and is, in fact, $Z = U \cap W$.

7.3. HOMOMORPHISMS AND ISOMORPHISMS

For the first time in a long time we are about to go into a territory that is totally new to us. What we shall talk about is the notion of a *homomorphism* of one vector space into another. This will be defined precisely below, but for the moment, suffice it to say that a *homomorphism is a mapping that preserves structure*.

The analogue of this concept occurs in every part of mathematics. For instance, the analog in analysis might very well be that of a continuous function. In every part of algebra such decent mappings are defined by algebraic relations they satisfy.

A special kind of homomorphism is an *isomorphism,* which is really a homomorphism that is 1-1 and onto. If there is an isomorphism from one vector space to another, the spaces are *isomorphic.* Isomorphic spaces are, in a very good sense, equal. Whatever is true in one space gets transferred, so it is also true in the other. For us, the importance will be that any *finite-dimensional* vector space V over F will turn out to be isomorphic to $F^{(n)}$ for some n. This will allow us to transfer *everything* we did in $F^{(n)}$ to finite-dimensional vector spaces. In a sense, $F^{(n)}$ is the universal model of a finite-dimensional vector space, and in treating $F^{(n)}$ and the various concepts in $F^{(n)}$, we lose nothing in generality. These notions and results will be transferred to V by an isomorphism.

Of course, all this now seems vague. It will come more and more into focus as we proceed here and in the coming sections.

Definition.   If V and W are vector spaces over F, then the mapping $\Phi: V \longrightarrow W$ is said to be a *homomorphism* if

1. $\Phi(v_1 + v_2) = \Phi(v_1) + \Phi(v_2)$
2. $\Phi(av) = a\Phi(v)$

for all $v, v_1, v_2 \in V$ and all $a \in F$.

Note that in $\Phi(v_1 + v_2)$, the addition $v_1 + v_2$ takes place in V, whereas the addition $\Phi(v_1) + \Phi(v_2)$ takes place in W.

We did run into this notation when we talked about linear transformations of $F^{(n)}$ (and of $F^{(n)}$ into $F^{(m)}$) in Chapters 3 and 6. Here the context is much broader.

Decent concepts deserve decent examples. We shall try to present some now.

EXAMPLES

1. Let $V = F^{(4)}$ and $W = F^{(2)}$. Define $\Phi: V \longrightarrow W$ by

$$\Phi \begin{bmatrix} a \\ b \\ c \\ d \end{bmatrix} = \begin{bmatrix} a + c - 3d \\ 4b - a \end{bmatrix}.$$

We claim that Φ is a homomorphism of V into W. Given $\begin{bmatrix} a \\ b \\ c \\ d \end{bmatrix}$ and

$\begin{bmatrix} a_1 \\ b_1 \\ c_1 \\ d_1 \end{bmatrix}$ in $F^{(4)}$; then

$$\Phi \begin{bmatrix} a \\ b \\ c \\ d \end{bmatrix} = \begin{bmatrix} a + c - 3d \\ 4b - a \end{bmatrix}, \Phi \begin{bmatrix} a_1 \\ b_1 \\ c_1 \\ d_1 \end{bmatrix} = \begin{bmatrix} a_1 + c_1 - 3d_1 \\ 4b_1 - a_1 \end{bmatrix},$$

while

$$\Phi \left(\begin{bmatrix} a \\ b \\ c \\ d \end{bmatrix} + \begin{bmatrix} a_1 \\ b_1 \\ c_1 \\ d_1 \end{bmatrix} \right) = \Phi \begin{bmatrix} a + a_1 \\ b + b_1 \\ c + c_1 \\ d + d_1 \end{bmatrix}$$

$$= \begin{bmatrix} a + a_1 + c + c_1 - 3(d + d_1) \\ 4(b + b_1) - (a + a_1) \end{bmatrix}$$

$$= \begin{bmatrix} a + c - 3d \\ 4b - a \end{bmatrix} + \begin{bmatrix} a_1 + c_1 - 3d_1 \\ 4b_1 - a_1 \end{bmatrix}$$

$$= \Phi \left(\begin{bmatrix} a \\ b \\ c \\ d \end{bmatrix} \right) + \Phi \left(\begin{bmatrix} a_1 \\ b_1 \\ c_1 \\ d_1 \end{bmatrix} \right).$$

Similarly,

$$\Phi \left(a \left(\begin{bmatrix} a \\ b \\ c \\ d \end{bmatrix} \right) \right) = a \Phi \left(\begin{bmatrix} a \\ b \\ c \\ d \end{bmatrix} \right)$$

for all $a \in F$. So Φ is a homomorphism of $F^{(4)}$ into $F^{(2)}$. It is actually *onto* $F^{(2)}$. (Prove!)

2. Let $V = F^{(n)}$ and $W = F$ be viewed as vector spaces over F. Let

$$\Phi\begin{bmatrix} a_1 \\ \vdots \\ a_n \end{bmatrix} = a_1 + a_2 + \cdots + a_n.$$

We claim that Φ is a homomorphism of $F^{(n)}$ *onto* F. First, why is it onto? That is easy, for given $a \in F$, then $a = \Phi\begin{bmatrix} a \\ 0 \\ \vdots \\ 0 \end{bmatrix}$, so Φ is onto. Why a homomorphism? By the definition.

$$\Phi\left(\begin{bmatrix} a_1 \\ \vdots \\ a_n \end{bmatrix} + \begin{bmatrix} b_1 \\ \vdots \\ b_n \end{bmatrix}\right) = \Phi\begin{bmatrix} a_1 + b_1 \\ \vdots \\ a_n + b_n \end{bmatrix}$$

$$= (a_1 + b_1) + \cdots + (a_n + b_n)$$

$$= (a_1 + \cdots + a_n) + (b_1 + \cdots + b_n)$$

$$= \Phi\begin{bmatrix} a_1 \\ \vdots \\ a_n \end{bmatrix} + \Phi\begin{bmatrix} b_1 \\ \vdots \\ b_n \end{bmatrix}.$$

Similarly,

$$\Phi\left(c\begin{bmatrix} a_1 \\ \vdots \\ a_n \end{bmatrix}\right) = c\Phi\begin{bmatrix} a_1 \\ \vdots \\ a_n \end{bmatrix}.$$

3. Let V be the set of all polynomials in x over $\mathbb{R}$. Define

$$F: V \longrightarrow V \quad \text{by} \quad \Phi(a_0 x^n + a_1 x^{n-1} + \cdots + a_n)$$
$$= na_0 x^{n-1} + (n-1)a_1 x^{n-2} + \cdots + a_{n-1}.$$

So Φ is just the derivative of the polynomial in question. Either by a direct check, or by remembrances from the calculus, we have that Φ is a homomorphism of V into V. Is it onto?

4. Again let V be the set of polynomials over $\mathbb{R}$ and let $\Phi\colon V \longrightarrow V$ be defined by

$$\Phi(p(x)) = \int_0^x p(t)\, dt.$$

Since we have that

$$\int_0^x (p(t) + q(t))\, dt = \int_0^x p(t)\, dt + \int_0^x q(t)\, dt$$

and

$$\int_0^x cp(t)\, dt = c \int_0^x p(t)\, dt \text{ for } c \in \mathbb{R}$$

we see that

$$\Phi(v_1 + v_2) = \Phi(v_1) + \Phi(v_2)$$

and

$$\Phi(cv_1) = c\Phi(v_1)$$

for all v_1, v_2 in V and all $c \in \mathbb{R}$. So Φ is a homomorphism of V into itself. Is it onto?

5. Let $V = F^{(5)}$ and $W = F^{(3)}$. Define $\Phi\colon V \longrightarrow W$ by

$$\Phi\begin{bmatrix} a_1 \\ a_2 \\ \vdots \\ a_5 \end{bmatrix} = \begin{bmatrix} a_1 \\ a_2 \\ a_3 \end{bmatrix}.$$

It is immediate that Φ is a homomorphism of V onto W.

6. Let V and W be vector spaces over F and let $X = V \oplus W$. Define $\Phi\colon X \longrightarrow V$ by

$$\Phi(v, w) = v.$$

We see here, too, that Φ is a homomorphism of X onto V.

7. Let V be the complex numbers, viewed as a vector space over $\mathbb{R}$. If $z = a + bi$, define

$$\Phi(z) = a - bi = \bar{z}.$$

Then Φ is a homomorphism of V onto itself. It is called *complex conjugation*.

8. Let $V = M_n(F)$ and $W = F$ and let $\Phi: V \longrightarrow W$ be defined by

$$\Phi(A) = \text{tr } (A).$$

Since tr $(A + B) = \text{tr } (A) + \text{tr } (B)$ and tr $(aA) = a$ tr (A), we get that Φ is a homomorphism of V onto W. It is the *trace function* on $M_n(F)$.

9. Let V be the set of polynomials in x of degree n or less over $\mathbb{R}$, and let W be the set of all polynomials in x over F. Define $\Phi: V \longrightarrow W$ by

$$\Phi(p(x)) = x^2 p(x)$$

for all $p(x) \in V$. It is easy to see that Φ is a homomorphism of V into W, but not onto.

From the examples given, one sees that it is not too difficult to construct homomorphisms of vector spaces.

What are the important attributes of homomorphisms? For one thing they carry linear combinations into linear combinations.

Lemma 7.3.1. If Φ is a homomorphism of V into W, then

$$\Phi(a_1 v_1 + \cdots + a_k v_k) = a_1 \Phi(v_1) + \cdots + a_k \Phi(v_k),$$

for any $v_1, \ldots, v_k$ in V and $a_1, \ldots, a_k$ in F.

Proof. By the additive property of homomorphisms we have that

$$\Phi(a_1 v_1 + \cdots + a_k v_k) = \Phi(a_1 v_1) + \cdots + \Phi(a_k v_k).$$

Since $\Phi(a_i v_i) = a_i \Phi(v_i)$ this top relation becomes the one asserted in the lemma. ∎

Another property is that homomorphisms carry subspaces into subspaces. This really is implicit in Lemma 7.3.1, but we state it as a lemma, leaving the proof to the reader.

Lemma 7.3.2. If V is a vector space, U a subspace of V and Φ a homomorphism of V into W, then $\Phi(U) = \{\Phi(u) \mid u \in U\}$ is a subspace of W.

Thus the image of V and any subspace of V under Φ is a subspace of W. Finally, another subspace crops up when dealing with homomorphisms.

Definition. If Φ is a homomorphism of V into W, then Ker (Φ), the *kernel* of Φ, is defined by

$$\text{Ker } (\Phi) = \{v \in V \mid \Phi(v) = 0\}.$$

Does this remind you of the nullspace of a linear transformation? It should, because it is. For Φ is merely a linear transformation from V to W.

Lemma 7.3.3. Ker (Φ) is a subspace of V.

Proof. Suppose that $u, v \in$ Ker (Φ) and $a \in F$. Then

$$\Phi(u + v) = \Phi(u) + \Phi(v) = 0$$

since $\Phi(u) = \Phi(v) = 0$. Also,

$$\Phi(au) = a\Phi(u) = a0 = 0.$$

Thus $u + v$ and au are also in Ker (Φ), hence Ker (Φ) is a subspace of V. ∎

Is there anything special that happens when Ker (Φ) is the particularly nice subspace consisting only of 0? Yes! Suppose that we have Ker $(\Phi) = \{0\}$ and $\Phi(u) = \Phi(v)$. Then

$$0 = \Phi(u) - \Phi(v) = \Phi(u - v),$$

hence $u - v$ is in Ker $(\Phi) = \{0\}$. Therefore, $u - v = 0$, and so $u = v$. What this says is that Φ is then 1-1. If Φ is 1-1, it is trivial that Ker $(\Phi) = \{0\}$, hence

Lemma 7.3.4. The homomorphism $\Phi: V \longrightarrow W$ is 1-1 if and only if Ker $(\Phi) = \{0\}$.

A homomorphism that maps V *onto* W in a 1-1 way is indeed something special. We give it a name.

Definition. The homomorphism $\Phi: V \longrightarrow W$ is called an *isomorphism* if Φ is 1-1 and onto W.

The importance of isomorphisms is that two vector spaces that are *isomorphic* (i.e., for which there is an isomorphism of one onto the other) are essentially the same. All that is different is the naming we give the elements. So if Φ is an isomorphism of V onto W, the renaming is given by calling $v \in V$ by the name $\Phi(v)$ in W. The fact that Φ preserves sums and products by scalars assures us that this renaming is consistent with the structure of V and W.

For practical purposes isomorphic vector spaces are the same. Anything true in one of these gets transferred to something true in the other. We shall see how the concept of isomorphism reduces the study of finite-dimensional vector spaces to that of $F^{(n)}$.

Definition.   We shall denote that there is an isomorphism of V onto W by $V \simeq W$.

PROBLEMS

NUMERICAL PROBLEMS

1. In Example 1 show that the mapping Φ is onto $F^{(2)}$.

2. Verify that the mapping Φ in Example 3 is a homomorphism of V into itself.

3. Verify that the mapping Φ in Example 5 is a homomorphism of $F^{(5)}$ *onto* $F^{(3)}$.

4. Verify that the mapping $\Phi: V \oplus W \longrightarrow V$ in Example 6 is a homomorphism.

5. Verify that the mapping Φ in Example 10 is a homomorphism of V into, but not onto, W.

Middle-Level Problems

6. In all Examples 1 through 9 given, find Ker (Φ). That is, express the form of the general element in Ker (Φ).

7. Prove that the identity mapping of a vector space V into itself is an isomorphism of V onto itself.

8. If $\Phi: V \longrightarrow W$ is defined by $\Phi(v) = 0$ for all $v \in V$, show that Φ is a homomorphism.

9. Prove Lemma 7.3.2.

10. In Examples 1 through 9, determine which Φ are isomorphisms.

MORE THEORETICAL PROBLEMS

Easier Problems

11. Prove for U, V, W vector spaces over F that:

   (a) $V \simeq V$.

   (b) $U \simeq V$ implies that $V \simeq U$.

   (c) $U \simeq V$, $V \simeq W$ implies that $U \simeq W$.

12. Let V be the vector space of all polynomials in x over $\mathbb{R}$ of degree m or less. Prove that $V \simeq F^{(m+1)}$.

13. If $m \leq n$ prove that there is a subspace V of $F^{(n)}$ such that $V \simeq F^{(m)}$.

14. If $\Phi: U \longrightarrow V$ and $\psi: V \longrightarrow W$ are homomorphisms of the vector spaces U, V, W over F, show that $\psi\Phi$ defined by $(\psi\Phi)(u) = \psi(\Phi(u))$ for every $u \in U$ is a homomorphism of U into W.

15. If $\Phi: V \longrightarrow W$ is a homomorphism and if $a \in F$, define

$$\psi(v) = a\Phi(v)$$

for every $v \in V$. Prove that ψ is a homomorphism of V into W.

16. Prove that $V \oplus W \simeq W \oplus V$.

Middle-Level Problems

17. Let $X = \{(v, v) \mid v \in V\}$. Show that $X \simeq V$.

18. If $\phi: M_n(F) \longrightarrow F$ is defined by $\Phi(A) = \operatorname{tr}(A)$, find Ker (Φ).

19. If Φ and ψ are homomorphisms of V into W, define $\Phi + \psi$ by $(\Phi + \psi)(v) = \Phi(v) + \psi(v)$ for every v in V. Prove that $\Phi + \psi$ is a homomorphism of V into W.

20. Combine the results of Problems 15 and 19 to show that the set of homomorphisms of V into W is a vector space over F.

21. Let $\Phi: V \oplus V \longrightarrow V$ be defined by $\Phi(v_1, v_2) = v_1 + v_2$. Find Ker (Φ) and show $V \oplus V = \{(v, v) \mid v \in V\} + $ Ker (Φ) (see Problem 13 of Section 7.2).

22. Suppose that U, W are subspaces of V. Define $\psi: U \oplus W \longrightarrow V$ by $\psi(u, w) = u + w$. Show that Ker $(\psi) \simeq U \cap W$.

23. If V is finite-dimensional over F and Φ is a homomorphism of V into W, show that $\phi(V) = \{\phi(v) \mid v \in V\}$ is finite-dimensional over F.

24. Let $V = F^{(n)}$ and $W = F^{(m)}$. Show that every homomorphism of V into W can be given by an $m \times n$ matrix over F.

25. Show that as vector spaces, $M_n(F) \simeq F^{(n^2)}$.

7.4 ISOMORPHISMS FROM V TO $F^{(n)}$

Recall that Φ is said to be an isomorphism of V *onto* W if Φ is a homomorphism of V *onto* W which is $1-1$. We also saw that Φ is $1-1$ if and only if Ker $(\Phi) = \{0\}$.

As we pointed out in Section 7.3, two vector spaces that are isomorphic are "essentially" equal. For one thing any result in one of these vector spaces is carried into an analogous result in the other one via the isomorphism.

This section has as its principal goal the theorem that if V is a finite-dimensional vector space over F, then $V \simeq F^{(n)}$ for some n. The dimension n of $F^{(n)}$ will become the dimension of V over F.

Every result proved in Chapters 3 and 4 for $F^{(n)}$ about linear independence, bases, dimension, and so on, becomes a theorem in V, established by use of the isomorphism of V onto $F^{(n)}$. We shall make a list of the principal results that transfer this way, referring back to the proofs given in $F^{(n)}$.

Suppose then that V is finite-dimensional over F. By the very definition of "finite-dimensional" we know that there is a finite set of elements, $v_1, \ldots, v_k$, in V that span V over F. Such a set of vectors is called a *generating set of V over F*. Since there is such a finite set, there is a finite set with the fewest number of elements that span V over F. We call such a smallest set, $u_1, \ldots, u_n$, a *minimal* generating set of V over F.

At this point we introduce concepts that we introduced and exploited heavily in Chapters 3 and 4, namely, linear independence and basis over F. As we shall soon see, a minimal generating set $u_1, \ldots, u_n$ of V over F will turn out to be a basis for V over F.

Definition.   The elements $v_1, \ldots, v_m$ in V are said to be *linearly independent over F* if $a_1 v_1 + \cdots + a_m v_m = 0$ only if $a_1 = 0, \ldots,$ $a_m = 0$. If $v_1, \ldots, v_m$ are *not* linearly independent over F, we say that they are *linearly dependent over F*.

Two small things should be pointed out.

To say that elements $v_1, \ldots, v_m$ of V are linearly independent is equivalent to saying that if v is in the span of $v_1, \ldots, v_m$, then v has a unique representation $v = a_1 v_1 + \cdots + a_m v_m$. That is, if v is also $v = b_1 v_1 + \cdots + b_m v_m$, then $a_1 = b_1, a_2 = b_2,$ $\ldots, a_m = b_m$.

Why? Suppose first that $v_1, \ldots, v_m$ are linearly independent and that an element v in their span can be written as

$$v = a_1 v_1 + \cdots + a_m v_m$$

and also as

$$v = b_1 v_1 + \cdots + b_m v_m.$$

Subtracting the first expression from the second, we get

$$0 = (b_1 - a_1)v_1 + \cdots + (b_m - a_m)v_m$$

so that

$$(b_1 - a_1) = 0, \ldots, (b_m - a_m) = 0$$

by the linear independence. Thus

$$a_1 = b_1, \ldots, a_m = b_m.$$

Conversely, suppose that each element v in the span of $v_1, \ldots, v_m$ has a unique expression $v = a_1 v_1 + \cdots + a_m v_m$. Then the condition $a_1 v_1 + \cdots + a_m v_m = 0$ implies that

$$a_1 v_1 + \cdots + a_m v_m = 0v_1 + \cdots + 0v_m,$$

so that

$$a_1 = 0, \ldots, a_m = 0$$

by the unicity of the representation.

> *If $v_1, \ldots, v_m$ are linearly independent over F, then none of $v_1, \ldots, v_m$ can be 0.*

For suppose that $v_1 = 0$; then $av_1 + 0v_2 + \cdots + 0v_m = 0$ for any $a \in F$. By the paragraph above, we see that $v_1, \ldots, v_m$ then cannot be linearly independent over F.

In V, the vector space of all polynomials in x over F, the elements

$$1 + x, 2 - x + x^2, x^3, \tfrac{1}{4}x^5$$

can easily be shown to be linearly independent over F. (Do it!) On the other hand, the elements

$$5, 1 + x, 2 + x, 7 + x + x^2$$

are linearly dependent over F, since

$$-\tfrac{1}{5}(5) + (-1)(1 + x) + 1(2 + x) + (0)(7 + x + x^2) = 0.$$

Definition. A *basis* for V is a set $v_1, \ldots, v_n$ of vectors of V such that

1. The vectors $v_1, \ldots, v_n$ are linearly independent;
2. The vectors $v_1, \ldots, v_n$ span V.

A generating set $v_1, \ldots, v_n$ for a finite-dimensional vector space V is a basis if and only if it is linearly independent. If $v_1, \ldots, v_n$ is a basis for V over F, then each element $v \in V$ has a *unique* representation $v = a_1 v_1 + \cdots + a_n v_n$, where the $a_1, \ldots, a_n$ are from F. Using this, we now prove the

Theorem 7.4.1. If $u_1, \ldots, u_n$ is a minimal generating set of finite-dimensional vector space V over F, then $u_1, \ldots, u_n$ is a basis for V over F.

Proof. Suppose to the contrary that $u_1, \ldots, u_n$ are linearly dependent, that is,

$$a_1 u_1 + \cdots + a_n u_n = 0,$$

where not all the a_i are 0. Since the numbering of $u_1, \ldots, u_n$ is not important, we may assume that $a_1 \neq 0$. Then

$$a_1 u_1 = -a_2 u_2 - \cdots - a_n u_n,$$

and since $a_1 \neq 0$, we have

$$u_1 = (-a_2/a_1)u_2 - \cdots - (a_n/a_1)u_n = b_2 u_2 + \cdots + b_n u_n,$$

where the $b_r = (-a_r/a_1)$ are in F. For $v \in V$, then, since $u_1, \ldots, u_n$ is a generating set of V over F, we have

$$
\begin{aligned}
v &= c_1 u_1 + \cdots + c_n u_n \\
  &= c_1(b_2 u_2 + \cdots + b_n u_n) + c_2 u_2 + \cdots + c_n u_n \\
  &= (c_1 b_2 + c_2)u_2 + \cdots + (c_1 b_n^- + c_n)u_n
\end{aligned}
$$

and since the $c_r b_s + c_s$ are in F, we get that $u_2, \ldots, u_n$ already span V over F. Thus $u_2, \ldots, u_n$ has fewer elements than $u_1, \ldots, u_n$, and $u_1, \ldots, u_n$ has been assumed to be a *minimal* generating set. This is a contradiction. So we are forced to conclude that all of $a_1, \ldots, a_n$ are 0. In other words, $u_1, \ldots, u_n$ are linearly independent over F. Since they span V, they are a basis. ■

Corollary 7.4.2. Every finite-dimensional vector space has a basis. We are now in a position to prove the principal results of this section.

Theorem 7.4.3. If V is a vector space with basis $v_1, \ldots, v_n$ over F, then there is an isomorphism Φ from V to $F^{(n)}$ such that

$$\Phi(a_1 v_1 + \cdots + a_n v_n) = \begin{bmatrix} a_1 \\ \vdots \\ a_n \end{bmatrix}$$

for all $a_1, \ldots, a_n$ in F.

Proof. Let $v_1, \ldots, v_n$ be a basis for V over F. Given $u \in V$, u has a representation as $u = a_1 v_1 + \cdots + a_n v_n$. Since the set of coefficients $a_1, \ldots, a_n$ are the only ones that will give us u, by the linear independence of the vectors $v_1, \ldots, v_n$, we can use them to

define $\Phi: V \to F^{(n)}$ by $\Phi(u) = \begin{bmatrix} a_1 \\ \vdots \\ a_n \end{bmatrix}$.

Clearly Φ maps V *onto* $F^{(n)}$, for given $\begin{bmatrix} c_1 \\ \vdots \\ c_n \end{bmatrix}$ in $F^{(n)}$, then

$$\begin{bmatrix} c_1 \\ \vdots \\ c_n \end{bmatrix} = \Phi(c_1 v_1 + \cdots + c_n v_n).$$

We claim that Φ is a homomorphism of V onto $F^{(n)}$. Why? If u, w are in V, then,

$$u = a_1 v_1 + \cdots + a_n v_n$$

and

$$w = b_1 v_1 + \cdots + b_n v_n.$$

Thus

$$u + w = (a_1 v_1 + \cdots + a_n v_n) + (b_1 v_1 + \cdots + b_n v_n)$$
$$= (a_1 + b_1) v_1 + \cdots (a_n + b_n) v_n,$$

whence, by the definition of Φ,

$$\Phi(u + w) = \begin{bmatrix} a_1 + b_1 \\ \vdots \\ a_n + b_n \end{bmatrix} = \begin{bmatrix} a_1 \\ \vdots \\ a_n \end{bmatrix} + \begin{bmatrix} b_2 \\ \vdots \\ b_n \end{bmatrix}.$$

However, we recognize $\begin{bmatrix} a_1 \\ \vdots \\ a_n \end{bmatrix}$ as $\Phi(u)$ and $\begin{bmatrix} b_1 \\ \vdots \\ b_n \end{bmatrix}$ as $\Phi(w)$. Putting all these pieces together, we get that $\Phi(u + w) = \Phi(u) + \Phi(w)$. A very similar argument shows that $\Phi(au) = a\Phi(u)$ for $a \in F$, $u \in V$. In short, Φ is a homomorphism of V onto $F^{(n)}$.

To finish the proof, all we need to show now is that Φ is $1-1$. By Lemma 7.3.4 it is enough to show that Ker $(\Phi) = (0)$. What is Ker (Φ)? If $z = c_1v_1 + \cdots + c_nv_n$ is in Ker (Φ), then, by the very definition of kernel, $\Phi(z) = \begin{bmatrix} 0 \\ \vdots \\ 0 \end{bmatrix}$. But we know precisely what Φ does to z, by the definition of Φ, that is, we know that $\Phi(z) = \begin{bmatrix} c_1 \\ \vdots \\ c_n \end{bmatrix}$. Comparing these two evaluations of $\Phi(z)$, we get $\begin{bmatrix} c_1 \\ \vdots \\ c_n \end{bmatrix} = \begin{bmatrix} 0 \\ \vdots \\ 0 \end{bmatrix}$, that is, $c_1 = c_2 = \cdots = c_n = 0$. Thus Ker (Φ) consists only of 0. Thus Φ is an isomorphism of V onto $F^{(n)}$. This completes the proof of Theorem 7.4.3. ∎

Since every finite-dimensional vector space has a basis, by Corollary 7.4.2, we have the

Corollary 7.4.4.   Any finite-dimensional vector space over F is isomorphic to $F^{(n)}$ for some positive integer n.

Theorem 7.4.3 opens a floodgate of results for us. By the isomorphism given in this theorem, we can carry over the whole corpus of results proved in Chapters 3 and 4 without the need to go into proof. Why? We can now use the following general *transfer principle*, thanks to Theorem 7.4.3 and Lemma 7.3.1:

> *Any phenomena expressed in $F^{(n)}$ in terms of addition, scalar multiplication, or linear combinations are transferred by any isomorphism ϕ from V to $F^{(n)}$ to corresponding phenomena expressed in V in terms of addition, scalar multiplication, or linear combinations.*

Let's be specific. In $F^{(n)}$, concepts of span, linear independence, basis, dimension, and so on, have been defined in such terms. In a finite-dimensional vector space V, we have defined span, linear independence along the same lines. Since $F^{(n)}$ has the standard basis $e_1, \ldots, e_n$ and any basis for $F^{(n)}$ has n elements by Corollary 3.6.4, $F^{(n)}$ has a basis and any two bases have the same number of elements. By our transfer principle, it follows that any finite-dimensional vector space has a basis and any two bases have the same number of elements.

We have now established the following theorems.

Theorem 7.4.5. A finite-dimensional vector space V over F has a basis.

Theorem 7.4.6. Any two bases of a finite-dimensional vector space V over F have the same number of elements.

We now can define dimension.

Definition. The *dimension* of a finite-dimensional vector space V over F is the number of elements in a basis. It is denoted by dim (V).

Of course, all results from Chapter 3 on $F^{(n)}$ that can be expressed in terms of linear transformations carry over by our transfer principle to finite-dimensional vector spaces. Here is a partial list corresponding to principal results:

1. $v_1, \ldots, v_n$ form a basis if and only if $v_1, \ldots, v_n$ is a minimal generating set.
2. $v_1, \ldots, v_n$ are linearly independent implies that $m \leq$ dim (V).
3. $v_1, \ldots, v_n$ form a basis if and only if $v_1, \ldots, v_n$ is a maximal linearly independent subset of V.
4. Any linearly independent subset of V is contained in a basis of V.
5. Any linearly independent set of n elements of an n-dimensional vector space V is a basis.

We shall return to linear independence soon, because we need this concept even if V is not finite-dimensional over F.

As we shall see as we progress, using Theorem 7.4.3 we will be able to carry over the results on inner product spaces to the general context of vector spaces.

PROBLEMS

MORE THEORETICAL PROBLEMS

Easier Problems

1. Make a list of all the results in Chapter 3 on $F^{(n)}$ as a vector space that can be carried over to a general finite-dimensional vector space by use of Theorem 7.4.3.

2. Prove the converse of Theorem 7.4.1—that any basis for V over F is a minimal generating set.

7.5. LINEAR INDEPENDENCE IN INFINITE-DIMENSIONAL VECTOR SPACES

We have described linear independence and dependence several times. Why, then, redo these things once again? The basic reason is that in all our talk about this we stayed in a finite-dimensional context. But these notions are important and meaningful even in the infinite-dimensional situation. This will be especially true when we pick up the subject of inner product spaces.

As usual, if V is any vector space over F, we say that the finite set $v_1, \ldots, v_n$ of elements in V is *linearly independent over* F if $a_1 v_1 + \cdots + a_n v_n = 0$, where $a_1, \ldots, a_n$ are in F, only if $a_1 = a_2 = \cdots = a_n = 0$. To avoid repetition of a phrase, we shall merely say "linearly independent," it being understood that this is over F. Of course, if $v_1, \ldots, v_n$ are not linearly independent, we say that they are *linearly dependent*.

But we can also speak about an infinite subset as a subset of linearly independent (or dependent) elements. We do so as follows:

Definition. If $S \subset V$ is any nonempty subset, we say that S is *linearly independent* if *any finite* subset of S consists of linearly independent elements.

For instance, if V is the vector space of polynomials in x over $\mathbb{R}$, then the set

$$S = \{1, x, x^2, \ldots, x^k, \ldots\}$$

is a linearly independent set of elements. If

$$x^{i_1}, \ldots, x^{i_n}$$

is any finite subset of S, where

$$0 \leq i_1 < i_2 < \cdots < i_n,$$

and if $a_1 x^{i_1} + \cdots + a_n x^{i_n} = 0$, then, by invoking the definition of when a polynomial is identically zero, we get that $a_1, \ldots, a_n$ are all 0. Thus S is a linearly independent set.

It is easy to construct other examples. For instance, if V is the vector space of all real-valued differential functions, then the functions

$$e^x, e^{2x}, \ldots, e^{nx}, \ldots$$

are linearly independent over $\mathbb{R}$. Perhaps it would be worthwhile to verify this statement in this case, for it may provide us with a technique we could use elsewhere. Suppose then that

$$a_1 e^{m_1 x} + \cdots + a_k e^{m_k x} \equiv 0, \ a_k \neq 0,$$

where

$$1 \leq m_1 < m_2 < \cdots < m_k$$

are integers. Suppose that this is the *shortest possible* such relation. We differentiate the expression to get

$$m_1 a_1 e^{m_1 x} + \cdots + m_k a_k e^{m_k x} \equiv 0.$$

Multiplying the first by m_k and subtracting the second we get

$$(m_k - m_1) a_1 e^{m_1 x} + \cdots + (m_k - m_{k-1}) a_{k-1} e^{m_{k-1} x} \equiv 0.$$

This is a *shorter* relation, so each $(m_k - m_j) a_j = 0$ for $1 \leq j \leq k - 1$. But since $m_k > m_j$ for $j \leq k - 1$, we end up with $a_j = 0$ for $j \leq k - 1$. This leaves us with $a_k m^{kx} \equiv 0$, hence $a_k = 0$, a contradiction. Therefore, the functions

$$e^x, e^{2x}, \ldots, e^{mx}, \ldots$$

are linearly independent over $\mathbb{R}$.

In the problem set there will be other examples. This is about all we wanted to add to the notion of linear independence, namely, to stress that the concept has significance even in the infinite-dimensional case.

PROBLEMS

NUMERICAL PROBLEMS

1. If V is the vector space of differentiable functions over $\mathbb{R}$, show that the following sets are linearly independent.

   (a) $\cos(x)$, $\cos(2x)$, $\ldots$, $\cos(nx)$, $\ldots$

   (b) $\sin(x)$, $\sin(2x)$, $\ldots$, $\sin(nx)$, $\ldots$

   (c) $\cos(x)$, $\sin(x)$, $\cos(2x)$, $\sin(2x)$, $\ldots$, $\cos(nx)$, $\sin(nx)$, $\ldots$

   (d) 1, $2 + x$, $3 + 2x^2$, $4 + 3x^3$, $\ldots$, $n + (n-1)x^{n-1}$, $\ldots$

2. Which of the following sets are linearly independent, and which are not?

   (a) V is the vector space of polynomials over F, and the elements are 1, $2 + x$, $3 + 2x$, $\ldots$, $n + (n-1)x$, $\ldots$

   (b) V is the set of all real valued functions on $[0, 1]$ and the elements are $\dfrac{1}{x + 1}$, $\dfrac{1}{x + 2}$, $\ldots$, $\dfrac{1}{x + n}$, $\ldots$

   (c) V is the set of polynomials over $\mathbb{R}$ and the elements are 1, x, $1 + x + x^2$, $1 + x + x^2 + x^3$, $\ldots$, $1 + x + \cdots + x^n$, $\ldots$

   (d) V is the set of polynomials over F, and the elements are 1, $(x + 2)$, $(x + 3)^2$, $\ldots$, $(x + n)^{n-1}$, $\ldots$

MORE THEORETICAL PROBLEMS

If S is in V, we say that the *subspace of V spanned by S over F* is the set of *all finite linear combinations* of elements of S over F.

Middle-Level Problems

3. Find the subspace of V spanned by the given elements:

   (a) V is the set of polynomials in x over F, and the elements are 1, $(x + 2)$, $\ldots$, $(x + n)^{n-1}$, $\ldots$

   (b) V is the set of polynomials in x over F, and the elements are 1, x, x^2, $\ldots$, x^n, $(x + 1)^n$, $\ldots$, $(x + k)^n$, $\ldots$

4. Show that there is no *finite* subset of the vector space of polynomials in x over F which span all this vector space.

5. Let V be the vector space of all complex-valued functions, as a vector space over $\mathbb{C}$. Let e^{ix} denote the function $\cos(x) + i\sin(x)$, where $i^2 = -1$.

(a) Show that $e^{\pm ix}$, $e^{\pm 2ix}$, $\ldots$, $e^{\pm inx}$, $\ldots$ are linearly independent over $\mathbb{C}$.

(b) From Part (a), show that $\cos(x)$, $\sin(x)$, $\ldots$, $\cos(nx)$, $\sin(nx)$, $\ldots$ are linearly independent over $\mathbb{C}$.

(c) Show that $e^{\pm ix}$, $\ldots$, $e^{\pm inx}$, $\ldots$, $\cos(x)$, $\cos(2x)$, $\ldots$, $\cos(nx)$, $\ldots$ are *not* linearly independent over $\mathbb{C}$.

Harder Problems

6. Let V be the vector space of all continuous real-valued functions on the real line and let S be the set of all differentiable functions on the real line. Prove that S does not span V over $\mathbb{R}$ by showing that $f(x) = |x|$ is *not* in the span of S.

7. If $f_0(x)$, $\ldots$, $f_n(x)$ are differentiable functions over $\mathbb{R}$, show that $f_0(x)$, $\ldots$, $f_n(x)$ are linearly independent over $\mathbb{R}$ if

$$\begin{vmatrix} f_0(x) & \cdots & f_n(x) \\ f_0'(x) & \cdots & f_n'(x) \\ \cdots & \cdots & \cdots \\ f_0^{(n)}(x) & \cdots & f_n^{(n)}(x) \end{vmatrix}$$

is not identically 0. [This determinant is called the *Wronskian* of $f_0(x)$, $\ldots$, $f_n(x)$.]

8. Use the result of Problem 7 to show that the set e^x, e^{2x}, $\ldots$, e^{nx} is linearly independent.

9. Use the result of Problem 8 to show that the set e^x, e^{2x}, $\ldots$, e^{nx}, $\ldots$ is a linearly independent set.

7.6. INNER PRODUCT SPACES

Once again we return to a theme that was expounded on at some length in Chapter 3, the notion of an *inner product space*. Recall that in $\mathbb{C}^{(n)}$

we defined the inner product of the vectors $v = \begin{bmatrix} a_1 \\ \vdots \\ a_n \end{bmatrix}$, $w = \begin{bmatrix} b_1 \\ \vdots \\ b_n \end{bmatrix}$ by

$(v, w) = \sum_{j=1}^{n} a_j \bar{b}_j$. We discussed many properties of this inner product and saw how useful a tool it was in deriving theorems about $F^{(n)}$ and about matrices.

In line with our general program we want to carry these ideas over to general vector spaces over $\mathbb{C}$ (or $\mathbb{R}$). In fact, for a large part we will not even insist that the vector space be finite-dimensional. When we do focus on finite-dimensional inner product spaces we shall see that they are isomorphic to $F^{(n)}$ by an isomorphism that preserves inner products. This will become more precise when we actually produce this isomorphism.

It is easy to define what we shall be talking about.

Definition.   A vector space V over $\mathbb{C}$ will be called an *inner product space* if there is defined a function from V to $\mathbb{C}$, denoted by (v, w), for pairs v, w in V, such that:

1. $(v, v) \geq 0$, and $(v, v) = 0$ only if $v = 0$.
2. $(v, w) = \overline{(w, v)}$, the complex conjugate of (v, w).
3. $(av, w) = a(v, w)$ for $a \in \mathbb{C}$.
4. $(v_1 + v_2, w) = (v_1, w) + (v_2, w)$.

Notice that (1)–(4) imply:

1. $(v, aw) = \bar{a}(v, w)$. For

$$(v, aw) = \overline{(aw, v)}$$
$$= \overline{a(w, v)} \text{ [by (3)]}$$
$$= \bar{a}\overline{(w, v)} = \bar{a}(v, w) \text{ [by (2)]}.$$

2. $(v, w_1 + w_2) = (v, w_1) + (v, w_2)$. For

$$(v, w_1 + w_2) = \overline{(w_1 + w_2, v)}$$
$$= \overline{(w_1, v) + (w_2, v)} \text{ [by (4)]}$$
$$= \overline{(w_1, v)} + \overline{(w_2, v)} = (v, w_1) + (v, w_2) \text{ [by (2)]}.$$

3.   If $(v, w) = 0$ for all $w \in V$, then $v = 0$. For in particular,

$$(v, v) = 0,$$

hence by (1), $v = 0$.

It might be illuminating to see some examples of inner product spaces.

EXAMPLES

1. We already saw one example, namely, $V = \mathbb{C}^n$ and

$$(v, w) = \sum_{j=1}^{n} a_j \bar{b}_j, \text{ where } v = \begin{bmatrix} a_1 \\ \vdots \\ a_n \end{bmatrix}, w = \begin{bmatrix} b_1 \\ \vdots \\ b_n \end{bmatrix}.$$

2. This will be an example of an inner product space over $\mathbb{R}$. Let V be the set of all continuous functions on the closed interval $[-1, 1]$. For $f(x)$, $g(x)$ in V define

$$(f, g) = \int_{-1}^{1} f(x)g(x)\, dx.$$

For instance, if $f = x$, $g = x^3$, then

$$(x, x^3) = \int_{-1}^{1} xx^3\, dx = \int_{-1}^{1} x^4\, dx = [x^5/5]_{-1}^{1} = \tfrac{2}{5}.$$

What is (x, e^x)?

We verify the various properties of an inner product.

1. $(f, f) = \int_{-1}^{1} f(x)f(x)\, dx \geq 0$ since $(f(x))^2 \geq 0$ and this integral is 0 if and only if $f(x) \equiv 0$.

2. $(f, g) = \int_{-1}^{1} f(x)g(x)\, dx = \int_{-1}^{1} g(x)f(x)\, dx = (g, f)$ [and $(g, f) = \overline{(f, g)}$ since everything in sight is real].

3. $(f_1 + f_2, g) = \int_{-1}^{1} [f_1(x) + f_2(x)]g(x)\, dx = \int_{-1}^{1} [f_1(x)g(x) + f_2(x)g(x)]\, dx = \int_{-1}^{1} f_1(x)g(x)\, dx + \int_{-1}^{1} f_2(x)g(x)\, dx = (f_1, g) + (f_2, g)$.

4. $(af, g) = \int_{-1}^{1} af(x)g(x)\, dx = a \int_{-1}^{1} f(x)g(x)\, dx = a(f, g)$ for $a \in \mathbb{R}$.

So V is an inner product space over $\mathbb{R}$.

3. Let V be the set of differentiable functions on $[-\pi, \pi]$ and let W be the linear span of

$$\cos(x), \cos(2x), \ldots, \cos(nx), \ldots$$

over $\mathbb{R}$. We make V into an inner product space over $\mathbb{R}$ by defining

$$(f, g) = \int_{-\pi}^{\pi} f(x)g(x) \, dx.$$

As in Example 2, V is an inner product space relative to this inner product. We want to compute the inner product of the elements $\cos (mx)$ which span W. For m, n we have

$$(\cos (mx), \cos (nx)) = \int_{-\pi}^{\pi} \cos (mx) \cos (nx) \, dx.$$

If $m = n \neq 0$, we have

$$(\cos (mx), \cos (mx)) = \int_{-\pi}^{\pi} \cos^2 (mx) \, dx = \pi,$$

and if $m \neq n$,

$$(\cos (mx), \cos (nx)) = \int_{-\pi}^{\pi} \cos (mx) \cos (nx) \, dx,$$

which equals

$$\left(\frac{1}{2(m + n)} \sin ([m + n]x) + \frac{1}{2(m - n)} \sin ([m - n]x) \right)_{-\pi}^{\pi} = 0.$$

Finally, if $m = n = 0$.

$$\int_{-\pi}^{\pi} \cos (mx) \cos (nx) \, dx = \int_{-\pi}^{\pi} dx = 2\pi.$$

The example is an interesting one. Relative to the inner product the functions

$$\cos (x), \quad \cos (2x), \quad \ldots, \quad \cos (nx), \quad \ldots$$

form an *orthogonal* set in V.

4. Let V be as in Example 3 and let S be the set

$$\{\cos (mx) | \text{ all integers } m \geq 0\} \cup \{\sin (nx) \mid \text{ all integers } n \geq 1\}.$$

As in Example 3, if we define the inner product

$$(f, g) = \int_{-\pi}^{\pi} f(x)g(x) \, dx.$$

Then S is a linearly independent set and $(f, g) = 0$ for $f \neq g$ in S.

This remark is the foundation on what is known as *Fourier series* rests.

We now go on to some properties of inner product spaces. The first result is one we proved earlier, in Chapter 1. But we redo it here, in exactly the same way.

Lemma 7.6.1. If $a > 0$, b, c are real numbers and $ax^2 + bx + c \geq 0$ for all real x, then $b^2 - 4ac \leq 0$.

Proof. Since $ax^2 + bx + c \geq 0$ for all real x, it is certainly true for the real number $-b/2a$:

$$a(-b/2a)^2 + b(-b/2a) + c \geq 0.$$

Hence $b^2/4a - b^2/2a + c \geq 0$, that is, $c - b^2/4a \geq 0$. Because $a > 0$, this leads us to $b^2 \leq 4ac$. ∎

This technical result has a famous inequality as a consequence. It is known as the *Schwarz Inequality*.

Lemma 7.6.2. In an inner product space V,

$$|(v, w)|^2 \leq (v, v)(w, w)$$

for all v, w in V.

Proof. If $v, w \in V$ and $x \in \mathbb{R}$ then, by Property 1 defining an inner product space, $(v + xw, v + xw) \geq 0$. Expanding this, we obtain

$$(v, v) + x((v, w) + (w, v)) + x^2(w, w) \geq 0.$$

We first settle the case in which (v, w) is real. Then $(v, w) = \overline{(v, w)} = (w, v)$. So the inequality above becomes

$$(w, w)x^2 + 2x(v, w) + (v, v) \geq 0$$

for all real x. Invoking Lemma 7.6.1, with

$$a = (w, w), \, b = 2(v, w), \, c = (v, v),$$

we get

$$4(v, w)^2 \leq 4 \, (v, v)(w, w),$$

so

$$|(v, w)|^2 \leq (v, v)(w, w),$$

the desired conclusion.

Suppose then that (v, w) is not real; then certainly $(v, w) \neq 0$. Let $z = v/(v, w)$; then

$$(z, w) = \left(\frac{v}{(v, w)}, w \right) = \frac{1}{(v, w)} (v, w) = 1.$$

Thus (z, w) is real. By what we did above,

$$1 = |(z, w)|^2 \leq (z, z)(w, w).$$

Now

$$(z, z) = \left(\frac{v}{(v, w)}, \frac{v}{(v, w)} \right) = \frac{1}{(v, w)} \frac{1}{\overline{(v, w)}} (v, v)$$

$$= \frac{1}{|(v, w)|^2} (v, v)$$

since $(v, w)\overline{(v, w)} = |(v, w)|^2$. So we obtain that

$$1 = |(z, w)|^2 \leq \frac{1}{|(v, w)|^2} (v, v)(v, v),$$

from which we get

$$|(v, w)|^2 \leq (v, v)(w, w).$$

This finishes the proof.   ■

If $F^{(n)}$ we saw that this lemma implies that

$$\left(\left| \sum_{j=1}^{n} a_j \overline{b_j} \right| \right)^2 \leq \left(\sum_{j=1}^{n} |a_j|^2 \right) \left(\sum_{j=1}^{n} |b_j|^2 \right)$$

using the definition $(v, w) = \sum_{j=1}^{n} a_j \overline{b_j}$, where $v = \begin{bmatrix} a \\ a_2 \\ \vdots \\ a_n \end{bmatrix}$ and

$w = \begin{bmatrix} b_1 \\ b_2 \\ \vdots \\ b_n \end{bmatrix}$.

Perhaps of greater interest is the case of the real continuous functions, as a vector space over $\mathbb{R}$, with the inner product defined, say, by

$$(f, g) = \int_{-1}^{1} f(x)g(x) \, dx.$$

The Schwarz inequality then tell us that

$$\left| \int_{-1}^{1} f(x)g(x) \, dx \right|^2 \leq \left(\int_{-1}^{1} f^2(x) \, dx \right)\left(\int_{-1}^{1} g^2(x) \, dx \right).$$

Of course, we could define the inner product using other limits of integration. These integral inequalities are very important in analysis in carrying out estimates of integrals. For example, we could use this to estimate integrals where we cannot carry out the integration in closed form.

We apply it, however, to a case where we could do the integration. Suppose, that we wanted some upper bound for $\int_{-\pi}^{\pi} x \sin (x) \, dx$. According to the Schwarz inequality,

$$\left| \int_{-\pi}^{\pi} x \sin (x) \, dx \right|^2 \leq \left(\int_{-\pi}^{\pi} x^2 \, dx \right)\left(\int_{-\pi}^{\pi} \sin^2 (x) \, dx \right)$$

$$= (\tfrac{2}{3} \pi^3)(\pi)$$

$$= \tfrac{2}{3} \pi^4.$$

We know that $\int x \sin (x) \, dx = \sin (x) - x \cos (x)$, so

$$\int_{-\pi}^{\pi} x \sin (x) \, dx = 2\pi.$$

Hence

$$\left(\int_{-\pi}^{\pi} x \sin (x) \, dx \right)^2 = (2\pi)^2 = 4\pi^2$$

So we see here that the upper bound $2\pi^4/3$ is crude; but not too crude.

In an inner product space we can introduce the concept of length.

Definition. If V is an inner product space, then the *length* of v in V, denoted by $\|v\|$, is defined by $\|v\| = \sqrt{(v, v)}$.

In $\mathbb{R}^{(n)}$ this coincides with our usual (and intuitive) notion of length. Note that $\|v\| = 0$ if and only if $v = 0$.

What properties does this length function enjoy? One of the basic ones is the *triangle inequality*.

Lemma 7.6.3.   For $v, w \in V$, $\|v + w\| \leq \|v\| + \|w\|$.

   Proof.   By definition, $\|v + w\| = \sqrt{(v + w, v + w)}$. Now

$$(v + w, v + w) = (v, v) + (w, w) + (v, w) + (w, v).$$

By Lemma 7.6.2, $|(v, w)| = |(w, v)| \leq \sqrt{(v, v)(w, w)}$. So

$$
\begin{aligned}
\|v + w\|^2 &= (v + w, v + w) \\
&= (v, v) + (w, w) + (v, w) + (w, v) \\
&\leq (v, v) + (w, w) + 2\sqrt{(v, v)(w, w)} \\
&= \|v\|^2 + \|w\|^2 + 2\|v\|\,\|w\|.
\end{aligned}
$$

This gives us

$$\|v + w\|^2 \leq (\|v\|^2 + \|w\|^2 + 2\|v\|\,\|w\|) = (\|v\| + \|w\|)^2.$$

Taking square roots, we end up with $\|v + w\| \leq \|v\| + \|w\|$, the desired result.   ∎

   In working in $\mathbb{C}^{(n)}$ we came up with a scheme whereby from a set $v_1, \ldots, v_k$ of linearly independent vectors we created a set $w_1, \ldots, w_k$ of orthogonal vectors. We want to do the same thing in general, especially when V is infinite-dimensional over $\mathbb{C}$ (or $\mathbb{R}$). But first we better say what is meant by "orthogonal."

Definition.   If v, w are in V, then v is said to be *orthogonal* to w if $(v, w) = 0$.

   Since $(v, w) = \overline{(w, v)}$ we see immediately that if v is orthogonal to w, then w is orthogonal to v.
   Let S be a set of non-zero orthogonal elements in V.

Lemma 7.6.4.   A set S of nonzero orthogonal elements in V is a linearly independent set.

   Proof.   Suppose for some finite set of elements $s_1, \ldots, s_k$ in S that $a_1 s_1 + \cdots + a_k s_k = 0$. Thus

$$
\begin{aligned}
0 &= (a_1 s_1 + \cdots + a_k s_k, s_j) \\
&= (a_1 s_1, s_j) + \cdots + (a_k s_k, s_j) \\
&= a_1(s_1, s_j) + \cdots + a_k(s_k, s_j).
\end{aligned}
$$

Because the elements of S are orthogonal, $(s_t, s_j) = 0$ for $t \neq j$. So the equation above reduces to $a_j(s_j, s_j) = 0$. Since $(s_j, s_j) \neq 0$, the outcome of all this is that each $a_j = 0$. In other words, any finite subset of S consists of linearly independent elements. Hence S is a linearly independent set. ∎

Definition. A set S in V is said to be an *orthonormal* set if S is an orthogonal set and each element in S is of length 1.

Note that from an orthogonal set of nonzero elements we readily produce an orthonormal set. This is the content of

Lemma 7.6.5. If S is an orthogonal set of nonzero elements of V, then $\tilde{S} = \left\{ \dfrac{s}{\|s\|} \mid s \in S \right\}$ is an orthonormal set.

Proof. Given $s \in S$, then $\left\| \dfrac{s}{\|s\|} \right\| = \dfrac{1}{\|s\|} \|s\| = 1$, so every element in $\tilde{S}$ has length 1. Also, if $s \neq t$ are in S, then $(s, t) = 0$, hence

$$\left(\frac{s}{\|s\|}, \frac{t}{\|t\|} \right) = \frac{1}{\|s\|} \frac{1}{\|t\|} (s, t) = 0.$$

In short, $\tilde{S}$ is an orthonormal set. ∎

In the next section we shall see a procedure—which we saw operate in $\mathbb{C}^{(n)}$—of how to produce orthonormal sets from sets of linearly independent elements.

PROBLEMS

NUMERICAL PROBLEMS

1. Compute the inner products (f, g) in V, the set of all real-valued continuous functions on $[-\pi, \pi]$, where $(f, g) = \int_{-1}^{1} f(x)g(x)\, dx$.

   (a) $f(x) = \cos(x)$, $g(x) = x$.

   (b) $f(x) = e^x$, $g(x) = \sin(x) + \cos(x)$.

(c) $f(x) = \cos(4x)$, $g(x) = \sin(x)$.

(d) $f(x) = e^x$, $g(x) = 1 - e^x$.

2. In V of Problem 1, find a vector w orthogonal to the given v.

(a) $v = \cos(x) + \sin(2x) + \cos(3x)$.

(b) $v = e^x$.

(c) $v = 1 + x$.

3. Show that $(0, w) = (v, 0)$ for all $v, w \in V$, where V is an inner product space.

4. In Problem 1 find the length of each of the given functions $g(x)$.

MORE THEORETICAL PROBLEMS

Easier Problems

5. If W is a subspace of V, let

$$W^\perp = \{v \in V \mid (v, w) = 0 \text{ for all } w \in W\}.$$

(a) Prove that $W^\perp$ is a subspace of V.

(b) Show that $(W^\perp)^\perp \supset W$.

Middle-Level Problems

6. In Problem 5 show that $((W^\perp)^\perp)^\perp = W^\perp$.

7. If V is the vector space of polynomials with real coefficients with inner product give by $(f, g) = \int_{-1}^{1} f(x)g(x)\, dx$, find $W^\perp$ for

(a) W is the linear span of 1, x, x^2.

(b) W is the linear span of $x + x^2$, $x^2 + x^3$.

8. Show that if W is a subspace of V, then $W \cap W^\perp = \{0\}$.

9. in V, an inner product space, define the *distance* between two elements v and w by $d(v, w) = \|v - w\|$. Prove:

(a) $d(v, w) = 0$ if and only if $v = w$.

(b) $d(v, w) = d(w, v)$.

(c) $d(v, w) + d(w, z) \geq d(v, z)$ (triangle inequality).

10. If V is the inner product space of continuous real-valued functions on $[-\pi, \pi]$ find $d(v, w)$ for

(a) $v = \cos(x)$, $w = \sin(x)$.

(b) $v = e^x$, $w = e^{-x}$.

(c) $v = \cos(x) + \cos(2x)$, $w = 3\sin(x) - 4\sin(2x)$.

Harder Problems

11. If W is a subspace of V, let

$$W + W^\perp = \{u + v \mid u \in W, \, v \in W^\perp\}.$$

Show that if z is orthogonal to every element of $W + W^\perp$, then $z = 0$.

12. Let $V = M_n(\mathbb{R})$ and define for $A, B \in V$, the inner product of A and B by $(A, B) = \text{tr}\,(AB')$, where B' is the transpose of B. Show that this is an inner product on V.

13. Let E_{rs} be that matrix all of whose entries are 0 except the (r, s)-entry, which is 1. Prove that relative to the inner product of Problem 12, the E_{rs} form an orthonormal basis of V.

14. In Problem 12 show that $(AB, C) = (B, A'C)$.

Very Hard Problems

15. If V is an inner product space and $v \in V$, let

$$v^\perp = \{w \in V \mid (v, w) = 0\}.$$

Prove that $V = Fv + v^\perp$.

16. If V is an inner product space and W is a 2-dimensional subspace of V, show that $V = W + W^\perp$.

17. In Problem 12, if W is the subspace of all diagonal matrices, find $W^\perp$ and verify that $W + W^\perp = V = M_n(\mathbb{R})$.

7.7. MORE ON INNER PRODUCT SPACES

Once again we shall repeat something that was done earlier in $\mathbb{C}^{(n)}$ and $\mathbb{R}^{(n)}$. But here V will be an arbitrary inner product space over $\mathbb{R}$ or $\mathbb{C}$. The results we obtain are of importance not only for the finite-dimensional case but possibly more so for the infinite-dimensional one. One by-product of what we shall do is the fact that studying finite-dimensional inner product spaces is no more nor less than studying $F^{(n)}$ as an inner product space. This will be done by exhibiting a particular isomorphism. From this we shall be able to transfer everything we did in Chapters 3 and 4 for the special case of $F^{(n)}$ to arbitrary finite-dimensional vector spaces over F.

The principal tool we need is what was earlier called the *Gram–Schmidt process*. The outcome of this process is that for any set

$S = \{s_r\}$ of linearly independent elements we can find an orthonormal set S_1 such that the linear span of S over F equals the linear span of S_1 over F.

Theorem 7.7.1 (Gram–Schmidt). Let $S = \{s_r\}$ be a nonempty set of linearly independent elements in V. Then there is an orthonormal set S_1, contained in the linear span of S, such that the linear span of S_1 equals that of S.

     Proof. By Lemma 7.6.5 it is enough for us to produce an orthogonal set S_2 contained in the linear span of S, such that S_2 has the same linear span as does S. To modify S_2 to an orthonormal set—as is shown in Lemma 7.6.5—is easy.

     Since S consists of linearly independent elements, we know, to begin with, that 0 is not in S. Thus $(s, s) \neq 0$ for every $s \in S$.

     Suppose that the elements of S are enumerated as $s_1, s_2, \ldots, s_n, \ldots$. Step by step we shall construct nonzero elements $t_1, t_2, \ldots, t_n, \ldots$ such that t_j is a linear combination of $s_1, \ldots, s_j$, and such that $(t_j, t_k) = 0$ for $j \neq k$, and finally, such that the linear span of $S_1 = \{t_1, t_2, \ldots, t_n, \ldots\}$ equals that of S.

     Where do we begin? We pick our first element t_1 in the easiest possible way, namely

$$t_1 = s_1.$$

How do we get t_2? Let $t_2 = as_1 + s_2 = at_1 + s_2$, where $a \in \mathbb{C}$ is to be determined. We want $(t_1, t_2) = 0$; that is, we want

$$0 = (as_1 + s_2, s_1) = (as_1, s_1) + (s_2, s_1) = a(s_1, s_1) + (s_2, s_1).$$

Since $(s_1, s_1) \neq 0$ we can solve for a as $a = -\dfrac{(s_2, s_1)}{(s_1, s_1)}, = -\dfrac{(s_2, t_1)}{(t_1, t_1)} t_1,$

getting

$$t_2 = s_2 - \frac{(s_2, t_1)}{(t_1, t_1)} t_1.$$

Now we do have $(t_1, t_2) = 0$. Note, too, that since t_2 is a nonzero linear combination of s_1 and s_2, which are linearly independent, $t_2 \neq 0$. Also note that both s_1 and s_2 are expressible as linear combinations of t_1 and t_2. So the linear span of t_1 and t_2 equals that of s_1 and s_2.

     We go on in this way, step by step. Suppose that we have constructed for $1 \leq j \leq k$ nonzero $t_1, t_2, \ldots, t_k$ which are orthogonal, where t_j is in the linear span of $s_1, \ldots, s_j$. We want to construct the next one, t_{k+1}. How? We want t_{k+1} to be a linear combination of $s_1, \ldots, s_{k+1}$, with $t_{k+1} \neq 0$, and most important, $(t_{k+1}, t_j) = 0$ for $j \leq k$. Let t_{k+1} be

$$s_{k+1} - \frac{(s_{k+1}, t_1)}{(t_1, t_1)} t_1 - \frac{(s_{k+1}, t_2)}{(t_2, t_2)} t_2 - \cdots - \frac{(s_{k+1}, t_k)}{(t_k, t_k)} t_k.$$

Because $t_1, \ldots, t_k$ are linear combinations of the s_j where $j < k + 1$, and since the s_i's are linearly independent, we see that $t_{k+1} \neq 0$. Also, t_{k+1} is in the linear span of $s_1, \ldots, s_{k+1}$. Finally, for $j \leq k$, the inner product (t_{k+1}, t_j) is

$$\left(s_{k+1} - \frac{(s_{k+1}, t_1)}{(t_1, t_1)} t_1 - \frac{(s_{k+1}, t_2)}{(t_2, t_2)} t_2 - \cdots - \frac{(s_{k+1}, t_k)}{(t_k, t_k)} t_k, t_j \right),$$

which equals

$$(s_{k+1}, t_j) - \frac{(s_{k+1}, t_1)}{(t_1, t_1)} (t_1, t_j) - \cdots - \frac{(s_{k+1}, t_j)}{(t_j, t_j)} (t_j, t_j)$$

$$- \cdots - \frac{(s_{k+1}, t_k)}{(t_k, t_k)} (t_k, t_j).$$

Now the elements $t_1, \ldots, t_k$ had already been constructed in such a way that $(t_i, t_j) = 0$ for $i \neq j$. So the only nonzero contributions to our sum expression for (t_{k+1}, t_j) come from (s_{k+1}, t_j) and $\frac{(s_{k+1}, t_j)}{(t_j, t_j)} (t_j, t_j)$. Thus this sum reduces to

$$(t_{k+1}, t_j) = (s_{k+1}, t_j) - \frac{(s_{k+1}, t_j)}{(t_j, t_j)} (t_j, t_j)$$

$$= (s_{k+1}, t_j) - (s_{k+1}, t_j) = 0.$$

So we have gone one more step and have constructed t_{k+1} from $t_1, \ldots, t_k$ and s_{k+1}. Continue with the process. Note that from the form of the construction s_{k+1} is a linear combination of $t_1, \ldots, t_{k+1}$. This is true for all k. So, the linear span of the $\{t_1, t_2, \ldots, t_n, \ldots\}$ is equal to the linear span of $S = \{s_1, \ldots, s_n, \ldots\}$.

This finishes the proof of this very important construction.   ■

Corollary 7.7.2. If V is a finite-dimensional inner product space, then V has an orthonormal basis.

What the theorem says is that given a linearly independent set $\{s_1, \ldots, s_n, \ldots\}$ we can modify it, using linear combinations of $s_1, \ldots, s_n, \ldots$ to achieve an orthonormal set that has the same span.

We have seen this theorem used in $F^{(n)}$, so it is important here to give example in an infinite-dimensional inner product space.

EXAMPLE

Let V be the inner product space of all polynomials in x with real coefficients, where we define

$$(f, g) = \int_{-1}^{1} f(x)g(x) \, dx.$$

How do we construct an orthonormal set

$$p_1(x), p_2(x), \ldots, p_n(x), \ldots$$

whose linear span is that of

$$s_1 = 1, s_2 = x, s_3 = x^2, \ldots, s_n = x^{n-1}, \ldots,$$

that is, all of V? The recipe for doing this is given in the proof of Theorem 7.7.1.

We first construct a suitable orthogonal set. Let $q_1(x) = 1$, $q_2(x) = s_2 + as_1 = x + a$, where

$$a = -\frac{(q_2, q_1)}{(q_1, q_1)} = -\int_{-1}^{1} (1)(x) \, dx \bigg/ \int_{-1}^{1} 1 \, dx = 0.$$

So $q_2(x) = x$. We go on to $q_3(x)$. What do we require of $q_3(x)$? Only that it be of the form

$$q_3 = s_3 + aq_1 + bq_2$$

with

$$a = -\frac{(s_3, q_1)}{(q_1, q_1)} \quad b = -\frac{(q_3, q_2)}{(q_2, q_2)} \quad \text{(as in the proof)}.$$

This ensures that $(q_3, q_1) = (q_3, q_2) = 0$. Since

$$(s_3, q_1) = \int_{-1}^{1} (x^2)(1) \, dx = [x^3/3]_{-1}^{1} = \tfrac{2}{3},$$

$$(q_1, q_1) = 2$$

$$(s_3, q_2) = \int_{-1}^{1} (x^2)(x) \, dx = [x^4/4]_{-1}^{1} = 0,$$

we get

$$q_3(x) = x^2 - \tfrac{1}{3}.$$

We go on in this manner to $q_4(x)$, $q_5(x)$, and so on. The set we get consists of orthogonal elements. To make them an ortho-normal set we merely must divide each $q_i(x)$ by its length

$$\sqrt{q_i(x)q_i(x)} = \sqrt{\int_{-1}^{1} (q_i(x))^2 \, dx}.$$

We leave the evaluation of these and of some of the other $q_i(x)$ to the problem set that follows.

This orthonormal set of polynomials constructed is a version of what are called the *Legendre polynomials*. They are of great importance in physics, chemistry, engineering, and mathematics.

In the finite-dimensional case of $F^{(n)}$ we exploited the result of the Gram–Schmidt construction to show that $F^{(n)} = W \oplus W^\perp$, where W is any subspace of $F^{(n)}$ and

$$W^\perp = \{v \in F^{(n)} \mid (v, w) = 0, \text{ all } w \in W\}$$

is the *orthogonal complement* of W. We want to do something similar in general.

In the problem sets we were introduced to the *direct sum* of vector spaces V and W. This was defined as

$$V \oplus W = \{(v, w) \mid v \in V, w \in W\},$$

where we added component-wise and $a(v, w) = (av, aw)$ for $a \in F$. We want to make a more formal introduction of this notion and some related ones.

Definition. Let V be a vector space over F and let U, W be subspaces of V. We say that V is the *direct sum* of U and W if every element z in V can be written in a *unique* way as $z = u + w$, where $u \in U$ and $w \in W$.

What does this have to do with the notion of $U \oplus W$ as ordered pairs described above? The answer is

Lemma 7.7.3. *If U, W are subspaces of V, then V is the direct sum of U and W if and only if*

1. $U + W = \{u + w \mid u \in U, w \in W\} = V$.
2. $U \cap W = \{0\}$.

Furthermore, in this case $V \simeq U \oplus W$, where

$$U \oplus W = \{(u, w) \mid u \in U, w \in W\}.$$

Proof. If $V = U + W$ then every element v in V is of the form $v = u + w$. If $U \cap W = \{0\}$, we claim that this representation of v is unique, for suppose that $v = u + w = u_1 + w_1$, where $u, u_1 \in U$ and $w, w_1 \in W$. So $u - u_1 = w_1 - w$; but $u - u_1 \in U$ and $w_1 - w \in W$ and since $u - u_1 = w_1 - w$, we get $u - u_1 \in U \cap W = \{0\}$. Hence $u = u_1$. Similarly, $w = w_1$. So the representation of v in the desired form is indeed unique. Hence V is the direct sum of U and W.

To show that if V is the direct sum of U and W, then $V = U + W$ and $U \cap W = \{0\}$, we leave as an exercise.

Now to the isomorphism of $U \oplus W$ and V. Define the mapping $\Phi \colon U \oplus W \to V$ by the rule

$$\Phi((u, w)) = u + w.$$

We leave to the reader the verification that Φ is a homomorphism of $U \oplus W$ into V. It is onto since $V = U + W$ (because V is the direct sum of U and W). To show that Φ is an isomorphism we need but show that the kernel

$$\mathrm{Ker}\,(\Phi) = \{(u, w) \in U \oplus W \mid \Phi\,((u, w)) = 0\}$$

consists only of 0. If $\Phi((u, w)) = 0$, knowing that $\Phi((u, w)) = u + w$ leads us to $u + w = 0$, hence $u = -w \in U \cap W = \{0\}$. Thus $u = w = 0$. Therefore, $\mathrm{Ker}\,(\Phi) = \{0\}$, and the lemma is proved. ∎

We are now able to prove the important analog of what we once did in $F^{(n)}$ for any inner product space.

Theorem 7.7.4. If V is an inner product space and W is a *finite-dimensional* subspace of V, then V is the direct sum of W and $W^\perp$. So $V \simeq W \oplus W^\perp$, the set of all ordered pairs (w, z), where $w \in W$, $z \in W^\perp$.

Proof. Since W is finite-dimensional over F, it has a finite basis over F. By Theorem 7.7.1 we know that W then has an orthonormal basis $w_1, \ldots, w_m$ over F. We claim that an element of z is in $W^\perp$ if and only if $(z, w_r) = 0$ for $r = 1, 2, \ldots, m$. Why? Clearly, if

$z \in W^{\perp}$, then $(z, w) = 0$ for *all* $w \in W$, hence certainly $(z, w_r) = 0$. In the other direction, if $(z, w_r) = 0$ for $r = 1, 2, \ldots, m$, we claim that $(z, w) = 0$ for all $w \in W$. Because $w_1, w_2, \ldots, w_m$ is a basis of W over F, $w = a_1 w_1 + a_2 w_2 + \cdots + a_m w_m$. Thus

$$(z, w) = (z, a_1 w_1 + \cdots + a_m w_m)$$
$$= (z, a_1 w_1) + \cdots + (z, a_m w_m)$$
$$= \bar{a}_1 (z, w_1) + \bar{a}_2 (z, w_2) + \cdots + \bar{a}_m (z, w_m) = 0.$$

Hence $z \in W^{\perp}$.

Let $v \in V$; consider the element

$$z = v - (v, w_1) - \cdots - (v, w_m) w_m.$$

We assert that $(z, w_r) = 0$ for $r = 1, 2, \ldots, m$. Computing, we get

$$(z, w_r) = (v - (v, w_1) w_1 - \cdots - (v, w_m) w_m, w_r)$$
$$= (v, w_r) - (v, w_1)(w_1, w_r) - \cdots - (v, w_m)(w_m, w_r).$$

But the $w_1, \ldots, w_m$ are an orthonormal set, hence $(w_s, w_r) = 0$ if $s \neq r$ and $(w_r, w_r) = 1$. So the only nonzero contribution above is from the rth term; in other words,

$$(z, w_r) = (v, w_r) - (v, w_r)(w_r, w_r) = (v, w_r) - (v, w_r) = 0.$$

We have shown that

$$z = v - (v, w_1) w_1 - \cdots - (v, w_m) w_m$$

is in $W^{\perp}$. But

$$v = z + ((v, w_i) w_1 + \cdots + (v, w_m) w_m)$$

and

$$(v, w_i) w_1 + \cdots + (v, w_m) w_m$$

is in W. Thus $v \in W + W^{\perp}$. But $W \cap W^{\perp} = \{0\}$; hence, by Lemma 7.7.3, V is the direct sum of W and $W^{\perp}$. By Lemma 7.7.3 again, $V \simeq W \oplus W^{\perp}$. ∎

    An important special situation in which Theorem 7.7.4 holds is that of V finite-dimensional over F, for in that case any subspace of V is also finite-dimensional over F. Thus

Theorem 7.7.5.   If V is a finite-dimensional inner product space, then for any subspace W of V, V is the direct sum of W and $W^{\perp}$.

Because the direct sum of W and $W^\perp$ as subspaces of V is V, and since it is isomorphic to $W \oplus W^\perp$, we shall denote the fact that V is the direct sum of W and $W^\perp$ by $V = W \oplus W^\perp$. In these terms Theorem 7.7.5 reads as:

If V is a finite-dimensional inner product space, then

$$V = W \oplus W^\perp$$

for any subspace W of V.

We already know that if V is a finite-dimensional inner product space, then V merely as a vector space is isomorphic to $F^{(n)}$, where $n = \dim(V)$. We might very well ask:

Is V isomorphic to $F^{(n)}$ as an inner product space?

What does this mean? It means precisely that there is an isomorphism Φ of V onto $F^{(n)}$ such that

$$(\Phi(u), \Phi(v)) = (u, v) \qquad \text{for all } u, v \in V.$$

Here the inner product $(\Phi(u), \Phi(v))$ is that of $F^{(n)}$, while the inner product (u, v) is that of V. *In other words, Φ also preserves inner products.*
The anser is "yes," as is shown by

Theorem 7.7.6. If V is a finite-dimensional inner product space, then V is isomorphic to $F^{(n)}$, where $n = \dim(V)$, as an inner product space.

Proof. By Theorem 7.7.1, V has an orthonormal basis $v_1, \ldots, v_n$ where $n = \dim(V)$. Given $v \in V$, then $v = a_1 v_1 + \cdots + a_n v_n$, with $a_1, \ldots, a_n \in F$ uniquely determined by v. Define $\Phi \colon V \to F^{(n)}$ by

$$\Phi(v) = \begin{bmatrix} a_1 \\ \vdots \\ a_n \end{bmatrix}.$$

We saw in Theorem 7.4.3, that Φ is an isomorphism of V onto $F^{(n)}$.
Does Φ preserve the inner product? Let $v = a_1 v_1 + \cdots + a_n v_n$ and $u = b_1 v_1 + \cdots + b_n v_n$ for any u, v in V. Thus

$$(u, v) = (b_1 v_1 + \cdots + b_n v_n, a_1 v_1 + \cdots + a_n v_n)$$

$$= \sum_{i=1}^{n} \sum_{j=1}^{n} b_i \bar{a}_j (v_i, v_j).$$

(Prove!) Now, since $v_1, \ldots, v_n$ are orthonormal, $(v_i, v_j) = 0$ if $i \neq j$ and $(v_i, v_i) = 1$. So the double sum for (u, v) reduces to

$$(u, v) = \sum_{j=1}^{n} b_i \bar{a}_i.$$

But $\sum_{j=1}^{n} b_i \bar{a}_i$ is precisely the inner product of the vectors $\Phi(u) = \begin{bmatrix} b_1 \\ \vdots \\ b_n \end{bmatrix}$ and $\Phi(v) = \begin{bmatrix} a_1 \\ \vdots \\ a_n \end{bmatrix}$ in $F^{(n)}$. In other words, $(\Phi(u), \Phi(v)) = (u, v)$.

This completes the proof. ∎

     In light of Theorem 7.7.6:

We can carry over everything *we did in Chapters* 3 *or* 4 *for* $F^{(n)}$ *as an inner product space to* all *finite-dimensional inner product spaces over* F.

     For instance, we get from Theorem 4.3.7 that

Theorem 7.7.7. If V is a finite-dimensional inner product space and W is a subspace of V then $(W^{\perp})^{\perp} = W$.

     Go back to Chapter 3 and see what else you can carry over to the general case by exploiting Theorem 7.7.6.
     One could talk about vector spaces over F for systems F other than $\mathbb{R}$ or $\mathbb{C}$. However, we have restricted ourselves to working over $\mathbb{R}$ or $\mathbb{C}$. This allows us to show that any finite-dimensional vector space over $\mathbb{R}$ or $\mathbb{C}$ is, in fact, an inner product space.

Theorem 7.7.8. If V is a finite-dimensional vector space over $\mathbb{R}$ or $\mathbb{C}$, then V is an inner product space.

     Proof. Because V is finite-dimensional over $\mathbb{R}$ (or $\mathbb{C}$) it has a basis $v_1, \ldots, v_n$ over $\mathbb{R}$ (or $\mathbb{C}$). We define an inner product on V by insisting that $(v_i, v_j) = 0$ for $i \neq j$ and $(v_i, v_i) = 1$. That is we define

$$(u, v) = \sum_{j=1}^{n} a_j \bar{b}_j,$$

where $u = \sum_{i=1}^{n} a_i v_i$ and $v = \sum_{i=1}^{n} b_i v_i$.

We leave it to the reader to complete the proof that this is a legitimate inner product on V.   ■

Because of Theorem 7.7.8:

We can carry over to general finite-dimensional V over F = R or C everything we did in Chapter 2 for $F^{(n)}$ as a vector space over F.

A sample result of this kind is:

If V is finite-dimensional over F and $u_1, \ldots, u_m$ in V are linearly independent over F, then we can fill $u_1, \ldots, u_m$ out to a basis of V over F. That is, we can find elements $w_1, \ldots, w_r$ in V such that $u_1, \ldots, u_m, w_1, \ldots, w_r$ is a basis of V over F (and so dim (V) = m + r).

PROBLEMS

NUMERICAL PROBLEMS

1. In the example, find $q_4(x)$, $q_5(x)$, $q_6(x)$.

2. In the example, what are the lengths of $q_1(x)$, $q_2(x)$, $q_3(x)$?

3. Make a list of all the results on inner product spaces that can be carried over to a general inner product space form $F^{(n)}$ by use of Theorem 7.7.5.

4. Let $V = M_3(\mathbb{R})$ with inner product given by $(A, B) = \text{tr } (AB')$, where $A, B \in M_3(\mathbb{R})$ and B' is the transpose of B. Find an orthonormal basis for V relative to this inner product.

5. If V is as in Problem 4 and $W = \left\{ \begin{bmatrix} a & b & c \\ 0 & 0 & 0 \\ 0 & 0 & 0 \end{bmatrix} \mid a, b, c \in \mathbb{R} \right\}$,

find $W^\perp$ and verify that $V = W \oplus W^\perp$.

6. Prove Theorem 7.7.7 using the results of Chapters 3 and 4 and Theorem 7.7.6.

7. Make a list of the results in Chapters 3 and 4 that hold for any finite-dimensional vector space over $\mathbb{R}$ or $\mathbb{C}$ using Theorems 7.7.6 and 7.7.8.

8. In the proof of Theorem 7.7.8, verify that $(u, v) = \sum_{j=1}^{n} a_j \bar{b}_j$.

MORE THEORETICAL PROBLEMS

Easier Problems

9. If $V = M_n(\mathbb{C})$, show that $(A, B) = \text{tr}(AB^*)$ for $A, B \in M_n(\mathbb{C})$ and where B^* is the Hermitian adjoint of B defines an inner product on V over $\mathbb{C}$.

Middle-Level Problems

10. In the V of Problem 9 find an orthonormal basis of V relative to the given inner product.

11. If in the V of Problem 9, W is the set of all diagonal matrices, find $W^{\perp}$ and verify that $V = W \oplus W^{\perp}$.

12. If W and U are finite-dimensional subspaces of V such that

    $$W^{\perp} = U^{\perp},$$

    prove that $W = U$.

Harder Problems

13. In the example show that $q_n(x)$ is a polynomial of degree $n - 1$ over $\mathbb{R}$.

14. For $q_n(x)$ as in the text material show:
    - (a) If n is odd, then $q_n(x)$ is a polynomial in even powers of x.
    - (b) If n is even, then $q_n(x)$ is a polynomial in odd powers of x. (*Hint:* Use induction.)

15. If V is finite-dimensional, show that for any subspace W of V, $\dim(V) = \dim(W) + \dim(W^{\perp})$.

C H A P T E R

8

LINEAR TRANSFORMATIONS

8.1. INTRODUCTION

Earlier in the book we encountered linear transformations in the setting of $F^{(n)}$. The point of the material in this chapter is to consider linear transformations in the wider context of an arbitrary vector space. Although almost everything we do will be in the case of a *finite-dimensional* vector space V over F, in the definitions, early results, and examples we also consider the infinite-dimensional situation.

Our basic strategy will be to use the isomorphism established in Theorem 7.4.3 for a finite-dimensional vector space V over F, with $F^{(n)}$, where $n = \dim (V)$. This will also be exploited for inner product spaces. As we shall see, everything done for matrices goes over to general linear transformations on an arbitrary finite-dimensional vector space. In a nutshell, we shall see that such a linear transformation can be represented as a matrix and that the operations of addition and multiplication of linear transformations coincides with those of the associated matrices. Thus all the material developed for $n \times n$ matrices immediately can be transferred to the general case. Therefore, we often will merely sketch a proof or refer the reader back to the appropriate proof carried out in Chapters 3 and 4.

We can also speak about linear transformations—they are merely homomorphisms—of one vector space into another one. We shall touch on this lightly at the end of the chapter. Our main concern, however, will be linear transformations of a vector space V over F into itself, which is the more interesting situation where more can be said.

8.2. DEFINITIONS, EXAMPLES, AND SOME PRELIMINARY RESULTS

Let V be a vector space over F, where $F = \mathbb{R}$ or $F = \mathbb{C}$.

Definition. A *linear transformation T of V into V over F* is a mapping $T: V \longrightarrow V$ such that

(1) $T(v + w) = T(v) + T(w)$

(2) $T(av) = aT(v)$

for all $v, w \in V$ and all $a \in F$.

Of course, properties (1) and (2) generalize to

$$T(a_1v_1 + \cdots + a_kv_k) = a_1T(v_1) + \cdots + a_kT(v_k)$$

for any finite positive integer k. (Prove!)

We consider several examples of linear transformations on a variety of vector spaces. We shall refer to them by their numbering from time to time later.

EXAMPLES

1. Recall first the basic example we studied in such detail in Chapter 2, namely, $V = F^{(n)}$ and $T = (a_{ij})$, where T acts on V by the rule

$$\begin{bmatrix} a_{11} & \cdots & a_{1n} \\ \vdots & & \vdots \\ a_{n1} & \cdots & a_{nn} \end{bmatrix}\begin{bmatrix} x_1 \\ \vdots \\ x_n \end{bmatrix} = \begin{bmatrix} y_1 \\ \vdots \\ y_n \end{bmatrix},$$

where $y_r = \sum_{s=1}^{n} a_{rs}x_s$. We saw that T acts on V as a transformation on V and, furthermore, *every linear transformation* on V over F can be realized in this way.

2. Let V be the set of all polynomials in x over F. Define $T: V \longrightarrow V$ by defining $T(1) = 1$, $T(x) = x + 1$, $T(x^2) = (x + 1)^2 = x^2 + 2x + 1$, $\ldots$, the general term being

$$T(x^n) = (x + 1)^n = \sum_{j=0}^{n} \binom{n}{j} x^j$$

$$\left(\text{recall that } \binom{n}{j} = \frac{n!}{j!(n - j!)} \right);$$

and then, for any $p(x) = a_0 + a_1 x + a_2 x^2 + \cdots + a_k x^k$, defining

$$T(p(x)) = a_0 T(1) + a_1 T(x) + a_2 T(x^2) + \cdots + a_k T(x^k).$$

Then, for instance,

$$\begin{aligned} T(5 - x + x^2 + 6x^3) &= 5T(1) - T(x) + T(x^2) + 6T(x^3) \\ &= 5 - (x + 1) + (x + 1)^2 \\ &\quad + 6(x + 1)^3 \\ &= 11 + 19x + 19x^2 + 6x^3. \end{aligned}$$

This mapping T is a linear transformation on V because we have *forced* it to be one of the very definition of T. Specifying what T does to $1, x, x^2, \ldots, x^n, \ldots$, which generate V over F, and then *insisting* that T acts on $p(x) = a_0 + a_1 x + \cdots + a_k x^k$ via

$$T(p(x)) = a_0 T(1) + a_1 T(x) + \cdots + a_k T(x^k)$$

guarantees that T will be a linear transformation on V.

If we pick a sequence of polynomials

$$p_0(x), p_1(x), \ldots, p_n(x), \ldots$$

and define

$$T(x^j) = p_j(x)$$

and

$$\begin{aligned} T(a_0 + a_1 x &+ \cdots + a_k x^k) \\ &= a_0 p_0(x) + a_1 p_1(x) + \cdots + a_k p_k(x), \end{aligned}$$

we will equally get a linear transformation on V over F.

So, for example, if we define

$$T(1) = x^5, \quad T(x) = 6, \quad T(x^2) = x^7 + x, \quad T(x^k) = 0$$

for $k > 2$, then T is a linear transformation on V when defined as above. Thus

$$T(a_0 + a_1 x + \cdots + a_k x^k)$$
$$= a_0 T(1) + a_1 T(x) + a_2 T(x^2) + \cdots + a_k T(x^k)$$
$$= a_0 x^5 + 6a_1 + a_2(x^7 + x).$$

3. Example 2 is quite illustrative. First notice that we never really used the fact that V was the set of polynomials over F. What we did was to exploit the fact that V is spanned by the set of all powers of x, and these powers of x are linearly independent over F. We can do something similar for any vector space that is spanned by some linearly independent set of elements. We make this more precise.

Let V be a vector space over F spanned by the linearly independent elements $v_1, \ldots, v_n, \ldots$. Let $w_1, w_2, \ldots, w_n, \ldots$ be *any* elements of V and define T by

$$T(v_j) = w_j$$

for all j. Given any element u in V, u has a *unique* expansion of the form

$$u = a_1 v_1 + \cdots + a_k v_k,$$

where $a_1, \ldots, a_k$ are in F. Define

$$T(u) = a_1 T(v_1) + \cdots + a_k T(v_k) = a_1 w_1 + \cdots + a_k w_k.$$

By the linear independence of $v_1, \ldots, v_n, \ldots$ this mapping T is well defined. It is easy to check that it is a linear transformation on V.

Note how important it is that the elements

$$v_1, v_2, \ldots, v_n, \ldots$$

are linearly independent. Take the example of the polynomials in x; the elements

$$1, x, 1 + x, x^2, \ldots$$

are *not* linearly independent. If we try to define a linear transformation T by requiring

$$T(1) = 1, \ T(x) = x, \ T(1 + x) = x^3, \ldots$$

we see that T *cannot* be a linear transformation, for otherwise $T(1 + x)$ would equal

$$T(1) + T(x) = 1 + x,$$

yet we tried to insist that

$$T(1 + x) = x^3 \neq 1 + x.$$

4. Let V be a finite-dimensional vector space over F and let $v_1, \ldots, v_n$ be a basis of V over F. If

$$u = a_1v_1 + \cdots + a_nv_n,$$

define

$$T(u) = a_1v_1 + a_3v_3.$$

It is routine to verify that T is a linear transformation on V. If W is the subspace spanned by v_1 and v_3 we call T the *projection* of V onto W.

It is the first order of business that we establish some simple computational facts about linear transformations.

Lemma 8.2.1.   If T is a linear transformation on V then

1. $T(0) = 0$.
2. $T(-v) = -T(v) = (-1)T(v)$.

Proof.   To see that $T(0) = 0$, note that $T(0 + 0) = T(0) + T(0)$. Yet $T(0 + 0) = T(0)$. Thus $T(0) + T(0) = T(0)$. There is an element $w \in V$ such that $T(0) + w = 0$, from the definition of a vector space. So

$$\begin{aligned}
0 &= T(0) + w \\
&= (T(0) + T(0)) + w \\
&= T(0) + (T(0) + w) \\
&= T(0) + 0 = T(0).
\end{aligned}$$

This yields the desired result that $T(0) = 0$.

We leave the proof of Part (2) to the reader.   ■

If V is any vector space over F, we can introduce two operations on the linear transformations on V. These are addition and multiplication by a scalar (i.e., by an element of F). We define both of these operations now.

Definition. If T_1, T_2 are two linear transformations of V into itself, then the *sum, $T_1 + T_2$,* of T_1 and T_2 is defined by

$$(T_1 + T_2)(v) = T_1(v) + T_2(v)$$

for all $v \in V$.

Definition. If T is a linear transformation on V and $a \in F$, then aT, the *multiplication of T by the scalar a,* is defined by

$$(aT)(v) = aT(v)$$

for all $v \in V$.

So, for instance, if V is the set of all polynomials in x over $\mathbb{R}$ and if T_1 is defined by

$$T_1(a_0 + a_1x + a_2x^2 + \cdots + a_nx^n) = 3a_0 + \tfrac{1}{2}a_1x,$$

while T_2 is defined by

$$T_2(b_0 + b_1x + b_2x^2 + \cdots + b_nx^n) = b_3x^3,$$

then

$$\begin{aligned}
(T_1 + T_2)(c_0 + c_1x + c_2x^2 + \cdots &+ c_nx^n) \\
= T_1(c_0 + c_1x + c_2x^2 + \cdots &+ c_nx^n) \\
+ T_2(c_0 + c_1x + c_2x^2 + \cdots &+ c_nx^n) \\
= 3c_0 + \tfrac{1}{2}c_1x + c_3x^3.&
\end{aligned}$$

Similarly, $15T_1$ acts on V by the action

$$(15T_1)(c_0 + c_1x + c_2x^2 + \cdots + c_nx^n)$$
$$= 15(3c_0 + \tfrac{1}{2}c_1x) = 45c_0 + \tfrac{15}{2}c_1x.$$

One would hope that combining linear transformations and scalars in these ways always leads us again to linear transformations on V. There is no need to hope; we easily verify this to be the case.

Lemma 8.2.2. If T_1, T_2 are linear transformations on V and if $a \in F$, then $T_1 + T_2$ and aT_1 are linear transformations on V.

Proof. What must we show? To begin with, if T_1, T_2 are linear transformations on V and if u, $v \in V$, we need that

$$(T_1 + T_2)(u + v) = (T_1 + T_2)(u) + (T_1 + T_2)(v).$$

To see this we expand according to the definition of the sum

$$(T_1 + T_2)(u + v) = (T_1)(u + v) + (T_2)(u + v)$$
$$= T_1(u) + T_1(v) + T_2(u) + T_2(v)$$
$$\text{(since } T_1, T_2 \text{ are linear transformations)}$$
$$= T_1(u) + T_2(u) + T_1(v) + T_2(v)$$
$$= (T_1 + T_2)(u) + (T_1 + T_2)(v)$$
$$\text{(again by the definition of } (T_1 + T_2)),$$

which is exactly what we wanted to show.

We also need to see that

$$(T_1 + T_2)(av) = a((T_1 + T_2)(v))$$

for $v \in V$, $a \in F$. But

$$(T_1 + T_2)(av) = T_1(av) + T_2(av) = aT_1(v) + aT_2(v)$$
$$= a(T_1(v) + T_2(v))$$
$$= a((T_1 + T_2)(v)),$$

as required.

So $T_1 + T_2$ is indeed a linear transformation on V. An analogous argument shows that bT_1 is a linear transformation on V for $b \in F$ and T_1 a linear transformation. ∎

Definition. If V is a vector space over F, then $L(V)$ is the set of all linear transformations on V over F.

What we showed was that if $T_1, T_2 \in L(V)$ and $a \in F$, then $T_1 + T_2$ and aT_1 are both in $L(V)$. We can now repeat the argument—especially after the next theorem—to show that if $T_1, \ldots, T_k \in L(V)$ and $a_1, \ldots, a_k \in F$, then $a_1T_1 + \cdots + a_kT_k$ is again in $L(V)$.

In the two operations defined in $L(V)$ we have imposed an algebraic structure on $L(V)$. How does this structure behave?

We first point out one significant element present in $L(V)$. Let's define $O: V \longrightarrow V$ by the rule

$$O(v) = 0$$

for all $v \in V$. Trivially, O is a linear transformation on V. Furthermore, if $v \in V$ and $T \in L(V)$, then

$$(T + O)(v) = T(v) + O(v) = T(v) + 0 = T(v),$$

from which we deduce that

$$T + O = T.$$

Similarly,

$$O + T = T.$$

We shall usually write O merely as 0 and refer to it as the *zero element* of $L(V)$.

If $T \in L(V)$, consider the element

$$(-1)T = S,$$

which we know also lies in $L(V)$. For any $v \in V$,

$$(T + S)(v) = T(v) + S(v) = T(v) + (-1)T(v)$$
$$= T(v + (-1)v) = T(0) = 0.$$

Hence

$$T + S = O = 0.$$

So $S = (-1)T$ acts as a negative for T in $L(V)$. We write it merely as $-T$.

With these preliminaries out of the way we have

Theorem 8.2.3. $L(V)$ is itself a vector space over F.

We leave the proof as a series of exercises. We must verify that each of the defining rules for a vector space holds in $L(V)$ under the operations we introduced.

PROBLEMS

NUMERICAL PROBLEMS

1. In Example 4 verify that the T defined is a linear transformation.

2. Show that $O: V \longrightarrow V$ defined by the rule $O(v) = 0$ for all $v \in V$ is a linear transformation on V.

3. If V is the set of all polynomials in x over $\mathbb{R}$, show that

$$d: V \longrightarrow V,$$

defined by $d(p(x)) = \dfrac{d}{dx} p(x)$, is a linear transformation on V.

4. If V is as in Problem 3 and $S: V \longrightarrow V$ is defined by

$$S(p(x)) = \int_0^x p(t) \, dt,$$

show that S is a linear transformation on V.

5. Calculate $d + S$ and $-\frac{1}{2}S$ on

   (a) $p(x) = x^5 - \frac{1}{4}x + \pi$.

   (b) $p(x) = \frac{1}{6}x^6 - \frac{1}{7}x^7$.

   (c) $p(x) = 4\pi - 17x^{20}$.

6. In $L(V)$ show

   (a) $T_1 + T_2 = T_2 + T_1$

   (b) $(T_1 + T_2) + T_3 = T_1 + (T_2 + T_3)$

   (c) $(a + b)T_1 = aT_1 + bT_1$

   (d) $a(T_1 + T_2) = aT_1 + aT_2$

   for all $a, b \in F$ and all $T_1, T_2, T_3 \in L(V)$.

7. If V is the linear span over $\mathbb{R}$ of $\cos(x)$, $\cos(2x)$, $\cos(3x)$ in the set of all real-valued functions on $[0, 1]$, show that

   (a) The functions $\cos(x)$, $\cos(2x)$, $\cos(3x)$ are linearly independent over $\mathbb{R}$.

   (b) The mapping T on V defined by

$$T(a\cos(x) + b\cos(2x) + c\cos(3x))$$
$$= c\cos(x) + a\cos(2x) + b\cos(3x)$$

   is a linear transformation.

   (c) Show that T is $1-1$ and onto.

MORE THEORETICAL PROBLEMS

Easier Problems

8. Show that aT, where $a \in F$ and $T \in L(V)$, defined by

$$(aT)(v) = a(T(v))$$

is a linear transformation on V.

Middle-Level Problems

9. If T is a linear transformation on V, show by mathematical induction that $T(a_1 v_1 + \cdots + a_k v_k) = a_1 T(v_1) + \cdots + a_k T(v_k)$ for all $v_1, \ldots, v_k \in V$ and all $a_1, \ldots, a_k \in F$.

10. If W is a subspace of V and $T \in L(V)$ is such that $T(v) = 0$ for all $v \notin W$, prove that $T = 0$, the zero element of $L(V)$.

11. If $T_1, \ldots, T_k \in L(V)$ and $v \in V$, show that the set

$$\{(a_1 T_1 + \cdots + a_k T_k)(v) \mid a_1, \ldots, a_k \in F\}$$

is a finite-dimensional subspace of V.

12. Prove that if S is a finite-dimensional subspace of $L(V)$ and W is a finite-dimensional subspace of V, then

  (a) $S(W) = \{T(w) \mid T \in S, w \in W\}$ is a subspace of V.

  (b) $S(W)$ is finite-dimensional over F.

  (c) $\dim (S(W)) \leq (\dim (W))(\dim (S))$.

13. If $T \in L(V)$, show that

  (a) $U = \{v \in V \mid T(v) = 0\}$ is a subspace of V.

  (b) $T(V) = W = \{T(v) \mid v \in V\}$ is a subspace of V.

14. If W is a subspace of V, let $K = \{T \in L(V) \mid T(W) = 0\}$. Show that K is a subspace of $L(V)$.

15. Define the mapping $\Phi \colon L(V) \longrightarrow V$ by $\Phi(T) = T(v_0)$, where $T \in L(V)$, v_0 is a *fixed* element of V. Prove that Φ is a homomorphism of $L(V)$ into V.

16. In Problem 15 show that the kernel of Φ cannot be merely $\{0\}$.

17. Let W be a subspace of V and let $T \in L(V)$. If

$$U = \{v \in V \mid T(v) \in W\},$$

show that U is a subspace of V. [Compare this with Part (a) of Problem 13.]

18. If V is a finite-dimensional vector space over F and $T \in L(V)$ is such that T maps V *onto* V, prove that T is $1-1$.

19. If V is a finite-dimensional vector space over F and $T \in L(V)$ is $1-1$ on V, prove that T maps V *onto* V.

8.3. PRODUCTS OF LINEAR TRANSFORMATIONS

Given any set S and f, g mappings of S into S, we had earlier in the book defined the product or composition of f and g by the rule

$$(fg)(s) = f(g(s))$$

for every $s \in S$. If T_1, T_2 are linear transformations on V over F, they are mappings of V into itself; hence their product, T_1T_2, as mappings makes sense. *We shall use this product as an operation in $L(V)$.*

The first question that then comes to mind is: If T_1, T_2 are in $L(V)$, is T_1T_2 also in $L(V)$?

The answer is "yes," as we see in

Lemma 8.3.1. If T_1, $T_2 \in L(V)$, then $T_1T_2 \in L(V)$.

Proof. We must check out two things: For all u, $v \in V$, $a \in F$,

1. Is $(T_1T_2)(u + v) = (T_1T_2)(u) + (T_1T_2)(v)$?
2. Is $(T_1T_2)(av) = a((T_1T_2)(v))$?

We look at these two in turn.
By definition,

$$
\begin{aligned}
(T_1T_2)(u + v) &= T_1(T_2(u + v)) \\
&= T_1(T_2(u) + T_2(v)) \text{ [since } T_2 \in L(V)] \\
&= T_1(T_2(u)) + T_1(T_2(v)) \text{ [since } T_1 \in L(V)] \\
&= (T_1T_2)(u) + (T_1T_2)(v).
\end{aligned}
$$

Thus (1) is established.

We leave the proof of (2) to the reader.   ■

Lemma 8.3.1 assures us that multiplying two elements of $L(V)$ throws us back into $L(V)$. Knowing this, we might ask what properties this product enjoys and how product, sum, and multiplication by a scalar interact.

There is a very special element, I, in $L(V)$ defined by

$$I(v) = v$$

for all $v \in V$. It is easy to show that this I is indeed in $L(V)$. As the identity mapping on V we know from before that

$$TI = IT = T$$

for all $T \in L(V)$. We also know, just from the property of mappings, that

$$(T_1T_2)T_3 = T_1(T_2T_3).$$

If $T \in L(V)$ is *both* $1-1$ and onto V, there is a mapping S such that

$$ST = TS = I.$$

Is this mapping S also in $L(V)$? Let's check it out. Since

$$T \in L(V), \; T(S(u) + S(v)) = (TS)(u) + (TS)(v)$$
$$= I(u) + I(v) = u + v,$$

we have

$$S(T(S(u) + S(v))) = S(u + v).$$

Since

$$S(T(S(u) + S(v))) = (ST)(S(u) + S(v))$$
$$= I(S(u) + S(v)) = S(u) + S(v),$$

we end up with the desired result that

$$S(u + v) = S(u) + S(v).$$

Similarly, we can show that

$$S(av) = aS(v) \text{ for } a \in F, v \in V.$$

Consequently, $S \in L(V)$. If such an S exists for T—and it needn't for every T—we write it as T^{-1}, in accordance with what we did for mappings. So we have

Definition. If $S \in L(V)$ and $ST = TS = I$, we call S the *inverse* of T and denote it as T^{-1}.

We have shown

Lemma 8.3.2. If $T \in L(V)$ is $1 - 1$ and onto, then T^{-1} is also in $L(V)$.

The properties of the product by itself of linear transformations behaves exactly as the product of mappings, that is, decently. How do product and sum intertwine?

Lemma 8.3.3. If S, T, Q are in $L(V)$, then

1. $S(T + Q) = ST + SQ$;
2. $(T + Q)S = TS + QS$.

Proof. To prove each part we must show only that both sides of the alleged equalities agree on every element v in V. So

$$(S(T + Q))(v) = S(T(v) + Q(v)) = S(T(v)) + S(Q(v))$$
$$= (ST)(v) + (SQ)(v) = (ST + SQ)(v).$$

Therefore,

$$S(T + Q) = ST + SQ,$$

establishing Part (1). The proof of Part (2) is similar.   ■

We leave the proof of the next lemma to the reader. It is similar in spirit to the one just done.

Lemma 8.3.4.   If S, $T \in L(V)$ and a, $b \in F$, then

1. $S(aT) = a(ST)$;
2. $(aS)(bT) = (ab)(ST)$.

For any $T \in L(V)$ we shall use the notation of exponents, that is, for $k \geq 0$, $T^0 = I$, $T^1 = T$, . . . , $T^k = T(T^{k-1})$. The usual rules of exponents prevail, namely,

$$T^k T^m = T^{k+m}$$

and

$$(T^k)^m = T^{km}.$$

If T^{-1} exists in $L(V)$, we say that T is *invertible* in $L(V)$. In this case, we define T^{-n} for $n \geq 0$ by

$$T^{-n} = (T^{-1})^n.$$

Again, here, the usual rules of exponents hold.

PROBLEMS

NUMERICAL PROBLEMS

1. If V is the set of polynomials, of degree n or less, in x over $\mathbb{R}$ and $D\colon V \longrightarrow V$ is defined by $D(p(x)) = \dfrac{d}{dx}(p(x))$, prove that $D^{n+1} = 0$.

2. If V is as in Problem 1 and $T\colon V \longrightarrow V$ is defined by

$$T(p(x)) = p(x + 1),$$

show that T^{-1} exists in $L(v)$.

3. If V is the set of all polynomials in x over $\mathbb{R}$, let $D: V \longrightarrow V$ by
$D(p(x)) = \dfrac{d}{dx}(p(x))$; and $S: V \longrightarrow V$ by $S(p(x)) = \displaystyle\int_0^x p(t)\,dt$.
Prove:

(a) $DS = I$.

(b) $SD \neq I$.

4. If V is as in Problem 3 and $T: V \longrightarrow V$ according to the rules
$T(1) = 1$, $T(x) = x$, $T(x^2) = x^2$, and $T(x^k) = 0$ if $k > 2$, show
that $T^2 = T$.

5. In Problem 4, if $S: V \longrightarrow V$ according to

$$T(1) = x, \quad T(x) = x^2, \quad T(x^2) = 1, \quad T(x^k) = x^k \text{ for } k > 2,$$

show that $T^3 = I$.

6. If V is the linear span over $\mathbb{R}$ in the set of all real-valued functions
on $[0, 1]$ of the functions

$$\cos(x),\ \sin(x),\ \cos(2x),\ \sin(2x),\ \cos(3x),\ \sin(3x),$$

and $D: V \longrightarrow V$ is defined by

$$D(v) = \frac{d}{dx}(v)$$

for $v \in V$, show that

(a) D maps V into V.

(b) D maps V onto V.

(c) D is a linear transformation on V.

(d) D^{-1} exists on V.

7. In Problem 6 find the explicit form of D^{-1}.

MORE THEORETICAL PROBLEMS

Easier Problems

8. Prove the Part (2) of Lemma 8.3.1.

9. In $L(V)$ prove that $(T + Q)S = TS + QS$.

10. Prove Lemma 8.3.4.

11. Prove that if $T \in L(V)$ and $a \in F$, then multiplication by the scalar
a coincides with the product of the two linear transformations
aI and T.

12. If $E \in L(V)$ satisfies $E^2 = E$ and E is invertible, what must E be?

13. If $F = \mathbb{C}$ and V is any vector space over $\mathbb{C}$, define $T: V \longrightarrow V$ by $T(v) = av$ for all $v \in V$, where $a \in F$. Show that

  (a) T is a linear transformation on V.

  (b) If $a \neq 0$, then T is invertible in $L(V)$.

  (c) Find T^{-1} if $a \neq 0$.

  (d) If $a^n = 1$, then $T^n = I$.

14. If $T \in L(V)$ and $T^2 = I$, suppose, for some $0 \neq v \in V$, that $T(v) = av$, where $a \in F$. What are the possible values of a?

15. If $T^2 = I$ in $L(V)$ and $T \neq I$, find an element $v \neq 0$ in V and an element $w \neq 0$ in V such that $T(v) = v$ and $T(w) = -w$.

16. If $T \in L(V)$ satisfies $T^2 = T$ and $T \neq 0$, $T \neq I$, find an element $v \neq 0$ in V and an element $w \neq 0$ in V such that $T(v) = v$ and $T(w) = 0$.

Middle-Level Problems

17. If $T^2 = I$ in $L(V)$ and $T \neq I$, let $U = \{v \in V \mid T(v) = v\}$ and $W = \{v \in V \mid T(v) = -v\}$. Show that

  (a) $U \cap W = \{0\}$.

  (b) U and W are nonzero subspaces of V.

  (c) $V = U \oplus W$.

18. If $T^2 = T$ and $T \neq 0$, $T \neq I$, let $U = \{v \in V \mid T(v) = v\}$ and let $W = \{v \in V \mid T(v) = 0\}$. Prove that

  (a) U and W are nonzero subspaces of V.

  (b) $U \cap W = \{0\}$.

  (c) $V = U \oplus W$.

  [***Hint:*** For Part (c), if $T^2 = T$, then $(I - T)^2 = I - T$.]

19. If $T \in L(V)$ is such that $T^2 - 2T + I = 0$, show that there is a $v \neq 0$ in V such that $T(v) = v$.

20. For T as in Problem 19, show that T is invertible in $L(V)$ and find T^{-1} explicitly.

Harder Problems

21. If $T \in L(V)$ satisfies $T^n + a_1 T^{n-1} + \cdots + a_{n-1}T + a_nI = 0$, where $a_1, \ldots, a_n$ are in F, prove:

  (a) T is invertible if $a_n \neq 0$.

  (b) If $a_n \neq 0$, find T^{-1} explicitly.

22. If $T \in L(V)$ is the T of Problem 20 and $v_0 \in V$, let

$U = \{p(T)(v_0) \mid p(x)$ is any polynomial in x over $F\}$. Prove:

(a) U is a subspace of V.

(b) U is finite-dimensional over F.

23. If $v_1, \ldots, v_n$ is a basis for V over F and if $T \in L(V)$ acts on V according to $T(v_1) = v_2$, $T(v_2) = v_3$, $\ldots$, $T(v_{n-1}) = v_n$, $T(v_n) = a_1 v_1 + \cdots + a_{n-1} v_{n-1}$, with $a_1, \ldots, a_{n-1} \in F$, show that there exist $b_1, \ldots, b_{n-1}$ in F such that

$$T^n + b_1 T^{n-1} + \cdots + b_{n-1} I = 0.$$

Can you describe the b_i explicitly?

8.4. LINEAR TRANSFORMATIONS AS MATRICES

In studying linear transformations on $V = F^{(n)}$ we showed that any linear transformation T in $L(V)$ can be realized as a matrix in $M_n(F)$. [And, conversely, every matrix in $M_n(F)$ defines a linear transformation on $F^{(n)}$.] Let's recall how this was done.

Let $v_1, \ldots, v_n$ be a basis of the n-dimensional vector space V over F. If $T \in L(V)$, to know what T does to any element it is enough to know what T does to each of the basis elements $v_1, \ldots, v_n$. This is true because, given $u \in V$, then $u = a_1 v_1 + \cdots + a_n v_n$ where $a_1, \ldots, a_n \in F$. Hence

$$T(u) = T(a_1 v_1 + \cdots + a_n v_n) = a_1 T(v_1) + \cdots + a_n T(v_n);$$

thus, knowing each of the $T(v_i)$ lets us know the value of $T(u)$.

So we only need to know $T(v_1), \ldots, T(v_n)$ in order to know T. Since $T(v_1), T(v_2), \ldots, T(v_n)$ are in V, and since $v_1, \ldots, v_n$ form a basis of V over F, we have that

$$T(v_s) = \sum_{r=1}^{n} t_{rs} v_r,$$

where the t_{rs} are in F. We associate with T the matrix

$$m(T) = (t_{rs}) = \begin{bmatrix} t_{11} & \cdots & t_{1n} \\ \vdots & & \vdots \\ t_{n1} & \cdots & t_{nn} \end{bmatrix}.$$

Notice that this matrix $m(T)$ depends heavily on the basis $v_1, \ldots, v_n$ used, as we take into account in

Definition.   The *matrix of $T \in L(V)$ in the basis $v_1, \ldots, v_n$* is the matrix (t_{rs}) determined by the equations $Tv_s = \sum_{r=1}^{n} t_{rs} v_r$.

Thus if we calculate $m(T)$ according to a basis $v_1, \ldots, v_n$ and you calculate $m(T)$ according to a basis $u_1, \ldots, u_n$, there is no earthly reason why we should get the same matrices. As we shall see, the matrices you get and we get are closely related but are not necessarily equal. In fact, we really should not write $m(T)$. Rather *we should indicate the dependence on $v_1, \ldots, v_n$ by some device such as* $m_{v_1, \ldots, v_n}(T)$. But this is clumsy, so we take the easy way out and write it simply as $m(T)$.

Let's look at an example. Let V be the set of all polynomials in x over $\mathbb{R}$ of degree n or less. Thus $1, x, \ldots, x^n$ is a basis of V over $\mathbb{R}$. Let $D: V \longrightarrow V$ be defined by

$$D(p(x)) = \frac{d}{dx} p(x);$$

that is,

$$D(1) = 0, \; D(x) = 1, \; D(x^2) = 2x, \; \ldots, \; D(x^i)$$
$$= ix^{i-1}, \; \ldots, \; D(x^n) = nx^{n-1}.$$

If we denote

$$v_1 = 1, \; v_2 = x, \; \ldots, \; v_k = x^{k-1}, \; \ldots, \; v_{n+1} = x^n,$$

then $D(v_k) = (k - 1)v_{k-1}$. So the matrix of D in this basis is

$$m(D) = \begin{bmatrix} 0 & 1 & 0 & \cdots & 0 \\ 0 & 0 & 2 & \cdots & 0 \\ \vdots & \vdots & \vdots & \ddots & \vdots \\ & & & & n \\ 0 & 0 & 0 & \cdots & 0 \end{bmatrix}.$$

For instance, if $n = 2$, then $m(D) = \begin{bmatrix} 0 & 1 & 0 \\ 0 & 0 & 2 \\ 0 & 0 & 0 \end{bmatrix}$.

What is $m(D)$ if we use the basis $1, 1 + x, x^2, \ldots, x^n$ of V over F? If

$$u_1 = 1, \quad u_2 = 1 + x, \quad u_3 = x^2, \ldots, u_{n+1} = x^n,$$

then, since

$$D(u_1) = 0, \; D(u_2) = 1 = u_1, \; D(u_3) = 2x = 2(u_2 - 1)$$
$$= 2u_2 - 2u_1, \; D(u_4) = 3x^2 = 3u_3, \ldots,$$

the matrix $m(D)$ in this basis is

$$m(D) = \begin{bmatrix} 0 & 1 & -2 & 0 & \cdots & 0 \\ 0 & 0 & 2 & 0 & \cdots & 0 \\ 0 & 0 & 0 & 3 & & \vdots \\ \vdots & \vdots & \vdots & \vdots & \ddots & \\ & & & & & n \\ 0 & 0 & 0 & 0 & \cdots & 0 \end{bmatrix}.$$

For instance, if $n = 2$, $m(D) = \begin{bmatrix} 0 & 1 & -2 \\ 0 & 0 & 2 \\ 0 & 0 & 0 \end{bmatrix}$.

Let V be an n-dimensional vector space over F and $v_1, \ldots, v_n$, $n = \dim(V)$, a basis of V. *This basis $v_1, \ldots, v_n$ will be the one we use in the discussion to follow.*

How do the matrices $m(T)$, for $T \in L(V)$, behave with regard to the operations in $L(V)$? The proofs of the statements we are going to make are *precisely* the same as the proofs given in Section 3.9, where we considered linear transformations on $F^{(n)}$.

Theorem 8.4.1. If $T_1, T_2, \in L(V)$, $a \in F$, then

1. $m(T_1 + T_2) = m(T_1) + m(T_2)$;
2. $m(aT_1) = am(T_1)$;
3. $m(T_1 T_2) = m(T_1)m(T_2)$.

Furthermore, given any matrix A in $M_n(F)$, then $A = m(T)$ for some $T \in L(V)$. Finally, $m(T_1) = m(T_2)$ implies that $T_1 = T_2$. So the mapping $m: L(V) \longrightarrow M_n(F)$ is $1-1$ and onto.

Corollary 8.4.2. $T \in L(V)$ is invertible if and only if $m(T) \in M_n(F)$ is invertible; and if T is invertible, then $m(T^{-1}) = m(T)^{-1}$.

Corollary 8.4.3. If $T \in L(V)$ and $S \in L(V)$ is invertible, then $m(S^{-1}TS) = m(S)^{-1}m(T)m(S)$.

What this theorem says is that m is a mechanism for carrying us from linear transformations to matrices in a way that preserves all properties. So we can argue back and forth readily using matrices and linear transformations interchangeably.

Corollary 8.4.4. If $T \in L(V)$, then the following conditions are equivalent:

1. The only solution to $Tx = 0$ is $x = 0$.
2. T is $1-1$.
3. T is onto.
4. T is invertible.

Proof. Letting A be the matrix $m(T)$ of T in a basis for V, we proved that conditions (1)–(4) with T replaced by A are equivalent. So by Theorem 8.4.1 and Corollary 8.4.2, conditions (1)–(4) for T are also equivalent. ∎

Corollary 8.4.5. If $T \in L(V)$ takes a basis $v_1, \ldots, v_n$ to a basis $Tv_1, \ldots, Tv_n$ of V, then T is invertible in $L(V)$.

Proof. Let $v_1, \ldots, v_n$ be a basis of V such that $Tv_1, \ldots, Tv_n$ is also a basis of V. Then, given $u \in V$, there are $a_1, \ldots, a_n \in F$ such that

$$u = a_1 Tv_1 + \cdots + a_n Tv_n = T(a_1 v_1 + \cdots + a_n v_n).$$

Thus T maps V *onto* V. Is T $1-1$? If $v = b_1 v_1 + \cdots + b_n v_n$ and $Tv = 0$, then

$$0 = T(b_1 v_1 + \cdots + b_n v_n) = b_1 Tv_1 + \cdots + b_n Tv_n.$$

So since $Tv_1, \ldots, Tv_n$ is a basis, the b_s are all 0 and $v = 0$. Thus T is invertible in $L(V)$ by Corollary 8.4.5. ∎

If $v_1, \ldots, v_n$ and $w_1, \ldots, w_n$ are two bases of V and T a linear transformation on V, we can compute, as above, the matrix $m_1(T)$ in the first basis and $m_2(T)$ in the second. How are these matrices related? If $m_1(T) = (t_{rs})$ and $m_2(T) = (\tau_{rs})$, then by the very definition of m_1 and m_2,

$$Tv_s = \sum_{r=1}^{n} t_{rs} v_r \tag{1}$$

$$Tw_s = \sum_{r=1}^{n} \tau_{rs} w_r \tag{2}$$

If S is the linear transformation defined by

$$w_s = Sv_s$$

for $s = 1, 2, \ldots, n$, then since S takes a basis into a basis, it must be invertible.

Equation (2) becomes

$$TSv_s = \sum_{r=1}^{n} \tau_{rs} Sv_r = S\left(\sum_{r=1}^{n} \tau_{rs} v_r \right). \tag{3}$$

$$S^{-1}TSv_s = \sum_{r=1}^{n} \tau_{rs} v_r. \tag{4}$$

What (4) says is that the matrix of $S^{-1}TS$ in the basis $v_1, \ldots, v_n$ is precisely $(\tau_{rs}) = m_2(T)$. So $m_1(S^{-1}TS) = m_2(T)$. Using Corollary 8.4.3 we get $m_2(T) = m_1(S)^{-1}m_1(T)m_1(S)$.

We have proved

Theorem 8.4.6. If $m_1(T)$ and $m_2(T)$ are the matrices of T in the bases $v_1, \ldots, v_n$ and $w_1, \ldots, w_n$, respectively, then

$$m_2(T) = m_1(S)^{-1}m_1(T)m_1(S),$$

where S is defined by $w_s = Sv_s$ for $s = 1, 2, \ldots, n$.

We illustrate the theorem with an example. Let V be the vector space of polynomials of degree 2 or less over F and let D be defined by

$$D(p(x)) = \frac{d}{dx} p(x).$$

Then D is in $L(V)$. In the basis $v_1 = 1, \quad v_2 = x, \quad v_3 = x^2$ of V, we saw that $m_1(D) = \begin{bmatrix} 0 & 1 & 0 \\ 0 & 0 & 2 \\ 0 & 0 & 0 \end{bmatrix}$, while in the basis

$$w_1 = 1, w_2 = 1 + x, w_3 = x^2,$$

we saw that $m_2(D) = \begin{bmatrix} 0 & 1 & -2 \\ 0 & 0 & 2 \\ 0 & 0 & 0 \end{bmatrix}.$

What does $m_1(S)$ look like where $w_1 = Sv_1, w_2 = Sv_2, w_3 = Sv_3$? We have

$$v_1 = w_1 = Sv_1,$$
$$v_1 + v_2 = 1 + x = w_2 = Sv_2,$$
$$v_3 = x^2 = w_3 = Sv_3.$$

Thus

$$m_1(S) = \begin{bmatrix} 1 & 1 & 0 \\ 0 & 1 & 0 \\ 0 & 0 & 1 \end{bmatrix}.$$

As is easily checked,

$$m_1(S)^{-1} = \begin{bmatrix} 1 & -1 & 0 \\ 0 & 1 & 0 \\ 0 & 0 & 1 \end{bmatrix}.$$

So we can calculate $m_2(D) = m_1(S)^{-1}m_1(D)m_1(S)$ as

$$\begin{bmatrix} 1 & -1 & 0 \\ 0 & 1 & 0 \\ 0 & 0 & 1 \end{bmatrix} \begin{bmatrix} 0 & 1 & 0 \\ 0 & 0 & 2 \\ 0 & 0 & 0 \end{bmatrix} \begin{bmatrix} 1 & 1 & 0 \\ 0 & 1 & 0 \\ 0 & 0 & 1 \end{bmatrix}$$

$$= \begin{bmatrix} 0 & 1 & -2 \\ 0 & 0 & 2 \\ 0 & 0 & 0 \end{bmatrix} \begin{bmatrix} 1 & 1 & 0 \\ 0 & 1 & 0 \\ 0 & 0 & 1 \end{bmatrix}$$

$$= \begin{bmatrix} 0 & 1 & -2 \\ 0 & 0 & 2 \\ 0 & 0 & 0 \end{bmatrix} = m_2(D),$$

as asserted in our theorem.

We saw earlier that in $M_n(F)$, tr $(C^{-1}AC) = $ tr (A) and det $(C^{-1}AC)$ $= $ det (A). So, in light of Corollary 8.4.3, *no matter what bases* we use to represent $T \in L(V)$ as a matrix, then $m_2(T) = m_1(S)^{-1}m_1(T)m_1(S)$; hence tr $(m_2(T)) = $ tr $(m_1(T))$ and det $(m_1(T)) = $ det $(m_2(T))$. So we can make meaningful

Definition. If $T \in L(V)$, then

1. tr $(T) = $ tr $(m(T))$
2. det $(T) = $ det $(m(T))$

for $m(T)$ the matrix of T in *any* basis of V.

We now can carry over an important result from matrices, the Cayley–Hamilton Theorem.

Theorem 8.4.7. If V is n-dimensional over F and $T \in L(V)$, then T satisfies $p_T(x) = \det(xI - T)$, where $m(T)$ is the matrix of T in any basis of V. So T satisfies a polynomial of degree n over F.

Proof. $m(xI - T) = xI - m(T)$; hence

$$\det(xI - T) = \det(m(xI - T)) = \det(xI - m(T))$$
$$= p_{m(T)}(m(T)) = 0$$

by the Cayley–Hamilton Theorem for the matrix $m(T)$. ∎

We call $p_T(x) = \det(xI - T)$ the *characteristic polynomial* of T.

Thus if $a \in F$ is such that $\det(aI - T) = 0$, then $m(aI - T) = aI - m(T)$ cannot be invertible. Since m is $1-1$ and onto, $aI - T$ *cannot* be invertible. So there exists a vector $v \neq 0$ in V such that $(aI - T)v = 0$, that is, $Tv = av$. Furthermore, if $b \in F$ is *not* a root of $p_T(x)$, then $\det(bI - T) \neq 0$. This implies that $(bI - T)$ is invertible, hence there is *no* $v \neq 0$ such that $Tv = bv$. Thus

Theorem 8.4.8. If a is a root of $p_T(x) = \det(xI - T)$, then $Tv = av$ for some $v \neq 0$ in V, and if b is not a root of $p_T(x)$, there is no such $v \neq 0$ in V.

The roots of $p_T(x)$ are called the *characteristic roots* of T, and these are precisely the elements $a_1, \ldots, a_n$ in F such that $Tv_r = a_r v_r$ for some $v_r \neq 0$ in V. Also, v_r is called the *characteristic vector* associated with a_r.

With this said,

> everything *we did for characteristic roots and characteristic vectors of matrices* carries over in its entirety *to linear transformations on finite-dimensional vector spaces over F.*

Readers should make a list of what strikes them as the most important results that can so be transferred.

A further word about this transfer.

> *Not only are the results on characteristic roots transferable to linear transformations, but all results proved for matrices have their exact analogs for linear transformations. The mechanism to do this is the one we used, namely, passing to the matrix of T in a given basis, and making use of the theorems proved about this.*

PROBLEMS

NUMERICAL PROBLEMS

1. Find the matrix of D defined on the set of polynomials in x over F of degree at most 2 by $D(p(x)) = \dfrac{d}{dx}p(x)$ in the bases

   (a) $v_1 = 1, \quad v_2 = 2 - x, \quad v_3 = x^2 + 1.$

   (b) $v_1 = 1 + x, \quad v_2 = x, \quad v_3 = 1 + 2x + x^2.$

   (c) $v_1 = \dfrac{1 - x}{2}, \quad v_2 = \dfrac{1 + x}{2}, \quad v_3 = x^2.$

2. In Problem 1 find C such that $C^{-1}m_1(D)C = m_2(D)$, where $m_1(D)$ is the matrix of D in the basis of Part (a), and $m_2(D)$ is that in the basis of Part (c).

3. Do Problem 2 for the bases in Parts (b) and (c).

4. What are tr (D) and det (D) for the D in Problem 1?

5. Let V be all functions of the form $ae^x + be^{2x} + ce^{3x}$. Then if $D: V \longrightarrow V$ is defined by $D(f(x)) = \dfrac{d}{dx} f(x)$, find:

   (a) The matrix of D in the basis

   $$v_1 = e^x, \qquad v_2 = e^{2x}, \qquad v_3 = e^{3x}.$$

   (b) The matrix of D in the basis

   $$v_1 = e^x + e^{2x}, \qquad v_2 = e^{2x} + e^{3x}, \qquad v_3 = e^x + e^{3x}.$$

   (c) Find a matrix C such that $C^{-1}m_1(D)C = m_2(D)$, where $m_1(D)$ is the matrix of D in the basis of Part (a) and $m_2(D)$ is that in the basis of Part (b).

   (d) Find det $(m_1(D))$ and det $(m_2(D))$, and tr $(m_1(D))$ and tr $(m_2(D))$.

6. In Problem 5, find the characteristic roots of D and their associated characteristic vectors.

7. Show that for D in Problem 5, det (D) is the product of the characteristic roots of D and tr (D) is the sum of the characteristic roots of D.

8. Let V be the set of all polynomials in T of degree 3 or less over F. Define $T: V \longrightarrow V$ by $T(f(x))) = \dfrac{1}{x} \displaystyle\int_0^x f(t)\, dt$ for $f(x) \neq 0$ and $T(0) = 0$. Is T a linear transformation on V? If not, why not?

9. Find the matrix of T defined on the V of Problem 8 by

$$T(1) = 1 + x, \quad T(x) = (1 + x)^2, \quad T(x^2)$$
$$= (1 + x)^3, \quad T(x^3) = x;$$

and find det (T),   tr (T), and the characteristic polynomial $p_T(x)$.

MORE THEORETICAL PROBLEMS

Easier Problems

10. In V the set of polynomials in x over F of degree n or less, find $m(T)$, where T is defined in the basis

$$v_1 = 1, \quad v_2 = x, \quad \ldots, \quad v_n = x^n$$

by

(a) $T(1) = x, \quad T(x) = 1, \quad T(x^k) = x^k$ for $1 < k \le n$.

(b) $T(1) = x, \quad T(x) = x^2, \quad \ldots, \quad T(x^k) = x^{k+1}$ for $1 \le k < n, \ T(x^n) = 1$.

(c) $T(1) = x, \quad T(x) = x^2, \quad T(x^2) = x^3, \quad \ldots, \quad T(x^k) = x^k$ for $2 < k \le n$.

11. For Parts (a), (b), and (c) of Problem 10, find det (T).

Middle-Level Problems

12. For Parts (a) and (c) of Problem 10, if $F = \mathbb{C}$, find the characteristic roots of T and the associated characteristic vectors.

13. Can $\begin{bmatrix} 1 & 2 & 5 \\ 0 & 1 & 6 \\ 1 & 0 & 1 \end{bmatrix}$ and $\begin{bmatrix} -1 & 0 & 1 \\ 0 & 4 & 2 \\ 0 & 1 & 3 \end{bmatrix}$ be the matrices of the same

linear transformation in two different bases?

14. Show that $\begin{bmatrix} 1 & 2 & 3 \\ 0 & 1 & 2 \\ 0 & 0 & 1 \end{bmatrix}$ and $\begin{bmatrix} 1 & 2 & 5 \\ 0 & 1 & 2 \\ 0 & 0 & 1 \end{bmatrix}$ are the matrices of a linear

transformation T on $F^{(3)}$ in two different bases.

Harder Problems

15. If V is an n-dimensional vector space over $\mathbb{C}$ and $T \in L(V)$, show that we can find a basis $v_1, \ldots, v_n$ of V, such that $T(v_r)$ is a linear combination of v_r and its predecessors $v_1, \ldots, v_{r-1}$ for $r = 1, 2, \ldots, n$ (see Theorem 4.7.1).

16. If $T \in L(V)$ is nilpotent, that is, $T^k = 0$ for some k, show that $T^n = 0$, where $n = \dim (V)$.

17. If $T \in L(V)$ is nilpotent, prove that 0 is the only characteristic root of T.

18. Prove that

$$m(T_1 + T_2) = m(T_1) + m(T_2)$$

and

$$m(T_1T_2) = m(T_1)m(T_2).$$

19. Prove that if V is n-dimensional over F, then m maps $L(V)$ *onto* $M_n(F)$ in a $1-1$ way.

20. If V is a vector space and W is a subspace of V such that $T(W) \subset W$, show that we can find a basis of V such that

$$m(T) = \left[\begin{array}{c|c} A & * \\ \hline 0 & B \end{array}\right], \quad \text{where } A \text{ is an } r \times r \text{ matrix}$$

$(r = \dim (W))$ and B is an $(n - r) \times (n - r)$ matrix. (We don't care what * is.)

21. If V is a vector space and U, W are subspaces of V such that $V = U \oplus W$, show that if $T \in L(V)$ is such that $T(U) \subset U$ and $T(W) \subset W$, then there is a basis of V such that the matrix of T in

that basis looks like $\left[\begin{array}{c|c} A & 0 \\ \hline 0 & B \end{array}\right]$, where A, B are $r \times r$,

$(n - r) \times (n - r)$ matrices, respectively, where $r = \dim (U)$.

22. In Problem 20, what interpretation can you give the matrix U?

23. In Problem 21, what interpretation can you give the matrices A and B?

24. Prove that $T \in L(V)$ is invertible if and only if $\det (T) \neq 0$. What is $\det (T^{-1})$ in this case?

8.5. HERMITIAN IDEAS

Since we are always working over $F = \mathbb{R}$ or $F = \mathbb{C}$, we saw that we can introduce an inner product structure on $F^{(n)}$. This depends heavily on the fact that $F = \mathbb{R}$ or $\mathbb{C}$. We could extend the notion of vector spaces over "number systems" F that are neither $\mathbb{R}$ nor $\mathbb{C}$. Depending on the nature of this "number system" the "vector space" might or might not be susceptible to the introduction of an inner product on it.

However, we are working only over $F = \mathbb{R}$ or $\mathbb{C}$, so we have no such worries. So given any finite-dimensional vector space V, we saw in Section 7.7 that we could make of V an inner product space. There we did it using the existence of orthornormal bases for V and for $F^{(n)}$.

We redo it here—to begin with—in a slightly more abstract way. It is healthy to see something from a variety of points of view. Given $F^{(n)}$ we had introduced on $F^{(n)}$ an inner product via the equation

$$(v, w) = \sum_{r=1}^{} a_r \bar{b}_r,$$

where $v = \begin{bmatrix} a_1 \\ \vdots \\ a_n \end{bmatrix}$ and $w = \begin{bmatrix} b_1 \\ \vdots \\ b_n \end{bmatrix}$. By Theorem 7.4.3 we know that there is an isomorphism Φ of V onto $F^{(n)}$, where $n = \dim(V)$. We exploit Φ to produce an inner product on V. How? Just define the function $[\cdot, \cdot]$ by the rule

$$[v, w] = (\Phi(v), \Phi(w)) \qquad \text{for } v, w \in V.$$

Since $\Phi(v)$ and $\Phi(w)$ are in $F^{(n)}$ and $(\cdot, \cdot)$ is defined on $F^{(n)}$, the definition given above for $[\cdot, \cdot]$ makes sense.

It makes sense, but is it an inner product? The answer is "yes." We check out the rules that spell out an inner product, in turn.

1. If $v \in V$, then $[v, v] = (\Phi(v), \Phi(v)) \geq 0$. If $[v, v] = 0$, then $(\Phi(v), \Phi(v)) = 0$; hence $\Phi(v) = 0$ since $(\,,\,)$ is an inner product on $F^{(n)}$. Because Φ is $1-1$ we end up with the desired result, $v = 0$.

2. Is $[v, w] = \overline{[w, v]}$ the complex conjugate of (v, w)? Well,

$$\begin{aligned} [v, w] &= \overline{(\Phi(v), \Phi(w))} \\ &= \overline{(\Phi(w), \Phi(v))} \\ &= \overline{[w, v]}. \end{aligned}$$

3. Given $a \in F$, is $[av, w] = a[v, w]$? Again,

$$\begin{aligned} [av, w] &= (\Phi(av), \Phi(w)) \\ &= (a\Phi(v), \Phi(w)) \\ &= a(\Phi(v), \Phi(w)) = a[v, w]. \end{aligned}$$

4. Finally, does $[u + v, w] = [u, w] + [v, w]$? Because

$$\begin{aligned} [u + v, w] &= (\Phi(u + v), \Phi(w)) \\ &= (\Phi(u) + \Phi(v), \Phi(w)) \\ &= (\Phi(u), \Phi(w)) + (\Phi(v), \Phi(w)) = [u, w] + [v, w]. \end{aligned}$$

We have shown that $[\cdot, \cdot]$ is a legitimate inner product on V. Hence V is an inner product space. Thus

Theorem 8.5.1. If V is finite-dimensional over $\mathbb{R}$ or $\mathbb{C}$, then V is an inner product space.

Utilizing the isomorphisms Φ of V onto $F^{(n)}$, we propose to carry over to the abstract space V everything we did for $F^{(n)}$ as an inner product space. When this is done we shall define the Hermitian adjoint of any element in $L(V)$ and prove the analogs for Hermitian linear transformations of the results proved for Hermitian matrices.

Recall that the inner product on V is defined by

$$[v, w] = (\Phi(v), \Phi(w)),$$

where $(\cdot, \cdot)$ indicates the inner product in $F^{(n)}$. Let $e_1, \ldots, e_n$ be the canonical basis of $F^{(n)}$; so $e_1, \ldots, e_n$ is an orthonormal basis of $F^{(n)}$. Since Φ is an isomorphism of V onto $F^{(n)}$ there exist elements $v_1, \ldots, v_n$ in V such that $e_r = \Phi(v_r)$ for $r = 1, 2, \ldots, n$. Thus

$$[v_r, v_s] = (\Phi(v_r), \Phi(v_s)) = (e_r, e_s) = 1$$

if $r = s$ and $= 0$ if $r \neq s$.

If we call $v, w \in V$ *orthogonal* if $[v, w] = 0$, then our basis $v_1, \ldots, v_n$ above consists of mutually orthogonal elements. Furthermore, if $\sqrt{[v, v]}$ is called the *length* of v, then each v_r is of length 1. So it is reasonable, using $F^{(n)}$ as a prototype, to call $v_1, \ldots, v_n$ an *orthonormal* basis of V.

What we did for V also holds for any subspace W of V. So W also has an orthonormal basis. This is

Theorem 8.5.2. Given a subspace W of V, then W has an orthonormal basis.

The analog of the *Gram–Schmidt process* discussed earlier also carries over. We feel it would be good for the reader to go back to where this was done in Section 4.3 and to carry over the proof there to the abstract space V.

So, according to the Gram–Schmidt process, given $u_1, \ldots, u_k$ in V, linearly independent over F, we can find an orthonormal bases $x_1, \ldots, x_k$ of $<u_1, \ldots, u_k>$ (or cite Theorem 8.5.2). Let

$$S = <u_1, \ldots, u_k>^{\perp}$$

that is,

$$S = \{w \in V \mid [v, w] = 0 \text{ for all } v \in <u_1, \ldots, u_k>\}.$$

Then $<u_1, \ldots, u_k>^{\perp}$ is a subspace of V. If $v \in V$, then

$$s = v - [v, x_1]x_1 - [v, x_2]x_2 - \cdots - [v, x_k]x_k$$

has the property $[s, x_r] = 0$ for $r = 1, 2, \ldots, k$. Since s is orthogonal to every element of a basis of $<u_1, \ldots, u_k>$, s must be orthogonal to every element of $<u_1, \ldots, u_k>$. In short, $s \in S = <u_1, \ldots, u_k>^{\perp}$. So

$$v = s + [v, x_1]x_1 + \cdots + [v, x_k]x_k,$$

whence $v \in S + <u_1, \ldots, u_k>$. Thus $S + <u_1, \ldots, u_k> = V$. Also, $S \cap <u_1, \ldots, u_k> = 0$ (Why?). Therefore, $V = S \oplus <u_1, \ldots, u_k>$. (Verify all the statements made above!)

Given any subspace W of V then W has a finite basis so W is the linear span of a finite set of linearly independent elements. Thus, if $W^{\perp} = \{s \in V \mid [w, s] = 0 \text{ for all } w \in W\}$, then by the above we have

Theorem 8.5.3. If W is a subspace of V, then $V = W \oplus W^{\perp}$.

Definition. $W^{\perp}$ is called the *orthogonal complement* of W.

If $U = (W^{\perp})^{\perp}$, then for $u \in U$, $[u, s] = 0$ for all $s \in W^{\perp}$. Since $[w, s] = 0$ for $w \in W$, we know that $W \subset (W^{\perp})^{\perp} = W^{\perp \perp}$. However, $V = W^{\perp} \oplus W^{\perp \perp}$, dim $(V) = $ dim $W^{\perp}) + $ dim $(W^{\perp \perp})$. (Prove!) But since $V = W \oplus W^{\perp}$, dim $(V) = $ dim $(W) + $ dim $(W^{\perp})$. The upshot of all this is that dim $(W) = $ dim $(W^{\perp \perp}$. Because W is a subspace of $W^{\perp \perp}$ and of the same dimension as $W^{\perp \perp}$, we must have $W = W^{\perp \perp}$.

We have proved

Theorem 8.5.4. If W is a subspace of V, then $W = W^{\perp \perp}$.

We now turn our attention to linear transformations on V and how they interact with the inner product $[\cdot, \cdot]$ on V.

Let $v_1, \ldots, v_n$ be an orthonormal basis of V; hence $[v_r, v_s] = 0$ if $r \neq s$, $[v_r, v_r] = 1$, for $r, s = 1, 2, \ldots, n$. Let $T \in L(V)$ be a linear transformation on V. Thus for any $s = 1, 2, \ldots, n$,

$$Tv_s = \sum_{r=1}^{n} t_{rs}v_r,$$

where the $t_{rs} \in F$. Define S by

$$Sv_s = \sum_{r=1}^{n} \sigma_{rs}v_r,$$

where $\sigma_{rs} = \bar{t}_{sr}$. We prove

Theorem 8.5.5. S is a linear transformation on V and for any

$$v, w \in V, \quad [Tv, w] = [v, Sw].$$

Proof. It is enough to show that $[Tv_r, v_s] = [v_r, Sv_s]$ for all r, s. (Why?) But

$$[T(v_r), v_s] = \left[\sum_{k=1}^{n} t_{kr}v_k, v_s \right] = \sum_{k=1}^{n} t_{kr}[v_k, v_s] = t_{sr}$$

since $[v_k, v_s] = 0$ if $k \neq s$, and $[v_s, v_s] = 1$.

Now for the calculation of $[v_r, Sv_s]$. Since $Sv_s = \sum_{r=1}^{n} \sigma_{rs}v_r$ where $\sigma_{rs} = \bar{t}_{sr}$,

$$[v_r, Sv_s] = \left[v_r, \sum_{k=1}^{n} \sigma_{ks}v_k \right] = \sum_{k=1}^{n} \bar{\sigma}_{ks}[v_r, v_k] \text{ (Why?)} = \bar{\sigma}_{rs},$$

since $[v_r, v_k] = 0$ if $k \neq r$, and $= 1$ if $k = r$. So

$$\bar{\sigma}_{rs} = [v_r, Sv_s] = [Tv_r, v_s] = t_{sr}.$$

This gives us the desired result, $\sigma_{rs} = \bar{t}_{sr}$, and proves the theorem. ∎

Definition. We call S the *Hermitian adjoint* of T and denote it by T^*.

Note that if $m(T)$ is the matrix of T in an orthonormal basis of V, then the proof shows that $m(T^*) = m(T)^*$, where $m(T)^* = \overline{m(T)}'$, the Hermitian adjoint of the matrix $m(T)$. (Prove!)

We know from Theorem 8.5.5 that $[Tv, w] = [v, T^*w]$ for all $v, w \in V$. Conversely, if $[Tv, w] = [v, Sw]$ for all $v, w \in V$, then $S = T^*$. Why? If R and S are linear transformations of V such that $[v, Rw] = [v, Sw]$ for all $v, w \in V$, then for any $w \in V$, the vector $u = Sw - Rw$ satisfies $[v, u] = 0$ for all $v \in V$. Taking $v = u$ gives $[u, u] = 0$, so that $u = 0$. This, in turn, implies that $Rw = Sw$. Since we can show this for any $w \in V$, R and S are equal. Taking R to be T^*, we have

$$[v, T^*w] = [Tv, w] = [v, Sw]$$

for all $v, w \in V$. So $S = R = T^*$. We can now prove

Lemma 8.5.6. For $S, T \in L(V)$ and $a \in F$:

1. $T^{**} = T$, where T^{**} denotes $(T^*)^*$;

2. $(aS + T)^* = \bar{a}S^* + T^*$;
3. $(ST)^* = T^*S^*$.

Proof. By Theorem 8.5.5, we have the following for all v, $w \in V$:

$[Tv, w] = [v, T^*w]$ for all $v, w \in V$;
$[T^*w, v] = [w, (T^*)^*(v)]$ for all $v, w \in V$;
$[(aS + T)v, w] = [v, (aS + T)^*w]$;
$[STv, w] = [v, (ST)^*w]$.

Taking conjugates of both sides of the second equation, we get

$[v, T^*w] = [(T^*)^*(v), w]$,

which, by the first equation, implies that

$[(T^*)^*(v), w] = [Tv, w]$

for all $v, w \in V$. So, $T = (T^*)^*$. This proves (1). For (2), we expand the left-hand side of the third equation, getting

$$\begin{aligned}[(aS + T)v, w] &= [aSv + Tv, w] \\ &= a[Sv, w] + [Tv, w] = a[v, S^*w] + [Tv, w] \\ &= [v, \bar{a}S^*w] + [v, T^*w] = [v, (\bar{a}S^* + T^*)w]\end{aligned}$$

for all $v, w \in V$. From this, it follows that $(aS + T)^* = \bar{a}S^* + T^*$, proving (2). The proof of (3) is similar, and we leave it to the reader as an exercise. ∎

Definition. We call $T \in L(V)$ *Hermitian* if $T^* = T$. We call T *skew-Hermitian* if $T^* = -T$. We call T *unitary* if $TT^* = I = T^*T$.

Everything we proved for these kinds of matrices translates directly into a theorem for that kind of linear transformation. We give two sample results. The method used in proving them is that which is used in proving all such results.

Theorem 8.5.7. If T is Hermitian, then all its characteristic roots are real.

Proof. Let a be a characteristic root of T and $v \in V$ an associated characteristic vector. Thus $Tv = av$; hence

$[Tv, v] = [av, v] = a[v, v]$.

But

$[Tv, v] = [v, T^*v] = [v, Tv] = [v, av] = \bar{a}[v, v]$.

Therefore, $a[v, v] = \bar{a}[v, v]$, whence since $[v, v] \neq 0$, $a = \bar{a}$. Thus a is real. ■

Theorem 8.5.8. $T \in L(V)$ is unitary if and only if $[Tv, Tw] = [v, w]$ for all $v, w \in V$.

Proof. Since T is unitary, $T^*T = I$. So

$$[v, w] = [Iv, w] = [T^*Tv, w] = [Tv, T^{**}w] = [Tv, Tw].$$

On the other hand, if $[Tv, Tw] = [v, w]$ for all $v, w \in V$, then

$$[T^*Tv, w] = [v, w];$$

thus $[T^*Tv - v, w] = 0$ for all w. This forces $T^*Tv = v$ for all $v \in V$. (Why?) So $T^*T = I$. We leave the proof that $TT^* = I$ to the reader. ■

The further properties of Hermitian, skew-Hermitian, and unitary linear transformations will be found in the exercises.

PROBLEMS

NUMERICAL PROBLEMS

1. For the following vector spaces V, find an isomorphism of V onto $F^{(n)}$ for appropriate n and use this isomorphism to define an inner product on V.

(a) $V =$ all polynomials of degree 3 or less over F.

(b) $V =$ all real-valued functions of the form

$$f(x) = a \cos (x) + b \sin (x).$$

(c) $V =$ all real-valued functions of the form

$$f(x) = ae^x + be^{2x} + ce^{3x}.$$

2. If V is the vector space of all polynomials of degree 4 or less and W is the subspace consisting of elements of the form

$$a + bx^2 + cx^4,$$

find via an isomorphism of V onto $F^{(5)}$ an inner product on V. Then find $W^{\perp}$.

3. For V as in Problem 2 find an orthonormal basis for $<v_1, v_2, v_3>$ where $v_1 = 1 + x$, $v_2 = x^2$, $v_3 = 1 + x + x^2 + x^3$.

4. For V as in Part (a) of Problem 1 define

$$(f(x), g(x)) = \int_0^1 f(x)\overline{g(x)} \, dx.$$

Prove this defines an inner product on V.

5. In Problem 4 find an orthonormal basis of V with respect to the inner product defined there.

6. For V as in Part (b) of Problem 1 define

$$(f(x), g(x)) = \int_0^\pi f(x)g(x) \, dx.$$

(a) Show that this defines an inner product on V.

(b) Find an orthonormal basis of V with respect to this inner product.

7. For V as in Problem 6, let $T \in L(V)$ be defined by

$$T(\cos(x)) = \sin(x), \; T(\sin(x)) = \cos(x) + \sin(x).$$

(a) Find the matrix of T in the orthonormal basis of Part (b) of Problem 6.

(b) Express T^* as a linear transformation on V. [That is, what is $T^*(\cos(x))$, $T^*(\sin(x))$?]

8. If V is the vector space in Problem 5, using the orthonormal basis you found, find the form of T^* if $T(x^k) = x^{k+1}$ for $0 \le k < 3$ and $T(x^3) = 1$.

9. If V, W are as in Problem 2, show that $W \oplus W^\perp = V$ and $W^{\perp\perp} = W$.

MORE THEORETICAL PROBLEMS

V is a vector space over $\mathbb{C}$.

Easier Problems

10. If $T \in L(V)$ and T is Hermitian, show that $iT = S$ is skew-Hermitian.

11. If S is skew-Hermitian, prove that $T = iS$ is Hermitian.

12. Prove that the characteristic roots of a skew-Hermitian T in $L(V)$ must be pure imaginaries.

13. Given $T \in L(V)$, show that $T = A + B$ where A is Hermitian, B is skew-Hermitian. Prove that A and B are unique.

14. Prove the following to be Hermitian:

(a) TT^*

(b) $T + T^*$

(c) $TS + S^*T^*$

for $S, T \in L(V)$.

Middle-Level Problems

15. If T is skew-Hermitian, prove that T^{2k} is Hermitian and T^{2k+1} is skew-Hermitian for all integers $k \geq 0$.

16. If T is unitary, prove that if a is a characteristic root of T, then $|a| = 1$.

17. Suppose that $W \subset V$ is a subspace of V such that $T(W) \subset W$, where $T^* = T$. Prove that $T(W^\perp) \subset W^\perp$.

18. If W and T are as in Problem 17, show that $m(T) = \begin{bmatrix} A & 0 \\ 0 & B \end{bmatrix}$, where

A is $r \times r$ and $r = \dim(W)$, for some suitable basis of V.

Harder Problems

19. Use Problem 18 and induction to prove that if $T^* = T$, then there exists an orthonormal basis of V such that $m(T)$ is a diagonal matrix in this basis.

20. If T is Hermitian and $T^k v = 0$, prove that $Tv = 0$.

21. If U in $L(V)$ is such that for some orthonormal basis $v_1, \ldots, v_n$ of V, the elements $w_1 = Uv_1, \ldots, w_n = Uv_n$ form an orthonormal basis of V, prove that U must be unitary.

22. If T is Hermitian, prove that both tr (T) and det (T) are real.

23. In terms of the characteristic roots of T, where $T^* = T$, what are tr (T) and det (T)?

8.6. LINEAR TRANSFORMATIONS FROM ONE SPACE TO ANOTHER

Our main emphasis in this book has been on linear transformations of a given vector space into itself. But there is also some interest and need to speak about linear transformations from one vector space to another.

In point of fact, we already have done so when we discussed homomorphisms of one vector space into another. In a somewhat camouflaged way we also discussed this in talking about $m \times n$ matrices where $m \neq n$. The unmasking of this camouflage will be done here, in the only theorem we shall prove in this section.

Let V and W be vector spaces of dimensions n and m, respectively, over F. We say that $T: V \longrightarrow W$ is a *linear transformation* from V to W if

1. $T(v_1 + v_2) = T(v_1) + T(v_2)$
2. $T(av_1) = aT(v_1)$

for all $v_1, v_2 \in V$, $a \in F$.

If you look back at the appropriate section, T is precisely what we called a homomorphism from V to W.

We prove, for the setup above

Theorem 8.6.1.   Given a linear transformation from V to W, then T can be represented by an $m \times n$ matrix with entries from F. Furthermore, any such $m \times n$ matrix can be used to define a linear transformation from V to W.

Proof.   Let $v_1, \ldots, v_n$ be a basis of V over F, and $w_1, \ldots, w_m$ a basis of W over F. Since $T(v_s) \in W$ for $s = 1, 2, \ldots, n$ and $w_1, \ldots, w_m$ is a basis of W, we have that

$$T(v_s) = \sum_{r=1}^{m} t_{rs} w_r.$$

Because any $v \in V$ is of the form

$$v = \sum_{s=1}^{n} a_s v_s,$$

with the $a_s \in F$, and since T is a linear transformation from V to W, we have

$$T(v_s) = T\left(\sum_{s=1}^{n} a_s v_s\right) = \sum_{s=1}^{n} a_s T(v_s).$$

Thus knowing $T(v_s)$ allows us to know exactly what $T(v)$ is for any $v \in V$. But all the information about $T(v_s)$ is contained in the $m \times n$ matrix

$$m(T) = \begin{bmatrix} t_{11} & t_{12} & \cdots & t_{1n} \\ t_{21} & t_{22} & \cdots & t_{2n} \\ \vdots & \vdots & & \vdots \\ t_{m1} & t_{m2} & \cdots & t_{mn} \end{bmatrix},$$

for we can read off $T(v_s)$ from this matrix. So T is represented by this matrix $m(T)$. It is easy to verify that under addition and multiplication of linear transformations by scalars, $m(T_1 + T_2) = m(T_1) + m(T_2)$, and $m(aT_1) = am(T_1)$, for T_1, T_2 linear transformations from V to W and $a \in F$. Of course, we are assuming that $T_1 + T_2$ is defined by

$$(T_1 + T_2)(v) = T_1(v) + T_2(v)$$

and

$$(aT_1)(v) = a(T_1(v))$$

for $v \in V$, $a \in F$.

Going the other way, given an $m \times n$ matrix

$$A = \begin{bmatrix} t_{11} & t_{12} & \cdots & t_{1n} \\ t_{21} & t_{22} & \cdots & t_{2n} \\ \vdots & \vdots & & \vdots \\ t_{m1} & t_{m2} & \cdots & t_{mn} \end{bmatrix},$$

with entries in F, define a linear transformation $T: V \longrightarrow W$, making use of the bases $v_1, \ldots, v_n$ of V and $w_1, \ldots, w_m$ of W by the equations

$$T(v_s) = \sum_{r=1}^{m} t_{rs} w_r.$$

We leave it to the reader to prove that T defined this way leads us to a linear transformation of V to W.

Note that our constructions above made heavy use of particular bases of V and W. Using other bases we would get other matrix representations of T. These representations are closely related, but we shall not go into this relationship here. ■

We give an example of the theorem. Let $V = F^{(2)}$ and $W = F^{(3)}$. Define $T: V \longrightarrow W$ by the rule

$$T\begin{bmatrix} a_1 \\ a_2 \end{bmatrix} = \begin{bmatrix} a_1 + a_2 \\ a_1 - a_2 \\ a_2 \end{bmatrix}.$$

It is easy to see that T defines a linear transformation of V to W. Using the bases $v_1 = \begin{bmatrix} 1 \\ 0 \end{bmatrix}$, $v_2 = \begin{bmatrix} 0 \\ 1 \end{bmatrix}$ of V and $w_1 = \begin{bmatrix} 1 \\ 0 \\ 0 \end{bmatrix}$, $w_2 = \begin{bmatrix} 0 \\ 1 \\ 0 \end{bmatrix}$,

$w_3 = \begin{bmatrix} 0 \\ 0 \\ 1 \end{bmatrix}$ of W, we have that

$$T(v_1) = T\begin{bmatrix} 1 \\ 0 \end{bmatrix} = \begin{bmatrix} 1 \\ 1 \\ 0 \end{bmatrix} = w_1 + w_2$$

and

$$T(v_2) = T\begin{bmatrix} 0 \\ 1 \end{bmatrix} = \begin{bmatrix} 1 \\ -1 \\ 1 \end{bmatrix} = w_1 - w_2 + w_3.$$

Thus the matrix in the given bases of T is given by

$$m(T) = \begin{bmatrix} 1 & 1 \\ 1 & -1 \\ 0 & 1 \end{bmatrix},$$

a 2×3 matrix over F. What would $m(T)$ be if we used the basis $v_1 = \begin{bmatrix} 1 \\ -1 \end{bmatrix}$ and $v_2 = \begin{bmatrix} 1 \\ 1 \end{bmatrix}$ of V over F, and w_1, w_2, w_3 as above?

Then

$$T(v_1) = \begin{bmatrix} 0 \\ 1 \\ -1 \end{bmatrix} = w_2 - w_3$$

and

$$T(v_2) = \begin{bmatrix} 2 \\ 0 \\ 1 \end{bmatrix} = 2w_1 + w_3.$$

Thus the matrix of T in these bases is

$$m(T) = \begin{bmatrix} 0 & 2 \\ 1 & 0 \\ -1 & 1 \end{bmatrix}.$$

PROBLEMS

NUMERICAL PROBLEMS

1. If V is the vector space of all polynomials in x of degree 3 or less over F and $W = F^{(2)}$, let $T: V \longrightarrow W$ be

$$T(a_0 + a_1x + a_2x^2 + a_3x^3) = \begin{bmatrix} a_3 \\ 0 \end{bmatrix}.$$

Using as basis for V the powers 1, x, x^2, x^3 of x, and as basis $\begin{bmatrix} 1 \\ 0 \end{bmatrix}$ and $\begin{bmatrix} 0 \\ 1 \end{bmatrix}$ for W, find the matrix of T.

2. In Problem 1, if we use $\begin{bmatrix} 1 \\ -1 \end{bmatrix}$, $\begin{bmatrix} 1 \\ 1 \end{bmatrix}$ as a basis of W, and the same basis of V, find the matrix of T.

3. Let V be the linear span of

$$\sin(x), \sin(2x), \sin(3x), \cos(x), \cos(2x)$$

over $\mathbb{R}$, and let W be the linear span of $\sin(x)$, $\sin(2x)$ over $\mathbb{R}$. Define $T: V \longrightarrow W$ by

$$T(a\sin(x) + b\sin(2x) + c\sin(3x) + d\cos(x) + e\cos(2x))$$
$$= a\sin(x) - 5c\sin(2x).$$

Using appropriate bases of V and W, find $m(T)$.

4. Let V be the real-valued functions of the form

$$a + be^x + ce^{-x} + de^{2x}.$$

Define $T: V \longrightarrow F^{(3)}$ by the rule

$$T(a + be^x + ce^{-x} + de^{2x}) = \begin{bmatrix} b \\ -c \\ 2d \end{bmatrix}.$$

Find $M(T)$ in suitable bases of V and $F^{(3)}$.

5. Let V be as in Problem 4 and let W be the set of all real-valued functions of the form $\alpha + \beta x + \gamma e^x + \delta e^{-x} + \mu e^{2x}$. Consider $T: V \longrightarrow W$ defined by $T(f(x)) = \int_0^x f(t)\, dt$.

   (a) Prove that T is a linear transformation from V to W.

   (b) Find the matrix of T in appropriate bases of V and W.

MORE THEORETICAL PROBLEMS

Easier Problems

6. Prove that $T_1 + T_2$ and aT_1 as defined are linear transformations from V to W.

7. If A is the $m \times n$ matrix (a_{rs}) in the canonical bases of $F^{(n)}$ and $F^{(m)}$ respectively,

   (a) Write down the linear transformation T defined by A using these canonical bases.

   (b) If $m(T)$ is the matrix of T in the canonical bases, prove that $m(T) = A$.

9

APPLICATIONS
(OPTIONAL)

9.1. FIBONACCI NUMBERS

The ancient Greeks attributed a mystical and aesthetic significance to what is called the *golden section*. The golden section is the division of a line segment into two parts such that the smaller one, a, is to the larger one, b, as b is to the total length of the segment. In mathematical terms this translates into

$$\frac{a}{b} = \frac{b}{a + b},$$

hence $a(a + b) = b^2$, and so $a^2 + ab - b^2 = 0$. In particular, if we normalize things by putting $b = 1$, we get $a^2 + a - 1 = 0$. Hence, by the quadratic formula,

$$a = \frac{-1 \pm \sqrt{1 + 4}}{2} = \frac{-1 \pm \sqrt{5}}{2}.$$

The particular value

$$a = \frac{\sqrt{5} - 1}{2}$$

is called the *golden mean*. This number also comes up in certain properties of regular pentagons.

To the Greeks, constructions where the ratio of the sides of a rectangle was the golden mean were considered to have a high aesthetic value. This fascination with the golden mean persists to this day. For example, in the book *A Maggot* by the popular writer John Fowles, he expounds on the golden mean, the Fibonacci numbers, and a hint that there is a relation between them. He even cites the Fibonacci numbers in relation to the rise of the rabbit population, which we discuss below.

We introduce the sequence of numbers known as the Fibonacci numbers. They derive their name from the Italian mathematician Leonardo Fibonacci, who lived in Pisa and is often known as Leonardo di Pisa. He lived from 1180 to 1250. To quote Boyer in his *A History of Mathematics:*

> *"Leonardo of Pisa was without doubt the most original and most capable mathematician of the medieval Christian world."*

Again quoting Boyer, Fibonacci in his book *Liber Abaci* posed the following problem:

> *"How many pairs of rabbits will be produced in a year, beginning with a single pair, if in every month each pair bears a new pair which becomes productive from the second month on?"*

As is easily seen, this problem gives rise to the sequence of numbers

$$a_0 = 0, \quad a_1 = 1, \quad a_2 = 1, \quad a_3 = 2, \quad a_4 = 3, \quad \ldots$$

where the $(n + 1)$st number, a_{n+1}, is determined by

$$a_{n+1} = a_n + a_{n-1}$$

for $n \geq 1$. In other words, the $(n + 1)$st member of the sequence is the sum of its two immediate predecessors. This sequence of numbers is called the *Fibonacci sequence,* and its terms are known as the *Fibonacci numbers.*

Can we find a nice simple formula that expresses a_n as a function of n? Indeed we can, as we shall demonstrate below. Interestingly enough, the answer we get involves the golden mean, which was mentioned above.

There are many approaches to the problem of finding a formula for a_n. The approach we take is by means of the 2×2 matrices, not that this is necessarily the shortest or best method of solving the problem.

What it does do is to give us a nice illustration of how one can exploit characteristic roots and their associated characteristic vectors. Also, this method lends itself to easy generalizations.

In $\mathbb{C}^{(2)}$ we write down a sequence of vectors built up from the Fibonacci numbers as follows:

$$v_0 = \begin{bmatrix} a_0 \\ a_1 \end{bmatrix} = \begin{bmatrix} 0 \\ 1 \end{bmatrix}, \quad v_1 = \begin{bmatrix} a_1 \\ a_2 \end{bmatrix} = \begin{bmatrix} 1 \\ 1 \end{bmatrix},$$

$$v_2 = \begin{bmatrix} a_2 \\ a_3 \end{bmatrix} = \begin{bmatrix} 1 \\ 2 \end{bmatrix}, \quad v_3 = \begin{bmatrix} a_3 \\ a_4 \end{bmatrix} = \begin{bmatrix} 2 \\ 3 \end{bmatrix}, \quad \ldots,$$

and generally,

$$v_n = \begin{bmatrix} a_n \\ a_{n+1} \end{bmatrix}$$

for all positive integers n.

Consider the matrix

$$A = \begin{bmatrix} 0 & 1 \\ 1 & 1 \end{bmatrix}.$$

How does A behave vis-à-vis the sequence of vectors $v_0, v_1, v_2, \ldots,$ $v_n, \ldots$? Note that

$$Av_0 = \begin{bmatrix} 0 & 1 \\ 1 & 1 \end{bmatrix}\begin{bmatrix} 0 \\ 1 \end{bmatrix} = \begin{bmatrix} 1 \\ 1 \end{bmatrix} = v_1,$$

$$Av_1 = \begin{bmatrix} 0 & 1 \\ 1 & 1 \end{bmatrix}\begin{bmatrix} 1 \\ 1 \end{bmatrix} = \begin{bmatrix} 1 \\ 2 \end{bmatrix} = v_2, \quad \ldots,$$

$$Av_n = \begin{bmatrix} 0 & 1 \\ 1 & 1 \end{bmatrix}\begin{bmatrix} a_n \\ a_{n+1} \end{bmatrix} = \begin{bmatrix} a_{n+1} \\ a_n + a_{n+1} \end{bmatrix} = \begin{bmatrix} a_{n+1} \\ a_{n+2} \end{bmatrix} = v_{n+1}.$$

Thus A carries each term of the sequence into its successor in the sequence. Going back to the beginning,

$$v_1 = Av_0, \quad v_2 = Av_1 = A(Av_0) = A^2v_0,$$
$$v_3 = Av_2 = A(A^2v_0) = A^3v_0,$$

and

$$v_{n+1} = Av_n = A(A^nv_0) = A^{n+1}v_0.$$

If we knew exactly what $A^n v_0 = v_n$ equals, then, by reading off the first component, we would have the desired formula for a_n.

We ask:

What are the characteristic roots of $A = \begin{bmatrix} 0 & 1 \\ 1 & 1 \end{bmatrix}$?

Since its characteristic polynomial is

$$p_A(x) = \det(xI - A) = \det \begin{bmatrix} x & -1 \\ -1 & x-1 \end{bmatrix} = x^2 - x - 1,$$

from the quadratic formula the roots of $p_A(x)$, that is, the characteristic roots of A, are

$$\frac{1 + \sqrt{5}}{2}, \frac{1 - \sqrt{5}}{2}.$$

Note that the second of these is just the negative of the golden mean.

We want to find explicitly the characteristic vectors w_1 and w_2 associated with $\dfrac{1 + \sqrt{5}}{2}$ and $\dfrac{1 - \sqrt{5}}{2}$, respectively. Because

$\left(\dfrac{1 + \sqrt{5}}{2} \right) \left(\dfrac{-1 + \sqrt{5}}{2} \right) = 1$ we see that if $w_1 =$

$\begin{bmatrix} (-1 + \sqrt{5})/2 \\ 1 \end{bmatrix}$, then

$$Aw_1 = \begin{bmatrix} 0 & 1 \\ 1 & 1 \end{bmatrix} \begin{bmatrix} (-1 + \sqrt{5})/2 \\ 1 \end{bmatrix} = \begin{bmatrix} 1 \\ 1 + (-1 + \sqrt{5})/2 \end{bmatrix}$$

$$= \begin{bmatrix} 1 \\ (1 + \sqrt{5})/2 \end{bmatrix} = (1 + \sqrt{5})/2 \begin{bmatrix} (-1 + \sqrt{5})/2 \\ 1 \end{bmatrix}.$$

Thus w_1 is a characteristic vector associated with the characteristic root $(1 + \sqrt{5})/2$. A similar computation shows that

$$w_2 = \begin{bmatrix} (-1 - \sqrt{5})/2 \\ 1 \end{bmatrix}$$

is a characteristic vector associated with $(1 - \sqrt{5})/2$, that is, $Aw_2 = ((1 - \sqrt{5})/2) \, w_2$.

From $Aw_1 = ((1 + \sqrt{5})/2)w_1$ we easily get, by successively multiplying by A, that

$$A^2w_1 = A(Aw_1) = A\left(\frac{1 + \sqrt{5}}{2}\right)w_1 = \left(\frac{1 + \sqrt{5}}{2}\right)Aw_1$$

$$= \left(\frac{1 + \sqrt{5}}{2}\right)^2 w_1,$$

$$\vdots$$

$$A^nw_1 = \left(\frac{1 + \sqrt{5}}{2}\right)^n w_1.$$

Similarly,

$$A^nw_2 = \left(\frac{1 - \sqrt{5}}{2}\right)^n w_2.$$

Can v_0 be expressed as a combination of w_1 and w_2? We write down this expression, which you can verify or find for yourself:

$$v_0 = \left(\frac{1 + \sqrt{5}}{2\sqrt{5}}\right)w_1 - \left(\frac{1 - \sqrt{5}}{2\sqrt{5}}\right)w_2.$$

From this expression, we find that

$$v_n = A^nv_0 = A^n\left(\left(\frac{1 + \sqrt{5}}{2\sqrt{5}}\right)w_1 - \left(\frac{1 - \sqrt{5}}{2\sqrt{5}}\right)w_2\right)$$

$$= \left(\frac{1 + \sqrt{5}}{2\sqrt{5}}\right)\left(\frac{1 + \sqrt{5}}{2}\right)^n w_1 - \left(\frac{1 - \sqrt{5}}{2\sqrt{5}}\right)\left(\frac{1 - \sqrt{5}}{2}\right)^n w_2$$

$$= \frac{1}{\sqrt{5}}\left[\begin{array}{c}\left(\frac{1 + \sqrt{5}}{2}\right)^n - \left(\frac{1 - \sqrt{5}}{2}\right)^n \\ \left(\frac{1 + \sqrt{5}}{2}\right)^{n+1} - \left(\frac{1 - \sqrt{5}}{2}\right)^{n+1}\end{array}\right] = \begin{bmatrix} a_n \\ a_{n+1} \end{bmatrix}.$$

Thus

$$a_n = \frac{1}{\sqrt{5}}\left(\left(\frac{1 + \sqrt{5}}{2}\right)^n - \left(\frac{1 - \sqrt{5}}{2}\right)^n\right).$$

(Verify all the arithmetic!) We have found the desired formula for a_n.

Note one little consequence of this formula for a_n. By its construction, a_n is an *integer;* therefore, the strange combination

$$\frac{1}{\sqrt{5}}\left(\left(\frac{1 + \sqrt{5}}{2}\right)^n - \left(\frac{1 - \sqrt{5}}{2}\right)^n\right)$$

is an integer. Can you find a direct proof of this, without resorting to the Fibonacci numbers?

There are several natural variants of the Fibonacci numbers that one could introduce. To begin with, we need not begin the proceedings with

$$a_0 = 0, \quad a_1 = 1, \quad a_2 = 1, \quad \ldots ;$$

we could start with

$$a_0 = a, \quad a_1 = b, \quad a_2 = a + b, \ldots, \quad a_{n+1} = a_n + a_{n-1}.$$

The same matrix $\begin{bmatrix} 0 & 1 \\ 1 & 1 \end{bmatrix}$, the same characteristic roots, and the same

characteristic vectors w_1 and w_2 arise. The only change we have to make

is to express $\begin{bmatrix} a \\ b \end{bmatrix}$ as a combination of w_1 and w_2. The rest of the argument

runs as above.

A second variant might be: Assume that

$$a_0 = 0, a_1 = 1, a_2 = c, \ldots, a_{n+1} = ca_n + da_{n-1}$$

where c, and d are fixed integers. The change from the argument above

would be that we use the matrix $B = \begin{bmatrix} 0 & 1 \\ d & c \end{bmatrix}$; note that

$$B\begin{bmatrix} a_{n-1} \\ a_n \end{bmatrix} = \begin{bmatrix} 0 & 1 \\ d & c \end{bmatrix}\begin{bmatrix} a_{n-1} \\ a_n \end{bmatrix} = \begin{bmatrix} a_n \\ da_{n-1} + ca_n \end{bmatrix} = \begin{bmatrix} a_n \\ a_{n+1} \end{bmatrix}.$$

If the characteristic roots of B are distinct, to find the formula for a_n we must find the characteristic roots and associated vectors of B, express

the first vector $\begin{bmatrix} 0 \\ 1 \end{bmatrix}$ as a combination of these characteristic vectors, and

proceed as before.

We illustrate this with a simple example. Suppose that

$$a_0 = 0, \quad a_1 = 1, \quad a_2 = 1, \quad a_3 = a_2 + 2a_1 = 3, \quad \ldots,$$
$$a_n = a_{n-1} + 2a_{n-2}$$

for $n \geq 2$. Thus our matrix B is merely

$$B = \begin{bmatrix} 0 & 1 \\ 2 & 1 \end{bmatrix}.$$

The characteristic roots of B are 2 and -1, and if

$$w_1 = \begin{bmatrix} 1 \\ 2 \end{bmatrix}, \quad w_2 = \begin{bmatrix} -1 \\ 1 \end{bmatrix},$$

then $Bw_1 = 2w_1$, $Bw_2 = -w_2$. Also,

$$\begin{bmatrix} 0 \\ 1 \end{bmatrix} = \tfrac{1}{3}w_1 + \tfrac{1}{3}w_2,$$

hence

$$v_n = B^n v_0 = \frac{1}{3} B^n w_1 + \frac{1}{3} B^n w_2 = \frac{2^n}{3} w_1 + \frac{(-1)^n}{3} w_2$$

$$= \begin{bmatrix} \dfrac{2^n}{3} + \dfrac{(-1)^{n+1}}{3} \\[2mm] \dfrac{2^{n+1}}{3} + \dfrac{(-1)^{n+2}}{3} \end{bmatrix}.$$

Thus

$$a_n = \frac{2^n + (-1)^{n+1}}{3}$$

for all $n \geq 0$.

The last variant we want to consider is one in which we don't assume that a_{n+1} is merely a fixed combination of its two predecessors. True, the Fibonacci numbers use only the two predecessors, but that doesn't make "two" sacrosanct. We might assume that $a_0, a_1, \ldots, a_{k-1}$ are given integers and for $n \geq k$,

$$a_n = c_1 a_{n-1} + c_2 a_{n-2} + \cdots + c_k a_{n-k},$$

where $c_1, \ldots, c_k$ are fixed integers. The solutions would involve the

k-tuples $\begin{bmatrix} a_{n+1} \\ a_{n+2} \\ \vdots \\ a_{n+k} \end{bmatrix}$ and the matrix

$$C = \begin{bmatrix} 0 & 1 & 0 & 0 & 0 & 0 & \cdots & 0 & 0 \\ 0 & 0 & 1 & 0 & 0 & 0 & \cdots & 0 & 0 \\ 0 & 0 & 0 & 1 & 0 & 0 & \cdots & 0 & 0 \\ \vdots & \vdots & \vdots & \vdots & \vdots & \vdots & & \vdots & \vdots \\ 0 & 0 & 0 & 0 & 0 & 0 & \cdots & 0 & 1 \\ c_k & c_{k-1} & c_{k-2} & c_{k-3} & c_{k-4} & c_{k-5} & \cdots & c_2 & c_1 \end{bmatrix}.$$

This would entail the roots of the characteristic polynomial

$$p_C(x) = x^k - c_1 x^{k-1} - c_2 x^{k-2} - \cdots - c_{k-1} x - c_k.$$

In general, we cannot find the roots of $p_C(x)$ exactly. However, if we could find the roots, and if they were distinct, we could find the asso-

ciated characteristic vectors $w_1, \ldots, w_k$ and express $\begin{bmatrix} a_0 \\ a_1 \\ \vdots \\ a_{k-1} \end{bmatrix}$ as a com-

bination of $w_1, \ldots, w_k$. We would then proceed as before, noting that

$$C \begin{bmatrix} a_n \\ a_{n+1} \\ \vdots \\ a_{n+k-1} \end{bmatrix} = \begin{bmatrix} a_{n+1} \\ a_{n+2} \\ \vdots \\ a_{n+k} \end{bmatrix}.$$

and

$$C^n \begin{bmatrix} a_0 \\ a_1 \\ \vdots \\ a_{k-1} \end{bmatrix} = \begin{bmatrix} a_n \\ a_{n+1} \\ \vdots \\ a_{n+k-1} \end{bmatrix}.$$

PROBLEMS

NUMERICAL PROBLEMS

1. Find a_n if
   (a) $a_0 = 0$, $a_1 = 4$, $a_2 = 4$, $\ldots$, $a_n = a_{n-1} + a_{n-2}$ for $n \geq 2$.
   (b) $a_0 = 1$, $a_1 = 1$, $\ldots$, $a_n = 4a_{n-1} + 3a_{n-2}$ for $n \geq 2$.

(c) $a_0 = 1$, $a_1 = 3$, $\ldots$, $a_n = 4a_{n-1} + 3a_{n-2}$ for $n \geq 2$.

(d) $a_0 = 1$, $a_1 = 1$, $\ldots$, $a_n = 2(a_{n-1} + a_{n-2})$ for $n \geq 2$.

MORE THEORETICAL PROBLEMS

2. For $v_0 = \begin{bmatrix} 0 \\ 1 \end{bmatrix}$ we found that $a_n = A^n v_0$. Show that for $\tilde{v}_0 = \begin{bmatrix} 1 \\ 0 \end{bmatrix}$ that $\tilde{a}_n = A^n \tilde{v}_0$ and $\tilde{a}_n = a_{n-1}$.

3. For $v_0 = \begin{bmatrix} x \\ y \end{bmatrix}$ and $a_n = A^n v_0$, use the result of Problem 2 to show that $a_n = ya_n + xa_{n-1}$.

4. Show directly that $\dfrac{1}{\sqrt{5}} (((1 + \sqrt{5})/2)^n - ((1 - \sqrt{5})/2)^n)$ is an integer and is positive.

5. Show that $(1 + \sqrt{5})^n + (1 - \sqrt{5})^n$ is an integer.

6. If w_1 and w_2 are as in the discussion of the Fibonacci numbers, show that $w_1 - w_2 = \sqrt{5} \begin{bmatrix} 0 \\ 1 \end{bmatrix}$ and therefore that

$$\begin{bmatrix} 0 \\ 1 \end{bmatrix} = \begin{bmatrix} 0 & 1 \\ 1 & 1 \end{bmatrix}\begin{bmatrix} 1 \\ 0 \end{bmatrix} = \left(\frac{1}{\sqrt{5}} [w_1 - w_2] \right)$$

$$= \frac{1 + \sqrt{5}}{2\sqrt{5}} w_1 - \frac{1 - \sqrt{5}}{2\sqrt{5}} w_2.$$

9.2. EQUATIONS OF CURVES

Determinants provide us with a tool to determine the equation of a curve of a given kind that passes through enough given points. We can immediately obtain the equation of the curve in determinant form. To get it more explicitly would require the expansion of the determinant. Of course, if the determinant is of high order, this is no mean job.

Be that as it may, we do get an effective method of determining curves through points this way. We illustrate what we mean with a few examples.

We all know that the equation of a straight line is of the form $ax + by + c = 0$, where a, b, c are constants. Furthermore, we know that a line is determined by two distinct points on it. Given

$$(x_1, y_1) \neq (x_2, y_2),$$

what is the equation of the straight line through them? Suppose the equation of the line to be $ax + by + c = 0$. Thus since (x_1, y_2) and (x_2, y_2) are on the line, we have

$$ax + by + c = 0$$
$$ax_1 + by_1 + c = 0$$
$$ax_2 + by_2 + c = 0.$$

In order that this system of equations have a nontrivial solution for

(a, b, c), we must have $\begin{vmatrix} x & y & 1 \\ x_1 & y_1 & 1 \\ x_2 & y_2 & 1 \end{vmatrix} = 0.$ (Why?) Hence, expanding

yields

$$x(y_1 - y_2) - y(x_1 - x_2) + (x_1y_2 - x_2y_1) = 0.$$

This is the equation of the straight line we are looking for. Of course, it is an easy enough matter to do this without determinants. But doing it with determinants illustrates what we are talking about.

Consider another situation, that of a circle. The equation of the general circle is given by $a(x^2 + y^2) + bx + cy + d = 0$. (If $a = 0$, the circle degenerates to a straight line.) Given three points, they determine a unique circle (if the points are collinear, this circle degenerates to a line). What is the equation of this circle? If the points are

$$(x_1, y_1), (x_2, y_2), (x_3, y_3)$$

we have the four equations

$$a(x^2 + y^2) + bx + cy + d = 0$$
$$a(x_1^2 + y_1^2) + bx_1 + cy_1 + d = 0$$
$$a(x_2^2 + y_2^2) + bx_2 + cy_2 + d = 0$$
$$a(x_3^2 + y_3^2) + bx_3 + cy_3 + d = 0.$$

Since not all of a, b, c, d are zero, we must have the determinant of their coefficients equal to 0. Thus

$$\begin{vmatrix} x^2 + y^2 & x & y & 1 \\ x_1^2 + y_1^2 & x_1 & y_1 & 1 \\ x_2^2 + y_2^2 & x_2 & y_2 & 1 \\ x_3^2 + y_3^2 & x_3 & y_3 & 1 \end{vmatrix} = 0$$

is the equation of the sought-after circle.

Consider a specific case the three points are $(1, -1)$, $(1, 2)$, $(2, 0)$. These are not collinear. The circle through them is given by

$$\begin{vmatrix} x^2 + y^2 & x & y & 1 \\ 1 + (-1)^2 & 1 & -1 & 1 \\ 1^2 + 2^2 & 1 & 2 & 1 \\ 2^2 + 0^2 & 2 & 0 & 1 \end{vmatrix} = 0.$$

Expanding and simplifying this, we get the equation

$$x^2 + y^2 - x - y - 2 = 0,$$

that is,

$$(x - \tfrac{1}{2})^2 + (y - \tfrac{1}{2})^2 = \tfrac{5}{2}.$$

This is a circle, center at $(\tfrac{1}{2}, \tfrac{1}{2})$ and radius $= \sqrt{(\tfrac{5}{2})}$.

We could go on with specific examples; however, we leave these to the exercises.

PROBLEMS

1. The equation of a conic having its axes of symmetry on the x- and y-axes is of the form $ax^2 + by^2 + cx + dy + e = 0$. Find the equation of the conic through the given points.

   (a) $(1, 2)$, $(2, 1)$, $(3, 2)$, $(2, 3)$.

   (b) $(0, 0)$, $(5, -4)$, $(1, -2)$, $(1, 1)$.

   (c) $(5, 6)$, $(0, 1)$, $(1, 0)$, $(0, 0)$.

   (d) $(-1, -1)$, $(2, 0)$, $(0, \tfrac{1}{2})$, $(-\tfrac{1}{2}, \tfrac{1}{3})$.

2. Identify the conics found in Problem 1.

3. Show from the determinant form of a circle why, if the three given points are collinear, the equation of the circle reduces to that of a straight line.

4. Find the equations of the curve of the form

$$a \sin (x) + b \sin (2y) + c$$

passing through $(\pi, \pi/2)$ and $(\pi/2, \pi/4)$.

5. Find the equation of the curve $ae^x + be^{-y} + ce^{2x} + de^{-2y}$ passing through $(1, 1)$, $(2, 2)$, and $(2, -1)$.

9.3. MARKOV PROCESSES

A country holding regular elections is always governed by one of two parties, A and B. At the time of any election, the probabilities for transition from one party to another are as follows:

A to B: 40% A to A: 60% (the remaining probability)
B to A: 30% B to B: 70% (the remaining probability)

Given these *transition probabilities,* we ask:

1. If party B was chosen in the first election, what is the probability that the party chosen in the fifth election is party A?

2. If $s_1^{(k)}$ is the probability of choosing party A in the kth election, and $s_2^{(k)}$ that of choosing party B, what are the probabilities $s_1^{(k+p)}$ of choosing party A in the $(k + p)$th election and $s_2^{(k+p)}$ that of choosing party B?

3. Is there a *steady state* for this transition; that is, are there nonnegative numbers s_1, s_2 which add up to 1 such that if s_1 is the probability of having chosen party A in the last election and s_2 that of having chosen party B, then the probabilities of choosing parties A and B in the next election are still s_1 and s_2, respectively?

To answer these questions, we let $S^{(k)}$ be the vector $\begin{bmatrix} s_1^{(k)} \\ s_2^{(k)} \end{bmatrix}$, where $s_1^{(k)}$ is the probability of choosing party A in the kth election and $s_2^{(k)}$ that of choosing party B. So the entries $s_1^{(k)}$, $s_2^{(k)}$ of $S^{(k)}$ are nonnegative, which add up to 1. We call $S^{(k)}$ the *state* after the kth election. If the current party is A, then $S^{(1)} = \begin{bmatrix} 1 \\ 0 \end{bmatrix}$ and $S^{(2)} = \begin{bmatrix} .6 \\ .4 \end{bmatrix}$ since the probabilities for transition to parties A and B from party A are 60% and 40%. Similarly, if the current party is B, then $S^{(1)} = \begin{bmatrix} 0 \\ 1 \end{bmatrix}$ and $S^{(2)} = \begin{bmatrix} .3 \\ .7 \end{bmatrix}$. So, in both

cases, the transition from $S^{(1)}$ to $S^{(2)}$ results from multiplying $S^{(1)}$ by the

transition matrix $T = \begin{bmatrix} .6 & .3 \\ .4 & .7 \end{bmatrix}$. Assuming that *transition is linear*, that

is, $S^{(k)} = s_1^{(k)} \begin{bmatrix} 1 \\ 0 \end{bmatrix} + s_2^{(k)} \begin{bmatrix} 0 \\ 1 \end{bmatrix}$ implies that

$$S^{(k+1)} = s_1^{(k)} \begin{bmatrix} .6 \\ .4 \end{bmatrix} + s_2^{(k)} \begin{bmatrix} .3 \\ .7 \end{bmatrix},$$

we get

$$S^{(k+1)} = s_1^{(k)} T \begin{bmatrix} 1 \\ 0 \end{bmatrix} + s_2^{(k)} T \begin{bmatrix} 0 \\ 1 \end{bmatrix} = TS^{(k)};$$

that is, $S^{(k+1)} = TS^{(k)}$. From this, it is apparent that $S^{(k+p)} = T^p S^{(k)}$ for
any positive integer p, and that $S^{(k)} = T^{k-1} S^{(1)}$, which provides us with
answers to our first two questions:

1. If party B was chosen in the first election, the probability that the
    party chosen in the 5th election is party A is the first entry of the

    second column $\begin{bmatrix} .6 & .3 \\ .4 & .7 \end{bmatrix}^4 \begin{bmatrix} 0 \\ 1 \end{bmatrix}$ of $\begin{bmatrix} .6 & .3 \\ .4 & .7 \end{bmatrix}^4$.

2. If $S^{(k)}$ is the state after the kth election, then the state after the
    $k + p$th election is $T^p S^{(k)}$.

For example, if the chances for choosing parties A and B, respectively, in the twelfth election are judged to be 20% and 80%, then the

state after the fifteenth election is $\begin{bmatrix} .6 & .3 \\ .4 & .7 \end{bmatrix}^3 \begin{bmatrix} .2 \\ .8 \end{bmatrix}$, that is, the first and

second entries of this vector give the probabilities of choosing parties A
and B, respectively, in the fifteenth election.

Let's now look for a steady state $S = \begin{bmatrix} s_1 \\ s_2 \end{bmatrix}$, that is, a vector S

whose entries are nonnegative and add up to 1 such that $TS = S$; that
is, $(T - 1I)S = 0$. To find S, it now suffices to find a nonzero vector

S in the nullspace of $T - 1 = \begin{bmatrix} -.4 & .3 \\ .4 & -.3 \end{bmatrix}$, having nonnegative entries. Now S must be in the nullspace of $T - 1I$. Since the second row is a multiple of the first, the $T - 1I$ row reduces to the matrix

$\begin{bmatrix} -.4 & .3 \\ 0 & 0 \end{bmatrix}$. So $\begin{bmatrix} 3 \\ 4 \end{bmatrix}$ is a basis for the nullspace of $T - 1I =$

$\begin{bmatrix} -.4 & .3 \\ .4 & -.3 \end{bmatrix}$. From this it follows that

There one and only one steady state $S = \begin{bmatrix} \frac{3}{7} \\ \frac{4}{7} \end{bmatrix}$.

We know that 1 is a characteristic root for T, and we can find the other characteristic root for T and use it to get a simple formula for $S^{(k)} = T^{k-1}S^{(k)}$. The other characteristic root of T turns out to be .3, and we get

$$\begin{bmatrix} 1 \\ -1 \end{bmatrix}$$

as a corresponding characteristic vector:

$$\begin{bmatrix} .6 & .3 \\ .4 & .7 \end{bmatrix}\begin{bmatrix} 1 \\ -1 \end{bmatrix} = .3\begin{bmatrix} 1 \\ -1 \end{bmatrix}.$$

For the characteristic root 1, we have the characteristic vector $\begin{bmatrix} \frac{3}{7} \\ \frac{4}{7} \end{bmatrix}$.

Multiplying by 7, we also have the characteristic vector $\begin{bmatrix} 3 \\ 4 \end{bmatrix}$. Letting

$C = \begin{bmatrix} 3 & 1 \\ 4 & -1 \end{bmatrix}$, we have $C^{-1}TC = \begin{bmatrix} 1 & 0 \\ 0 & .3 \end{bmatrix}$ or

$$T = C\begin{bmatrix} 1 & 0 \\ 0 & .3 \end{bmatrix}C^{-1}.$$

So we have the formula

$$S^{(k)} = T^{k-1}S^{(1)} = C\begin{bmatrix} 1 & 0 \\ 0 & (.3)^{k-1} \end{bmatrix}C^{-1}S^{(1)}.$$

One calculates that $C^{-1} = (\frac{1}{7})\begin{bmatrix} 1 & 1 \\ 4 & -3 \end{bmatrix}$, so that

$$S^{(k)} = (\frac{1}{7})\begin{bmatrix} 3 & 1 \\ 4 & -1 \end{bmatrix}\begin{bmatrix} 1 & 0 \\ 0 & (.3)^{k-1} \end{bmatrix}\begin{bmatrix} 1 & 1 \\ 4 & -3 \end{bmatrix}S^{(1)}.$$

From the formula for $S^{(k)}$, we see that the limit $S^{(\infty)}$ of $S^{(k)}$ as k goes to ∞ exists and is

$$S^{(\infty)} = (\tfrac{1}{7}) \begin{bmatrix} 3 & 1 \\ 4 & -1 \end{bmatrix} \begin{bmatrix} 1 & 0 \\ 0 & 0 \end{bmatrix} \begin{bmatrix} 1 & 1 \\ 4 & -3 \end{bmatrix} S^{(1)}$$

$$= (\tfrac{1}{7}) \begin{bmatrix} 3 & 0 \\ 4 & 0 \end{bmatrix} \begin{bmatrix} 1 & 1 \\ 4 & -3 \end{bmatrix} S^{(1)} = (\tfrac{1}{7}) \begin{bmatrix} 3 & 3 \\ 4 & 4 \end{bmatrix} S^{(1)}$$

$$= \begin{bmatrix} \tfrac{3}{7} & \tfrac{3}{7} \\ \tfrac{4}{7} & \tfrac{4}{7} \end{bmatrix} S^{(1)}.$$

Since the sum of the entries of $S^{(1)}$ is 1, it follows that $S^{(\infty)}$ is

$$S^{(\infty)} = \begin{bmatrix} \tfrac{3}{7} \\ \tfrac{4}{7} \end{bmatrix}$$

for any initial state $S^{(1)}$. Since this vector is the steady state S, this means that

if k is large, then s_k is close to the steady state $\begin{bmatrix} \tfrac{3}{7} \\ \tfrac{4}{7} \end{bmatrix}$, *regardless*

of the results of the first election.

The process of passing from $S^{(k)}$ to $S^{(k+1)}$ ($k = 1, 2, 3, \ldots$) described above is an example of a Markov process. What is a Markov process in general?

Definition.   A *transition matrix* is a matrix $T \in M_n(\mathbb{R})$ such that each entry of T is nonnegative and the entries of each column add up to 1.

Definition.   A *state* is any element $S \in \mathbb{R}^{(n)}$ whose entries are all non-negative and add up to 1.

Any transition matrix $T \in M_n(\mathbb{R})$ has the property that TS is a state for any state $S \in \mathbb{R}^{(n)}$ (Prove!). With this terminology, a *Markov process* is any process involving transition from one of several mutually exclusive conditions $C_1, \ldots, C_n$ to another such that for some transition matrix T, the following is true:

If at any stage in the process, the probabilities of the conditions
C_r are the entries S_r of the state S, then the probability of
condition C_r at the next stage in the process is entry r of TS for
all r.

For any such Markov process, T is called the transition matrix of the process and its entries are called the *transition probabilities*. The transition probabilities can be described conveniently by giving a *transition diagram,* which in our example with $T = \begin{bmatrix} .6 & .3 \\ .4 & .7 \end{bmatrix}$ is

Note that the probabilities for remaining at nodes A or B are not given, since they can be determined from the others.

PROBLEMS

NUMERICAL PROBLEMS

1. Find a steady state for $T = \begin{bmatrix} .2 & .3 \\ .8 & .7 \end{bmatrix}$, and find an expression for the state $S^{(304)}$ if $S^{(1)}$ is $\begin{bmatrix} .5 \\ .5 \end{bmatrix}$.

2. Suppose that companies R and T compete for the same market, which 10 years ago was held completely by company T. Suppose that at the end of each of these 10 years, each company lost a percentage of its market to the other company as follows:

   R to T: 40%

   T to R: 30%

   Describe the process whereby the distribution of the market changes as a Markov process and give an expression for the percentage of the market that will be held by company T at the end of 2 more years. Find a steady state for the distribution of the market.

MORE THEORETICAL PROBLEMS

Easier Problems

3. In Problem 2, give a formula for the percentage of the market which will be held by company T at the end of p more years and determine the limit of this percentage as p goes to infinity.

9.4. INCIDENCE MODELS

Models used to determine *local prices* of goods transported among various cities, and certain other models used to determine *electrical potentials* at nodes of a network of electrical currents and to determine *displacements* at nodes of a mechanical structure under stress, all have one thing in common. When these models are stripped of the trappings that go with the particular model, what is left is an *incidence diagram,* such as the one shown here.

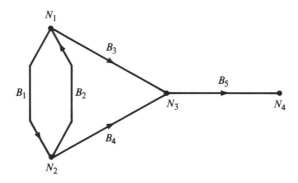

What, exactly, is an incidence diagram? It consists of a certain number m of *nodes* and a certain number n of oriented *branches,* each of which begins and ends at different nodes, such that every node is the *beginning* or *ending* of some branch. How is an incidence diagram represented in a mathematical model? Corresponding to each incidence diagram is an *incidence matrix,* that is, a matrix such that each column has one entry whose value is 1, one entry whose value is -1, and 0 for all other entries. To form this matrix, let there be one row for each node N_r and one column for each branch B_s. If node N_r is the beginning of branch B_s, let the (r, s) entry of the matrix be -1. If N_r is the ending of branch B_s, let the (r, s) entry of the matrix be 1. Otherwise, let the (r, s) entry be 0, indicating that N_r is neither the beginning nor ending of branch B_s. For example, the matrix of the foregoing incidence diagram is

$$\begin{bmatrix} -1 & 1 & -1 & 0 & 0 \\ 1 & -1 & 0 & -1 & 0 \\ 0 & 0 & 1 & 1 & -1 \\ 0 & 0 & 0 & 0 & 1 \end{bmatrix}.$$

Conversely, given an $m \times n$ incidence matrix T, there is a corresponding incidence diagram with m nodes $N_1, \ldots, N_m$ corresponding to the rows of T and n oriented branches $B_1, \ldots, B_n$ corresponding to the columns of T. If row r has -1 in column s, then node N_r is the *beginning* of branch B_s. On the other hand, if row r has 1 in column s, then N_r is the *end* of branch B_s. For example, the matrix

$$T = \begin{bmatrix} 1 & 1 & -1 & 0 & 0 \\ 1 & -1 & 0 & -1 & 0 \\ 0 & 0 & 1 & 1 & -1 \\ 0 & 0 & 0 & 0 & 1 \end{bmatrix}$$

is an incidence matrix whose incidence diagram is the one illustrated here.

EXAMPLE

In a transportation model, the nodes N_r of our incidence diagram represent cities and the branches B_s transportation routes between the cities. Suppose that beer is produced and consumed in the cities N_r and transported along the routes B_s so that:

1. For each city N_r, if we add the rates F_s at which beer (measured in cans of beer per hour) is transported on routes B_s heading into city N_r and then subtract the rates F_s for routes B_s heading out of city N_r, we get the difference G_r between the rate of production and the rate of consumption of beer in city N_r. So for city N_2, since route B_1 heads in, whereas routes B_2 and B_4 head out, $F_1 - F_2 - F_4$ equals G_2.

2. If we let P_r denote the price of beer (per can) in city N_r and E_s the price at the end of the route B_s minus the price at the beginning of the route B_s, there is a positive constant value R_s such that $R_s F_s = E_s$. This constant R_s reflects distance and other *resistance to flow* along route B_s that will increase the price difference E_s. For $s = 3$, the price at the beginning of B_3 is P_1 and the price at the end of B_3 is P_3, so that $R_3 F_3 = E_3$, where $E_3 = P_3 - P_1$.

In terms of the incidence matrix, we now have

         Kirchhoff's First Law: $TF = G$
         Kirchhoff's Second Law: $T'P = E$
         Ohm's Law: $RF = E,$

$$\text{where} \quad R = \begin{bmatrix} R_1 & & 0 \\ & \ddots & \\ 0 & & R_n \end{bmatrix}.$$

For example, for the incidence matrix above, the equation $T'P = E$ is

$$\begin{bmatrix} -1 & 1 & 0 & 0 \\ 1 & -1 & 0 & 0 \\ -1 & 0 & 1 & 0 \\ 0 & -1 & 1 & 0 \\ 0 & 0 & -1 & 1 \end{bmatrix} \begin{bmatrix} P_1 \\ P_2 \\ P_3 \\ P_4 \end{bmatrix} = \begin{bmatrix} P_2 - P_1 \\ P_1 - P_2 \\ P_3 - P_1 \\ P_4 - P_2 \\ P_5 - P_3 \end{bmatrix} = \begin{bmatrix} E_1 \\ E_2 \\ E_3 \\ E_4 \\ E_5 \end{bmatrix},$$

and the equation $RF = E$ is

$$\begin{bmatrix} R_1 & & & & 0 \\ & R_2 & & & \\ & & R_3 & & \\ & & & R_4 & \\ 0 & & & & R_5 \end{bmatrix} \begin{bmatrix} F_1 \\ F_2 \\ F_3 \\ F_4 \\ F_5 \end{bmatrix} = \begin{bmatrix} R_1 F_1 \\ R_2 F_2 \\ R_3 F_3 \\ R_4 F_4 \\ R_5 F_5 \end{bmatrix} = \begin{bmatrix} E_1 \\ E_2 \\ E_3 \\ E_4 \\ E_5 \end{bmatrix}.$$

If a vector P satisfies the conditions for a price vector, how do we find it, given T, R, and G? Substituting $T'P$ for E in $RF = E$, we find that $F = R^{-1}T'P$, so that P is a solution of the matrix equation

$$(TR^{-1}T')P = G.$$

It can be shown that

if the total production equals the total consumption of beer, then this equation has an exact solution P.

Otherwise, we can use the methods of Chapter 10 to find the closest approximation P to an exact solution.

PROBLEMS

NUMERICAL PROBLEMS

1. Draw the incidence diagram for $T = \begin{bmatrix} 1 & -1 & 0 & 1 \\ -1 & 0 & 1 & 0 \\ 0 & 1 & -1 & -1 \end{bmatrix}.$

9.5. DIFFERENTIAL EQUATIONS

Finding the set of all solutions to the homogeneous differential equation

$$\frac{d^n v}{dt^n} + a_{n-1} \frac{d^{n-1} v}{dt^{n-1}} + \cdots + a_0 v = 0$$

is really the same as finding the nullspace of a corresponding linear transformation. Why? The mapping T that maps a function v of a real variable t to its derivative $Tv = \dfrac{dv}{dt}$ is linear on the vector space F_n of n times differentiable complex-valued functions v of t. Here, the derivative of the function

$$v(t) = a(t) + b(t)i$$

is defined as

$$\frac{dv}{dt} = \frac{da}{dt} + \frac{db}{dt} i.$$

So if we let $f(x)$ denote the polynomial

$$f(x) = x^n + a_{n-1} x^{n-1} + \cdots + a_0,$$

$f(T)$ is linear on F_n and we can express the set of solutions to

$$\frac{d^n v}{dt^n} + b_{n-1} \frac{d^{n-1} v}{dt^{n-1}} + \cdots + b_0 v = 0$$

as the nullspace

$$W = \{v \in F_n \mid f(T)v = 0\}$$

of $f(T)$.

To determine the nullspace W of $f(T)$, note that $TW \subseteq W$, since $f(T)v = 0$ implies that

$$0 = Tf(T)v = f(T) T(v).$$

Then take any $v \in W$ and form the subspace V of W spanned over $\mathbb{C}$ by the functions

$$v, \quad Tv, \quad \ldots, \quad T^{n-1}v.$$

Since $f(T)v = 0$, $T^n v$ is a linear combination of $v, Tv, \ldots, T^{n-1}v$, so is contained in V. It follows that T maps $v, Tv, \ldots, T^{n-1}v$ into V, that is, $TV \subseteq V$.

Now we can regard T as a linear transformation of the finite dimensional vector space V. Since $f(T)V = 0$, we can apply the

Theorem 9.5.1. Let T be a linear transformation of a finite-dimensional vector space V over the field F of complex numbers and let $f(T)V = 0$, where $f(x) = (x - a_1)^{m_1} \cdots (x - a_k)^{m_k}$. Then V is the direct sum

$$V = V_{a_1}(T) \oplus \cdots \oplus V_{a_k}(T),$$

where the $V_{a_r}(T)$ are the *generalized characteristic subspaces*

$$V_{a_r}(T) = \{v \in V \mid (T - a_r I)^e v = 0 \text{ for some } e\}.$$

Proof. We assume that $V \neq \{0\}$, take b to be a characteristic root of T, and note that $x - b$ divides $f(x)$. Letting $S = T - bI$, we have $V = X \oplus Y$, where X is the *generalized nullspace*

$$X = \{v \in V \mid S^e v = 0 \text{ for some } e\}$$

of S and $Y = \cap_{e=1}^{\infty} S^e(V)$. To prove this we use the equations

$$X = \{v \in V \mid S^e v = 0\}, \qquad Y = S^e V$$

for any $e \geq \dim (V)$. (Prove!) Since $S^e X = 0$ and $S^e Y = S^{2e} V = S^e V = Y$ for such e, the mapping defined by S^e on X is 0 and the mapping defined by S^e on Y is $1-1$ and onto. So $X \cap Y = \{0\}$. Now let $e = \dim (V)$ and take any $v \in V$. Choose $w \in Y$ such that $S^e v = S^e w$. Then $v = (v - w) + w$. Since $S^e(v - w) = S^e v - S^e w = 0$, we get that $v \in X + Y$. Since we know that $X \cap Y = \{0\}$, we conclude that $V = X \oplus Y$.

The subspace X of V is nonzero since b is a characteristic root of T. If $X = V$, then $V = V_b(T)$ (Prove!) and we are done since $x - b$ divides $f(x)$. Otherwise, both X and Y are nonzero subspaces of lower dimension than V. Since T maps the subspaces X and Y into themselves and $f(T)X = 0$, $f(T)Y = 0$, we get the desired decomposition

$$Y = Y_{a_1}(T) \oplus \cdots \oplus Y_{a_k}(T)$$

for Y, by induction. Moreover, we also have $X = V_b(T)$, where $x - b$ divides $f(x)$ and where $Y_b(T) = \{0\}$ (since $S = T - bI$ is $1 - 1$ on Y). Renumbering the a_r so that $b = a_1$, we get

$$V = X \oplus Y = V_{a_1}(T) \oplus Y_{a_2}(T) \oplus \cdots \oplus Y_{a_k}(T)$$
$$= V_{a_1}(T) \oplus V_{a_2}(T) \oplus \cdots \oplus V_{a_k}(T). \qquad \blacksquare$$

Writing v as $v = v_1 + \cdots + v_k$ with $v_r \in V_{a_r}(T)$ for all r, we have

$$(T - a_r I)^{m_r} v_r = 0.$$

Let's fix r and denote $a_r = a$, $v_r = w$ and $m_r = m$. Since $T(e^{at}u) = ae^{at}u + e^{at}Tu$ for any differentiable u, we have

$$(T - aI)(e^{at}u) = e^{at}Tu.$$

Applying $(T - aI)$ in this fashion $m - 1$ more times, we also have

$$(T - aI)^m(e^{at}u) = e^{at}T^m u.$$

Replacing u by $e^{-at}w$, we have

$$(T - aI)^m(e^{at}e^{-at}w) = e^{at}T^m(e^{-at}w),$$

that is,

$$(T - aI)^m w = e^{at}T^m(e^{-at}w).$$

So, our condition $(T - aI)^m w = 0$ is equivalent to the condition $T^m(e^{-at}w) = 0$, which in turn is equivalent to the condition that $e^{-at}w$ is a polynomial $p(t)$ of degree less than m. But then $w = p(t)e^{at}$, that is, w is a linear combination of the functions e^{at}, te^{at}, $\ldots$, $t^{m-1}e^{at}$. Conversely, it is easy to check that the functions e^{at}, te^{at}, $\ldots$, $t^{m-1}e^{at}$ are solutions to the differential equation $(T - aI)^m = 0$. Since $(x - a)^m = (x - a_r)^{m_r}$ is a factor of $f(x)$, they are also solutions to the differential equations $f(T)v = 0$.

Since each solution $v \in W$ is a sum of functions v_r each of which is a linear combination of $e^{a_r t}$, $te^{a_r t}$, $\ldots$, $t^{m_r - 1}e^{a_r t}$, we conclude that

$$\{t^{n_r}e^{a_r t} \mid 1 \leq r \leq k, 0 \leq n_r \leq m_r - 1\}$$

spans W. So W is finite dimensional. Regarding T as a linear transformation of W, we have $f(T)W = 0$ and

$$W = W_{a_1}(T) \oplus \cdots \oplus W_{a_k}(T),$$

where $W_{a_1}(T), \ldots, W_{a_k}(T)$ are the generalized characteristic subspaces introduced in Theorem 9.5.1. Since the functions

$$e^{a_r t}, \quad te^{a_r t}, \quad \ldots, \quad t^{m_r - 1}e^{a_r t}$$

are linearly independent elements of $W_{a_r}(T)$ for each r (Prove!), the set

$$\{t^{n_r}e^{a_r t} \mid 1 \leq r \leq k, 0 \leq n_r \leq m_r - 1\}$$

is linearly independent and we have

Theorem 9.5.2.   The space W of n times differentiable complex-valued
solutions v to $\dfrac{d^n v}{dt^n} + b_{n-1}\dfrac{d^{n-1}v}{dt^{n-1}} + \cdots + b_0 v = 0$ is n-dimensional
over $\mathbb{C}$. A basis for W is $\{t^{n_r}e^{a_r t} \mid 1 \le r \le k, 0 \le n_r \le m_r - 1\}$, where
$x^n + b_{n-1}x^{n-1} + \cdots + b_0$ factors as $(x - a_1)^{m_1} \cdots (x - a_k)^{m_k}$.

Can we get a solution f to

$$\frac{d^n v}{dt^n} + b_{n-1}\frac{d^{n-1}v}{dt^{n-1}} + \cdots + b_0 v = 0$$

satisfying initial conditions

$$f^{(r)}(t_0) = v_r \ (0 \le r \le n - 1)$$

at a specified value t_0 of t? Yes: we can give such a desired solution
explicitly using the exponential function e^{tT} introduced in Section 4.8.
For this we consider the system

$$u_1' = u_2$$
$$u_2' = u_3$$
$$\vdots$$
$$u_n' = -b_0 u_1 - \cdots - b_{n-1}u_n$$

of n linear differential equations in n unknowns represented by the matrix
equation $u' = Tu$, where T is

$$\begin{bmatrix} 0 & & & & -a_0 \\ 1 & & & & \vdots \\ 0 & & & & \\ \vdots & & \ddots & & \vdots \\ 0 & \cdots & 0 & 1 & -a_{n-1} \end{bmatrix},$$

the *companion matrix* to the polynomial $b_0 + \cdots + b_{n-1}x^{n-1} + x^n$
with the same coefficients as the differential equation. How does this
system of linear differential equations relate to the differential equation

$$\frac{d^n v}{dt^n} + b_{n-1}\frac{d^{n-1}v}{dt^{n-1}} + \cdots + b_0 v = 0?$$

The condition $u' = Tu$ implies that $v = u_1$ satisfies the conditions

$$u_1 = v^{(0)}, u_2 = v^{(1)}, \ldots, u_n = v^{(n-1)}.$$

But then

$$v^{(n)} = u_n' = -b_0 u_1 - \cdots - b_{n-1} u_n,$$

that is,

$$v^{(n)} = -b_0 v^{(0)} - \cdots - b_{n-1} v^{(n-1)}.$$

So Theorem 4.8.2 now gives us

Theorem 9.5.3. For any t_0 and $v_0, \ldots, v_{n-1}$, the differential equation

$$\frac{d^n v}{dt^n} + b_{n-1} \frac{d^{n-1} v}{dt^{n-1}} + \cdots + b_0 v = 0$$

has one and only one solution v such that $v^{(r)}(t_0) = v_r$ for $0 \le r \le$

$n - 1$, namely, $v = u_1$, where $u = e^{(t-t_0)T} \begin{bmatrix} v_0 \\ \vdots \\ v_{n-1} \end{bmatrix}$.

How can we compute e^{tT} efficiently? To give a definitive answer to this question would take us far afield. Instead, we take a single step in the direction of an answer by giving an expression for e^{tT} in terms of simpler exponential functions. In Section 4.8 we noted that the exponential function $e^{t(aI + N)}$ could be expressed as $e^{at}e^{tN}$. To get the most out of this, letting T be any $n \times n$ matrix and viewing T as a linear transformation of $V = \mathbb{C}^{(n)}$, we use Theorem 9.5.1, to express V as the direct sum

$$V = V_{a_1}(T) \oplus \cdots \oplus V_{a_k}(T),$$

where $f(x) = (x - a_1)^{m_1} \cdots (x - a_k)^{m_k}$ is the characteristic polynomial of T and the $V_{a_r}(T)$ are the subspaces

$$V_{a_r}(T) = \{v \in V \mid (T - a_r I)^e v = 0 \text{ for some } e\}.$$

For each r we let T_r be the linear transformation of $V_{a_r}(T)$ defined by $T_r v = Tv$ for $v \in V_{a_r}(T)$ and take a matrix C_r whose columns are a basis for the subspace $V_{a_r}(T)$ in which the matrix A_r of T_r is upper triangular, that is, $A_r = a_r I_r + N_r$, where I_r is the identity matrix and N_r is an upper triangular nilpotent matrix. Letting C be the matrix $C =$

$[C_1, \ldots, C_k]$, the columns of C are a basis for V in which the matrix A of T is

$$A = \begin{bmatrix} a_1 I_1 + N_1 & & 0 \\ & \ddots & \\ 0 & & a_k I_k + N_k \end{bmatrix}.$$

Since $A = C^{-1}TC$, we can write T as CAC^{-1}. For the same reason that $f(CAC^{-1}) = Cf(A)C^{-1}$ for polynomials $f(x)$, and taking convergence considerations into account, we can show that

$$e^{tT} = e^{C^{-1}tAC} = C^{-1}e^{tA}C.$$

So we can get an expression for e^{tT} in terms of simpler exponentials:

$$e^{tT} = C^{-1}e^{t\begin{bmatrix} a_1 I_1 + N_1 & & 0 \\ & \ddots & \\ 0 & & a_k I_k + N_k \end{bmatrix}}C$$

$$= C^{-1}\begin{bmatrix} e^{t(a_1 I_1 + N_1)} & & 0 \\ & \ddots & \\ 0 & & e^{t(a_k I_k + N_k)} \end{bmatrix}C$$

$$= C^{-1}\begin{bmatrix} e^{a_1 t}e^{tN_1} & & 0 \\ & \ddots & \\ 0 & & e^{a_k t}e^{N_k} \end{bmatrix}C.$$

PROBLEMS

NUMERICAL PROBLEMS

1. Find a basis over $\mathbb{C}$ for solution space of $\dfrac{d^3v}{dv^3} - 2\dfrac{d^2v}{dt^2} + \dfrac{dv}{dt} = 0$.

2. In Problem 1, find a real solution v such that $v(0) = 3$, $v'(0) = 3$, $v''(0) = 4$, where $v'(t)$ denotes the derivative of $v(t)$.

MORE THEORETICAL PROBLEM

Easier Problem

3. Use the formula $T(e^{at}u) = ae^{at}u + e^{at}Tu$ to determine the matrix of T on the space W introduced in the discussion preceding Theorem 9.5.2 with respect to the basis

$$\{t^{n_r}e^{a_r t} \mid 1 \leq r \leq k, 0 \leq n_r \leq m_r - 1\}.$$

10

LEAST SQUARES METHODS
(OPTIONAL)

10.1. APPROXIMATE SOLUTIONS OF SYSTEMS OF LINEAR EQUATIONS

Since the system of m linear equations in n variables with matrix equation $Ax = y$ has a solution x if and only if y is in the column space of A, many systems of linear equations have no solutions. What shall we do with an equation $Ax = y$ such as

$$\begin{bmatrix} 6 & 3 \\ 2 & 1 \end{bmatrix}\begin{bmatrix} x_1 \\ x_2 \end{bmatrix} = \begin{bmatrix} 2 \\ 2 \end{bmatrix}$$

which has no solution? We can *always* find an *approximate solution* x by replacing y by the vector y' in the column space of A nearest to y and solving $Ax = y'$ instead. In the case of the equation

$$\begin{bmatrix} 6 & 3 \\ 2 & 1 \end{bmatrix}\begin{bmatrix} x_1 \\ x_2 \end{bmatrix} = \begin{bmatrix} 2 \\ 2 \end{bmatrix},$$

where A and y are $\begin{bmatrix} 6 & 3 \\ 2 & 1 \end{bmatrix}$ and $\begin{bmatrix} 2 \\ 2 \end{bmatrix}$, the column space of A is the span

$$W = \mathbb{R} \begin{bmatrix} 3 \\ 1 \end{bmatrix}$$

of $\begin{bmatrix} 3 \\ 1 \end{bmatrix}$. So y' is the multiple $y' = t \begin{bmatrix} 3 \\ 1 \end{bmatrix}$ of $\begin{bmatrix} 3 \\ 1 \end{bmatrix}$ such that the length

$$\|y - y'\| = \left\| \begin{bmatrix} 2 \\ 2 \end{bmatrix} - t \begin{bmatrix} 3 \\ 1 \end{bmatrix} \right\|$$

is as small as possible. We can represent y, y', and W pictorially as follows:

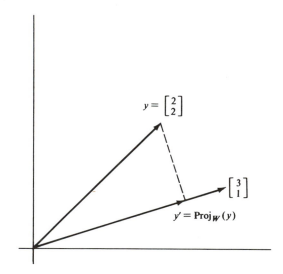

To get an explicit expression for the vector y' in the column space W of A nearest to y, we need

Definition.   Let W be a subspace of $\mathbb{R}^{(n)}$, so that $\mathbb{R}^{(n)} = W \oplus W^{\perp}$, and write $y \in \mathbb{R}^{(n)}$ as $y_1 + y_2$, where $y_1 \in W$ and $y_2 \in W^{\perp}$. Then y_1 is called the *projection* of y on W and is denoted $y_1 = \text{Proj}_W(y)$.

For the equation

$$\begin{bmatrix} 6 & 3 \\ 2 & 1 \end{bmatrix} \begin{bmatrix} x_1 \\ x_2 \end{bmatrix} = \begin{bmatrix} 2 \\ 2 \end{bmatrix},$$

we take W to be the span $\mathbb{R}\begin{bmatrix} 3 \\ 1 \end{bmatrix}$ of $\begin{bmatrix} 3 \\ 1 \end{bmatrix}$. Then $W^\perp$ is the span of

$\begin{bmatrix} -1 \\ 3 \end{bmatrix}$. So we get that

$$\begin{bmatrix} 2 \\ 2 \end{bmatrix} = (\tfrac{4}{5})\begin{bmatrix} 3 \\ 1 \end{bmatrix} + (\tfrac{2}{5})\begin{bmatrix} -1 \\ 3 \end{bmatrix},$$

the first term of which is

$$\text{Proj}_W \begin{bmatrix} 2 \\ 2 \end{bmatrix} = (\tfrac{4}{5})\begin{bmatrix} 3 \\ 1 \end{bmatrix}.$$

To minimize the length

$$\|y - y'\| = \left\| \begin{bmatrix} 2 \\ 2 \end{bmatrix} - t\begin{bmatrix} 3 \\ 1 \end{bmatrix} \right\|$$

$$= \left\| (\tfrac{4}{5})\begin{bmatrix} 3 \\ 1 \end{bmatrix} + (\tfrac{2}{5})\begin{bmatrix} -1 \\ 3 \end{bmatrix} - t\begin{bmatrix} 3 \\ 1 \end{bmatrix} \right\|$$

$$= \left\| ((\tfrac{4}{5}) - t)\begin{bmatrix} 3 \\ 1 \end{bmatrix} + (\tfrac{2}{5})\begin{bmatrix} -1 \\ 3 \end{bmatrix} \right\|.$$

since $\begin{bmatrix} 3 \\ 1 \end{bmatrix}$ and $\begin{bmatrix} -1 \\ 3 \end{bmatrix}$ are orthogonal we must take $t = \tfrac{4}{5}$ to eliminate

the term involving $\begin{bmatrix} 3 \\ 1 \end{bmatrix}$. (Prove!) So the element y' of W nearest to

$\begin{bmatrix} 2 \\ 2 \end{bmatrix}$ is

$$y' = \text{Proj}_W \begin{bmatrix} 2 \\ 2 \end{bmatrix} = (\tfrac{4}{5})\begin{bmatrix} 3 \\ 1 \end{bmatrix}.$$

In general, we have the *theorem of Pythagoras* that

if $y = y_1 + y_2$, where y_1 and y_2 are orthogonal, then

$$\|y\|^2 = \|y_1\|^2 + \|y_2\|^2,$$

since

$$\|y\|^2 = (y_1 + y_2, y_1 + y_2)$$
$$= (y_1, y_1) + (y_1, y_2) + (y_2, y_1) + (y_2, y_2)$$
$$= (y_1, y_1) + (y_2, y_2) = \|y_1\|^2 + \|y_2\|^2.$$

Using this, we get

Theorem 10.1.1.   Let W be a subspace of $\mathbb{R}^{(n)}$. Then the element y' of W nearest to $y \in \mathbb{R}^{(n)}$ is $y' = \text{Proj}_W(y)$.

Proof.   Write $y = y_1 + y_2$ with $y_1 \in W$ and $y_2 \in W^\perp$, so that $y_1 = \text{Proj}_W(y)$. Then for $w \in W$, the squared distance $\|y - w\|^2$ is $\|(y_1 - w) + y_2\|^2$. By the theorem of Pythagoras, this is

$$\|y_1 - w\|^2 + \|y_2\|^2$$

which is minimal if and only if $w = y_1 = \text{Proj}_W(y)$.   ■

Given a vector y and subspace W, the method of going from y to the vector $y' = \text{Proj}_W(y)$ in W nearest to y is sometimes called the *method of least squares* since the sum of squares of the entries of $y - y'$ is thereby minimized.

How do we compute $y' = \text{Proj}_W(y)$? Let's first consider the case where W is the span $\mathbb{R}w$ of a single nonzero vector w. Then we have $V = W \oplus W^\perp$ and get simple formula

$$\text{Proj}_W(y) = \frac{(y,\, w)}{(w,\, w)}\, w$$

since

1. $\dfrac{(y,\, w)}{(w,\, w)}\, w$ is in W.

2. $y - \dfrac{(y,\, w)}{(w,\, w)}\, w$ is in $W^\perp$. (Prove!)

3. $y = \dfrac{(y,\, w)}{(w,\, w)}\, w + \left(y - \dfrac{(y,\, w)}{(w,\, w)}\, w \right)$.

So if W is spanned by a single element w, we get a formula for $\text{Proj}_W(y)$ involving only y and w. In the case where W is the span $W = \mathbb{R}\begin{bmatrix} 3 \\ 1 \end{bmatrix}$ of $\begin{bmatrix} 3 \\ 1 \end{bmatrix}$, our formula gives us

$$\text{Proj}_W \begin{bmatrix} 2 \\ 2 \end{bmatrix} = \frac{2 \cdot 3 + 2 \cdot 1}{3 \cdot 3 + 1 \cdot 1} \begin{bmatrix} 3 \\ 1 \end{bmatrix} = (\tfrac{4}{5}) \begin{bmatrix} 3 \\ 1 \end{bmatrix}.$$

which is what we got when we computed it directly.

What happens for an arbitrary subspace W? If we are given an orthogonal basis for W, we get a formula for $\text{Proj}_W(y)$ which involves only y and that orthogonal basis, as we now show in

Theorem 10.1.2.   Let W be a subspace of $\mathbb{R}^{(n)}$ with orthogonal basis $w_1, \ldots, w_k$ and let $y \in \mathbb{R}^{(n)}$. Then

$$\text{Proj}_W(y) = \frac{(y, w_1)}{(w_1, w_1)} w_1 + \cdots + \frac{(y, w_k)}{(w_k, w_k)} w_k.$$

Proof.   Let

$$y_1 = \frac{(y, w_1)}{(w_1, w_1)} w_1 + \cdots + \frac{(y, w_k)}{(w_k, w_k)} w_k.$$

Then we have

$$(y - y_1, w_j) = (y, w_j) - \frac{(y, w_j)}{(w_j, w_j)} (w_j, w_j) = 0$$

for $1 \leq j \leq k$. (Prove!) So $y - y_1 \in W^\perp$ and $y_1 = \text{Proj}_W(y)$. ∎

A nice geometric interpretation of this theorem is that

the projection of y on the span W of mutually orthogonal vectors $w_1, \ldots, w_k$ equals the sum of the projections of y on the lines $\mathbb{R}w_1, \ldots, \mathbb{R}w_k$.

For example, if $n = 3$ and $k = 2$, we can represent the projection $y' = \text{Proj}_W(y)$ of a vector y onto the span W of nonzero orthogonal vectors w_1, w_2 pictorially as follows, where y_1 and y_2 are the projections of y on $\mathbb{R}w_1$ and $\mathbb{R}w_2$:

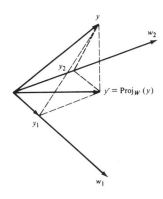

If W is a subspace of $\mathbb{R}^{(n)}$ with orthonormal basis $w_1, \ldots, w_k$ and $y \in \mathbb{R}^{(n)}$. Then the formula in our theorem simplifies to

$$\text{Proj}_W(y) = (y, w_1)w_1 + \cdots + (y, w_k)w_k.$$

EXAMPLE

Let's find the element of W nearest $\begin{bmatrix} 1 \\ 2 \\ 3 \end{bmatrix}$, where W is the span of vectors

$$\begin{bmatrix} 1 \\ 2 \\ -1 \end{bmatrix}, \begin{bmatrix} -1 \\ 4 \\ 1 \end{bmatrix}.$$

Using the Gram–Schmidt process, we get an orthonormal basis

$$1/\sqrt{6} \begin{bmatrix} 1 \\ 2 \\ -1 \end{bmatrix}, \quad 1/\sqrt{3} \begin{bmatrix} -1 \\ 1 \\ 1 \end{bmatrix}$$

for W. Then the projection of $\begin{bmatrix} 1 \\ 2 \\ 3 \end{bmatrix}$ on W is

$$\frac{1}{\sqrt{6}} (1 + 4 - 3) \frac{1}{\sqrt{6}} \begin{bmatrix} 1 \\ 2 \\ -1 \end{bmatrix} + \frac{1}{\sqrt{3}} (-1 + 2 + 3) \frac{1}{\sqrt{3}} \begin{bmatrix} -1 \\ 1 \\ 1 \end{bmatrix}$$

$$= (\tfrac{1}{3}) \begin{bmatrix} 1 \\ 2 \\ -1 \end{bmatrix} + (\tfrac{4}{3}) \begin{bmatrix} -1 \\ 1 \\ 1 \end{bmatrix} = \begin{bmatrix} -1 \\ 2 \\ 1 \end{bmatrix}.$$

So, we have

$$\begin{bmatrix} 1 \\ 2 \\ 3 \end{bmatrix} = \begin{bmatrix} -1 \\ 2 \\ 1 \end{bmatrix} + \begin{bmatrix} 2 \\ 0 \\ 2 \end{bmatrix}$$

in $W \oplus W^\perp$ and the element of W closest to $\begin{bmatrix} 1 \\ 2 \\ 3 \end{bmatrix}$ in W is

$$\text{Proj}_W \begin{bmatrix} 1 \\ 2 \\ 3 \end{bmatrix} = \begin{bmatrix} -1 \\ 2 \\ 1 \end{bmatrix}.$$

Now we have a way to compute $\text{Proj}_W(y)$. First, find an orthonormal basis for W (e.g., by the Gram–Schmidt orthogonalization process) and then use the formula for $\text{Proj}_W(y)$ given in Theorem 10.1.2.

EXAMPLE

Suppose that we have a supply of 5000 units of S, 4000 units of T and 2000 units of U, materials used in manufacturing products P and Q. We ask:

If each unit of P uses 2 units of S, 0 units of T and 0 units of U, and each unit of Q uses 3 units of S, 4 units of T and 1 unit of U, how many units p and q of P and Q should we make if we want to use up the entire supply?

The system is represented by the equation

$$\begin{bmatrix} 2 & 3 \\ 0 & 4 \\ 0 & 1 \end{bmatrix} \begin{bmatrix} p \\ q \end{bmatrix} = \begin{bmatrix} 5000 \\ 4000 \\ 2000 \end{bmatrix}.$$

Since the vector $\begin{bmatrix} 5000 \\ 4000 \\ 2000 \end{bmatrix}$ is not a linear combination

$$p\begin{bmatrix} 2 \\ 0 \\ 0 \end{bmatrix} + q\begin{bmatrix} 3 \\ 4 \\ 1 \end{bmatrix}$$

of the columns of $\begin{bmatrix} 2 & 3 \\ 0 & 4 \\ 0 & 1 \end{bmatrix}$, *there is no exact solution* $\begin{bmatrix} p \\ q \end{bmatrix}$. So we

get an approximate solution $\begin{bmatrix} p \\ q \end{bmatrix}$ by finding those values of p and

q for which the distance from $p\begin{bmatrix} 2 \\ 0 \\ 0 \end{bmatrix} + q\begin{bmatrix} 3 \\ 4 \\ 1 \end{bmatrix}$ to $\begin{bmatrix} 5000 \\ 4000 \\ 2000 \end{bmatrix}$ is as small

as possible. To do this, we first find the vector in the space W of

linear combinations $p\begin{bmatrix} 2 \\ 0 \\ 0 \end{bmatrix} + q\begin{bmatrix} 3 \\ 4 \\ 1 \end{bmatrix}$ that is closest to $\begin{bmatrix} 5000 \\ 4000 \\ 2000 \end{bmatrix}$.

We've just seen that this vector is the projection $\mathrm{Proj}_W\left(\begin{bmatrix} 5000 \\ 4000 \\ 2000 \end{bmatrix}\right)$

of $\begin{bmatrix} 5000 \\ 4000 \\ 2000 \end{bmatrix}$ on W. Computing this by the formula $\mathrm{Proj}_W(v) =$

$(v, w_1)w_1 + (v, w_2)w_2$ of Theorem 10.1.2, where w_1, w_2 is the orthonormal basis

$$w_1 = \begin{bmatrix} 1 \\ 0 \\ 0 \end{bmatrix}, \quad w_2 = \frac{1}{c}\begin{bmatrix} 0 \\ 4 \\ 1 \end{bmatrix} \quad (c = \sqrt{17})$$

of W, we get

$$\left(\begin{bmatrix} 5000 \\ 4000 \\ 2000 \end{bmatrix}, \begin{bmatrix} 1 \\ 0 \\ 0 \end{bmatrix}\right)\begin{bmatrix} 1 \\ 0 \\ 0 \end{bmatrix} + \left(\begin{bmatrix} 5000 \\ 4000 \\ 2000 \end{bmatrix}, \frac{1}{c}\begin{bmatrix} 0 \\ 4 \\ 1 \end{bmatrix}\right)\frac{1}{c}\begin{bmatrix} 0 \\ 4 \\ 1 \end{bmatrix},$$

that is,

$$\mathrm{Proj}\left(\begin{bmatrix} 5000 \\ 4000 \\ 2000 \end{bmatrix}\right) = (5000)\begin{bmatrix} 1 \\ 0 \\ 0 \end{bmatrix} + \left(\frac{1}{c}\,16{,}000 + \frac{1}{c}\,2000\right)\frac{1}{c}\begin{bmatrix} 0 \\ 4 \\ 1 \end{bmatrix}$$

$$= (5000)\begin{bmatrix} 1 \\ 0 \\ 0 \end{bmatrix} + \frac{18{,}000}{17}\begin{bmatrix} 0 \\ 4 \\ 1 \end{bmatrix}.$$

To get p and q amounts to expressing

$$\mathrm{Proj}_W\left(\begin{bmatrix} 5000 \\ 4000 \\ 2000 \end{bmatrix}\right) = (5000)\begin{bmatrix} 1 \\ 0 \\ 0 \end{bmatrix} + \frac{18000}{17}\begin{bmatrix} 0 \\ 4 \\ 1 \end{bmatrix}$$

as a linear combination

$$\left(2500 - \left(\frac{3}{2}\right)\left(\frac{18000}{17}\right)\right)\begin{bmatrix} 2 \\ 0 \\ 0 \end{bmatrix} + \frac{18000}{17}\begin{bmatrix} 3 \\ 4 \\ 1 \end{bmatrix}$$

of $\begin{bmatrix} 2 \\ 0 \\ 0 \end{bmatrix}$ and $\begin{bmatrix} 3 \\ 4 \\ 1 \end{bmatrix}$. So we get

$$p = 2500 - \left(\frac{3}{2}\right)\left(\frac{18,000}{17}\right) = 911.76,$$

$$q = \frac{18,000}{17} = 1058.82$$

and our approximate solution is

$$\begin{bmatrix} p \\ q \end{bmatrix} = \begin{bmatrix} 911.76 \\ 1058.82 \end{bmatrix}.$$

By making 911.76 units of P and 1058.82 units of Q, the vector representing supplies used is

$$911.76 \begin{bmatrix} 2 \\ 0 \\ 0 \end{bmatrix} + 1058.82 \begin{bmatrix} 3 \\ 4 \\ 1 \end{bmatrix} = \text{Proj}_W\left(\begin{bmatrix} 5000 \\ 4000 \\ 2000 \end{bmatrix}\right)$$

$$= (5000)\begin{bmatrix} 1 \\ 0 \\ 0 \end{bmatrix} + \frac{18000}{17}\begin{bmatrix} 0 \\ 4 \\ 1 \end{bmatrix}$$

$$= \begin{bmatrix} 5000 \\ 4235.29 \\ 1058.82 \end{bmatrix}$$

So we use exactly 5000 units of S as desired, 4235.29 units of T (we need 235.29 units of T), and 1058.82 units of U (we have 941.18 units of U left over).

In the example, we found an approximate solution in the following sense.

Definition. For any $m \times n$ matrix A with real entries, an *approximate solution* x to an equation $Ax = y$ is a solution to the equation $Ax = \text{Proj}_{A(\mathbb{R}^{(n)})}(y)$.

In the example, W is the column space $A(\mathbb{R}^{(2)})$ of A and we found

$$\text{Proj}_{A(\mathbb{R}^{(2)})}\left(\begin{bmatrix} 5000 \\ 4000 \\ 2000 \end{bmatrix}\right) \text{ by using the formula}$$

$$\text{Proj}_{A(\mathbb{R}^{(2)})}(v) = (v, w_1)w_1 + (v, w_2)w_2$$

of Theorem 10.1.2, where w_1, w_2 was an orthonormal basis for $A(\mathbb{R}^{(2)})$. We were then able to solve the equation

$$\begin{bmatrix} 2 & 3 \\ 0 & 4 \\ 0 & 1 \end{bmatrix} x = \text{Proj}_{A(\mathbb{R}^{(2)})}\left(\begin{bmatrix} 5000 \\ 4000 \\ 2000 \end{bmatrix}\right).$$

Until we have a better method, we can use this same method for any $m \times n$ matrix A with real entries:

1. Compute the projection $\text{Proj}_{A(\mathbb{R}^{(n)})}(y)$ of y onto the column space of A.

2. Find one solution $x = v$ to the equation $Ax = \text{Proj}_{A(\mathbb{R}^{(n)})}(y)$.

3. The set of approximate solutions x to $Ax = y$ is then the set $v + \text{Nullspace}(A)$ of solutions to the equation $Ax = \text{Proj}_{A(\mathbb{R}^{(n)})}(y)$.

How can we find the *shortest* approximate solution x to $Ax = y$? We use

Corollary 10.1.3. Let W be a subspace of $\mathbb{R}^{(n)}$ and let $y \in \mathbb{R}^{(n)}$. Then the element of $y + W = \{y + w \mid w \in W\}$ of shortest length is $y - \text{Proj}_W(y)$.

Proof. The length $\|y + w\|$ of $y + w$ is the distance from y to $-w$. To minimize this distance over all $w \in W$, we simply take $-w = \text{Proj}_W(y)$, by Theorem 10.1.1. Then $y + w = y - \text{Proj}_W(y)$. ∎

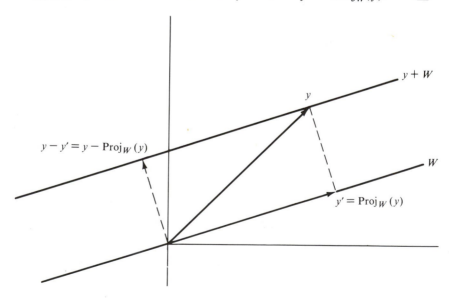

By this corollary, we get the shortest element of v + Nullspace(A) by replacing v by $v - \text{Proj}_N(v) = \text{Proj}_{N^\perp}(v)$ where N = Nullspace (A). Since $N^\perp$ is the column space of the transpose A' of A (Prove!), $\text{Proj}_{N^\perp}(v)$ is just the projection $\text{Proj}_{A'(\mathbb{R}^{(m)})}(v)$ of v on the column space of A'. So we can find the shortest approximate solution x to $Ax = y$ as follows:

1. Find one approximate solution v to the equation $Av = y$ by any method (e.g., by the one given above).

2. Replace the approximate solution v by $x = \text{Proj}_{A'(\mathbb{R}^{(m)})}(v)$.

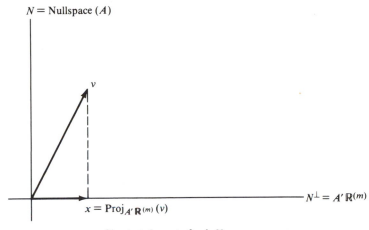

N = Nullspace (A)

v

$x = \text{Proj}_{A'\,\mathbb{R}^{(m)}}(v)$

$N^\perp = A'\,\mathbb{R}^{(m)}$

x = Shortest element of $v + N$

EXAMPLE

Let's find the shortest approximate solution to

$$\begin{bmatrix} 2 & 3 & 5 \\ 0 & 4 & 4 \\ 0 & 1 & 1 \end{bmatrix} x = \begin{bmatrix} 5000 \\ 4000 \\ 2000 \end{bmatrix}.$$

Since the column space of $A = \begin{bmatrix} 2 & 3 & 5 \\ 0 & 4 & 4 \\ 0 & 1 & 1 \end{bmatrix}$ is the same as the

column space W of the matrix $\begin{bmatrix} 2 & 3 \\ 0 & 4 \\ 0 & 1 \end{bmatrix}$ in our earlier example, the approximate solution $\begin{bmatrix} p \\ q \end{bmatrix} = \begin{bmatrix} 911.76 \\ 1058.82 \end{bmatrix}$ to the equation

$$\begin{bmatrix} 2 & 3 \\ 0 & 4 \\ 0 & 1 \end{bmatrix} \begin{bmatrix} p \\ q \end{bmatrix} = \begin{bmatrix} 5000 \\ 4000 \\ 2000 \end{bmatrix},$$

which satisfies the equation

$$\begin{bmatrix} 2 & 3 \\ 0 & 4 \\ 0 & 1 \end{bmatrix} \begin{bmatrix} p \\ q \end{bmatrix} = \mathrm{Proj}_{A(\mathbb{R}^3)} \left(\begin{bmatrix} 5000 \\ 4000 \\ 2000 \end{bmatrix} \right),$$

leads to the approximate solution

$$v = \begin{bmatrix} p \\ q \\ 0 \end{bmatrix} = \begin{bmatrix} 911.76 \\ 1058.82 \\ 0 \end{bmatrix}$$

to the equation

$$\begin{bmatrix} 2 & 3 & 5 \\ 0 & 4 & 4 \\ 0 & 1 & 1 \end{bmatrix} v = \begin{bmatrix} 5000 \\ 4000 \\ 2000 \end{bmatrix},$$

since it satisfies the equation

$$\begin{bmatrix} 2 & 3 & 5 \\ 0 & 4 & 4 \\ 0 & 1 & 1 \end{bmatrix} x = \mathrm{Proj}_{A(\mathbb{R}^{(3)})} \left(\begin{bmatrix} 5000 \\ 4000 \\ 2000 \end{bmatrix} \right).$$

So to get the shortest approximate solution, we replace v by $v - \mathrm{Proj}_N(v)$, where

$$v = \begin{bmatrix} 911.76 \\ 1058.82 \\ 0 \end{bmatrix}, N = \mathrm{Nullspace} \left(\begin{bmatrix} 2 & 3 & 5 \\ 0 & 4 & 4 \\ 0 & 1 & 1 \end{bmatrix} \right).$$

Since N is the set of vectors rw, where $w = \begin{bmatrix} -1 \\ -1 \\ 1 \end{bmatrix}$ (verify!), we

replace v by

$$v = \text{Proj}_N(v) = v - \frac{(v, w)}{(w, w)}\, w$$

$$= \begin{bmatrix} 911.76 \\ 1058.82 \\ 0 \end{bmatrix} - \frac{-911.76 - 1058.82}{3} \begin{bmatrix} -1 \\ -1 \\ 1 \end{bmatrix}$$

$$= \begin{bmatrix} 911.76 \\ 1058.82 \\ 0 \end{bmatrix} - \begin{bmatrix} 656.86 \\ 656.86 \\ -656.86 \end{bmatrix} = \begin{bmatrix} 254.90 \\ 401.96 \\ 656.86 \end{bmatrix}.$$

The approximate solution $\begin{bmatrix} 911.76 \\ 1058.82 \\ 0 \end{bmatrix}$ has length 1397.29. Our new

approximate solution $\begin{bmatrix} 254.90 \\ 401.96 \\ 656.86 \end{bmatrix}$ has length 811.18, a substantial

decrease in length.

We've seen that there is a shortest approximate solution v to the equation $Ax = y$ and that it satisfies the two conditions

1. $Av = \text{Proj}_{A(\mathbb{R}^{(n)})}(y)$;
2. v is in the column space of A'.

Suppose that both u and v satisfy these two conditions. Then the vector $w = v - u$ is in the nullspace N of A, since

$$Aw = Av - Au = \text{Proj}_{A(\mathbb{R}^{(n)})}(y) - \text{Proj}_{A(\mathbb{R}^{(n)})}(y) = 0.$$

Moreover, $w = v - u$ is orthogonal to N, since u and v are in the column space of A'. Since w is in N, w is orthogonal to w, that is, w is 0. But then $u = v$. This proves

Theorem 10.1.4.   Let A be an $m \times n$ matrix with real entries and let $y \in \mathbb{R}^{(m)}$. Then $Ax = y$ has one and only one shortest approximate solution x. Necessary and sufficient conditions that x be the shortest approximate solution to $Ax = y$ are that

1. $Ax = \text{Proj}_{A(\mathbb{R}^{(n)})}(y)$; and
2. x is in the column space of A'.

PROBLEMS

NUMERICAL PROBLEMS

1. Find all approximate solutions to $\begin{bmatrix} 2 & 1 & 5 \\ 0 & 2 & 2 \\ 0 & 0 & 0 \end{bmatrix} x = \begin{bmatrix} 2 \\ 0 \\ 2 \end{bmatrix}$.

2. Find the shortest approximate solution to $\begin{bmatrix} 2 & 1 & 5 \\ 0 & 2 & 2 \\ 0 & 0 & 0 \end{bmatrix} x = \begin{bmatrix} 2 \\ 0 \\ 2 \end{bmatrix}$.

3. Find the shortest approximate solution to $\begin{bmatrix} 1 & 1 \\ 0 & 2 \\ 1 & 1 \end{bmatrix} x = \begin{bmatrix} 2 \\ 1 \\ 0 \end{bmatrix}$.

MORE THEORETICAL PROBLEMS

Easier Problems

4. If $y' = \text{Proj}_W(y)$, show that $y - y' = \text{Proj}_{W^\perp}(y)$.

5. Using Problem 4, show that if W is the plane in $\mathbb{R}^{(3)}$ orthogonal to a nonzero vector $N \in \mathbb{R}^{(3)}$, then $\text{Proj}_W(y) - \dfrac{(y, N)}{(N, N)} N$.

10.2. THE APPROXIMATE INVERSE OF AN $m \times n$ MATRIX

By Corollary 6.3.7,

> *any $m \times n$ matrix A such that the equation $Ax = y$ has one and only one solution $x \in \mathbb{R}^{(n)}$ for every $y \in \mathbb{R}^{(m)}$ is an invertible $n \times n$ matrix.*

Now that we know from the preceding section that

> *for* any $m \times n$ *matrix A with real entries the equation $Ax = y$ has one and only one shortest approximate solution $x \in \mathbb{R}^{(n)}$ for every $y \in \mathbb{R}^{(m)}$,*

what, then, should we be able to say?

> *We denote the shortest approximate solution to $Ax = y$ by $A^-(y)$ for all y in $\mathbb{R}^{(m)}$.*

By Theorem 10.1.4, this means that

$A^-(y)$ is the unique vector $v \in \mathbb{R}^{(n)}$ such that

1. $Av = \text{Proj}_{A(\mathbb{R}^{(n)})}(y)$;
2. v is in the column space of A'.

We claim that the mapping A^- from $\mathbb{R}^{(m)}$ to $\mathbb{R}^{(n)}$ is a linear mapping. Letting $y, z \in \mathbb{R}^{(m)}$ and $c \in \mathbb{R}$, we have

1. $A(A^-(y) + A^-(z)) = AA^-(y) + AA^-(z) = \text{Proj}_{A(\mathbb{R}^{(n)})}(y) + \text{Proj}_{A(\mathbb{R}^{(n)})}(z) = \text{Proj}_{A(\mathbb{R}^{(n)})}(y + z)$ (Prove!);
2. $A^-(y) + A^-(z)$ is in the column space of A', since $A^-(y)$ and $A^-(z)$ are.

It follows that $A^-(y) + A^-(z) = A^-(y + z)$. Similarly, we have

1. $A(cA^-(y)) = cAA^-(y) = c\,\text{Proj}_{A(\mathbb{R}^{(n)})}(y) = \text{Proj}_{A(\mathbb{R}^{(n)})}(cy)$ (Prove!);
2. $cA^-(y)$ is in the column space of A', since $A^-(y)$ is.

From this, it follows that $A^-(cy) = cA^-(y)$. So, the mapping A^- is linear. We have now proved

Theorem 10.2.1.   The mapping A^- from $\mathbb{R}^{(m)}$ to $\mathbb{R}^{(n)}$ is linear.

By Theorems 10.2.1 and 6.1.1 we get that A^- is an $n \times m$ matrix.

Definition.   For any $m \times n$ matrix A with real entries, we call the $n \times m$ matrix A^- the *approximate inverse* of A, since A^-y is the shortest approximate solution x to $Ax = y$ for all y.

The approximate inverse is also called the *pseudoinverse*.
How do we find the $n \times m$ matrix A^-? Its columns are

$$A^-(e_1), \ldots, A^-(e_m),$$

which we can compute as the shortest approximate solutions to the equations $Ax = e_1, \ldots, Ax = e_m$, where $e_1, \ldots, e_m$ is the standard basis for $\mathbb{R}^{(m)}$.

EXAMPLE

Let's find the approximate inverse A^- of the 3×2 matrix

$$A = \begin{bmatrix} 2 & 3 \\ 0 & 4 \\ 0 & 1 \end{bmatrix}$$

and use it to find the shortest approximate solution $A^- \begin{bmatrix} 5000 \\ 4000 \\ 2000 \end{bmatrix}$ of

$$\begin{bmatrix} 2 & 3 \\ 0 & 4 \\ 0 & 1 \end{bmatrix} \begin{bmatrix} p \\ q \end{bmatrix} = \begin{bmatrix} 5000 \\ 4000 \\ 2000 \end{bmatrix}.$$

The projections of $\begin{bmatrix} 1 \\ 0 \\ 0 \end{bmatrix}$, $\begin{bmatrix} 0 \\ 1 \\ 0 \end{bmatrix}$, $\begin{bmatrix} 0 \\ 0 \\ 1 \end{bmatrix}$ on the column space of

$$\begin{bmatrix} 2 & 3 \\ 0 & 4 \\ 0 & 1 \end{bmatrix} \text{ are}$$

$$\begin{bmatrix} 1 \\ 0 \\ 0 \end{bmatrix}, \begin{bmatrix} 0 \\ \frac{16}{17} \\ \frac{4}{17} \end{bmatrix}, \begin{bmatrix} 0 \\ \frac{4}{17} \\ \frac{1}{17} \end{bmatrix}$$

and the solutions to the equations

$$\begin{bmatrix} 2 & 3 \\ 0 & 4 \\ 0 & 1 \end{bmatrix} x = \begin{bmatrix} 1 \\ 0 \\ 0 \end{bmatrix}, \quad \begin{bmatrix} 2 & 3 \\ 0 & 4 \\ 0 & 1 \end{bmatrix} x = \begin{bmatrix} 0 \\ \frac{16}{17} \\ \frac{4}{17} \end{bmatrix}, \quad \begin{bmatrix} 2 & 3 \\ 0 & 4 \\ 0 & 1 \end{bmatrix} x = \begin{bmatrix} 0 \\ \frac{4}{17} \\ \frac{1}{17} \end{bmatrix}$$

are $\begin{bmatrix} \frac{1}{2} \\ 0 \end{bmatrix}$, $\begin{bmatrix} -\frac{6}{17} \\ \frac{4}{17} \end{bmatrix}$, $\begin{bmatrix} -\frac{3}{34} \\ \frac{1}{17} \end{bmatrix}$. Since column space of the transpose

of $\begin{bmatrix} 2 & 3 \\ 0 & 4 \\ 0 & 1 \end{bmatrix}$ is $\mathbb{R}^{(2)}$, the projections of

$$\begin{bmatrix} \frac{1}{2} \\ 0 \end{bmatrix}, \begin{bmatrix} -\frac{6}{17} \\ \frac{4}{17} \end{bmatrix}, \begin{bmatrix} -\frac{3}{34} \\ \frac{1}{17} \end{bmatrix}$$

on the column space of the transpose of $\begin{bmatrix} 2 & 3 \\ 0 & 4 \\ 0 & 1 \end{bmatrix}$ are just

$$\begin{bmatrix} \frac{1}{2} \\ 0 \end{bmatrix}, \quad \begin{bmatrix} -\frac{6}{17} \\ \frac{4}{17} \end{bmatrix}, \quad \begin{bmatrix} -\frac{3}{34} \\ \frac{1}{17} \end{bmatrix}$$

themselves. So, the approximate inverse of $\begin{bmatrix} 2 & 3 \\ 0 & 4 \\ 0 & 1 \end{bmatrix}$ is

$$\begin{bmatrix} \frac{1}{2} & -\frac{6}{17} & -\frac{3}{34} \\ 0 & \frac{4}{17} & \frac{1}{17} \end{bmatrix}$$

and the shortest approximate solution of

$$\begin{bmatrix} 2 & 3 \\ 0 & 4 \\ 0 & 1 \end{bmatrix} \begin{bmatrix} p \\ q \end{bmatrix} = \begin{bmatrix} 5000 \\ 4000 \\ 2000 \end{bmatrix}$$

is

$$\begin{bmatrix} p \\ q \end{bmatrix} = \begin{bmatrix} \frac{1}{2} & -\frac{6}{17} & -\frac{3}{34} \\ 0 & \frac{4}{17} & \frac{1}{17} \end{bmatrix} \begin{bmatrix} 5000 \\ 4000 \\ 2000 \end{bmatrix} = \begin{bmatrix} 911.76 \\ 1058.82 \end{bmatrix},$$

the value that we got in the second example of Section 10.1.

We close this section by interrelating the four matrices

A,

A^-,

$\text{Proj}_{A'(\mathbb{R}^{(n)})}$ (projection from $\mathbb{R}^{(m)}$ to $A(\mathbb{R}^{(n)})$, an $n \times m$ matrix),

$\text{Proj}_{A'(\mathbb{R}^{(m)})}$ (projection from $\mathbb{R}^{(n)}$ to $A'\mathbb{R}^{(m)}$, an $m \times n$ matrix).

Theorem 10.2.2. $AA^- = \text{Proj}_{A(\mathbb{R}^{(n)})}$ and $A^-A = \text{Proj}_{A'(\mathbb{R}^{(m)})}$.

Proof. Since A^- maps y to the shortest x such that $Ax = \text{Proj}_{A'\mathbb{R}^{(m)}} y$, AA^- maps y to $\text{Proj}_{A(\mathbb{R}^{(n)})} y$. And since A maps x or y where-upon A^- maps y to $\text{Proj}_{A'(\mathbb{R}^{(m)})} x$, A^-A maps x to $\text{Proj}_{A'(\mathbb{R}^{(m)})} x$. ∎

PROBLEMS

NUMERICAL PROBLEMS

1. Find the approximate inverse of $A = \begin{bmatrix} 1 \\ 0 \\ 1 \end{bmatrix}$ and use it to find approximate solutions to the equation $Ax = \begin{bmatrix} 1 \\ 1 \\ 1 \end{bmatrix}$ and $Ax = \begin{bmatrix} 1 \\ 2 \\ 1 \end{bmatrix}$.

2. Find the approximate inverse of $A = \begin{bmatrix} 1 & 1 \\ 0 & 2 \\ 1 & 1 \end{bmatrix}$ and use it to find an

approximate solution to $Ax = \begin{bmatrix} 2 \\ 1 \\ 1 \end{bmatrix}$.

3. Find the approximate inverse of $A = \begin{bmatrix} 1 & 0 & 1 \\ 1 & 2 & 1 \end{bmatrix}$ and use it to find

an approximate solution to $Ax = \begin{bmatrix} 1 \\ 3 \end{bmatrix}$.

MORE THEORETICAL PROBLEM

4. Prove that $\mathrm{Proj}_W(y)$ is linear in y, that is,

$$\mathrm{Proj}_W(y + z) = \mathrm{Proj}_W(y) + \mathrm{Proj}_W(z),$$
$$\mathrm{Proj}_W(cy) = c\,\mathrm{Proj}_W(y).$$

10.3. SOLVING A MATRIX EQUATION USING ITS NORMAL EQUATION

Up to now, we have found approximate solutions to $Ax = y$ and computed the approximate inverse of A directly from the definitions. Are there better methods? Fortunately, we can find the approximate solutions x to $Ax = y$ by finding the solutions x to the corresponding *normal equation* $A'Ax = A'y$. In most applications, finding the solutions to $A'Ax = A'y$ is easier than finding the approximate solutions to $Ax = y$ directly. One reason for this is that $A'A$ is an $n \times n$ matrix when A is an $m \times n$ matrix. So if $n < m$, which is true in most applications. $A'A$ being $n \times n$ is of smaller size than A which is $m \times n$. Another is that $A'A$ is symmetric, so that it is similar to a diagonal matrix.

Why can we find the approximate solutions x to $Ax = y$ by finding the solutions x to the corresponding normal equation $A'Ax = A'y$? The condition $Ax = \mathrm{Proj}_{A(\mathbb{R}^{(n)})}(y)$ on the element Ax of the column space A is that $y - Ax$ be orthogonal to the column space of A, that is, that

$A'(y - Ax) = 0$. But this is just the condition that x be a solution of the equation $A'Ax = A'y$. To require further that x be the shortest approximate solution to $Ax = y$ is, by Theorem 10.1.4, equivalent to requiring that x be in the column space of A'. This proves

Theorem 10.3.1.   Let $y \in \mathbb{R}^{(m)}$. Then

1. The approximate solutions x to $Ax = y$ are just the solutions x to the corresponding normal equation $A'Ax = A'y$; and

2. $x = A^{-}y$ if and only if $A'Ax = A'y$ and x is the column space of A'.

Before we discuss the general case further, let's first look at the special case where the columns of A are linearly independent. For this, we need

Theorem 10.3.2.   If the columns of A are linearly independent, then the matrix $A'A$ is invertible.

Proof.   Since $A'A$ is a square matrix, it is invertible if and only if its nullspace is 0. So letting x be any vector such that $A'Ax = 0$, it suffices to show that $x = 0$. Multiplying by x', we have $0 = x'A'Ax = (Ax)'(Ax)$. This implies that the length of Ax is 0, so that Ax is 0. Since the columns of A are linearly independent, it follows that x is 0. (Prove!)   ∎

EXAMPLE

Since the columns of $A = \begin{bmatrix} 2 & 3 \\ 0 & 4 \\ 0 & 1 \end{bmatrix}$ are linearly independent,

the matrix $A'A = \begin{bmatrix} 2 & 0 & 0 \\ 3 & 4 & 1 \end{bmatrix} \begin{bmatrix} 2 & 3 \\ 0 & 4 \\ 0 & 1 \end{bmatrix} = \begin{bmatrix} 4 & 6 \\ 6 & 26 \end{bmatrix}$ is invertible,

with inverse $(\frac{1}{34}) \begin{bmatrix} 13 & -3 \\ -3 & 2 \end{bmatrix}$.

If the columns of A are linearly independent, we know from Theorem 10.3.2 that $A'A$ is invertible. Since, by Theorem 10.3.1, x is an approximate solution of $Ax = y$ if and only if x is a solution of $A'Ax =$

$A'y$, it follows that x is an approximate solution of $Ax = y$ if and only if $x = (A'A)^{-1}A'y$, which proves

Corollary 10.3.3. Suppose that the columns of A are linearly independent. Then for any y in the $\mathbb{R}^{(m)}$, there is one and only one approximate solution x to $Ax = y$, namely $x = (A'A)^{-1}A'y$.

Theorem 10.3.4. If the columns of A are linearly independent, then the approximate inverse of A is $A^- = (A'A)^{-1}A'$.

EXAMPLE

Since the columns of $A = \begin{bmatrix} 2 & 3 \\ 0 & 4 \\ 0 & 1 \end{bmatrix}$ are linearly independent,

the approximate inverse of A is

$$A^- = \left(\begin{bmatrix} 2 & 0 & 0 \\ 3 & 4 & 1 \end{bmatrix} \begin{bmatrix} 2 & 3 \\ 0 & 4 \\ 0 & 1 \end{bmatrix} \right)^{-1} \begin{bmatrix} 2 & 0 & 0 \\ 3 & 4 & 1 \end{bmatrix}$$

$$= \begin{bmatrix} 4 & 6 \\ 6 & 26 \end{bmatrix}^{-1} \begin{bmatrix} 2 & 0 & 0 \\ 3 & 4 & 1 \end{bmatrix}$$

$$= (\tfrac{1}{34}) \begin{bmatrix} 13 & -3 \\ -3 & 2 \end{bmatrix} \begin{bmatrix} 2 & 0 & 0 \\ 3 & 4 & 1 \end{bmatrix}$$

$$= \begin{bmatrix} \frac{1}{2} & -\frac{6}{17} & -\frac{3}{34} \\ 0 & \frac{4}{17} & \frac{1}{17} \end{bmatrix},$$

which agrees with our calculation of A^- in the corresponding example of Section 10.2.

It is very useful to have the explicit formula $A^- = (A'A)^{-1}A'$ for the approximate inverse of A in the case that the columns of A are linearly independent. Is there such a formula in general? Yes! The shortest approximate solution x to $Ax = y$ is the shortest solution (and therefore shortest approximate solution) x to the normal equation $(A'A)x = A'y$. This x is just $x = (A'A)^-(A'y)$, where $(A'A)^-$ is the approximate inverse of $A'A$. This proves

Theorem 10.3.5. The approximate inverse of any $m \times n$ matrix A with real entries is $A^- = (A'A)^-A'$.

Of course, to make this theorem work for us in finding A^-, we need to find $(A'A)^-$. This is easier than finding A^- itself, since $A'A$ is a *symmetric* matrix and, in most applications, $A'A$ is much smaller than A.

PROBLEMS

NUMERICAL PROBLEMS

1. Using the formula $A^- = (A'A)^{-1}A'$, find the approximate inverse

of $A = \begin{bmatrix} 1 \\ 0 \\ 1 \end{bmatrix}$ and compare with your answer to Problem 1 of Section

10.2.

2. Using the formula $A^- = (A'A)^{-1}A'$, find the approximate inverse

of $A = \begin{bmatrix} 1 & 1 \\ 0 & 2 \\ 1 & 1 \end{bmatrix}$ and compare with your answer to Problem 2 of

Section 10.2.

3. Using the formula $A^- = (A'A)^-A'$, find the approximate inverse

of $A = \begin{bmatrix} 1 & 2 \\ 0 & 0 \\ 1 & 2 \end{bmatrix}$. Then calculate the projection $AA^- = A(A'A)^-A'$

onto the column space of A. Finally, use it to find approximate

solutions to the equations $Ax = \begin{bmatrix} 1 \\ 1 \\ 1 \end{bmatrix}$ and $Ax = \begin{bmatrix} 2 \\ 1 \\ 1 \end{bmatrix}$.

4. Verify that the matrix $\begin{bmatrix} 1 & 0 & 0 \\ 0 & \frac{16}{17} & \frac{4}{17} \\ 0 & \frac{4}{17} & \frac{1}{17} \end{bmatrix}$, which we calculated as the

projection onto the column space of a matrix, does actually coincide with its square.

MORE THEORETICAL PROBLEMS

Easier Problems

5. Viewing real problems d as 1×1 matrices, show directly that the approximate inverse of a real number d is d^-, where d^- is $1/d$ if d is nonzero and d^- is 0 if d is 0.

10.4. FINDING FUNCTIONS THAT APPROXIMATE DATA

In an experiment having input variable x and output variable y, we get output values $y_0, \ldots, y_m$ corresponding to input values $x_0, \ldots, x_m$ from data generated in the experiment. We then seek to find a useful *functional approximation* to the mapping $y_r = f(x_r)$ $(0 \leq r \leq m)$, that is, we want to find a function such as $y = ax^2 + bx + c$ such that y_r and $ax_r^2 + bx_r + c$ are equal or nearly equal for $0 \leq r \leq m$. If we seek a functional approximation *of order* n, that is, a function of the form $y = p(x) = c_0 + c_1 x + \cdots + c_n x^n$, how do we choose the coefficients c_r? We write down the equations

$$c_0 + c_1 x_0 + \cdots + c_n x_0^n = y_0$$
$$\vdots$$
$$c_0 + c_1 x_m + \cdots + c_n x_m^n = y_m,$$

viewing the x_r^s as the entries of the coefficient matrix A and the c_r as the unknowns. Then we find the approximate solution $\begin{bmatrix} c_0 \\ \vdots \\ c_n \end{bmatrix}$ to

$$\begin{bmatrix} 1 & x_0 & \cdots & x_0^n \\ \vdots & \vdots & & \vdots \\ 1 & x_m & \cdots & x_m^n \end{bmatrix} \begin{bmatrix} c_0 \\ \vdots \\ c_n \end{bmatrix} = \begin{bmatrix} y_0 \\ \vdots \\ y_m \end{bmatrix},$$

for example by calculating the approximate inverse A^- of $A =$

$$\begin{bmatrix} 1 & x_0 & \cdots & x_0^n \\ \vdots & \vdots & & \vdots \\ 1 & x_m & \cdots & x_m^n \end{bmatrix} \text{ and letting } \begin{bmatrix} c_0 \\ \vdots \\ c_n \end{bmatrix} = A^- \begin{bmatrix} y_0 \\ \vdots \\ y_m \end{bmatrix}.$$

The functional approximation

$$y = p(x) = c_0 + c_1 x + \cdots + c_n x^n$$

for the mapping $y_r = f(x_r)$ that we obtain in this way is the polynomial $p(x)$ of degree n for which the sum of squares

$$(y_0 - p(x_0))^2 + \cdots + (y_m - p(x_m))^2$$

is as small as possible. This method of finding a functional approximation is often called *the method of least squares*.

EXAMPLE

In a time study experiment that we conduct to find a functional relationship between the duration x of the coffee break (in minutes) and the value y (in thousands of dollars) of the work performed the same day by a group of employees, coffee breaks of

$$x_0 = 10, x_1 = 15, x_2 = 21, x_3 = 5$$

minutes duration and values

$$y_0 = 10, y_1 = 14, y_2 = 13, y_3 = 10$$

of thousands of dollars' worth of work performed were observed on the four successive days of the experiment. We decide to analyze the data two ways, namely, to use it to get a first-order approximation, and then to use it to get a second order approximation:

1. To get the first-order approximation, we first get the general approximate solution to

$$
\begin{bmatrix} 1 & x_0 \\ \vdots & \vdots \\ 1 & x_m \end{bmatrix}
\begin{bmatrix} c_0 \\ c_1 \end{bmatrix}
=
\begin{bmatrix} y_0 \\ \vdots \\ y_m \end{bmatrix}.
$$

This is

$$
\begin{bmatrix} c_0 \\ c_1 \end{bmatrix}
= A^{-}
\begin{bmatrix} y_0 \\ \vdots \\ y_m \end{bmatrix}
= (A'A)^{-1} A'
\begin{bmatrix} y_0 \\ \vdots \\ y_m \end{bmatrix},
$$

where $A = \begin{bmatrix} 1 & x_0 \\ \vdots & \vdots \\ 1 & x_m \end{bmatrix}$, that is,

$$
\begin{bmatrix} c_0 \\ c_1 \end{bmatrix}
= \left(
\begin{bmatrix} 1 & \cdots & 1 \\ x_0 & \cdots & x_m \end{bmatrix}
\begin{bmatrix} 1 & x_0 \\ \vdots & \vdots \\ 1 & x_m \end{bmatrix}
\right)^{-1}
\begin{bmatrix} 1 & \cdots & 1 \\ x_0 & \cdots & x_m \end{bmatrix}
\begin{bmatrix} y_0 \\ \vdots \\ y_m \end{bmatrix}
$$

$$
=
\begin{bmatrix} m+1 & 1 \cdot x \\ x \cdot 1 & x \cdot x \end{bmatrix}^{-1}
\begin{bmatrix} 1 \cdot y \\ x \cdot y \end{bmatrix}
\text{ or }
\begin{bmatrix} c_0 \\ c_1 \end{bmatrix}
$$

$$
=
\begin{bmatrix} m+1 & 1 \cdot x \\ x \cdot 1 & x \cdot x \end{bmatrix}^{-1}
\begin{bmatrix} 1 \cdot y \\ x \cdot y \end{bmatrix},
$$

where $\mathbf{1} = \begin{bmatrix} 1 \\ \vdots \\ 1 \end{bmatrix}$ and $u \cdot v$ denotes (u, v). In the time study, m is

3 and we have

$$x \cdot \mathbf{1} = 10 + 15 + 21 + 5 = 51$$
$$x \cdot x = 100 + 225 + 441 + 25 = 791$$
$$\mathbf{1} \cdot y = 10 + 14 + 13 + 10 = 47$$
$$x \cdot y = 100 + 210 + 273 + 50 = 633.$$

So

$$\begin{bmatrix} c_0 \\ c_1 \end{bmatrix} = \begin{bmatrix} m + 1 & \mathbf{1} \cdot x \\ x \cdot \mathbf{1} & x \cdot x \end{bmatrix}^{-1} \begin{bmatrix} \mathbf{1} \cdot y \\ x \cdot y \end{bmatrix} = \begin{bmatrix} c_0 \\ c_1 \end{bmatrix}$$

$$= \begin{bmatrix} 4 & 51 \\ 51 & 791 \end{bmatrix}^{-1} \begin{bmatrix} 47 \\ 633 \end{bmatrix}$$

$$= (\tfrac{1}{563}) \begin{bmatrix} 791 & -51 \\ -51 & 4 \end{bmatrix} \begin{bmatrix} 47 \\ 633 \end{bmatrix} = \begin{bmatrix} 8.69 \\ .24 \end{bmatrix}.$$

and we find that the linear function

$$y = 8.69 + .24x$$

is the approximation of first order. Let's see how well it approximates:

x		10	15	21	
actual y		10	14	13	10
approximating $y = 8.69 + .24x$	11.57	12.29	13.73	9.89

2. For the second order approximation, the general solution to

$$\begin{bmatrix} 1 & x_0 & x_0^2 \\ \vdots & \vdots & \vdots \\ 1 & x_m & x_m^2 \end{bmatrix} \begin{bmatrix} c_0 \\ c_1 \\ c_2 \end{bmatrix} = \begin{bmatrix} y_0 \\ \vdots \\ y_m \end{bmatrix}$$

is

$$\begin{bmatrix} c_0 \\ c_1 \\ c_2 \end{bmatrix} = A^{-} \begin{bmatrix} y_0 \\ \vdots \\ y_m \end{bmatrix} = (A'A)^{-1}A' \begin{bmatrix} y_0 \\ \vdots \\ y_m \end{bmatrix},$$

where $A = \begin{bmatrix} 1 & x_0 & x_0^2 \\ \vdots & \vdots & \vdots \\ 1 & x_m & x_m^2 \end{bmatrix}$, that is, $\begin{bmatrix} c_0 \\ c_1 \\ c_2 \end{bmatrix}$ is given by

$$\left(\begin{bmatrix} 1 & \cdots & 1 \\ x_0 & \cdots & x_m \\ x_0^2 & \cdots & x_m^2 \end{bmatrix} \begin{bmatrix} 1 & x_0 & x_0^2 \\ \vdots & \vdots & \vdots \\ 1 & x_m & x_m^2 \end{bmatrix} \right)^{-1} \begin{bmatrix} 1 & \cdots & 1 \\ x_1 & \cdots & x_m \\ x_1^2 & \cdots & tx_m^2 \end{bmatrix} \begin{bmatrix} y_0 \\ \vdots \\ y_m \end{bmatrix}.$$

But then

$$\begin{bmatrix} c_0 \\ c_1 \\ c_2 \end{bmatrix} = \begin{bmatrix} m+1 & \mathbf{1} \cdot x & \mathbf{1} \cdot x^2 \\ x \cdot \mathbf{1} & x \cdot x & x \cdot x^2 \\ x^2 \cdot \mathbf{1} & x^2 \cdot x & x^2 \cdot x^2 \end{bmatrix}^{-1} \begin{bmatrix} \mathbf{1} \cdot y \\ x \cdot y \\ x^2 \cdot y \end{bmatrix},$$

where $\mathbf{1} = \begin{bmatrix} 1 \\ \vdots \\ 1 \end{bmatrix}$. So

$$\begin{bmatrix} c_0 \\ c_1 \\ c_2 \end{bmatrix} = \begin{bmatrix} 4 & 51 & 791 \\ 51 & 791 & 13{,}761 \\ 791 & 13{,}761 & 255{,}231 \end{bmatrix}^{-1} \begin{bmatrix} 47 \\ 633 \\ 10{,}133 \end{bmatrix}.$$

For these c_0, c_1, c_2, the approximation of second order is the quadratic polynomial

$$y = c_0 + c_1 x + c_2 x^2.$$

In order to use the formula $A^- = (A'A)^{-1}A'$ for the approximate inverse of $A = \begin{bmatrix} 1 & x_0 & \cdots & x_0^n \\ \vdots & \vdots & & \vdots \\ 1 & x_m & \cdots & x_m^n \end{bmatrix}$ in finding the approximating function of order n for the mapping $y_r = f(x_r)$ ($1 \le r \le m$), we need to know that the columns of A are linearly independent. Assuming that the x_r are all different, this is true if $m \ge n$. Why? For $m \ge n$, A has an invertible $n \times n$ submatrix by the following theorem.

Theorem 10.4.1. The matrix $A = \begin{bmatrix} 1 & x_0 & \cdots & x_0^n \\ \vdots & \vdots & & \vdots \\ 1 & x_n & \cdots & x_n^n \end{bmatrix}$ is invertible if

and only the $x_0, \ldots, x_n$ are all different.

Proof. If two of the x_r are the same, then A has two identical rows, so it is not invertible. Suppose, conversely, that A is not invertible. Then the system of equations

$$c_0 + c_1 x_0 + \cdots + c_n x_0^n = 0$$

$$\vdots \qquad\qquad\qquad \vdots$$

$$c_0 + c_1 x_n + \cdots + c_n x_n^n = 0$$

has a nonzero solution $\begin{bmatrix} c_0 \\ \vdots \\ c_n \end{bmatrix}$; that is, there is a nonzero polynomial

$p(x) = c_0 + c_1 x + \cdots + c_n x^n$ of degree n which vanishes at all of the $n + 1$ numbers $x_0, \ldots, x_n$. Since a polynomial $p(x)$ of degree n has at most n roots, two of the x_r are equal. ∎

PROBLEMS

NUMERICAL PROBLEMS

1. Find an approximating function of order 1 for the function that maps 2 to 4, 3 to 5, and 4 to 6.

2. Find an approximating function of order 1 for the function that maps 2 to 5, 3 to 10, and 4 to 17.

3. Give an expression for an approximating function of order 2 for the function that maps 2 to 5, 3 to 10, and 4 to 17.

MORE THEORETICAL PROBLEMS

Easier Problems

4. Using the formula $A^- = (A'A)^{-1}A'$ for the approximate inverse of

A, give the formula for $A^- \begin{bmatrix} y_0 \\ \vdots \\ y_m \end{bmatrix}$ for $A = \begin{bmatrix} 1 & x_0 & \cdots & x_0^n \\ \vdots & \vdots & & \vdots \\ 1 & x_m & \cdots & x_m^n \end{bmatrix}$ that

generalizes the formula

$$A^- \begin{bmatrix} y_0 \\ \vdots \\ y_m \end{bmatrix} = \begin{bmatrix} m+1 & 1\cdot x & 1\cdot x^2 \\ x\cdot 1 & x\cdot x & x\cdot x^2 \\ x^2\cdot 1 & x^2\cdot x & x^2\cdot x^2 \end{bmatrix}^{-1} \begin{bmatrix} 1\cdot y \\ x\cdot y \\ x^2\cdot y \end{bmatrix}$$

for $A = \begin{bmatrix} 1 & x_0 & x_0^2 \\ \vdots & \vdots & \vdots \\ 1 & x_m & x_m^2 \end{bmatrix}$, and show that $x^r \cdot x^s = 1 \cdot x^{r+s}$ for all

r, s.

5. Show that if $f(x)$ is a polynomial of degree n, $x_0, \ldots, x_n$ are all different and $y_r = f(x_r)$ for $1 \le r \le n$, then the approximating function of order n is just the polynomial $y = f(x)$ itself.

11

LINEAR ALGORITHMS
(OPTIONAL)

11.1. INTRODUCTION

Now that we have the theory and applications behind us, we ask:

> *How can we instruct a computer to carry out the computations?*

So that we can get into the subject enough to get a glimpse of what it is about, we restrict ourselves to a single computational problem—but one of great importance—the problem of solving a matrix-vector equation

$$Ax = y$$

for x exactly or approximately, where the entries of A, x, and y are to be real.

To instruct a computer to find x for a given A and y,

> *we simply find a mathematical description for x and devise an unambiguous recipe or step by step process for computing x from the expression.*

We call such a process an *algorithm*.

How do we choose a particular algorithm to compute x? Many factors may be involved, depending on the uses that will be made of the algorithm. Here we want to be able to solve $Ax = y$ without knowing anything in advance about the size or nature of the matrix A or vectors y that we are given as input. So we take the point of view that

1. We want the algorithm to work for all rectangular matrices A and right-hand-side vectors y, and we want it to indicate whether the solution is exact or approximate.
2. We want it to require as little memory as possible for computing the solution.
3. We want the computation to take as little time as possible.
4. We want the algorithm to be simple and easy to translate into language understandable by a computer.

The algorithm that plays the most central role in this chapter, the *row reduction algorithm* discussed in Section 11.3, may come about as close to satisfying all four of the foregoing properties as one could wish. It is simple and works for all matrices. At the same time it is very fast and since it uses virtually no memory except the memory that held the original matrix, it makes very efficient use of memory. This algorithm, which row reduces data representing an ordinary matrix into data representing that matrix in a special *factored format,* has another very desirable feature. It is reversible; that is, there is a corresponding *inverse* algorithm to *undo* row reduction. This means that we can restore a matrix in its factored format back to its original unfactored format. In fact, the inverse algorithm is also very fast, so this restoration is very efficient.

To solve the equation $Ax = y$ exactly, we use the row reduction algorithm to replace A in memory by three very special matrices L, D, U whose product, in the case where no row interchanges are needed, is

$A = LDU.$

If r is the rank of A, L is an $m \times r$ matrix which is essentially lower triangular with 1's on the diagonal (its rows may have to be put in another order to make L truly lower triangular), D is an invertible $r \times r$ matrix, and U is an echelon $r \times n$ matrix. The entries of these factors, except for the known 0's and 1's, are nicely stored together in the memory that had been occupied by A. So that we can get them easily whenever we need them, we mark the location of the r diagonal entries of D.

EXAMPLE

When we apply the row reduction algorithm to the matrix

$$A = \begin{bmatrix} 6 & 12 & 18 & 0 \\ 2 & 9 & 6 & 2 \\ 3 & 7 & 10 & 7 \end{bmatrix},$$

we get it in its factored format,

$$\begin{bmatrix} \mathbf{6} & 2 & 3 & 0 \\ \frac{1}{3} & \mathbf{5} & 0 & 2 \\ \frac{1}{2} & \frac{1}{5} & \mathbf{1} & 5 \end{bmatrix},$$

from which we extract the matrices

$$L = \begin{bmatrix} 1 & 0 & 0 \\ \frac{1}{3} & 1 & 0 \\ \frac{1}{2} & \frac{1}{5} & 1 \end{bmatrix}, \quad D = \begin{bmatrix} \mathbf{6} & 0 & 0 \\ 0 & \mathbf{5} & 0 \\ 0 & 0 & \mathbf{1} \end{bmatrix} \quad U = \begin{bmatrix} 1 & 2 & 3 & 0 \\ 0 & 1 & 0 & 2 \\ 0 & 0 & 1 & 5 \end{bmatrix}$$

whose product LDU is A.

Solving $Ax = y$ for x now is reduced to solving $LDUx = y$ for x. To get x, we simply solve the equations

$$Lv = y, \quad Dw = v, \quad Ux = w.$$

These are easy to solve (or to determine to be unsolvable) because of the special nature of the matrices L, D, U. So in this way, we get our solution x to $Ax = y$, if a solution exists.

When we are done, we can reverse the row reduction to restore the matrix A. This amounts to using the inverse algorithm to multiplying the factors L, D, U together in a very efficient way, getting back $A = LDU$. This is of great importance, since we may, within a small time frame, want to go back and forth many times between the unfactored and factored formats of the matrix.

After discussing this, in Sections 11.2 through 11.4 we go on to develop an algorithm for solving $Ax = y$ approximately. This algorithm also finds exact solutions, if they exist, but not as efficiently as does the algorithm for finding exact solutions. Here the algorithm replaces the matrix A by a factorization

$$A^- = J'K$$

of its approximate inverse, where J', in the case where no row interchanges are needed, is upper triangular. Solving $Ax = y$ approximately for x now is reduced to solving $x = J'Ky$, which amounts to solving the two equations

$z = Ky$ and $x = J'z$.

Finally, we illustrate how these algorithms can be used to build a computer program for solving $Ax = y$ and finding the exact or approximate inverse of a matrix A.

PROBLEMS

NUMERICAL PROBLEMS

1. Let $A = LDU$, where $L = \begin{bmatrix} 1 & 0 & 0 \\ 2 & 1 & 0 \\ 3 & 4 & 1 \\ 1 & 1 & 1 \end{bmatrix}$, $D = \begin{bmatrix} 1 & 0 & 0 \\ 0 & 2 & 0 \\ 0 & 0 & 3 \end{bmatrix}$, and

$U = \begin{bmatrix} 0 & 0 & 1 & 1 & 2 & 0 & 3 \\ 0 & 0 & 0 & 0 & 1 & 4 & 2 \\ 0 & 0 & 0 & 0 & 0 & 1 & 1 \end{bmatrix}$. Then

(a) Solve the equation $Ax = \begin{bmatrix} 1 \\ 2 \\ 3 \\ 1 \end{bmatrix}$ for x by solving the equations

$$\begin{bmatrix} 1 & 0 & 0 \\ 2 & 1 & 0 \\ 3 & 4 & 1 \\ 1 & 1 & 1 \end{bmatrix} v = \begin{bmatrix} 1 \\ 2 \\ 3 \\ 1 \end{bmatrix},$$

$$\begin{bmatrix} 1 & 0 & 0 \\ 0 & 2 & 0 \\ 0 & 0 & 3 \end{bmatrix} w = v,$$

$$\begin{bmatrix} 0 & 0 & 1 & 1 & 2 & 0 & 3 \\ 0 & 0 & 0 & 0 & 1 & 4 & 2 \\ 0 & 0 & 0 & 0 & 0 & 1 & 1 \end{bmatrix} x = w.$$

(b) Calculate x by computing A and row reducing $[A, y]$.

2. Using the factors L, D, U in Problem 1 as a road map, find a row reduction of the matrix A of Problem 1 that leads to the echelon form U.

MORE THEORETICAL PROBLEMS

Easier Problems

3. If the reduction of a matrix A to its echelon form U by the method of Chapter 2 does not involve any interchanges, and if $E_1, \ldots, E_k$ denote the elementary matrices corresponding to the elementary row operations used in the reduction, show that $U = NA$, where $N = E_k \ldots E_1$ is a lower triangular matrix. Using this, show that A can be factored as $A = LDU$, where L is a lower triangular $m \times m$ matrix with 1's on the diagonal and D is a diagonal $m \times m$ matrix.

4. Show that the matrix

$$\begin{bmatrix} 1 & 0 & 0 & 0 \\ f & 1 & 0 & 0 \\ e & c & 1 & 0 \\ d & b & a & 1 \end{bmatrix}$$

equals

$$\begin{bmatrix} 1 & 0 & 0 & 0 \\ f & 1 & 0 & 0 \\ 0 & 0 & 1 & 0 \\ 0 & 0 & 0 & 1 \end{bmatrix}\begin{bmatrix} 1 & 0 & 0 & 0 \\ 0 & 1 & 0 & 0 \\ e & 0 & 1 & 0 \\ 0 & 0 & 0 & 1 \end{bmatrix}\begin{bmatrix} 1 & 0 & 0 & 0 \\ 0 & 1 & 0 & 0 \\ 0 & 0 & 1 & 0 \\ d & 0 & 0 & 1 \end{bmatrix}\begin{bmatrix} 1 & 0 & 0 & 0 \\ 0 & 1 & 0 & 0 \\ 0 & c & 1 & 0 \\ 0 & 0 & 0 & 1 \end{bmatrix}\begin{bmatrix} 1 & 0 & 0 & 0 \\ 0 & 1 & 0 & 0 \\ 0 & 0 & 1 & 0 \\ 0 & b & 0 & 1 \end{bmatrix}\begin{bmatrix} 1 & 0 & 0 & 0 \\ 0 & 1 & 0 & 0 \\ 0 & 0 & 1 & 0 \\ 0 & 0 & a & 1 \end{bmatrix}.$$

11.2. THE *LDU* FACTORIZATION OF *A*

For any $m \times n$ matrix A and vector y in $\mathbb{R}^{(m)}$, solving the matrix-vector equation $Ax = y$ for $x \in \mathbb{R}^{(n)}$ by the methods given in Chapter 2 amounts to reducing the augmented matrix $[A, y]$ to an echelon matrix $[U, z]$ and solving $Ux = z$ instead. Let's look again at the reason for this, so that we can improve our methods. If M is the product of the inverses of the elementary matrices used during the reduction, a matrix that we can build and store during the reduction, then M is an invertible $m \times m$ matrix and $[MU, Mz] = [A, y]$. Since $Ux = z$ if and only if $MUx = Mz$, x is a solution of $Ux = z$ if and only if x is a solution of $Ax = y$.

From this comes something quite useful. If we reduce A to its echelon form U, gaining M during the reduction, then $A = MU$ and for any right-hand-side vector y that we may be given, we can solve $Ax = y$ in two steps as follows:

1.   Solve $Mz = y$ for z.

2.   Solve $Ux = z$ for x.

Putting these two steps together, we see that the x we get satisfies

$$Ax = MUx = Mz = y.$$

If no interchanges were needed during the reduction, M is an invertible lower triangular matrix. So since U is an echelon matrix, both equations $Mz = y$ and $Ux = z$ are easy to solve, provided that solutions exist. We have already seen how to solve $Ux = z$ for x by back substitution, given z. And we can get z from $Mz = y$, for a given y, using a reversed version of back substitution which we call *forward substitution*. Of course, if interchanges are needed during the reduction, we must also keep track of them and take them into account.

How do we find and store M so that $A = MU$? Let's look at

EXAMPLE

Let's take

$$A = \begin{bmatrix} 6 & 12 & 18 & 0 \\ 2 & 9 & 6 & 10 \\ 3 & 7 & 10 & 7 \end{bmatrix} \quad \left(\begin{array}{l} \text{unfactored format} \\ \text{for the matrix A} \end{array} \right),$$

and now reduce it to an echelon matrix U. As we reduce A, we keep track of certain nonzero *pivot entries* a_{pq} and for each of them, we store each *multiplier* a_{tq}/a_{pq} $(t > p)$ as entry (t, q) after the (t, q) entry has been changed to zero as a result of performing the operation Add $(t, p; -a_{tq}/a_{pq})$. The row operations Add $(2, 1; -\frac{1}{3})$, Add $(3, 1; -\frac{1}{2})$, Add $(3, 2; -\frac{1}{5})$ reduce A to the upper triangular matrix

$$V = \begin{bmatrix} 6 & 12 & 18 & 0 \\ 0 & 5 & 0 & 10 \\ 0 & 0 & 1 & 5 \end{bmatrix}.$$

If we write the pivot entries used during the reduction in boldface, the successive matrices encountered in the reduction are

$$
\begin{bmatrix} \mathbf{6} & 12 & 18 & 0 \\ 2 & 9 & 6 & 10 \\ 3 & 7 & 10 & 7 \end{bmatrix},
\quad
\begin{bmatrix} \mathbf{6} & 12 & 18 & 0 \\ \frac{1}{3} & 5 & 0 & 10 \\ 3 & 7 & 10 & 7 \end{bmatrix},
$$

$$
\begin{bmatrix} \mathbf{6} & 12 & 18 & 0 \\ \frac{1}{3} & \mathbf{5} & 0 & 10 \\ \frac{1}{2} & 1 & 1 & 7 \end{bmatrix},
\quad
\begin{bmatrix} \mathbf{6} & 12 & 18 & 0 \\ \frac{1}{3} & \mathbf{5} & 0 & 10 \\ \frac{1}{2} & \frac{1}{5} & \mathbf{1} & 5 \end{bmatrix}
$$

These matrices successively displace A in memory. The upper triangular part of the last one,

$$
\begin{bmatrix} \mathbf{6} & 12 & 18 & 0 \\ \frac{1}{3} & \mathbf{5} & 0 & 10 \\ \frac{1}{2} & \frac{1}{5} & \mathbf{1} & 5 \end{bmatrix}
\quad (LV \text{ factored format for } A),
$$

holds

$$
V = \begin{bmatrix} \mathbf{6} & 12 & 18 & 0 \\ 0 & \mathbf{5} & 0 & 10 \\ 0 & 0 & \mathbf{1} & 5 \end{bmatrix},
$$

whereas the lower part of it holds the lower entries of the *matrix of multipliers*

$$
L = \begin{bmatrix} 1 & 0 & 0 \\ \frac{1}{3} & 1 & 0 \\ \frac{1}{2} & \frac{1}{5} & 1 \end{bmatrix}.
$$

We claim that $A = LV$. Of course, it is easy to compute the product to check that $A = LV$. To see *why* $A = LV$, however, note also that if we were to apply the same operations Add $(2, 1; -\frac{1}{3})$, Add $(3, 1; -\frac{1}{2})$, Add $(3, 2; -\frac{1}{5})$ to L, we would get

$$\text{Add } (3, 2; -\tfrac{1}{5}) \text{ Add } (3, 1; -\tfrac{1}{2}) \text{ Add } (2, 1; -\tfrac{1}{3})L = I,$$

which implies that L can be gotten by applying their inverses in the opposite order to I. So

$$L = \text{Add } (2, 1; \tfrac{1}{3}) \text{ Add } (3, 1; \tfrac{1}{2}) \text{ Add } (3, 2; \tfrac{1}{5})I$$

and

$$LV = \text{Add } (2, 1; \tfrac{1}{3}) \text{ Add } (3, 1; \tfrac{1}{2}) \text{ Add } (3, 2; \tfrac{1}{5})V = A.$$

Going on, we can factor $V = \begin{bmatrix} 6 & 12 & 18 & 0 \\ 0 & 5 & 0 & 10 \\ 0 & 0 & 1 & 5 \end{bmatrix}$ by taking the

matrix of pivots

$$D = \begin{bmatrix} 6 & 0 & 0 \\ 0 & 5 & 0 \\ 0 & 0 & 1 \end{bmatrix}$$

and the echelon matrix

$$U = D^{-1}V = \begin{bmatrix} 1 & 2 & 3 & 0 \\ 0 & 1 & 0 & 2 \\ 0 & 0 & 1 & 5 \end{bmatrix}.$$

So we get the factorizations $V = DU$ and $A = LDU$:

$$\begin{bmatrix} 6 & 12 & 18 & 0 \\ 0 & 5 & 0 & 10 \\ 0 & 0 & 1 & 5 \end{bmatrix} = \begin{bmatrix} 6 & 0 & 0 \\ 0 & 5 & 0 \\ 0 & 0 & 1 \end{bmatrix}\begin{bmatrix} 1 & 2 & 3 & 0 \\ 0 & 1 & 0 & 2 \\ 0 & 0 & 1 & 5 \end{bmatrix}$$

$$\begin{bmatrix} 6 & 12 & 18 & 0 \\ 2 & 9 & 6 & 10 \\ 3 & 7 & 10 & 7 \end{bmatrix} = \begin{bmatrix} 1 & 0 & 0 \\ \frac{1}{3} & 1 & 0 \\ \frac{1}{2} & \frac{1}{5} & 1 \end{bmatrix}\begin{bmatrix} 6 & 0 & 0 \\ 0 & 5 & 0 \\ 0 & 0 & 1 \end{bmatrix}\begin{bmatrix} 1 & 2 & 3 & 0 \\ 0 & 1 & 0 & 2 \\ 0 & 0 & 1 & 5 \end{bmatrix}.$$

Of course, we could do this directly, starting from where we left off with

$$\begin{bmatrix} 6 & 12 & 18 & 0 \\ \frac{1}{3} & 5 & 0 & 10 \\ \frac{1}{2} & \frac{1}{5} & 1 & 5 \end{bmatrix} \quad (LV \text{ factored format for } A).$$

Simply divide the upper entries above the main diagonal of V, row by row, by the pivot entry of the same row to get the upper entries of U. We then end up with

$$\begin{bmatrix} 6 & 2 & 3 & 0 \\ \frac{1}{3} & 5 & 0 & 2 \\ \frac{1}{2} & \frac{1}{5} & 1 & 5 \end{bmatrix} \quad (LDU \text{ factored format for } A).$$

So not only have we factored $A = MU$, but our M comes to us in the factored form $M = LD$, enabling us to store it in factored form by storing L and D.

From our example we see how to find and store M so that $A = MU$. In fact, M comes to us in a factored form $M = LD$, so that $A = LDU$, and the factors L, D, U are stored efficiently during the reduction. In the general case, things go the same way. If no interchanges are needed, we can reduce A to an upper triangular matrix V using only the elementary row operations Add $(t, p; -a_{tq}/a)$. At the stage where we have a nonzero entry a in row p and column q, the *pivot entry,* and use the operation Add $(t, p; -a_{tq}/a)$, to make the (t, q) entry 0 for $t > p$, the (t, q) entry becomes available to us for *storing* the multiplier a_{tq}/a. Letting L be the corresponding lower triangular *matrix of multipliers,* consisting of the multipliers a_{tq}/a (with $t > p$) used in the reduction, below the diagonal, and 1's on the diagonal, we get $A = LV$. Why? In effect, to get V we are multiplying A by the elementary matrices corresponding to Add $(t, p; -a_{tq}/a)$; and we are multiplying I by their inverses, *in reverse order,* to get L. To see this, just compute the product of their inverses in reverse order, which has the same effect as writing the same multipliers in the same places, but starting from the other end and working forward. For example, if L is $\begin{bmatrix} 1 & 0 & 0 \\ 3 & 1 & 0 \\ 4 & 5 & 1 \end{bmatrix}$, it is the product

$$\begin{bmatrix} 1 & 0 & 0 \\ 3 & 1 & 0 \\ 4 & 5 & 1 \end{bmatrix} = \begin{bmatrix} 1 & 0 & 0 \\ 3 & 1 & 0 \\ 0 & 0 & 1 \end{bmatrix}\begin{bmatrix} 1 & 0 & 0 \\ 0 & 1 & 0 \\ 4 & 0 & 1 \end{bmatrix}\begin{bmatrix} 1 & 0 & 0 \\ 0 & 1 & 0 \\ 0 & 5 & 1 \end{bmatrix}.$$

So if we get $V = E_k \cdots E_1 A$, then we get $A = E_1^{-1} \cdots E_k^{-1} V = LV$. After we get $A = LV$, we go on to factor V as $V = DD^{-1}V = DU$, where D is the diagonal *matrix of pivots* whose diagonal entry in row p is 1 if row p of V is 0 and a if a is the first nonzero entry of row p of V, and where U is the echelon matrix $D^{-1}V$. Then we can rewrite the product $A = LV$ as $A = LDU = MD$, where $M = LD$. This proves

Theorem 11.2.1. If no interchanges take place in the row reduction of an $m \times n$ matrix A to an echelon matrix U, then $A = LDU$, where L is the lower triangular matrix of multipliers, D is the matrix of pivots, and U is the echelon matrix.

We can further simplify the factorization $A = LDU$ by throwing away parts of the matrices that are not needed.

Letting r be the rank of A, we throw away all but the first r columns of L, all but the first r rows and columns of D, and all but the first r rows of U.

Then we still have $A = LDU$, but now

L is a lower triangular m × r matrix with 1's on the diagonal, D is an invertible diagonal r × r matrix, and U is an r × n echelon matrix.

EXAMPLE

If *LDU* is the product

$$
\begin{bmatrix} 1 & 0 & 0 & 0 \\ 2 & 1 & 0 & 0 \\ 3 & 0 & 1 & 0 \\ 2 & 0 & 5 & 1 \end{bmatrix}
\begin{bmatrix} 2 & 0 & 0 & 0 \\ 0 & 5 & 0 & 0 \\ 0 & 0 & 3 & 0 \\ 0 & 0 & 0 & 1 \end{bmatrix}
\begin{bmatrix} 1 & 2 & 3 & 4 & 3 & 3 & 3 & 3 \\ 0 & 0 & 1 & 2 & 2 & 2 & 2 & 2 \\ 0 & 0 & 0 & 0 & 0 & 0 & 1 & 6 \\ 0 & 0 & 0 & 0 & 0 & 0 & 0 & 0 \end{bmatrix},
$$

then it equals the product

$$
\begin{bmatrix} 1 & 0 & 0 \\ 2 & 1 & 0 \\ 3 & 0 & 1 \\ 2 & 0 & 5 \end{bmatrix}
\begin{bmatrix} 2 & 0 & 0 \\ 0 & 5 & 0 \\ 0 & 0 & 3 \end{bmatrix}
\begin{bmatrix} 1 & 2 & 3 & 4 & 3 & 3 & 3 & 3 \\ 0 & 0 & 1 & 2 & 2 & 2 & 2 & 2 \\ 0 & 0 & 0 & 0 & 0 & 0 & 1 & 6 \end{bmatrix}
$$

because the effect of the diagonal matrix D is to multiply the rows of U by scalars, and the last column of L has no effect, since the last row of the product DU is 0.

Given the factorization $A = LDU$, whenever we are given a vector $y \in \mathbb{R}^{(m)}$, we solve $Ax = y$ for x by taking apart the problem of solving $LDUx = y$ for x as follows:

1. Solve $Lv = y$ for v by forward substitution.
2. Next solve $Dw = v$ for w by stationary substitution.
3. Finally, solve $Ux = w$ for x by back substitution.

Putting this all back together, we then get

$$Ax = LDUx = LDw = Lv = y,$$

which shows that x is the desired solution.

When there are no interchanges, the resulting vectors v, w, x are:

1. For $p = 1$ to the rank of A:

$$v_p = y_p - \sum_{j=1}^{p-1} L_{pj} v_j \qquad \text{(forward substitution)}$$

2. For $p = 1$ to the rank of A:

$$w_p = v_p / D_{pp} \qquad \text{(stationary substitution)}$$

3. For $q = n$ down to 1:
    For q a column containing a (p, q) pivot entry:

$$x_q = w_p - \sum_{j=q+1}^{n} U_{pj} x_j \quad \text{(back substitution)}$$

    For any other q (x_q is then an independent variable):

$$x_q = \text{any desired value}$$

EXAMPLE

Let's consider the case where $A = LDU$ is the product

$$\begin{bmatrix} 1 & 0 & 0 \\ 2 & 1 & 0 \\ 3 & 0 & 1 \\ 2 & 0 & 5 \end{bmatrix} \begin{bmatrix} 2 & 0 & 0 \\ 0 & 5 & 0 \\ 0 & 0 & 3 \end{bmatrix} \begin{bmatrix} 1 & 2 & 3 & 4 & 3 & 3 & 3 & 3 \\ 0 & 0 & 1 & 2 & 2 & 2 & 2 & 2 \\ 0 & 0 & 0 & 0 & 0 & 0 & 1 & 6 \end{bmatrix}$$

of the example above and try to solve $LDUx = y$ for the values

$$\begin{bmatrix} 2 \\ 9 \\ 6 \\ 3 \end{bmatrix}, \begin{bmatrix} 2 \\ 9 \\ 6 \\ 4 \end{bmatrix}$$

for y. Taking

$$y = \begin{bmatrix} 2 \\ 9 \\ 6 \\ 3 \end{bmatrix}$$

first, we try to solve

$$
\begin{bmatrix} 1 & 0 & 0 \\ 2 & 1 & 0 \\ 3 & 0 & 1 \\ 2 & 0 & 5 \end{bmatrix}
\begin{bmatrix} v_1 \\ v_2 \\ v_3 \end{bmatrix}
=
\begin{bmatrix} 2 \\ 9 \\ 6 \\ 3 \end{bmatrix}.
$$

The forward substitution formula gives us

$$
\begin{bmatrix} v_1 \\ v_2 \\ v_3 \end{bmatrix}
=
\begin{bmatrix} 2 \\ 9 - 2 \cdot 2 \\ 6 - 3 \cdot 2 - 0 \end{bmatrix}
=
\begin{bmatrix} 2 \\ 5 \\ 0 \end{bmatrix}.
$$

Testing, we find that

$$
\begin{bmatrix} 1 & 0 & 0 \\ 2 & 1 & 0 \\ 3 & 0 & 1 \\ 2 & 0 & 5 \end{bmatrix}
\begin{bmatrix} 2 \\ 5 \\ 0 \end{bmatrix}
=
\begin{bmatrix} 2 \\ 9 \\ 6 \\ 4 \end{bmatrix}.
$$

So, for $y = \begin{bmatrix} 2 \\ 9 \\ 6 \\ 3 \end{bmatrix}$, *there is no exact solution.* Next, let's try to solve

$$
LDUx = \begin{bmatrix} 2 \\ 9 \\ 6 \\ 4 \end{bmatrix}.
$$

This time, we can solve

$$
Lv = \begin{bmatrix} 2 \\ 9 \\ 6 \\ 4 \end{bmatrix}
$$

by forward substitution, getting $v = \begin{bmatrix} 2 \\ 5 \\ 0 \end{bmatrix}$. We then solve

$$
\begin{bmatrix} 2 & 0 & 0 \\ 0 & 5 & 0 \\ 0 & 0 & 3 \end{bmatrix} w = \begin{bmatrix} 2 \\ 5 \\ 0 \end{bmatrix},
$$

getting $w = \begin{bmatrix} 1 \\ 1 \\ 0 \end{bmatrix}$. Finally, we solve

$$\begin{bmatrix} 1 & 2 & 3 & 4 & 3 & 3 & 3 & 3 \\ 0 & 0 & 1 & 2 & 2 & 2 & 2 & 2 \\ 0 & 0 & 0 & 0 & 0 & 0 & 1 & 6 \end{bmatrix} x = \begin{bmatrix} 1 \\ 1 \\ 0 \end{bmatrix}$$

using the back substitution formula, getting

$$x = \begin{bmatrix} 1-3 \\ 0 \\ 1-0 \\ 0 \\ 0 \\ 0 \\ 0-0 \\ 0 \end{bmatrix} = \begin{bmatrix} -2 \\ 0 \\ 1 \\ 0 \\ 0 \\ 0 \\ 0 \\ 0 \end{bmatrix}.$$

PROBLEMS

NUMERICAL PROBLEMS

1. Find L, D, U for the matrices $\begin{bmatrix} 1 & 0 & 0 \\ 2 & 1 & 0 \\ 3 & 4 & 1 \end{bmatrix}$, $\begin{bmatrix} 1 & 0 \\ 1 & 2 \\ 0 & 0 \end{bmatrix}$, $\begin{bmatrix} 1 & 0 & 0 \\ 1 & 2 & 0 \\ 0 & 0 & 3 \end{bmatrix}$, $\begin{bmatrix} 1 & 2 & 3 \\ 2 & 1 & 3 \end{bmatrix}$.

MORE THEORETICAL PROBLEMS

Easier Problems

2. If A is invertible, show that if $A = LDU$ and $A = MEV$, where L and M are lower triangular with 1's on the diagonal, D and E are diagonal, and U and V are upper triangular with 1's on the diagonal, then $L = M$, $D = E$ and $U = V$.

3. Show that the following algorithm delivers the inverse of a lower triangular matrix L with 1's on the diagonal:
1. Perform the following row operations successively on I for each value of q from 1 to $n - 1$: For each value of p from $q + 1$ to n, Add $(p, q; -a_{pq})$.
2. L^{-1} is the resulting matrix.

4. Show that some invertible matrices A cannot be factored as $A = LDU$, by showing that there are no values a, b, c, d such that

$$\begin{bmatrix} 0 & 1 \\ 1 & 0 \end{bmatrix} = \begin{bmatrix} 1 & 0 \\ a & 1 \end{bmatrix}\begin{bmatrix} b & 0 \\ 0 & c \end{bmatrix}\begin{bmatrix} 1 & d \\ 0 & 1 \end{bmatrix}.$$

Middle-Level Problem

5. Given the factorization $A = LDU$ of an invertible matrix A:

(a) Show that L, D, U are invertible.

(b) Find algorithms for inverting L, D, and U in their own memory, given at most one extra column for work space.

(c) Give an algorithm for using the matrices L^{-1}, D^{-1}, U^{-1} to solve $Ax = y$ for x without explicitly calculating the product $A^{-1} = U^{-1}D^{-1}L^{-1}$.

Harder Problems

6. Show that if A is an $m \times n$ matrix of rank r and $A = LDU$ and $A = MEV$, where L' and M' are echelon $r \times m$ matrices, D and E are invertible diagonal $r \times r$ matrices and U and V are echelon $r \times n$ matrices, then $L = M$, $D = E$ and $U = V$.

11.3. THE ROW REDUCTION ALGORITHM AND ITS INVERSE

In Chapter 2, we gave a method for reducing a matrix to an echelon matrix. We now describe an algorithm similar to that method, but with important differences.

In order to present the algorithm, we first describe how to store an $m \times n$ matrix A in memory and how to perform and keep track of row interchanges. Of course, we must have an $m \times n$ array Memory (R, C) of real numbers in the memory of the computer, to hold the entries of A. By doing this, we do not need to actually move entries when we perform a row or column interchange, we just keep track of the rows and columns by making and updating lists Row and Col of their rows and columns in memory. So if we load the 4×6 matrix

$$A = \begin{bmatrix} 0 & 1 & 0 & 3 & 4 & 5 \\ 1 & 3 & 4 & 5 & 5 & 5 \\ 3 & 3 & 3 & 3 & 3 & 3 \\ 2 & 2 & 2 & 2 & 2 & 2 \end{bmatrix}$$

into memory with Row = [1, 2, 3, 4] and Col = [1, 2, 3, 4, 5, 6], we can keep track of Row, Col, and entries, held in the array Memory (R, C), by the following 4 × 6 *matrix structure* **A**:

$$\mathbf{A} = \begin{array}{c} \\ 1 \\ 2 \\ 3 \\ 4 \end{array} \begin{array}{cccccc} 1 & 2 & 3 & 4 & 5 & 6 \\ \left[\begin{array}{cccccc} 0 & 1 & 0 & 3 & 4 & 5 \\ 1 & 3 & 4 & 5 & 5 & 5 \\ 3 & 3 & 3 & 3 & 3 & 3 \\ 2 & 2 & 2 & 2 & 2 & 2 \end{array}\right] \end{array}.$$

So, we load the 4 × 6 matrix A in the computer by setting up the two lists

Row = [1, 2, 3, 4]
Col = [1, 2, 3, 4, 5, 6]

and putting the entries of A into the array Memory (R, C) according to the lists Row and Col. In this case, Row and Col indicate that the usual order should be used, so the entries occur in the array Memory (R, C) in the same order as they occur in A. So giving **A** is the same as giving the lists Row and Col and the array Memory (R, C). After loading A in memory in this way, suppose that we first interchange rows 3 and 4, then rows 4 and 2, then columns 2 and 4. The matrix A undergoes the following changes:

$$\begin{bmatrix} 0 & 1 & 0 & 3 & 4 & 5 \\ 1 & 3 & 4 & 5 & 5 & 5 \\ 3 & 3 & 3 & 3 & 3 & 3 \\ 2 & 2 & 2 & 2 & 2 & 2 \end{bmatrix}, \quad \begin{bmatrix} 0 & 1 & 0 & 3 & 4 & 5 \\ 1 & 3 & 4 & 5 & 5 & 5 \\ 2 & 2 & 2 & 2 & 2 & 2 \\ 3 & 3 & 3 & 3 & 3 & 3 \end{bmatrix},$$

$$\begin{bmatrix} 0 & 1 & 0 & 3 & 4 & 5 \\ 3 & 3 & 3 & 3 & 3 & 3 \\ 2 & 2 & 2 & 2 & 2 & 2 \\ 1 & 3 & 4 & 5 & 5 & 5 \end{bmatrix}, \quad \begin{bmatrix} 0 & 3 & 0 & 1 & 4 & 5 \\ 3 & 3 & 3 & 3 & 3 & 3 \\ 2 & 2 & 2 & 2 & 2 & 2 \\ 1 & 5 & 4 & 3 & 5 & 5 \end{bmatrix}.$$

To make corresponding changes in the matrix structure **A**, we simply keep updating the lists:

Row = [1, 2, 4, 3]   (after   interchanging rows 3 and 4)
Row = [1, 4, 2, 3]   (after then interchanging rows 4 and 2)
Col = [1, 4, 3, 2, 5, 6] (after then interchanging columns
                           2 and 4).

Let's look at the matrix structure **A** which represents A as it undergoes the corresponding transformations:

$$
\begin{array}{c}
\begin{array}{cccccc} 1 & 2 & 3 & 4 & 5 & 6 \end{array} \\
\begin{array}{c} 1 \\ 2 \\ 3 \\ 4 \end{array}
\begin{bmatrix}
0 & 1 & 0 & 3 & 4 & 5 \\
1 & 3 & 4 & 5 & 5 & 5 \\
3 & 3 & 3 & 3 & 3 & 3 \\
2 & 2 & 2 & 2 & 2 & 2
\end{bmatrix},
\qquad
\begin{array}{cccccc} 1 & 2 & 3 & 4 & 5 & 6 \end{array} \\
\begin{array}{c} 1 \\ 2 \\ 4 \\ 3 \end{array}
\begin{bmatrix}
0 & 1 & 0 & 3 & 4 & 5 \\
1 & 3 & 4 & 5 & 5 & 5 \\
3 & 3 & 3 & 3 & 3 & 3 \\
2 & 2 & 2 & 2 & 2 & 2
\end{bmatrix},
\end{array}
$$

$$
\begin{array}{c}
\begin{array}{cccccc} 1 & 2 & 3 & 4 & 5 & 6 \end{array} \\
\begin{array}{c} 1 \\ 4 \\ 2 \\ 3 \end{array}
\begin{bmatrix}
0 & 1 & 0 & 3 & 4 & 5 \\
1 & 3 & 4 & 5 & 5 & 5 \\
3 & 3 & 3 & 3 & 3 & 3 \\
2 & 2 & 2 & 2 & 2 & 2
\end{bmatrix},
\qquad
\begin{array}{cccccc} 1 & 4 & 3 & 2 & 5 & 6 \end{array} \\
\begin{array}{c} 1 \\ 4 \\ 2 \\ 3 \end{array}
\begin{bmatrix}
0 & 1 & 0 & 3 & 4 & 5 \\
1 & 3 & 4 & 5 & 5 & 5 \\
3 & 3 & 3 & 3 & 3 & 3 \\
2 & 2 & 2 & 2 & 2 & 2
\end{bmatrix}.
\end{array}
$$

Having set up this labeling system, it is time to make things precise.

Definition. An $m \times n$ *matrix structure* **A** consists of Row, Col, and Memory (R, C), where Row is a $1-1$ onto function from $\{1, \ldots, m\}$ to itself, Col is a $1-1$ onto function from $\{1, \ldots, n\}$ to itself, and Memory (R, C) is an $m \times n$ array of real numbers.

When we write

Row $= [1, 4, 2, 3]$,

we mean that Row is the mapping

Row $(1) = 1$, Row $(4) = 2$, Row $(2) = 3$, Row $(3) = 4$

from $\{1, 2, 3, 4\}$ to itself. Similarly, writing

Col $= [1, 4, 3, 2, 5, 6]$

means that Col is the mapping from $\{1, 2, 3, 4, 5, 6]$ to itself such that Col $(s) = t$, where s is in position t in the list $[1, 4, 3, 2, 5, 6]$. So since 2 is in position 4, Col $(2) = 4$.

The 4×6 matrix structure consisting of Row $= [1, 4, 2, 3]$, Col $= [1, 4, 3, 2, 5, 6]$, and our 4×6 array Memory (R, C) is just

$$
\mathbf{A} =
\begin{array}{c}
\begin{array}{cccccc} 1 & 4 & 3 & 2 & 5 & 6 \end{array} \\
\begin{array}{c} 1 \\ 4 \\ 2 \\ 3 \end{array}
\begin{bmatrix}
0 & 1 & 0 & 3 & 4 & 5 \\
1 & 3 & 4 & 5 & 5 & 5 \\
3 & 3 & 3 & 3 & 3 & 3 \\
2 & 2 & 2 & 2 & 2 & 2
\end{bmatrix}.
\end{array}
$$

Matrices get put in, or taken from, matrix structures according to

Definition.   The $m \times n$ matrix A corresponding to the $m \times n$ data structure **A** consisting of the lists Row, Col, and the $m \times n$ array Memory (R, C) is the $m \times n$ matrix A whose (r, s) entry A_{rs} is given by the formula

$$A_{rs} = \text{Memory (Row } (r), \text{ Col } (s)).$$

For example, the 4×6 matrix A corresponding to the 4×6 matrix structure **A** described earlier is the matrix

$$A = (A_{rs}) = (\text{Memory (Row } (r), \text{ Col } (s)),$$

which can easily be read from **A** when we write it out to look at:

$$\mathbf{A} = \begin{array}{c} \\ 1 \\ 4 \\ 2 \\ 3 \end{array}\begin{array}{c} 1\ \ 4\ \ 3\ \ 2\ \ 5\ \ 6 \\ \begin{bmatrix} 0 & 1 & 0 & 3 & 4 & 5 \\ 1 & 3 & 4 & 5 & 5 & 5 \\ 3 & 3 & 3 & 3 & 3 & 3 \\ 2 & 2 & 2 & 2 & 2 & 2 \end{bmatrix}\end{array}, \quad A = \begin{bmatrix} 0 & 3 & 0 & 1 & 4 & 5 \\ 3 & 3 & 3 & 3 & 3 & 3 \\ 2 & 2 & 2 & 2 & 2 & 2 \\ 1 & 5 & 4 & 3 & 5 & 5 \end{bmatrix}.$$

For example, A_{43} = Memory (Row (4), Col (3)) = Memory (2, 3) is read from the matrix structure by going to the row of memory marked 4 (which is row 2 of memory) and to the column of memory marked 3 (which is column 3 of memory) and getting the entry $A_{43} = 4$ in that row and column.

Of course, our objective in all of this has been to represent $m \times n$ matrices by $m \times n$ matrix structures and row and column interchanges on $m \times n$ matrices by corresponding operations on $m \times n$ matrix structures. We now do the latter in

Definition.   To interchange rows (respectively, columns) p and q of an $m \times n$ matrix structure **A** consisting of Row, Col, and Memory, just interchange the values of Row (p) and Row (q) [respectively, Col (p), Col (q)].

In our example above, we interchanged rows 4 and 2 when Row was Row = [1, 2, 4, 3]. The result there was that the values Row (4) = 3 and Row (2) = 2 of Row = [1, 2, 4, 3] were interchanged, resulting in the new list Row = [1, 4, 2, 3], the new values 2 and 3 for Row (4) and Row (2) having been obtained by interchanging the old ones, 3 and 2.

We can now give the row reduction algorithm. This algorithm is similar to the method given in Chapter 2 for reducing a matrix to an echelon matrix, but we've made some important changes and added some new features:

1. Our operations are performed on an $m \times n$ matrix structure rather than an $m \times n$ matrix, to make it easy to perform them and keep track of row interchanges.

2. Where "0" occurs in the method in Chapter 2, we now say "less in absolute value than Epsilon" (where Epsilon is a fixed small positive value which depends on the computer to be used).

3. Instead of looking for the "first nonzero value if any" in the rest of a given column, we look for the "first value that is the largest in absolute value" in rest of that column.

4. As we reduce to the echelon matrix U, we keep track of the pivot entries and use the freed memory on and below them to store the entries of D and L. In particular, we record the number of pivot entries in the variable Rank A.

Of these changes, (2) and (3) lead to increased *numerical stability*. In other words, these changes are important if we prefer not to divide by numbers so small as to lead to serious errors in the computations. The others enable us to construct, store, and retrieve the factors L, D, and U of A. They also enable us to reverse the algorithm and restore A.

Of course, the operations on the entries A_{rs} of A performed in this algorithm are really performed as operations on Row, Col, and the array Memory (R, C), the correspondence of entries being

$A_{rs} =$ Memory (Row (r), Col (s)).

This algorithm does not involve column interchanges and, in fact, neither do the other algorithms considered in this chapter. So, henceforth, we take Col to be the identity list Col $(s) = s$ and we do not label columns of a matrix structure.

Algorithm to row reduce an $m \times n$ matrix A to an echelon matrix U

Starting with (p, q) equal to $(1, 1)$ and continuing as long as $p \leq m$ and $q \leq n$, do the following:

1. *Get the first largest (in absolute value) (p', q) entry*

$$a = A_{p'q} = \text{Memory (Row } (p'), \text{ Col } (q)) \text{ of } A$$

for $p' \geq p$.

2. *If its absolute value is less than* Epsilon, *then we decrease the value of p by 1 (so later we increase p and q by 1 we try again in the same row and next column) but otherwise we call it a* pivot entry *and we do the following:*

   (a) *We record that the (p, q) entry is the pivot entry in row p; we do this using a function* PivotList *by setting*

   $$\text{PivotList } (p) = q.$$

   (b) *If $p' > p$, we interchange rows p and p' (by interchanging the values of* Row (p) *and* Row (p')*).*

   (c) *For each row t with $t > p$, we perform the elementary row operation*

   $$\text{Add } (t, p; -A_{tq}/a) \text{ (add } -A_{tq}/a \text{ times row p to row t),}$$

   where A_{tq} is the current (t, q) entry of A; in doing this, we do not disturb the area in which we have already stored multipliers; we then record the operation by writing the multiplier A_{tq}/a as entry (t, q) of A; (since we know that there should be a 0 there, we lose no needed information when we take over this entry as storage for our growing record).

   (d) *We perform the elementary row operation*

   $$\text{Multiply}(p; 1/a) \text{ (divide the entries of row p by a);}$$

   we then record the operation by writing the divisor $d_p = a$ as entry (p, q) of A; (since we know there should be a 1 there, we lose no needed information when we take over this entry for our growing record).

3. *We increase the values of p and q by 1 (on to the next row and column . . .).*

After all this has been done, we record that row $p - 1$ was the last nonzero row by setting the value RankA equal to $p - 1$.

This algorithm changes a matrix structure representing A in unfactored format to a matrix structure representing A in LDU factored format. Since we keep track of the pivots and the number $r = \text{Rank}A$ of pivots, we can get the entries of the echelon matrix U, the diagonal matrix D, and the lower triangular matrix L:

1. L is the $m \times r$ matrix whose (p, q) entry is

$$A_{p\text{PivotList}(q)} = \text{Memory (Row } (p), \text{ Col (PivotList } (q)))$$

for $p > q$, 1 for $p = q$, and 0 for $p < q$;

2. D is the $r \times r$ diagonal matrix with (q, q) entry

$$A_{q\text{PivotList}(q)} = \text{Memory (Row } (q), \text{ Col (PivotList } (q)))$$

for $1 \le q \le r$;

3. U is the $r \times n$ echelon matrix whose (p, q) entry is

$$A_{p\text{PivotList}(q)} = \text{Memory (Row } (p), \text{ Col (PivotList } (q)))$$

for $p < q$, 1 for $p = q$ and 0 for $p > q$.

The $A = LDU$ factorization of Section 13.2 is then replaced by a factorization $A = PLDU$, where P is the *permutation matrix corresponding to the list* Row, defined by

4. P is the $m \times m$ matrix whose (p, q) entry is 1 if $p = \text{Row } (q)$ and 0 otherwise.

EXAMPLE

$$\text{Let } A = \begin{bmatrix} 3 & 7 & 10 & 7 \\ 6 & 12 & 18 & 0 \\ 2 & 9 & 6 & 10 \end{bmatrix} \text{ be represented by the matrix}$$

structure

$$A = \begin{matrix} 1 \\ 2 \\ 3 \end{matrix} \begin{bmatrix} 3 & 7 & 10 & 7 \\ 6 & 12 & 18 & 0 \\ 2 & 9 & 6 & 10 \end{bmatrix} \quad \text{(unfactored format for } A\text{)},$$

with Row $= [1, 2, 3]$. Then the row operations

Interchange $(1, 2)$,
Add $(2, 1; -\frac{1}{3})$,
Add $(3, 1; -\frac{1}{2})$,
Interchange $(2, 3)$,
Add $(3, 2, -\frac{1}{5})$

reduce the matrix structure A to

$$V = \begin{matrix} 3 \\ 1 \\ 2 \end{matrix} \begin{bmatrix} 0 & 0 & 1 & 5 \\ 6 & 12 & 18 & 0 \\ 0 & 5 & 0 & 10 \end{bmatrix},$$

which represents the upper triangular matrix

$$V = \begin{bmatrix} 6 & 12 & 18 & 0 \\ 0 & 5 & 0 & 10 \\ 0 & 0 & 1 & 5 \end{bmatrix}.$$

How do we get this, and what is the multiplier matrix? Writing the pivots in boldface, as in the earlier example, the successive matrices encountered in the reduction are:

$$\begin{matrix} \mathbf{1} \\ \mathbf{2} \\ \mathbf{3} \end{matrix}\begin{bmatrix} 3 & 7 & 10 & 7 \\ \mathbf{6} & 12 & 18 & 0 \\ 2 & 9 & 6 & 10 \end{bmatrix}, \quad \begin{matrix} \mathbf{2} \\ \mathbf{1} \\ \mathbf{3} \end{matrix}\begin{bmatrix} 3 & 7 & 10 & 7 \\ \mathbf{6} & 12 & 18 & 0 \\ 2 & 9 & 6 & 10 \end{bmatrix},$$

$$\begin{matrix} \mathbf{2} \\ \mathbf{1} \\ \mathbf{3} \end{matrix}\begin{bmatrix} \frac{1}{2} & 1 & 1 & 7 \\ \mathbf{6} & 12 & 18 & 0 \\ 2 & 9 & 6 & 10 \end{bmatrix}, \quad \begin{matrix} \mathbf{2} \\ \mathbf{1} \\ \mathbf{3} \end{matrix}\begin{bmatrix} \frac{1}{2} & 1 & 1 & 7 \\ \mathbf{6} & 12 & 18 & 0 \\ \frac{1}{3} & 5 & 0 & 10 \end{bmatrix},$$

$$\begin{matrix} \mathbf{3} \\ \mathbf{1} \\ \mathbf{2} \end{matrix}\begin{bmatrix} \frac{1}{2} & 1 & 1 & 7 \\ \mathbf{6} & 12 & 18 & 0 \\ \frac{1}{3} & 5 & 0 & 10 \end{bmatrix}, \quad \begin{matrix} \mathbf{3} \\ \mathbf{1} \\ \mathbf{2} \end{matrix}\begin{bmatrix} \frac{1}{2} & \frac{1}{5} & 1 & 5 \\ \mathbf{6} & 12 & 18 & 0 \\ \frac{1}{3} & 5 & 0 & 10 \end{bmatrix}.$$

Now **A** has been reduced to

$$V = \begin{matrix} \mathbf{3} \\ \mathbf{1} \\ \mathbf{2} \end{matrix}\begin{bmatrix} 0 & 0 & 1 & 5 \\ \mathbf{6} & 12 & 18 & 0 \\ 0 & 5 & 0 & 10 \end{bmatrix}$$

and A to

$$V = \begin{bmatrix} \mathbf{6} & 12 & 18 & 0 \\ 0 & \mathbf{5} & 0 & 10 \\ 0 & 0 & \mathbf{1} & 5 \end{bmatrix};$$

and the matrix structure of multipliers is

$$\mathbf{L} = \begin{matrix} \mathbf{3} \\ \mathbf{1} \\ \mathbf{2} \end{matrix}\begin{bmatrix} \frac{1}{2} & \frac{1}{5} & 1 \\ 1 & 0 & 0 \\ \frac{1}{3} & 1 & 0 \end{bmatrix},$$

with corresponding matrix

$$L = \begin{bmatrix} 1 & 0 & 0 \\ \frac{1}{3} & 1 & 0 \\ \frac{1}{2} & \frac{1}{5} & 1 \end{bmatrix}.$$

The matrix P is obtained by listing rows 1, 2, 3 of the identity matrix as the rows 3, 1, 2 of memory, that is, P is the matrix

$$P = \begin{bmatrix} 0 & 0 & 1 \\ 1 & 0 & 0 \\ 0 & 1 & 0 \end{bmatrix}$$

obtained from the matrix structure

$$\mathbf{L} = \begin{matrix} 3 \\ 1 \\ 2 \end{matrix} \begin{bmatrix} \frac{1}{2} & \frac{1}{5} & 1 \\ 1 & 0 & 0 \\ \frac{1}{3} & 1 & 0 \end{bmatrix}$$

by removing the row labels and the multipliers.

Suppose that we now perform the product of the matrices

$$P = \begin{bmatrix} 0 & 0 & 1 \\ 1 & 0 & 0 \\ 0 & 1 & 0 \end{bmatrix}, L = \begin{bmatrix} 1 & 0 & 0 \\ \frac{1}{3} & 1 & 0 \\ \frac{1}{2} & \frac{1}{5} & 1 \end{bmatrix}, V = \begin{bmatrix} 6 & 12 & 18 & 0 \\ 0 & 5 & 0 & 10 \\ 0 & 0 & 1 & 5 \end{bmatrix}.$$

For LV we get

$$LV = \begin{bmatrix} 6 & 12 & 18 & 0 \\ 2 & 9 & 6 & 10 \\ 3 & 7 & 10 & 7 \end{bmatrix}.$$

Multiplying this by $P = \begin{bmatrix} 0 & 0 & 1 \\ 1 & 0 & 0 \\ 0 & 1 & 0 \end{bmatrix}$ then rearranges the rows,

giving us back our

$$A = \begin{bmatrix} 3 & 7 & 10 & 7 \\ 6 & 12 & 18 & 0 \\ 2 & 9 & 6 & 10 \end{bmatrix}.$$

If we want to factor V further as DU, we continue with the matrix structure where we left off,

$$\begin{matrix} 3 \\ 1 \\ 2 \end{matrix} \begin{bmatrix} \frac{1}{2} & \frac{1}{5} & 1 & 5 \\ 6 & 12 & 18 & 0 \\ \frac{1}{3} & 5 & 0 & 10 \end{bmatrix} \quad (LV \text{ factored format for } A),$$

and reduce it further so that it contains all three factors L, D, U as in the example above. The only change in the matrix structure is that the upper entries of V are changed to the upper entries of U, by dividing them row by row by the pivots:

$$\begin{matrix} 3 \\ 1 \\ 2 \end{matrix} \begin{bmatrix} \frac{1}{2} & \frac{1}{5} & 1 & 5 \\ 6 & 2 & 3 & 0 \\ \frac{1}{3} & 5 & 0 & 2 \end{bmatrix} \quad (LDU \text{ factored format for } A).$$

Of course, since $A = PLV$ and $V = DU$, we can get A back as before as

$$A = PLDU.$$

Since this reconstruction of A from L, D, U and the permutation matrix corresponding to the list Row built during reduction works the same way for any matrix A, we have

Theorem 11.3.1.   Suppose that A is represented by the matrix structure A with Row $= [1, \ldots, m]$, which is reduced to its LDU format with list Row updated during reduction to record the affect of interchanges. Then A is obtained by performing the product $PLDU$, where P, L, D, U are as described above.

If we have used the reduction algorithm to change a matrix structure representing A in unfactored format to a matrix structure representing A in LDU factored format, we can reverse the algorithm and return the matrix structure to matrix format by

Algorithm to undo row reduction of *A* to an echelon matrix *U*

Starting with p $=$ RankA, and continuing as long as p $\geq$ 1, do the following:

1. *Let q $=$ PivotList (p) and let d_p be the current (p, q) entry of A.*
2. *Multiply row p of U by d_p, ignoring the storage area on and below the pivot entries.*
3. *For each row t $> $ p of A, perform the elementary row operation Add (t, p; m_{tq}), where m_{tq} is the multiplier stored in row t below the pivot d_p of column q. (When doing this, first set the current (t, q) entry to 0.)*
4. *Decrease p by 1.*

Rearrange the rows by setting Row $=$ [1, ..., n].

EXAMPLE

The matrix structures encountered successively when the row reduction algorithm is applied to the matrix A in the above example are

$$
\begin{matrix}
1 \\ 2 \\ 3
\end{matrix}
\begin{bmatrix}
3 & 7 & 10 & 7 \\
6 & 12 & 18 & 0 \\
2 & 9 & 6 & 10
\end{bmatrix}
\quad
\begin{matrix}
2 \\ 1 \\ 3
\end{matrix}
\begin{bmatrix}
3 & 7 & 10 & 7 \\
6 & 12 & 18 & 0 \\
2 & 9 & 6 & 10
\end{bmatrix}
\quad
\begin{matrix}
2 \\ 1 \\ 3
\end{matrix}
\begin{bmatrix}
\frac{1}{2} & 1 & 1 & 7 \\
6 & 12 & 18 & 0 \\
2 & 9 & 6 & 10
\end{bmatrix}
$$

PivotList (1) = 1

$$
\begin{matrix}
2 \\ 1 \\ 3
\end{matrix}
\begin{bmatrix}
\frac{1}{2} & 1 & 1 & 7 \\
6 & 12 & 18 & 0 \\
\frac{1}{3} & 5 & 0 & 10
\end{bmatrix}
\quad
\begin{matrix}
2 \\ 1 \\ 3
\end{matrix}
\begin{bmatrix}
\frac{1}{2} & 1 & 1 & 7 \\
6 & 2 & 3 & 0 \\
\frac{1}{3} & 5 & 0 & 10
\end{bmatrix}
\quad
\begin{matrix}
3 \\ 1 \\ 2
\end{matrix}
\begin{bmatrix}
\frac{1}{2} & 1 & 1 & 7 \\
6 & 2 & 3 & 0 \\
\frac{1}{3} & 5 & 0 & 10
\end{bmatrix}
$$

PivotList (32) = 2

$$
\begin{matrix}
3 \\ 1 \\ 2
\end{matrix}
\begin{bmatrix}
\frac{1}{2} & \frac{1}{5} & 1 & 5 \\
6 & 2 & 3 & 0 \\
\frac{1}{3} & 5 & 0 & 10
\end{bmatrix}
\quad
\begin{matrix}
3 \\ 1 \\ 2
\end{matrix}
\begin{bmatrix}
\frac{1}{2} & \frac{1}{5} & 1 & 5 \\
6 & 2 & 3 & 0 \\
\frac{1}{3} & 5 & 0 & 2
\end{bmatrix}
\quad
\begin{matrix}
3 \\ 1 \\ 2
\end{matrix}
\begin{bmatrix}
\frac{1}{2} & \frac{1}{5} & 1 & 5 \\
6 & 2 & 3 & 0 \\
\frac{1}{3} & 5 & 0 & 2
\end{bmatrix}
$$

PivotList (3) = 3
RankA = 3

The same matrix structures are encountered in reverse order when the algorithm to undo row reduction of A is applied. Since RankA = 3, the algorithm starts with $p = 3$, $q = $ PivotList (3) = 3 and $d_3 = 1$. It then goes to $p = 3 - 1 = 2$, $q = $ PivotList (2) = 2, $d_2 = 5$ and multiplies the rest of row 2 by 5. It sets the (3, 2) entry $\frac{1}{5}$ to 0 and performs Add (3, 2; $\frac{1}{5}$) (ignoring the entries still holding multipliers). Finally, it goes to $p = 2 - 1 = 1$, $q = $ PivotList (1) = 1, $d_1 = 6$ and multiplies the rest of row 1 by 6. It sets the (2, 1) entry $\frac{1}{3}$ to 0 and performs Add (2, 1; $\frac{1}{3}$). Then it sets the (3, 1) entry $\frac{1}{2}$ to 0 and performs Add (3, 1; $\frac{1}{2}$). Finally, it resets Row = [1, 2, 3].

PROBLEMS

NUMERICAL PROBLEMS

1. Find L, D, U, and Row for the matrix

$$
\begin{bmatrix}
0 & 0 & 3 & 0 & 0 \\
0 & 4 & 0 & 0 & 0 \\
5 & 0 & 0 & 0 & 0 \\
0 & 0 & 0 & 0 & 6 \\
0 & 0 & 0 & 7 & 0
\end{bmatrix}.
$$

of PivotList (p), $1 \le p \le 5$?

2. If the matrix A is symmetric, that is, $A = A'$, and no interchanges take place when the row reduction algorithm is applied to A, show that in the resulting factorization $A = LDU$, L is the transpose of U.

3. In Problem 2, show that the assumption that no interchanges take place is necessary.

11.4. BACK AND FORWARD SUBSTITUTION. SOLVING $Ax = y$

Now that we can use the row reduction algorithm to go from the $m \times n$ matrix A to the matrices L, D, U and the list Row, which was built from the interchanges during the reduction, we ask: How do we use P, L, D, U, and Row to solve $Ax = y$? By Theorem 11.8.1, $A = PLDU$. So we can break up the problem of solving $Ax = y$ into parts, namely solving $Pu = y$, $Lv = u$, $Dw = v$, and $Ux = w$ for u, v, w, x. The only thing that is new here is solving $Pu = y$ for u. So let's look at this in the case of the preceding example. There, we have Row $= [3, 1, 2]$ and P is the corresponding permutation matrix

$$P = \begin{bmatrix} 0 & 0 & 1 \\ 1 & 0 & 0 \\ 0 & 1 & 0 \end{bmatrix}.$$

So, solving

$$\begin{bmatrix} 0 & 0 & 1 \\ 1 & 0 & 0 \\ 0 & 1 & 0 \end{bmatrix} \begin{bmatrix} u_1 \\ u_2 \\ u_3 \end{bmatrix} = \begin{bmatrix} y_1 \\ y_2 \\ y_3 \end{bmatrix}$$

for $\begin{bmatrix} u_1 \\ u_2 \\ u_3 \end{bmatrix}$, we get

$$\begin{bmatrix} u_1 \\ u_2 \\ u_3 \end{bmatrix} = \begin{bmatrix} y_2 \\ y_3 \\ y_1 \end{bmatrix} = \begin{bmatrix} y_{\text{Row}(1)} \\ y_{\text{Row}(2)} \\ y_{\text{Row}(3)} \end{bmatrix}$$

What this means is that we need make only one alteration in our earlier solution of $Ax = y$, namely, replace $\begin{bmatrix} y_1 \\ y_2 \\ y_3 \end{bmatrix}$ by $\begin{bmatrix} y_{Row(1)} \\ y_{Row(2)} \\ y_{Row(3)} \end{bmatrix}$. So, we now

have the

Algorithm for solving $Ax = y$ for x exactly

Use the row reduction algorithm to get the matrices L, D, U, and the list Row. Given a particular y in the column space of A, do the following:

1. Solve $Lv = y$ by the *forward substitution equations*

$$v_p = y_{Row(p)} - \sum_{j=1}^{p-1} L_{pj}v_j$$

for $p = 1$ to the rank of A.
2. Solve $Dw = v$ by the formula

$$w_p = v_p/D_{pp}$$

for $p = 1$ to the rank of A.
3. Solve $Ux = v$ by the *back substitution equations:*

$$x_q = w_p - \sum_{j=q+1}^{n} U_{pj}x_j \text{ if column } q \text{ contains a pivot entry;}$$

$$x_q = 0 \qquad\qquad \text{if column } q \text{ contains no pivot entry}$$

for $1 \le q \le n$.

When y is not in the column space of A, the above algorithm for solving $Ax = y$ exactly cannot be used. Instead, we can solve $Ax = y$ approximately by solving the normal equation $A'Ax = A'y$ exactly by the above algorithm. Since $A'y$ is in the column space of $A'A$, by our earlier discussion of the normal equation, this is always possible. So, we have

Algorithm for solving $Ax = y$ for x approximately

Use the algorithm for solving $A'Ax = A'y$ for x exactly.

If we need to solve $Ax = y$ approximately for many different vectors y, it is more efficient first to find A^- and then to use it to get $x = A^-y$ for each y. In the next section, we give an algorithm for finding A^- for any A.

We now turn to the important special case when the columns of A are linearly independent. In this case, we can solve the normal equation $A'Ax = A'y$ and find A^- efficiently by a simple algorithm involving the Gram–Schmidt Orthogonalization Process described in Section 4.3.

From the columns $v_1, \ldots, v_k$ of A, the Gram–Schmidt Orthogonalization Process gives us orthogonal vectors $w_1, \ldots, w_k$, where

$$w_s = v_s - \frac{(v_s, w_{s-1})}{(w_{s-1}, w_{s-1})} w_{s-1} - \cdots - \frac{(v_s, w_1)}{(w_1, w_1)} w_1$$

or

$$v_s = w_s + \frac{(v_s, w_{s-1})}{(w_{s-1}, w_{s-1})} w_{s-1} + \cdots + \frac{(v_s, w_1)}{(w_1, w_1)} w_1$$

for $1 \le s \le k$. Letting

$$u_1 = (1/|w_1|)w_1, \ldots, u_k = (1/|w_k|)w_k,$$

and setting

$$b_{ss} = |w_s|,$$

$$b_{rs} = \frac{(v_s, w_r)}{(w_r, w_r)} |w_r| \text{ for } r < s, \ 1 \le s \le k,$$

we can rewrite this as

$$v_s = b_{1s}u_1 + \cdots + b_{s-1,s}u_{s-1} + b_{ss}u_s$$

for $1 \le s \le k$. Letting Q be the $m \times k$ matrix whose columns are the orthonormal vectors $u_1, \ldots, u_k$ and R be the $k \times k$ matrix whose (r, s) entry is b_{rs} for $r \le s$ and 0 for $r > s$, these equations imply that

$$A = QR.$$

(Prove!) This is the so-called *QR factorization* of A as product of a matrix Q with orthonormal columns and an invertible upper triangular matrix R. (We leave it as an exercise for the reader to show that there is only one such factorization of A). So

applying the Gram–Schmidt Orthogonalization Process to the columns of A to get orthonormal vectors $u_1, \ldots, u_k$ in the manner above gives us the QR factorization $A = QR$.

For example, if $A = \begin{bmatrix} 1 & 2 \\ 3 & 4 \end{bmatrix}$, we get the orthogonal vectors

$$w_1 = \begin{bmatrix} 1 \\ 3 \end{bmatrix}$$

$$w_2 = \begin{bmatrix} 2 \\ 4 \end{bmatrix} - \frac{1 \cdot 2 + 3 \cdot 4}{1 \cdot 1 + 3 \cdot 3}\begin{bmatrix} 1 \\ 3 \end{bmatrix} = \begin{bmatrix} \frac{3}{5} \\ -\frac{1}{5} \end{bmatrix}$$

whose lengths are $|w_1| = \sqrt{10}$ and $|w_2| = (\frac{1}{5})\sqrt{10}$. From these, we get

the orthonormal vectors $u_1 = (1/\sqrt{10})\begin{bmatrix} 1 \\ 3 \end{bmatrix}$ and $u_2 = (1/\sqrt{10})\begin{bmatrix} 3 \\ -1 \end{bmatrix}$.

The equations $v_s = b_{1s}u_1 + \cdots + b_{ss}u_s$ $(1 \le s \le 2)$ are then

$$\begin{bmatrix} 1 \\ 3 \end{bmatrix} = \sqrt{10}u_1$$

$$\begin{bmatrix} 2 \\ 3 \end{bmatrix} = (\tfrac{7}{5})\sqrt{10}\ u_1 + (\tfrac{1}{5})\sqrt{10}\ u_2,$$

the matrices Q, R are

$$Q = \left(\frac{1}{\sqrt{10}}\right)\begin{bmatrix} 1 & 3 \\ 3 & -1 \end{bmatrix}$$

and

$$R = \sqrt{10}\begin{bmatrix} 1 & \frac{7}{5} \\ 0 & \frac{1}{5} \end{bmatrix},$$

and the *QR* factorization of $\begin{bmatrix} 1 & 2 \\ 3 & 4 \end{bmatrix}$ is

$$\begin{bmatrix} 1 & 2 \\ 3 & 4 \end{bmatrix} = \left(\left(\frac{1}{\sqrt{10}}\right)\begin{bmatrix} 1 & 3 \\ 3 & -1 \end{bmatrix}\right)\left(\sqrt{10}\begin{bmatrix} 1 & \frac{7}{5} \\ 0 & \frac{1}{5} \end{bmatrix}\right).$$

Given the *QR* factorization $A = QR$ for a matrix A with independent columns, the normal equation

$$A'Ax = A'y$$

can be solved for x easily and efficiently. Replacing A by QR in the normal equation $A'Ax = A'y$, we get $R'Q'QRx = R'Q'y$. Since R and R' are invertible and the columns of Q are orthonormal, this simplifies to

$$Rx = Q'y$$

and

$$x = R^{-1}Q'y.$$

(Prove!) Since R is upper triangular and invertible, the equation $Rx = Q'y$ can be solved for x using the back-substitution equations

$$b_{qq}x_q = z_q - \sum_{j=q+1}^{k} b_{qj}x_j,$$

where z_q is entry q of $Q'y$ ($1 \le q \le k$). Moreover, since R is invertible, *there is only one such solution x.* So x is the shortest approximate solution to $Ax = y$, from which it follows that $A^- = R^{-1}Q'$.

Computing the inverse R^{-1} can be done very easily, since R is upper triangular. One simply finds the columns $c_1, \ldots, c_k$ of R^{-1} as the solutions c_s to the equations $Rc_s = e_s$ (column s of the $k \times k$ identity matrix) using back-substitution equations:

$$b_{qq}c_{qs} = 0 - \sum_{j=q+1}^{k} b_{qj}c_{js} \qquad \text{for } q \ne s$$

$$b_{ss}c_{ss} = 1 - \sum_{j=q+1}^{k} b_{sj}c_{js}$$

for each s with $1 \le s \le k$. As it turns out, $c_{js} = 0$ for $j > s$. So the back-substitution equations above simplify to the equations

$$b_{qq}c_{qs} = -\sum_{j=q+1}^{s} b_{qj}c_{js} \qquad \text{for } q < s$$
$$b_{ss}c_{ss} = 1$$
$$c_{qs} = 0 \qquad\qquad\qquad \text{for } q > s$$

for $1 \le s \le k$. We leave it as an exercise for the reader to prove directly that the entries c_{qs} of the inverse R^{-1} satisfy these equations and are completely determined by them.

We summarize all of this by formulating the following algorithms.

Algorithm for solving $Ax = y$ for x approximately when the columns of A are linearly independent

1. *Use the QR factorization $A = QR$ to get the equation $Rx = Q'y$ (which replaces the normal equation $A'Ax = A'y$).*

2. *Solve $Rx = Q'y$ for x by the back-substitution equations*

$$b_{qq}x_q = z_q - \sum_{j=q+1}^{k} b_{qj}x_j,$$

where z_q is entry q of $Q'y$ for $1 \le q \le k$.

Algorithm for finding R^{-1} for an invertible upper triangular matrix R

Letting b_{rs} denote the (r, s) entry of an invertible upper triangular $k \times k$ matrix R, the entries c_{rs} of R^{-1} are determined by the back-substitution equations

$$b_{rr}c_{rs} = -\sum_{j=r+1}^{s} b_{rj}c_{js} \qquad \text{for } r < s$$

$$b_{ss}c_{ss} = 1$$

$$c_{rs} = 0 \qquad\qquad \text{for } r > s$$

for $1 \le s \le k$.

Algorithm for finding A^- when the columns of A are linearly independent

1. *Use the QR factorization $A = QR$ to get Q and R.*
2. *Find R^{-1} by the algorithm above.*
3. *Then $A^{-1} = R^{-1}Q'$.*

PROBLEMS

NUMERICAL PROBLEMS

1. Suppose that the row reduction algorithm, applied to A, gives us

$$L = \begin{bmatrix} 1 & 0 & 0 \\ 2 & 1 & 0 \\ 3 & 0 & 1 \\ 2 & 0 & 5 \end{bmatrix},$$

$$D = \begin{bmatrix} 2 & 0 & 0 \\ 0 & 5 & 0 \\ 0 & 0 & 3 \end{bmatrix},$$

$$U = \begin{bmatrix} 1 & 2 & 3 & 4 & 3 & 3 & 3 & 3 \\ 0 & 0 & 1 & 2 & 2 & 2 & 2 & 2 \\ 0 & 0 & 0 & 0 & 0 & 0 & 1 & 6 \end{bmatrix},$$

and Row $= [2, 3, 4, 1]$. Then solve $Ax = y$ (or show that there is no solution) for the vectors

$$\begin{bmatrix} 1 \\ 0 \\ 1 \\ 1 \end{bmatrix}, \begin{bmatrix} 1 \\ 0 \\ 1 \\ 5 \end{bmatrix}$$

by modifying the discussion in the related example in Section 11.2.

2. Find the inverse of $\begin{bmatrix} 2 & 3 & 4 \\ 0 & 5 & 6 \\ 0 & 0 & 7 \end{bmatrix}$ by the method described in this section.

3. Find the QR decomposition of the matrix $\begin{bmatrix} 1 & 3 \\ 2 & 4 \end{bmatrix}$ and compare it with the example given in this section.

4. Using the QR decomposition, find the approximate inverse of the matrix $\begin{bmatrix} 1 & 2 \\ 0 & 0 \\ 3 & 4 \end{bmatrix}$.

MORE THEORETICAL PROBLEMS

Easier Problems

5. For an invertible upper triangular $k \times k$ matrix R with entries b_{rs}, show that the entries c_{rs} of R^{-1} satisfy the equations

$$b_{qq}c_{qs} = -\sum_{j=q+1}^{s} b_{qj}c_{js} \qquad \text{for } q < s$$

$$b_{ss}c_{ss} = 1$$

$$c_{qs} = 0 \qquad\qquad \text{for } q > s$$

for $1 \le s \le k$, and that they are completely determined by them.

Middle-level Problems

6. Suppose that vectors $v_1, \ldots, v_k \in F^{(m)}$ are expressed as linear combinations

$$v_s = b_{1s}u_1 + \cdots + b_{ks}u_k \qquad (1 \le s \le k)$$

of vectors $u_1, \ldots, u_k \in F^{(m)}$. Letting A be the $m \times k$ matrix whose columns are $v_1, \ldots, v_k$, Q be the $m \times k$ matrix whose columns are $u_1, \ldots, u_k$, and R be the $k \times k$ matrix whose (r, s) entry is b_{rs} for $1 \le r, s \le k$, show that $A = QR$.

Harder Problems

7. Show that if $QR = ST$, where Q and S are $m \times k$ matrices whose columns are orthonormal and R and T are invertible upper triangular $k \times k$ matrices, then $Q = S$ and $R = T$.

11.5. APPROXIMATE INVERSE AND PROJECTION ALGORITHMS

In Chapter 10 we saw how to find the approximate solution x to an equation $Ax = y$, where A is an $m \times n$ real matrix. To do this efficiently for each of a larger number of different y, we should first get A^- and then compute x as $x = A^-y$. How do we get A^-? In principle, we can get A^- by using the methods of Chapter 10 to calculate each of its columns A^-e_s (where e_s is column s of the identity matrix) as the approximate solution x_s to the equation $Ax_s = e_s$. However, there are more efficient methods, which are based on Theorem 10.3.5 and diagonalization of a symmetric matrix. Unfortunately, however, these methods are also somewhat complicated.

To avoid the complications, we have worked out an efficient new method for finding A^- which uses only elementary row operations. This method, a variation of the method in Chapter 3 for finding the inverse of an invertible matrix, is based on two facts. the first of these is that the matrix $\text{Proj}_{A'(\mathbb{R}^{(m)})}$ is just $J'J$, where J is gotten by row reducing $A'A$ to an orthonormalized matrix in the sense of

Definition. An *orthonormalized matrix* is an $m \times n$ matrix J satisfying the following conditions:

1. Each nonzero row of J has length 1.
2. Any two different nonzero rows of J are orthogonal.

We can always row reduce a matrix to an orthonormalized upper triangular matrix. How? First, reduce it to an echelon matrix. Then orthonormalize the nonzero rows in the reverse order $r, \ldots, 1$ by performing the following row operations on A for each value of k from r down to 1:

1. Multiply $(k; 1/u_k)$, where u_k is the current length of row k.
2. For each value of q from $k - 1$ down to 1, Add $(q, k; -v_{kq})$, where v_{kq} is the inner product of the current rows k and q.

How do we show that $\text{Proj}_{A'(\mathbb{R}^{(m)})}$ equals $J'J$? First, we need some preliminary tools.

Definition. A matrix $P \in M_n(\mathbb{R})$ is a *projection* if $P = P'$ and $P^2 = P$.

Theorem 11.5.1. If P is a projection, then $P = \text{Proj}_{P(\mathbb{R}^{(n)})}$.

Proof. Let's denote the column space of P by W. Let $v \in \mathbb{R}^{(n)}$ and write $v = v_1 + v_2$, where $v_1 \in W$, $v_2 \in W^\perp$. Then $v_1 = Pu$ for some u, so that $Pv_1 = P^2u = Pu = v_1$. Thus $Pv_1 = v_1$. Letting u now represent an arbitrary element of $\mathbb{R}^{(n)}$, we find that Pu is in W, so that Pu and v_2 are orthogonal. This implies that

$$0 = (Pu)'v_2 = u'P'v_2 = u'Pv_2.$$

But then Pv_2 is orthogonal to u for all $u \in \mathbb{R}^{(n)}$, which implies that $Pv_2 = 0$. It follows that

$$Pv = Pv_1 + Pv_2 = v_1 + 0 = v_1 = \text{Proj}_W(v)$$

for all v, that is, $P = \text{Proj}_W(v)$. ∎

The nullspaces of $A'A$ and A are equal, since the condition $A'Ax = 0$ implies that

$$(Ax)'(Ax) = x'A'Ax = 0$$

which in turn implies that Ax has length 0 and so is 0. So since the matrices $A'A$ and A both have n columns and the rank plus the nullity adds up to n for both of them, by Theorem 6.3.6, the ranks of $A'A$ and A are equal. Using this, we can prove something even stronger, namely

Theorem 11.5.2. The column spaces of $A'A$ and A' are the same.

Proof. Certainly, the column space of $A'A$ is contained in the column space of A'. Since the dimensions of the column spaces of $A'A$ and A' are the ranks of $A'A$ and A, respectively, and since these are equal as we just saw, it follows that the column spaces of $A'A$ and A' are equal. ■

We now can prove two theorems that give us row reduction algorithms to compute $\text{Proj}_{A'(\mathbb{R}^{(m)})}$, the nullspace of A and A^-.

Theorem 11.5.3. Let A be a real $m \times n$ matrix. Then for any orthonormalized matrix J that is row equivalent to A, $\text{Proj}_{A'(\mathbb{R}^{(m)})} = J'J$ and the columns of $I - J'J$ span the nullspace of A.

Proof. Since J is an orthonormalized matrix, we get

$$(JJ')J = J.$$

But then $J'JJ'J = J'J$. Since $(J'J)' = J'J$ and $(J'J)^2 = J'J$, $J'J$ is a projection and $J'J = \text{Proj}_{J'J(\mathbb{R}^{(n)})}$. Since J and A are row equivalent, J' and A' have the same column spaces. So, by Theorem 11.5.2, $J'J$, J', A' have the same column spaces. But then

$$J'J = \text{Proj}_{J'J(\mathbb{R}^{(n)})} = \text{Proj}_{A'(\mathbb{R}^{(m)})}.$$

It follows that the columns of $I - J'J$ span the nullspace of A, since:

1. The nullspace of A is $(A'\mathbb{R}^{(m)})^{\perp}$ (Prove!).
2. $A'\mathbb{R}^{(m)})^{\perp} = (J'J\mathbb{R}^{(n)})^{\perp} = (I - J'J)\mathbb{R}^{(n)}$ [Prove, using $(J'J)^2 = J'J$!] ■

Lemma 11.5.4. Let $P \in M_n(\mathbb{R})$ satisfy the equation $PP' = P'$. Then P is a projection.

Proof. Since $P' = PP'$, $P = P'$. But then

$$P = P' = PP' = PP = P^2,$$

so P is a projection. ■

Theorem 11.5.5. Let A be an $m \times n$ real matrix, and let M be an invertible $m \times m$ matrix such that $J = MA'A$ is orthonormalized. Then $A^- = J'MA'$.

Proof. Let $B = J'MA'$. We claim first that

1. $BA = J'J = \text{Proj}_{A'(\mathbb{R}^{(m)})}$.

   2. $BAB = B$.
   3. $AB = \mathrm{Proj}_{A(\mathbb{R}^{(n)})}$.

Since $BA = J'MA'A = J'J$, and since $A'A$ and A' have the same column space by Theorem 11.5.2, (1) follows from Theorem 11.5.3. For (2), we first use the equation $JJ'J = J$ from the proof of Theorem 11.5.3 to get the equation $J'JJ' = J'$. Then

$$BAB = J'JB = J'JJ'MA' = J'MA' = B.$$

For (3), we first show that AB is a projection. By Lemma 11.5.4 it suffices to show that $(AB)(AB)' = (AB)'$, which follows from the equations

$$\begin{aligned}AB(AB)' &= (AJ'MA')(AM'JA') = AJ'(MA'A)M'JA' \\ &= AJ'JM'JA' = (AJ'J)M'JA' = AM'JA' = (AB)'.\end{aligned}$$

Here we use the fact that since $J'JA' = A'$, by Theorem 11.5.3, $AJ'J = A$. (Prove!) Finally, the column space of A contains $AB\mathbb{R}^{(m)}$, which in turn contains $ABA\mathbb{R}^{(n)} = A(BA\mathbb{R}^{(n)})$, which in turn is $A(A'\mathbb{R}^{(m)})$ by (1). Since AA' and A have the same column space by Theorem 11.5.2, it follows that all these spaces are actually equal; that is,

$$A\mathbb{R}^{(n)} = AB\mathbb{R}^{(m)} = ABA\mathbb{R}^{(n)} = A(BA\mathbb{R}^{(n)}) = A(A'\mathbb{R}^{(m)}) = A\mathbb{R}^{(n)}.$$

So A and AB have the same column spaces. But then the projection AB is just $AB = \mathrm{Proj}_{A(\mathbb{R}^{(m)})}$.

To show that $B = A^-$, let $x = By$. Then from (3) we get that

$$Ax = ABy = \mathrm{Proj}_{A(\mathbb{R}^{(m)})}y,$$

and from (1) and (2) we get that

$$\mathrm{Proj}_{A'(\mathbb{R}^{(m)})}x = BAx = BABy = By = x.$$

So x is the shortest solution to $Ax = \mathrm{Proj}_{A(\mathbb{R}^{(n)})}y$ and $x = A^-y$. Since this is true for all y, we get that $B = A^-$. ■

From these theorems and Theorem 10.2.2, we get the following methods, each of which builds toward the later ones:

Algorithm to compute the projection
$\mathrm{Proj}_{A'(\mathbb{R}^{(m)})}$ for a real $m \times n$ matrix A

   1. *Row reduce $A'A$ to an orthonormalized matrix $J = MA'A$.*
   2. *Then $\mathrm{Proj}_{A'(\mathbb{R}^{(m)})}$ is $J'J$.*

Algorithm to compute the nullspace of a real $m \times n$ matrix A

3. *Then the columns of $I - J'J$ span nullspace of A.*

Algorithm to compute the approximate inverse of a real $m \times n$ matrix A

4. *Then A^- is $J'K$, where $K = MA'$.*

Algorithm to compute the projection $\text{Proj}_{A(\mathbb{R}^{(n)})}$ for a real $m \times n$ matrix A

5. *Then $\text{Proj}_{A(\mathbb{R}^{(n)})}$ is AA^-.*

Algorithm to find all approximate solutions of $Ax = y$

6. *The shortest approximate solution to $Ax = y$ is $x = A^-y$, which we have by (4).*
7. *Every approximate solution is $x + w$, where $w \in (I - J'J)\mathbb{R}^{(n)}$, by (3).*

It is instructive to look at some examples.

EXAMPLES

1. To compute the projection of $\mathbb{R}^{(2)}$ onto the column space of $\begin{bmatrix} 1 \\ 2 \end{bmatrix}$, row reduce

$$\begin{bmatrix} 1 \\ 2 \end{bmatrix}[1, 2] = \begin{bmatrix} 1 & 2 \\ 2 & 4 \end{bmatrix}$$

to its orthonormalized echelon form

$$J = \begin{bmatrix} 1/\sqrt{5} & 2/\sqrt{5} \\ 0 & 0 \end{bmatrix}.$$

Then the projection is $J'J = \begin{bmatrix} \frac{1}{5} & \frac{2}{5} \\ \frac{2}{5} & \frac{4}{5} \end{bmatrix}$.

2. Let's find A^- for $A = \begin{bmatrix} 2 & 3 \\ 0 & 4 \\ 0 & 1 \end{bmatrix}$. Since

$$A'A = \begin{bmatrix} 2 & 0 & 0 \\ 3 & 4 & 1 \end{bmatrix} \begin{bmatrix} 2 & 3 \\ 0 & 4 \\ 0 & 1 \end{bmatrix} = \begin{bmatrix} 4 & 6 \\ 6 & 26 \end{bmatrix},$$

we row reduce

$$\begin{bmatrix} 4 & 6 & 2 & 0 & 0 \\ 6 & 26 & 3 & 4 & 1 \end{bmatrix}$$

to the echelon form

$$\begin{bmatrix} 1 & 1.5 & .5 & 0 & 0 \\ 0 & 1 & 0 & \frac{4}{17} & \frac{1}{17} \end{bmatrix}.$$

We then apply the operation Add $(1, 2, -v_{21})$ where v_{21} is the inner product 1.5 of $(1, 1.5)$ and $v = (0, 1)$, getting

$$\begin{bmatrix} 1 & 0 & .5 & -\frac{6}{17} & -1.5/17 \\ 0 & 1 & 0 & \frac{4}{17} & 1/17 \end{bmatrix}.$$

Since $J = I$, A^- is

$$IK = K = \begin{bmatrix} .5 & -\frac{6}{17} & -1.5/17 \\ 0 & \frac{4}{17} & 1/17 \end{bmatrix},$$

the same answer that we got when we computed A^- directly in Section 10.2.

Example 2 illustrates the fact that in the case where the columns of A are linearly independent, the row reduction is the same as reduction of $A'A$ to reduced echelon form I, with the operations applied to the augmented matrix. In this case, $A^- = K$.

EXAMPLE

Let's next find A^- for the matrix $A = \begin{bmatrix} 2 & 3 & 5 \\ 0 & 4 & 4 \\ 0 & 1 & 1 \end{bmatrix}$, to see what

happens when the columns of A are linearly dependent. Since

$$A'A = \begin{bmatrix} 2 & 0 & 0 \\ 3 & 4 & 1 \\ 5 & 4 & 1 \end{bmatrix} \begin{bmatrix} 2 & 3 & 5 \\ 0 & 4 & 4 \\ 0 & 1 & 1 \end{bmatrix} = \begin{bmatrix} 5 & 6 & 10 \\ 6 & 26 & 32 \\ 10 & 32 & 42 \end{bmatrix},$$

we row reduce

$$\begin{bmatrix} 4 & 6 & 10 & 2 & 0 & 0 \\ 6 & 26 & 32 & 3 & 4 & 1 \\ 10 & 32 & 42 & 5 & 4 & 1 \end{bmatrix}$$

successively to the matrices

$$\begin{bmatrix} 2 & 3 & 5 & 1 & 0 & 0 \\ 6 & 26 & 32 & 3 & 4 & 1 \\ 10 & 32 & 42 & 5 & 4 & 1 \end{bmatrix}, \begin{bmatrix} 2 & 3 & 5 & 1 & 0 & 0 \\ 0 & 17 & 17 & 0 & 4 & 1 \\ 0 & 17 & 17 & 0 & 4 & 1 \end{bmatrix},$$

$$\begin{bmatrix} 2 & 3 & 5 & 1 & 0 & 0 \\ 0 & 17 & 17 & 0 & 4 & 1 \\ 0 & 0 & 0 & 0 & 0 & 0 \end{bmatrix}, \begin{bmatrix} 2 & 3 & 5 & 1 & 0 & 0 \\ 0 & 1 & 1 & 0 & \frac{4}{17} & \frac{1}{17} \\ 0 & 0 & 0 & 0 & 0 & 0 \end{bmatrix}.$$

To avoid further fractions, we orthogonalize $(2, 3, 5)$ and $(0, 1, 1)$ directly, by the operation Add $(1, 2; -4)$, where 4 was chosen as the inner product 8 of $(2, 3, 5)$ and $(0, 1, 1)$ divided by the inner product 2 of $(0, 1, 1)$ and $(0, 1, 1)$. We then get

$$\begin{bmatrix} 2 & -1 & 1 & 1 & -\frac{16}{17} & -\frac{4}{17} \\ 0 & 1 & 1 & 0 & \frac{4}{17} & \frac{1}{17} \\ 0 & 0 & 0 & 0 & 0 & 0 \end{bmatrix}.$$

To normalize the orthogonal vectors $(2, -1, 1)$, $(0, 1, 1)$ to vectors of length 1, we apply the operations Multiply $(1, 1/\sqrt{6})$ and Multiply $(2, 1/\sqrt{2})$, getting

$$\begin{bmatrix} .8165 & -.4082 & .4082 & .4082 & -.3842 & -.0961 \\ 0 & .7071 & .7071 & 0 & .1664 & .0416 \\ 0 & 0 & 0 & 0 & 0 & 0 \end{bmatrix}.$$

So $A^- = J'K$ is the matrix

$$\begin{bmatrix} .8165 & 0 & 0 \\ -.4082 & .7071 & 0 \\ .4082 & .7071 & 0 \end{bmatrix} \begin{bmatrix} .4082 & -.3842 & -.0961 \\ 0 & .1664 & .0416 \\ 0 & 0 & 0 \end{bmatrix},$$

which we multiply out to

$$A^- = \begin{bmatrix} .3333 & -.3137 & -.0784 \\ -.1667 & .2745 & .0686 \\ .1667 & -.0391 & -.0098 \end{bmatrix}.$$

We can check to see if this does what it is supposed to do,

by checking whether $A^-\begin{bmatrix} 5000 \\ 4000 \\ 2000 \end{bmatrix}$ is the approximate inverse

$$\begin{bmatrix} 254.90 \\ 401.96 \\ 656.86 \end{bmatrix}$$

which we calculated in Section 11.5. We find that

$$A^-\begin{bmatrix} 5000 \\ 4000 \\ 2000 \end{bmatrix} = \begin{bmatrix} .3333 & -.3137 & -.0784 \\ .1667 & .2745 & .0686 \\ .1667 & -.0391 & -.0098 \end{bmatrix}\begin{bmatrix} 5000 \\ 4000 \\ 2000 \end{bmatrix}$$

$$= \begin{bmatrix} 254.90 \\ 402.96 \\ 657.86 \end{bmatrix},$$

which is the same.

In using the algorithms discussed above, note that:

1. To find approximate solutions x to $Ax = y$ for each of a large number of y, the most efficient way to proceed, using the algorithm above, is probably to keep A^- in its factored form $A^- = J'K$ and get $x = A^-y$ for each y by getting first $v = Ky$, then $x = J'v$. Here J' is lower triangular and the number of nonzero rows of K equals the rank of A. (Prove!)

2. The algorithm requires only enough memory to hold the original matrix A and the matrix $A'A$. Then $J'J$ will occupy the memory that had held $A'A$, and K occupies the memory that had held A.

3. We can carry out the multiplication $J'K$ to get A^- without using additional memory, by multiplying J' times K column by column until done. As a result of doing this, A will be replaced by A^- in the memory that had held it.

4. We can recover A^- by repeating the process, replacing A^- by $(A^-)^-$. Then the resulting matrix is just A (see Problem 4).

5. When memory is limited, we can assume without loss of efficiency that $m \geq n$, since A^- is the transpose of A'^- (see Problem 5).

PROBLEMS

NUMERICAL PROBLEM

1. Compute A', A'^-, A^-, $(A^-)^-$ for the following matrices.

(a) $\begin{bmatrix} 1 & 0 & 1 & 1 \\ 1 & 2 & 0 & 1 \end{bmatrix}$.

(b) $\begin{bmatrix} 1 & 2 & 2 \\ 2 & 2 & 2 \\ 3 & 2 & 2 \end{bmatrix}$.

(c) $\begin{bmatrix} 1 & 2 & 3 \\ 2 & 2 & 3 \\ 3 & 2 & 1 \end{bmatrix}$.

MORE THEORETICAL PROBLEMS

Easier Problems

2. Show that if $B = A^-$, then $ABA = A$, $BAB = B$, $(AB)' = AB$, $(BA)' = BA$.

Middle-Level Problems

3. Show that for any $m \times n$ matrix A, there is exactly one $n \times m$ matrix B such that $ABA = A$, $BAB = B$, $(AB)' = AB$, $(BA)' = BA$, namely $B = A^-$.

Harder Problems

4. Using Problem 3, show that $(A^-)^- = A$.
5. Using Problem 3, show that $(A^-)' = (A')^-$.
6. Using Problem 3, show that $(A'A)^- = A^- A'^-$.
7. Using Problem 3, show that $J^- = J'$.

Very Hard Problems

8. Using Problem 3, show that if A is given in its LDU factored format $A = LDU$, where A is $m \times n$ of rank r, L' is echelon $r \times m$, D is nonsingular $r \times r$, and U is echelon $r \times n$, then $A^- = U^- D^- L^-$.

9. Show that when A^- is factored as $A^- = J'K$ as described above, then the first $r = r(A)$ rows of K and nonzero and the others are 0.

11.6. A COMPUTER PROGRAM FOR FINDING EXACT AND APPROXIMATE SOLUTIONS

Having studied the basic algorithms related to solving equations, we now can use them to build many different kinds of computer programs. Listed below is a program for finding exact and approximate solutions to any matrix equation, which we wrote to illustrate how to use the algorithms studied in this chapter as models for procedures in such programs.

The program is written in TURBO PASCAL,* to make it easy to use on microcomputers. It is a simple matter to change it to standard PASCAL by bypassing a few special features of TURBO PASCAL used in the program.

This program enables you to load, use, and save *matrix files*. Here, a matrix file is a file whose first row gives the row and column degrees of a matrix, as positive integers, and whose subsequent lines give the entries, as real numbers. For example, the 2×3 matrix

$$A = \begin{bmatrix} 1.000 & -2.113 & 4.145 \\ 3.112 & 4.212 & 5.413 \end{bmatrix}$$

is represented by the contents of the matrix file

```
  2        3
1.000   -2.113  4.145
3.112    4.212  5.413
```

To load a matrix file, the program instructs the computer to get the row and column degrees m and n from the first line, then to load the entries into an $m \times n$ matrix structure. So, in the example, it gets the degrees 2 and 3 and loads the entries, as real numbers, into a 2×3 matrix structure.

While running the program, you have the following options, which you can select any number of times in any order by entering the commands L, S, W, E, A, D, U, I, and X and supplying appropriate data or choices when prompted to supply it:

L   Load a matrix A from a matrix file.

       (You will be asked to supply the file name for a matrix file. If you wish, you can supply CON, the special file name for the keyboard, in which case the computer loads A as you enter it

---

*TURBO PASCAL is a trademark of Borland International.

from the keyboard appropriately—with two positive integers m and n on the first line, and n real numbers on each of the next m lines.)

S   Save a matrix A to a matrix file.

(You will be asked to supply the file name. If you wish, you can supply LPT1, the special file name for the printer, in which case the computer prints A on the printer.)

W   Display a window of entries of the current matrix A.

E   Solve $Ax = y$ for x exactly, if x exists, and approximately if not.

(You will be asked to enter y's for which x's are desired until you indicate that you are done.)

A   Find the shortest approximate solution x to $Ax = y$ for any given matrix A.

(You will be asked to enter y's for which x's are desired until you indicate that you are done.)

D   Decompose A, given in matrix format, into its LDU factored format.

U   Undo the LDU decomposition to recover A, given in LDU factored format, into its matrix format.

I   Find the approximate inverse of A, which is the exact inverse when A is invertible.

X   Exit from the program.

```
PROGRAM LinAlg;  {Version 1.0.           An instructional aid.}

{***************************** DECLARATIONS ******************************}

CONST
   Epsilon = 0.1E-6; MaxDegree = 16;Format = 3;

TYPE
   IntegerVector   = ARRAY [1..MaxDegree            ]  OF INTEGER;
   RealVector      = ARRAY [1..MaxDegree            ]  OF REAL;
   RealMatrix      = ARRAY [1..MaxDegree,1..MaxDegree ]  OF REAL;
   StringType      = STRING [80];

VAR
   TurnedOn                      {Turns on the program loop} : BOOLEAN;
   RowDegA,ColDegA,RankA,PivotRowA,RowDegB,ColDegB          : INTEGER;
   ZeroList, IdentityList, RowListA, ColListA, PivotListA,
   InvRwListA , InvClListA                                  : IntegerVector;
   MatrixA, MatrixB                                         : RealMatrix;
   MatrixEntry                                              : REAL;
   v,w,x,y      {Solve Lv=y, Dw=v, Ux=w so Ax=LDUx=LDw=Lv=y} : RealVector;
   InputFile,OutputFile                                     : TEXT;
   InputFileN, OutputFileN, AnySt                           : StringType;

{*********************** LDU DECOMPOSITION OF A **************************}

PROCEDURE InterchangeRow (p,q: INTEGER);

{ Interchange rows p and q in both matrices A and B. }

VAR   CopyOfRow: INTEGER;
BEGIN
   CopyOfRow:=RowListA[q]; RowListA[q] := RowListA[p]; RowListA[p]:=CopyOfRow;
END;

PROCEDURE  AddRow(p,q :INTEGER;u:REAL;TruncationColA,ClDgB:INTEGER);

{ Add u times row q to row p in both matrices A and B, skipping the storage
  area in A determined by TruncationColA. }

VAR   s : INTEGER;
BEGIN
   FOR s := TruncationColA TO ColDegA DO        { Skipping the storage area, }
   MatrixA [RowListA[p],ColListA[s]]            { add u times row q to row p. }
      := MatrixA[RowListA[p],ColListA[s]] + u*MatrixA[RowListA[q],ColListA[s]];
   FOR s := 1 TO ClDgB DO                       { Operate in parallel on MatrixB. }
   MatrixB [RowListA[p],ColListA[s]]
      := MatrixB[RowListA[p],ColListA[s]] + u*MatrixB[RowListA[q],ColListA[s]];
END;
```

```
PROCEDURE GetPivotRow (p,q : INTEGER; VAR PivotRow: INTEGER);

{ Get the PivotRow, when at the stage of row p and column q. }

VAR  r:INTEGER;
BEGIN
  PivotRow := p;   { We record our best guess, at the outset, for PivotRow. }
  FOR r := p+1 TO RowDegA DO
    BEGIN       { We then look for a row with a bigger entry of column q. }
      IF (ABS(MatrixA [RowListA[r],ColListA[q]])
      > ABS(MatrixA [RowListA[PivotRow],ColListA[q]]))
      THEN PivotRow := r;       { If we find one, we update our last guess. }
    END;
END;

PROCEDURE ReduceAndStoreMultipliers (VAR p,q: INTEGER; VAR PivotEntry: REAL;
                        ClDgB: INTEGER; ReduceUtoDU, StoreL: BOOLEAN);

{ For PivotEntry in row p and column q, use the multipliers to reduce the
  subsequent rows and, if StoreL, store them below the PivotEntry as entries
  of L. }

VAR  t: INTEGER; Multiplier: REAL;
BEGIN
  FOR t := p+1 TO RowDegA DO
    BEGIN
      Multiplier := MatrixA[RowListA[t],ColListA[q]]/PivotEntry;
      AddRow(t,p,-Multiplier,q,ClDgB);         { Reduce MatrixA, MatrixB. }
      IF StoreL THEN MatrixA [RowListA[t],ColListA[q]] := Multiplier
      ELSE  MatrixA [RowListA[t],ColListA[q]] := 0;
    END;

{ If ReduceUtoDU, further reduce the factorizations A = LU, B = LV to
  A = LDU, B = LDV. }

  IF ReduceUtoDU THEN begin
    FOR t := q+1 TO ColDegA DO
      MatrixA [RowListA[p],ColListA[t]]
        := MatrixA[RowListA[p],ColListA[t]]/PivotEntry;
    FOR t := 1 TO ClDgB DO
      MatrixB [RowListA[p],ColListA[t]]
        := MatrixB[RowListA[p],ColListA[t]]/PivotEntry;
  END;
END;

PROCEDURE DoLDUdecompose (ClDgB: INTEGER; ReduceUtoDU, StoreL: BOOLEAN);

{ Starting with PivotRowA = 1, PivotColA = 1, keep getting the next Pivots
  and decompose Matrix A to its LDU factored format. }

VAR  PivotRow, PivotColA :INTEGER; VAR PivotEntry:REAL;
BEGIN
  IF PivotRowA<>0 THEN          { Do Nothing - the matrix is in LDU Format. }
  ELSE BEGIN                    { Begin to put the matrix in LDU Format. }
```

```
      WRITELN('DoLDUdecompose...');
      RowListA:=IdentityList; ColListA:=IdentityList;
      PivotRowA:=1; PivotColA:=1;                    { Start in upper left hand corner }
      WHILE ((PivotRowA<=RowDegA) AND (PivotColA<=ColDegA)) DO BEGIN
        GetPivotRow(PivotRowA,PivotColA,PivotRow);          { and get PivotRow. }
        PivotEntry := MatrixA [RowListA[PivotRow],ColListA[PivotColA]];
        IF (ABS(PivotEntry) < Epsilon) THEN PivotRowA:=PivotRowA-1 {Try again.}
        ELSE                                      { PivotColA is a PivotColumn. }
        BEGIN
          PivotListA[PivotRowA]:=ColListA[PivotColA];{Save PivotColA for UnDo.}
          IF PivotRow <> PivotRowA THEN InterchangeRow(PivotRowA,PivotRow);
          ReduceAndStoreMultipliers (PivotRowA,PivotColA,PivotEntry,ClDgB,
                                 ReduceUtoDU, StoreL);
        END;
        PivotRowA:=PivotRowA+1;PivotColA:=PivotColA+1; { Move on to the next. }
      END;
      PivotRowA:=PivotRowA-1;RankA:=PivotRowA; { Maintain PivotRowA for UnDo. }
   END;
END;


PROCEDURE RetrieveMultipliersAndRebuild (VAR p,q:INTEGER;ClDgB:INTEGER;
                                     ReduceUtoDU:Boolean);


{ Retrieve the stored Multipliers from L, to use to UnDo the decomposition
  of Matrix A to its LDU factored format and rebuild A. }


VAR  t:INTEGER;  Multiplier, PivotEntry:REAL;
BEGIN
   q:=PivotListA[p];                                { Get the pivot column q. }
   PivotListA[p]:=0 ;                               { Undo PivotListA too. }
   PivotEntry:=MatrixA[RowListA[p],ColListA[q]];      {Undo U in Matrix A. }
   IF ReduceUtoDU THEN BEGIN                          { to Undo DU to U. }
     FOR t := q+1 TO ColDegA DO
       MatrixA [RowListA[p],ColListA[t]]
         := MatrixA[RowListA[p],ColListA[t]]*PivotEntry;
     FOR t := 1 TO ClDgB DO                          { Undo in Matrix B too. }
       MatrixB [RowListA[p],ColListA[t]]
         := MatrixB[RowListA[p],ColListA[t]]*PivotEntry;
   END;
   FOR t:= RowDegA DOWNTO p+1 DO                               { Undo U to A. }
     BEGIN
       Multiplier := MatrixA[RowListA[t],ColListA[q]];{UnDo StoreMultipliers.}
       MatrixA [RowListA[t],ColListA[q]] := 0;
       AddRow(t,p, Multiplier,q,ClDgB);                        { UnDo Reduce A. }
     END;
END;


PROCEDURE UnDoLDUdecompose(ClDgB:INTEGER; ReduceUtoDU:Boolean);


{ UnDo the decomposition of Matrix A to its LDU factored format, restoring
  it to its unfactored format A. }


VAR  i, PivotCol, Row, Col: INTEGER; PivotEntry: REAL;
```

```
BEGIN                  { Note that if PivotRowA already 0, the procedure is idle. }
  IF PivotRowA <> 0 THEN WRITELN('UnDoLDUdecompose...');
  WHILE PivotRowA>=1 DO                  { Starting with PivotRowA, get PivotCol }
  BEGIN                                  { and rebuild in it; undo PivotRowA.    }
    RetrieveMultipliersAndRebuild (PivotRowA,PivotCol,ClDgB,ReduceUtoDU);
    PivotRowA:=PivotRowA-1;                            { Undo the PivotRow. }
  END;
  FOR i:=1 TO RowDegA DO RowListA[i]:=i;                 { Undo RowListA. }
END;

{*********************** APPROXIMATE INVERSE OF A *************************}

{ FOR EFFICIENCY: Modify according to the outline given in the chapter
  LINEAR ALGORITHMS, Section 5. }

PROCEDURE ApproximateInverse;

{ Start with Matrix A and replace it with its approximate inverse. }

VAR MatrixC: RealMatrix; RowDegC, ColDegC: INTEGER; r,s,t: INTEGER;

PROCEDURE TransposeAtoB(MatA: RealMatrix; VAR MatB: RealMatrix;
          RDegA, CdegA: INTEGER; VAR RdegB, CdegB: INTEGER);

{ Put the transpose of A in B. }

VAR i,j:INTEGER;
BEGIN
  RdegB := CdegA; CdegB := RdegA;
  FOR i := 1 TO RdegB DO
    FOR j := 1 TO CdegB DO
      MatB[i,j] := MatA[RowListA[j],i];      { Sometimes, RowList is needed. }
END;

PROCEDURE BtimesBprimeToA(MatB: RealMatrix;    VAR MatA    : RealMatrix;
                          RDegB,CDegB: INTEGER;VAR RdegA,CdegA: INTEGER);

{ Put B times its transpose in A. }

VAR  r,s,t :INTEGER;
BEGIN
  RdegA:=RdegB;CdegA:=RdegB;
  FOR r:= 1 TO RdegB DO
  FOR t:= 1 TO RdegB DO
    BEGIN
      MatA[r,t] := 0.0;
      FOR s:= 1 TO CdegB DO MatA[r,t]:=MatA[r,t]+MatB[r,s]*MatB[t,s];
    END
END;

PROCEDURE OrthogonalizeUp(RkA:INTEGER);

{ Orthogonalize the nonzero rows of A in reverse order by the Gram-Schmidt
  process. }

VAR dt,lngth : REAL; i,j,k : INTEGER;
```

```
FUNCTION Dot(i,k:INTEGER):REAL;

BEGIN
  dt:=0.0;
  FOR j:= 1 TO ColDegA DO dt
    :=dt + MatrixA[RowListA[i],j]*MatrixA[RowListA[k],j];
  Dot:=dt;
END;


FUNCTION Length(i:INTEGER):REAL;

BEGIN
  Lngth:=Dot(i,i); length:=sqrt(lngth);
END;

BEGIN {* OrthogonalizeUp *}
  FOR i:=RkA DOWNTO 1 DO
  BEGIN
    lngth:=Length(i);
    FOR j:= 1  TO ColDegA DO MatrixA[RowListA[i],j]
      :=MatrixA[RowListA[i],j]/lngth;
    FOR j:= 1  TO ColDegB DO MatrixB[RowListA[i],j]
      :=MatrixB[RowListA[i],j]/lngth;
    FOR k:= i-1 DOWNTO 1 DO AddRow(k,i,-Dot(i,k),1,ColDegB);
  END;
END;


BEGIN  {* ApproximateInverse *}
  IF PivotRowA<>0 THEN UnDoLDUdecompose(0,True);  { UnDo LDU decomposition. }
  WRITELN('ApproximateInverse...');
  TransposeAtoB(MatrixA,MatrixB, { Could be avoided by loading Aprime in B. }
    RowDegA,ColDegA,RowDegB,ColDegB);              { Get(A,Aprime). }
  BtimesBprimeToA(MatrixB,MatrixA,
            RowDegB,ColDegB,RowDegA,ColDegA);   { Get(AprimeA,Aprime). }
  DoLDUdecompose(ColDegB, False, False);    { LU decompose to get (J1,K1). }
  OrthogonalizeUp(RankA);                  { Go on to get(J,K) from (J1,K1). }
  TransposeAtoB(MatrixA,MatrixC,
            RowDegA,ColDegA,RowDegC,ColDegC);  { Get Jprime from (J,K). }
  ColDegA:=ColDegB;   { Give MatrixA its new column degree, that of Aprime. }
  FOR r:=1 TO RowDegA DO    { Get JprimeK in MatrixA from Jprime and (J,K). }
    BEGIN
      FOR t:=1 TO ColDegA DO
        BEGIN
          MatrixA[r,t]:=0.0;
          FOR s:= 1 TO {ColDegA}RowDegA DO
            MatrixA[r,t]:=MatrixA[r,t]+MatrixC[r,s]*MatrixB[RowListA[s],t];
        END;
    END;                               { The approximate inverse is JprimeK. }
  FOR r:=1 TO RowDegA DO BEGIN RowListA[r]:=r; PivotListA[r]:=r;END;
  PivotRowA:=0;  { Return A to matrix format now that it is a matrix again. }
END;
```

```
{****************** SOLVE Ax=y EXACTLY or APPROXIMATELY ******************}

PROCEDURE Multiply( VAR MatAminus:RealMatrix;RowDg,ColDg:INTEGER;
                    VAR y:RealVector);

{ Solve Ax = y by multiplying y by the approximate inverse of  A  to get x. }

VAR   i,j:INTEGER;
BEGIN
  FOR i:= 1 TO RowDg DO
    BEGIN x[i]:=0; FOR j:=1 TO ColDg DO x[i]:=x[i]+MatAminus[i,j]*y[j]; END;
END;

PROCEDURE ForwardSubstitute( VAR MatL:RealMatrix; VAR y,v:RealVector);

VAR  p,j:INTEGER;
BEGIN
  FOR p:=1 TO RowDegA DO
    BEGIN
      v[p]:=y[RowListA[p]];
      FOR j:=1 TO p-1        DO v[p]:=v[p]-MatL[RowListA[p],j]*v[j]
    END
END;

PROCEDURE BackSubstitute(VAR w,x:RealVector;VAR MatU:RealMatrix);

VAR p,q,j:INTEGER;
BEGIN
  FOR j:=1 TO ColDegA DO x[j]:= 0;  { Decide on values of independent x[j]. }
                                    { If j is pivot column, x[j] redefined later. }
  FOR p:= RankA DOWNTO 1 DO
    BEGIN
      q := PivotListA[p];
      x[q] := w[p];
      FOR j:=q+1 TO ColDegA DO x[q]:=x[q]-MatU[RowListA[p],j]*x[j]
    END
END;

PROCEDURE EnterVector (VAR v:RealVector;RwDgA:INTEGER);

{ Prompt user to enter the vector v. }

VAR i : INTEGER;
BEGIN  WRITELN;
  FOR i := 1 TO RwDgA DO
    BEGIN WRITE (' Enter y[',i:1,'] = '); READLN (v[i]); END;
  WRITELN;
END;

PROCEDURE WriteVector(z:RealVector;ClDgA:INTEGER);

{ Write the vector z on the screen. }

VAR j : INTEGER;
```

```
BEGIN
  FOR j:=1 TO ClDgA DO
    BEGIN WRITE('        x[',j:1,'] = '); WRITELN(z[j]:6:2) END;
  WRITELN;
END;

PROCEDURE GetRightHandSide(SolveItExactly:BOOLEAN;RwDgA,ClDgA:INTEGER);

{ Call EnterVector to get right hand side vector y from user. Solve
  Ax = y exactly using ForwardSubstitute and BackSubstitute;  or
  approximately using Multiply. }

VAR i: Integer; Ch: Char;
BEGIN
  Ch := 'Y'; WRITELN('  Right Hand Side Vector y: ');
  WHILE UPCASE(Ch) ='Y' DO BEGIN                        { to get y and solve for x. }
    EnterVector(y,RwDgA);
    IF SolveItExactly THEN
      BEGIN        { to solve Lv = y, Dw = v, Ux = w so LDUx = LDw = Lv = y. }
        ForwardSubstitute(MatrixA,y,v);                           { Get v. }
        FOR i:=1 TO RankA DO w[i]:=v[i]/MatrixA[RowListA[i],i];   { Get w. }
        BackSubstitute(w,x,MatrixA);                              { Get x. }
      END
    ELSE                                    { we are solving it approximately. }
      Multiply(MatrixA,ClDgA,RwDgA,y);   { where Matrix A holds inverse of A. }
    WRITELN('Final Solution Vector x: ');
    WriteVector(x,ClDgA);
    WRITELN ('Would you like to enter a different vector b? (Y/N):');
    READ (KBD,Ch); WRITELN(UPCASE(Ch));
  END
END;


PROCEDURE SolveApproximately;

{ Solve Ax = y for its shortest approximate solution x. }

VAR Ch: Char;
BEGIN
  ApproximateInverse;                        { Put the inverse of A in Matrix A. }
  GetRightHandSide(False,ColDegA,RowDegA);{ Get y and deliver x until done. }
  WRITELN ('Restore the matrix? (Y/N):'); READ (KBD,Ch);
  IF UPCASE(Ch)='N' THEN {Do Nothing} ELSE ApproximateInverse;
END;


PROCEDURE SolveExactly;
VAR Singular:Boolean; Ch: Char;
BEGIN
  Singular:=False;
  IF RowDegA <> ColDegA THEN Singular:=True ELSE DoLDUdecompose(0,True,True);
  IF RowDegA<>RankA THEN Singular:= True;
  IF NOT Singular THEN BEGIN                              { to solve exactly. }
    GetRightHandSide(True,RowDegA,ColDegA);
    WRITELN ('Restore to matrix format? (Y/N):');
    READ (KBD,Ch); WRITELN(UPCASE(Ch));
    IF UPCASE(Ch)='N' THEN {Do Nothing} ELSE UnDoLDUdecompose(0,True);
  END
```

```
    ELSE BEGIN                                  { to solve approximately or exit to menu. }
      WRITELN('Matrix singular. Solve Approximately? (Y/N) ');
      Read(KBD,Ch); IF UPCASE(Ch) = 'Y' THEN SolveApproximately;
    END;
END;

{***************************** WRITEMATRIX *****************************}

PROCEDURE WAIT; BEGIN WRITELN('Press a key ...');Repeat until KeyPressed END;

PROCEDURE InvertList( VAR RwList,ClList: IntegerVector; m, n:INTEGER);

{ Invert the row list. }

VAR i: Integer; AnyList: IntegerVector;
BEGIN  FOR i:= 1 TO m DO AnyList[RowListA[i]]:=i; RwList:= AnyList;
       FOR i:= 1 TO n DO AnyList[ColListA[i]]:=i; ClList:= AnyList; END;

PROCEDURE WriteMatrix (Mat:RealMatrix;m,n:INTEGER);

{ Writes RowListA, ColListA and matrix to screen. }

VAR
  i,j : INTEGER;
BEGIN
  InvertList(InvRwListA,InvClListA,m,n); WRITELN; WRITE('      ');
  FOR j:= 1 TO n DO WRITE('C',InvClListA[j]:2,'  ');WRITELN;WRITELN;
  FOR I:= 1 TO m DO
    BEGIN
      WRITE('R',InvRwListA[i]:2,' ');
      FOR j:= 1 TO n DO WRITE(Mat[i ,j]:6:2);
      WRITELN; WRITELN;
END;END;

PROCEDURE Window(Msg:StringType);

{ Display a message and write from the matrix to a window on the screen. }

BEGIN
  ClrScr; WRITELN(Msg); WRITELN('RowListA......Entries in memory:');
  WriteMatrix(MatrixA,RowDegA,ColDegA); Write('Rank = ',RankA,'  ');
  IF PivotRowA=0 THEN WRITELN('Matrix Format') else WRITELN('LDU Format');
  Wait;
END;

{************************* INPUT/OUTPUT FILES *************************}

PROCEDURE ReadFileMatrix ;

{ The input Matrix A is read into memory. }

VAR  i , j : INTEGER;
```

```
BEGIN
  RowListA:=IdentityList;ColListA:=IdentityList; PivotRowA:=0;{ A is matrix.}
  READLN (InputFile, RowDegA , ColDegA );
  IF RowDegA>MaxDegree Then RowDegA:=MaxDegree;
  IF ColDegA>MaxDegree Then ColDegA:=MaxDegree;
  FOR i := 1 TO RowDegA DO BEGIN
    FOR j := 1 TO ColDegA DO  READ (InputFile, MatrixA [i,j]);
    READLN (InputFile);
  END;
END;


PROCEDURE WriteFileMatrix ;

{ The Matrix B is written to file memory. }

VAR  i , j : INTEGER;
BEGIN
  WRITELN (OutputFile, RowDegA:10 ,ColDegA:10 );
  FOR i := 1 TO RowDegA DO BEGIN
    FOR j := 1 TO ColDegA DO
      WRITE (OutputFile, MatrixA [RowListA[i],ColListA[j]]:3*Format:Format);
    WRITELN (OutputFile);
  END;
END;


PROCEDURE GetMatrix;

{ Gets degrees, entries. }

VAR
  FileExists : BOOLEAN;
BEGIN
  Write('Loading '+InputFileN+' (or enter new File Name): ');READLN (AnySt);
  IF NOT(AnySt = '') THEN InputFileN:=AnySt;
  ASSIGN (InputFile,InputFileN);
  {$I-}  RESET (InputFile); {$I+};
  FileExists := (IOresult = 0);
  IF FileExists THEN BEGIN
    ReadFileMatrix; CLOSE (InputFile); RankA:=0; Window(InputFileN+' :  ')
  END
  ELSE BEGIN WRITELN('NO '+InputFileN+' exists'); Wait END;
END;


PROCEDURE PutMatrix;

{ Gets filename and writes degrees, entries to file. }

VAR  i,j  : INTEGER; AnySt:StringType;
BEGIN
  Write('Saving '+OutputFileN+' (or enter new File Name): '); READLN (AnySt);
  IF NOT(AnySt = '') THEN OutputFileN:=AnySt;                { Get file name. }
  ASSIGN (OutputFile,OutPutFileN);
  REWRITE (OutPutFile);
  IF pivotRowA <>0 THEN UnDoLDUdecompose(0,True);  { First make A a matrix. }
  WriteFileMatrix ;  CLOSE (OutputFile);          { Then save it.          }
END;
```

```
{*********************** INITIALIZATION OF VARIABLES ***********************}

PROCEDURE Initialize;

VAR i: Integer;
BEGIN
  TurnedOn := True;                             { Turns on the main program loop. }
  MatrixA[1,1]:=1; RowDegA:=1;ColDegA:=1;PivotRowA:=0;RankA:=0; { Matrix A. }
  FOR i:= 1 TO MaxDegree DO BEGIN IdentityList[i]:=i; ZeroList[i]:=0; END;
  RowListA:=IdentityList; InvRwListA:=IdentityList; ColListA:=IdentityList;
  PivotListA:=ZeroList;
  RowDegB:=0; ColDegB:=0;                                     { Matrix B. }
  InputFileN:='M1.MAT ';OutputFileN:='N1.MAT';               { Files   . }
END;

{***************************** MENU ********************************}

PROCEDURE MenuSelections;

BEGIN
  ClrScr;
  WRITELN('< To enter a matrix from console, use CON as the LOAD file >');
  WRITELN('< To print a matrix, use LPT1 as the SAVE file          >');
  WRITELN('< Please enter the upper case letter of your selection   >');
  WRITELN;
  WRITELN('    Loadmatrix:'+InputFileN+'           Savematrix:'+OutputfileN);
  WRITELN('    Decompose/Undecompose.            Window.');
  WRITELN('    Exactly solve/Approximately solve Ax=y for x.  ');
  WRITELN('    Invert exactly or approximately.              eXit');
END;

Procedure Menu (VAR TurnedOn: Boolean);
VAR Answer: Char;
BEGIN
  MenuSelections;
REPEAT Read(KBD,Answer); WRITELN(UPCASE(Answer));GotoXY(WhereX,WhereY-1);
UNTIL UPCASE(Answer) in  ['L','S','W','E','A','D','U','I','X'];
  CASE Answer of
'L','l': GetMatrix ;
'S','s': PutMatrix ;
'W','w': Window('CurrentEntries');
'E','e':BEGIN ClrScr; SolveExactly;                            END;
'A','a':BEGIN ClrScr; SolveApproximately;                      END;
'D','d':BEGIN ClrScr; DoLDUdecompose(0,True,True); Window('Decomposed.');END;
'U','u':BEGIN ClrScr; UnDoLDUdecompose(0,True); Window('UnDecomposed.'); END;
'I','i':BEGIN ClrScr; ApproximateInverse;
  IF ((RowDegB=ColDegB)AND(ColDegB=RankA)) THEN Window(' Exact Inverse: ')
  ELSE Window(' Approximate inverse: ');                       END;
'X','x': TurnedOn :=False ;
  END;END;

{***************************** MAIN PROGRAM ***************************}

BEGIN  Initialize; REPEAT Menu(TurnedOn) UNTIL Not TurnedOn; ClrScr   END.
```

PROBLEMS

Note: In Problems 1 to 9, you are asked to write various procedures to add to the program above. As you write them, integrate them into the program and expand the menu accordingly.

1. Write a procedure for computing the inverse of an invertible upper or lower triangular matrix, using the back-substitution equations of Section 11.4.

2. Write a procedure that uses the LDU factorization and the procedure in Problem 1 to compute the inverse of any invertible matrix A in its factored format $A^{-1} = U^{-1}D^{-1}L^{-1}$.

3. Write a procedure for determining whether the columns of a given matrix A are linearly independent.

4. Write a procedure for determining whether a given vector is in the column space of the current matrix.

5. Write a procedure for computing the matrices Q and R of the QR factorization of a matrix A with linearly independent columns.

6. Write a procedure for computing the approximate inverse of A using the QR factorization of A, when the columns of A are linearly independent, by solving the equations $Ax = e_1, \ldots, Ax_n = e_n$ approximately.

7. Write a procedure for computing, for a given matrix A, a matrix N whose columns form a basis for the nullspace of A.

8. Write a procedure to compute the following matrices: $\text{Proj}_{A(\mathbb{R}^{(n)})}$ and $\text{Proj}_{A'(\mathbb{R}^{(m)})}$.

9. Write a set of procedures and add to the menu an option EDIT to enable the user to create and edit matrices. EDIT should enable you to:

(a) Use the cursor keys to control what portion of a large current matrix is displayed on the screen. Alternatively, control what portion of a large current matrix is displayed on the screen by specifying a row and a column.

(b) Use the cursor keys to move the cursor to a desired entry displayed on the screen.

(c) Change the entry under the cursor.

(d) Copy to a matrix file all or part of the current matrix defined by specifying ranges for rows and columns.

(e) Copy from a matrix file into a specified portion of the current matrix.

10. Rewrite the program listed above in standard PASCAL.

INDEX